IMMOBILIZED ENZYME: PRINCIPLES AND APPLICATIONS

Immobilized Enzyme: Principles and Applications

Editor:

Nguyen Minh Nhut

Delve Publishing

www.delvepublishing.com

Immobilized Enzyme: Principles and Applications

Editor: **Nguyen Minh Nhut**

708 3rd Avenue, 6th Floor

New York, NY 10017

United States of America

www.delvepublishing.com

ISBN: 978-1-68095-745-7

Library of Congress Control Number: 2016951504

About the Editor

Nguyen Minh Nhut

Nhut received his Bachelor Degree in Advanced Biotechnology from Can Tho University (VN) and Michigan University (USA) (2011) and extended his research work at BOKU, Austria in "Safety in the Foodchain "Master course (2012). He pursues general food safety management, antibiotics resistance and drug development. Currently he is working at a start-up in the food supply chain as a manager.

Contents

List of Contributors....xiii
Preface....*xxi*

Chapter 1 Enzyme Immobilization: An Overview on Techniques and Support Materials....1
Abstract....1
1. Introduction....1
2. Different Techniques Used for Immobilization....2
3. Materials Used for Fabrication of Immobilization Supports....3
4. Applications and Scope....5
Conclusion....5
Acknowledgments....6
References....6

Chapter 2 Enzyme Immobilization by Microencapsulation: Methods, Materials, and Technological Applications....11
1. Introduction....11
2. Microencapsulation In Semipermeable Membranes....12
3. Entrapment in Network Matrices....17
4. Sol–Gels....19
5. Other Methods....23
6. Organic/Inorganic Hybrid Beads....24
7. New Materials....25
Conclusions....26
References....26

Chapter 3 Improvement of Thermal-Stability of Enzyme Immobilized onto Mesoporous Zirconia....33
Abstract....33
1. Introduction....33
2. Experimental Procedure....34
3. Stability of FDH Under Various Conditions....35
4. Results and Discussion....35
Conclusions....40
Acknowledgment....40
References....40

Chapter 4 Biodecolorization of Textile Dyes by Immobilized Enzymes in a Vertical Bioreactor System....43
Abstract....43
1. Introduction....43
2. Materials and Method....44
3. Results and Discussion....45
Conclusion....47
Acknowledgement....47
References....47

Chapter 5 Enzymatic Extract Containing Lignin Peroxidase Immobilized on Carbon Nanotubes: Potential Biocatalyst in Dye Decolourization....49
Abstract....49
1. Introduction....49

2. Materials And Methods 50
3. Results And Discussion 51
Conclusion 56
Acknowledgments 56
References 56

Chapter 6 **Preparation and Characterization of K-Carrageenase Immobilized onto Magnetic Iron Oxide Nanoparticles 59**
Abstract 59
1. Introduction 60
2. Materials and Methods 60
3. Results and Discussions 61
Conclusions 65
Conflict Of Interest 65
Financial Support 65
References 65

Chapter 7 **High Operational Stability of Invertase from Saccharomyces Cerevisiae Immobilized on Chitosan Nanoparticles 67**
Abstract 67
1. Introduction 67
2. Materials and Methods 68
3. Results and Discussion 69
Conclusions 72
References 73

Chapter 8 **Electrochemical Detection of Carbamate Pesticides in Fruit and Vegetables with a Biosensor Based on Acetylcholinesterase Immobilised on a Composite of Polyaniline–Carbon Nanotubes 75**
Abstract 75
1. Introduction 75
2. Experimental 76
3. Results and Discussion 77
Conclusions 81
Acknowledgements 81
References 81

Chapter 9 **Enzymatic Conversion of Glycerol to Glyceric Acid with Immobilised Laccase in Na-Alginate Matrix 83**
Abstract 83
1. Introduction 83
2. Materials and Methods 84
3. Results and Discussions 85
Conclusion 86
References 86

Chapter 10 **Laccase Immobilization by Chelated Metal Ion Coordination Chemistry 89**
Abstract 89
1. Introduction 89
2. Experimental Section 90
3. Results and Discussion 91
Conclusions 96
Acknowledgments 96

Author Contributions 96
References 96

Chapter 11 **Enzymes Encapsulation Within Calcium Alginate-Clay Beads: Characterization and Application for Cassava Slurry Saccharification 99**
Abstract 99
1. Introduction 99
2. Materials and Methods 100
3. Results and Discussion 101
Conclusions 102
Acknowledgements 103
References 103

Chapter 12 **Continuous Degradation of Maltose by Enzyme Entrapment Technology Using Calcium Alginate Beads as a Matrix 105**
Abstract 105
1. Introduction 105
2. Materials and Methods 106
3. Results and Discussion 107
Conclusions 111
Acknowledgments 111
References 111

Chapter 13 **Characterization And Multi-Step Transketolase-Ω-Transaminase Bioconversions in an Immobilized Enzyme Microreactor (IEMR) With Packed Tube 113**
Abstract 113
1. Introduction 114
2. Materials and Methods 115
3. Results and Discussion 117
Conclusion 120
Acknowledgements 121
References 121

Chapter 14 **Evaluation of Immobilized Lipases on Poly-Hydroxybutyrate Beads to Catalyze Biodiesel Synthesis 123**
Abstract 123
1. Introduction 123
2. Materials and Methods 125
3. Results and Discussion 126
Conclusions 131
Acknowledgments 131
References 131

Chapter 15 **Improved Activity of Lipase Immobilized In Microemulsion-Based Organogels for (R, S)-Ketoprofen Ester Resolution: Long-Term Stability and Reusability 135**
Abstract 135
1. Introduction 135
2. Materials and Methods 136
3. Results and Discussion 137
Conclusions 142
Acknowledgments 142
References 142

Chapter 16 **Improved Activity of Lipase Immobilized in Microemul Impact of Physicochemical Character of Hydrophobic orous Carrier on Reactivity of Immobilized Lipase Progressing Toward Higher Reaction Rate and High Yield in Repeated ... 145**

Abstract ... 145
1. Introduction ... 146
2. Material and Methods ... 146
3. Hydrolysis of Lipid Using Immobilized Lipase in Accurel Particles ... 146
4. Result and Discussion ... 147

Conclusion ... 150
Acknowledgements ... 151
References ... 151

Chapter 17 **Evealuation of Inorganic Matrixes as Supports for Immobilization of Microbial Lipase ... 153**

Abstract ... 153
1. Introduction ... 153
2. Materials and Methods ... 154
3. Results and Discussion ... 155

Conclusions ... 157
Acknowledgments ... 157
References ... 157

Chapter 18 **Glucose Oxidase Directly Immobilized onto Highly Porous Gold Electrodes for Sensing and Fuel Cell Applications ... 159**

Abstract ... 159
1. Introduction ... 159
2. Experimental ... 160
3. Results And Discussion ... 161

Conclusions ... 165
Acknowledgments ... 165
References ... 165

Chapter 19 **High Sensitivity Electrochemical Cholesterol Sensor Utilizing A Vertically Aligned Carbon Nanotube Electrode With Electropolymerized Enzyme Immobilization ... 169**

Abstract ... 169
1. Introduction ... 169
2. Experimental Section ... 170
3. Results and Discussion ... 171

Conclusions ... 174
References and Notes ... 174

Chapter 20 **Comparative Biochemical Characterization of Soluble and Chitosan Immobilized β-Galactosidase from Kluyveromyces Lactis Nrrl Y1564 ... 175**

Abstract ... 175
1. Introduction ... 175
2. Materials and Methods ... 176
3. Results and Discussion ... 179

Conclusions ... 185
Acknowledgements ... 185
References ... 185

Chapter 21 **High Stability of Immobilized β-D-Galactosidase for Lactose Hydrolysis and Galactooligosaccharides Synthesis** ... 189
Abstract ... 189
1. Introduction ... 189
2. Materials and Methods ... 190
3. Results and Discussion ... 191
Conclusions ... 194
Acknowledgements ... 194
References ... 194

Chapter 22 **Novel High-Performance Metagenome β-Galactosidases for Lactose Hydrolysis in the Dairy Industry** ... **197**
Abstract ... 197
1. Materials And Methods ... 198
2. Results ... 202
3. Discussion ... 205
Conclusions ... 206
Acknowledgments ... 206
References ... 206

Chapter 23 **Study of Immobilized Candida Rugosa Lipase for Biodiesel Fuel Production from Palm Oil Bby Flow Microcalorimetry** ... **209**
Abstract ... 209
1. Introduction ... 209
2. Methods ... 211
3. Results and Discussion ... 213
Conclusions ... 215
Acknowledgements ... 215
References ... 215

Index ... **217**

List of Contributors

Sumitra Datta
School of Chemical and Biotechnology, Shanmuga Arts, Science, Technology and Research Academy (SASTRA) University, Tirumalaisamudram, Thanjavur 613401, Tamilnadu, India

L. Rene Christena
School of Chemical and Biotechnology, Shanmuga Arts, Science, Technology and Research Academy (SASTRA) University, Tirumalaisamudram, Thanjavur 613401, Tamilnadu, India

Yamuna Rani Sriramulu Rajaram
School of Chemical and Biotechnology, Shanmuga Arts, Science, Technology and Research Academy (SASTRA) University, Tirumalaisamudram, Thanjavur 613401, Tamilnadu, India

Christina Rother
Graz University of Technology, Graz, Austria

Bernd Nidetzky
Graz University of Technology, Graz, Austria
Austrian Centre of Industrial Biotechnology, Graz, Austria

Yuichi Masuda
Materials Chemistry Course, Graduate School of Engineering, Aichi Institute of Technology, 1247 Yachigusa, Yakusa-cho, Toyota 470-0392, Japan
National Institute of Advanced Industrial Science and Technology (AIST), 2266-98, Anagahora, Shimosidami, Moriyama-ku, Nagoya 463-8510, Japan

Shin-ichi Kugimiya
Materials Chemistry Course, Graduate School of Engineering, Aichi Institute of Technology, 1247 Yachigusa, Yakusa-cho, Toyota 470-0392, Japan
Katsuya Kato [2]
National Institute of Advanced Industrial Science and Technology (AIST), 2266-98, Anagahora, Shimosidami, Moriyama-ku, Nagoya 463-8510, Japan

Dede Heri Yuli Yanto
The United Graduate School of Agricultural Sciences, Ehime University. 3-5-7 Tarumi, Matsuyama, Ehime 790-8566, Japan
Research and Development Unit for Biomaterials, Indonesian Institute of Science. Jl. Raya Bogor Km. 46, Cibinong, Bogor, 16911 Indonesia.

Sanro Tachibana
Department of Applied Bioscience, Faculty of Agriculture, Ehime University. 3-5-7 Tarumi, Matsuyama, Ehime, 790-8566,

Kazutaka Itoh
Department of Applied Bioscience, Faculty of Agriculture, Ehime University. 3-5-7 Tarumi, Matsuyama, Ehime, 790-8566,

Sabrina Feliciano Oliveira
Department of Microbiology, Universidade Federal de Minas Gerais, Av. Anto^nio Carlos, 6627 – Pampulha, 31270-901 Belo Horizonte, Minas Gerais, Brazil

Jose´ Maria Rodrigues da Luz
Department of Microbiology, Universidade Federal de Vic¸ osa, Av. Peter Henry Rolfs, s/n° – Campus Universita´rio, Vic¸ osa, Minas Gerais, Brazil

Maria Catarina Megumi Kasuya
Department of Microbiology, Universidade Federal de Vic¸ osa, Av. Peter Henry Rolfs, s/n° – Campus Universita´rio, Vic¸ osa, Minas Gerais, Brazil

Luiz Orlando Ladeira
Department of Microbiology, Universidade Federal de Minas Gerais, Av. Anto^nio Carlos, 6627 – Pampulha, 31270-901 Belo Horizonte, Minas Gerais, Brazil

Ary Correa Junior
Department of Microbiology, Universidade Federal de Minas Gerais, Av. Anto^nio Carlos, 6627 – Pampulha, 31270-901 Belo Horizonte, Minas Gerais, Brazil

Anfeng Xiao
College of Food and Biological Engineering, Jimei University, Xiamen 361021, China
Fujian Provincial Key Laboratory of Food Microbiology and Enzyme Engineering, Xiamen, Fujian Province 361021, China
Key Laboratory of Systemic Utilization and In-depth Processing of Economic Seaweed, Xiamen Southern Ocean Technology Center of China, Xiamen, Fujian Province 361021, China

Caiyun Xu
College of Food and Biological Engineering, Jimei University, Xiamen 361021, China

Yan Lin
College of Chengyi, Jimei University, Xiamen 361021, China

Hui Ni
College of Food and Biological Engineering, Jimei University, Xiamen 361021, China
Fujian Provincial Key Laboratory of Food Microbiology and Enzyme Engineering, Xiamen, Fujian Province 361021, China
Key Laboratory of Systemic Utilization and In-depth Processing of Economic Seaweed, Xiamen Southern Ocean Technology Center of China, Xiamen, Fujian Province 361021, China

Yanbing Zhu
College of Food and Biological Engineering, Jimei University, Xiamen 361021, China
Fujian Provincial Key Laboratory of Food Microbiology and Enzyme Engineering, Xiamen, Fujian Province 361021, China
Key Laboratory of Systemic Utilization and In-depth Processing of Economic Seaweed, Xiamen Southern Ocean Technology Center of China, Xiamen, Fujian Province 361021, China

Huinong Cai
College of Food and Biological Engineering, Jimei University, Xiamen 361021, China
Fujian Provincial Key Laboratory of Food Microbiology and Enzyme Engineering, Xiamen, Fujian Province 361021, China
Key Laboratory of Systemic Utilization and In-depth Processing of Economic Seaweed, Xiamen Southern Ocean Technology Center of China, Xiamen, Fujian Province 361021, China

Sheila G. Valerio
Enzymology Lab, Institute of Food Science and Technology, Federal University of Rio Grande do Sul, Av. Bento Gonc¸ alves 9500, ZC 91501-970 Porto Alegre, RS, Brazil

Joana S. Alves
Enzymology Lab, Institute of Food Science and Technology, Federal University of Rio Grande do Sul, Av. Bento Gonc¸ alves 9500, ZC 91501-970 Porto Alegre, RS, Brazil

Manuela P. Klein
Enzymology Lab, Institute of Food Science and Technology, Federal University of Rio Grande do Sul, Av. Bento Gonc¸ alves 9500, ZC 91501-970 Porto Alegre, RS, Brazil

Rafael C. Rodrigues
Biocatalysis and Enzyme Technology Lab, Institute of Food Science and Technology, Federal University of Rio Grande do Sul, Av. Bento Gonc¸ alves 9500, ZC 91501-970 Porto Alegre, RS, Brazil

Plinho F. Hertz
Enzymology Lab, Institute of Food Science and Technology, Federal University of Rio Grande do Sul, Av. Bento Gonc¸ alves 9500, ZC 91501-970 Porto Alegre, RS, Brazil

Ivana Cesarino
Instituto de Química de São Carlos, Universidade de São Paulo, São Carlos, São Paulo CEP 13566-590, Brazil

Fernando C. Moraes
Instituto de Química de São Carlos, Universidade de São Paulo, São Carlos, São Paulo CEP 13566-590, Brazil

Marcos R.V. Lanza
Instituto de Química de São Carlos, Universidade de São Paulo, São Carlos, São Paulo CEP 13566-590, Brazil

Sergio A.S. Machado
Instituto de Química de São Carlos, Universidade de São Paulo, São Carlos, São Paulo CEP 13566-590, Brazil

Chi Shein Hong
Faculty of Chemical & Natural Resources Engineering, Universiti Malaysia Pahang, 26300 Gambang, Pahang, MALAYSIA

Sim Yee Chin
Faculty of Chemical & Natural Resources Engineering, Universiti Malaysia Pahang, 26300 Gambang, Pahang, MALAYSIA

Chin Kui Cheng
Faculty of Chemical & Natural Resources Engineering, Universiti Malaysia Pahang, 26300 Gambang, Pahang, MALAYSIA

Mahmud Mohd Sabri
Faculty of Chemical & Natural Resources Engineering, Universiti Malaysia Pahang, 26300 Gambang, Pahang, MALAYSIA

Gek Kee Chua
Faculty of Chemical & Natural Resources Engineering, Universiti Malaysia Pahang, 26300 Gambang, Pahang, MALAYSIA

Qingqing Wang
Key Laboratory of Eco-Textiles, Ministry of Education, Jiangnan University, Wuxi 214122, China

Jing Cui
Key Laboratory of Eco-Textiles, Ministry of Education, Jiangnan University, Wuxi 214122, China

Guohui Li
Key Laboratory of Eco-Textiles, Ministry of Education, Jiangnan University, Wuxi 214122, China

Jinning Zhang
Key Laboratory of Eco-Textiles, Ministry of Education, Jiangnan University, Wuxi 214122, China

Fenglin Huang
Key Laboratory of Eco-Textiles, Ministry of Education, Jiangnan University, Wuxi 214122, China

Qufu Wei
Key Laboratory of Eco-Textiles, Ministry of Education, Jiangnan University, Wuxi 214122, China

Siti Noraida Abd Rahim
Tropical Agro-Biomass Research Group, Faculty of Plantation and Agrotechnology, Universiti Teknologi MARA, 40450 Shah Alam, Malaysia

Alawi Sulaiman
Tropical Agro-Biomass Research Group, Faculty of Plantation and Agrotechnology, Universiti Teknologi MARA, 40450 Shah Alam, Malaysia

Fazlena Hamzah
Faculty of Chemical Engineering, Universiti Teknologi MARA, 40450 Shah Alam, Malaysia

Ku Halim Ku Hamid
Faculty of Chemical Engineering, Universiti Teknologi MARA, 40450 Shah Alam, Malaysia

Miradatul Najwa Muhd Rodhi
Faculty of Chemical Engineering, Universiti Teknologi MARA, 40450 Shah Alam, Malaysia

Mohibbah Musa
Faculty of Chemical Engineering, Universiti Teknologi MARA, 40450 Shah Alam, Malaysia

Nurul Aini Edama
Tropical Agro-Biomass Research Group, Faculty of Plantation and Agrotechnology, Universiti Teknologi MARA, 40450 Shah Alam, Malaysia

Muhammad Asif Nawaz
The Karachi Institute of Biotechnology and Genetic Engineering (KIBGE), University of Karachi, Karachi 75270, Pakistan

Haneef Ur Rehman
The Karachi Institute of Biotechnology and Genetic Engineering (KIBGE), University of Karachi, Karachi 75270, Pakistan

Zainab Bibi
The Karachi Institute of Biotechnology and Genetic Engineering (KIBGE), University of Karachi, Karachi 75270, Pakistan

Afsheen Aman
The Karachi Institute of Biotechnology and Genetic Engineering (KIBGE), University of Karachi, Karachi 75270, Pakistan

Shah Ali Ul Qader
The Karachi Institute of Biotechnology and Genetic Engineering (KIBGE), University of Karachi, Karachi 75270, Pakistan

Amanatuzzakiah Abdul Halim
The Advanced Centre for Biochemical Engineering, University College London, Torrington Place, London WC1E 7JE, United Kingdom
Biotechnology Engineering Department, International Islamic University Malaysia (IIUM), 50728 Kuala Lumpur, Malaysia

Nicolas Szita
The Advanced Centre for Biochemical Engineering, University College London, Torrington Place, London WC1E 7JE, United Kingdom

Frank Baganz
The Advanced Centre for Biochemical Engineering, University College London, Torrington Place, London WC1E 7JE, United Kingdom

Adriano A. Mendes
Laboratory of Biocatalysis, Federal University of São João del Rei, 35701-970, Sete Lagoas, MG, Brazil
Department of Chemical Engineering, Federal University of São Carlos, 13565-905, São Carlos, SP, Brazil

Pedro C. Oliveira
Department of Chemical Engineering, Engineering School of Lorena, University of São Paulo, 12602-810, Lorena, SP, Brazil

Ana M. Vélez
Department of Chemical Engineering, Federal University of São Carlos, 13565-905, São Carlos, SP, Brazil

Roberto C. Giordano
Department of Chemical Engineering, Federal University of São Carlos, 13565-905, São Carlos, SP, Brazil

Raquel de L.C. Giordano
Department of Chemical Engineering, Federal University of São Carlos, 13565-905, São Carlos, SP, Brazil

Heizir F. de Castro
Department of Chemical Engineering, Engineering School of Lorena, University of São Paulo, 12602-810, Lorena, SP, Brazil

Wei-Wei Zhang
Key Laboratory of Green Chemistry & Technology, Ministry of Education, College of Chemistry, Sichuan University, Chengdu 610064, PR China

Jun-Qi Jia
Key Laboratory of Green Chemistry & Technology, Ministry of Education, College of Chemistry, Sichuan University, Chengdu 610064, PR China

Na Wang
Key Laboratory of Green Chemistry & Technology, Ministry of Education, College of Chemistry, Sichuan University, Chengdu 610064, PR China

Cheng-Li Hu
Key Laboratory of Green Chemistry & Technology, Ministry of Education, College of Chemistry, Sichuan University, Chengdu 610064, PR China

Sheng-Yong Yang
State Key Laboratory of Biotherapy and Cancer Center, West China Hospital, West China Medical School, Sichuan University, Chengdu 610064, PR China

Xiao-Qi Yu
Key Laboratory of Green Chemistry & Technology, Ministry of Education, College of Chemistry, Sichuan University, Chengdu 610064, PR China

M. Naya, M. Imai
Course in Bioresource Utilization Sciences, Graduate School of Bioresource Sciences, Nihon University, 1866 Kameino, Fujisawa, Kanagawa 252-0880, Japan

H.F.Castro
Department of Chemical Engineering, School of Chemical Engineering of Lorena,

M.L.C.P.Silva
Department of Chemical Engineering, School of Chemical Engineering of Lorena,

G.L.J.P Silva
Department of Chemical Engineering, School of Chemical Engineering of Lorena,

Hendrik du Toit
University of Bath, Department of Chemical Engineering, Bath, BA2 7AY, UK

Mirella Di Lorenzo
University of Bath, Department of Chemical Engineering, Bath, BA2 7AY, UK

Anurat Wisitsoraat
Nanoelectronics and MEMS Laboratory, National Electronics and Computer Technology Center, 112 Thailand Science Park, Pahol Yothin Rd., Klong Luang, Pathumthani 12120, Thailand

Chanpen Karuwan
Nanoelectronics and MEMS Laboratory, National Electronics and Computer Technology Center, 112 Thailand Science Park, Pahol Yothin Rd., Klong Luang, Pathumthani 12120, Thailand

Krongkamol Wong-ek
Nanoelectronics and MEMS Laboratory, National Electronics and Computer Technology Center, 112 Thailand Science Park, Pahol Yothin Rd., Klong Luang, Pathumthani 12120, Thailand

Ditsayut Phokharatkul
Nanoelectronics and MEMS Laboratory, National Electronics and Computer Technology Center, 112 Thailand Science Park, Pahol Yothin Rd., Klong Luang, Pathumthani 12120, Thailand

Pornpimol Sritongkham
Nanoelectronics and MEMS Laboratory, National Electronics and Computer Technology Center, 112 Thailand Science Park, Pahol Yothin Rd., Klong Luang, Pathumthani 12120, Thailand

Adisorn Tuantranont
Nanoelectronics and MEMS Laboratory, National Electronics and Computer Technology Center, 112 Thailand Science Park, Pahol Yothin Rd., Klong Luang, Pathumthani 12120, Thailand

Ariosvana Fernandes Lima
Instituto Federal de Educac¸ ão, Ciência e Tecnologia do Ceará, Campus Limoeiro do Norte, Brazil

Kenia Franco Cavalcante
Departamento de Engenharia Química, Universidade Federal do Ceará, Campus do Pici, Bloco 709, CEP 60455-760, Fortaleza, CE, Brazil

Maria de Fátima Matos de Freitas
Departamento de Engenharia Química, Universidade Federal do Ceará, Campus do Pici, Bloco 709, CEP 60455-760, Fortaleza, CE, Brazil

Tigressa Helena Soares Rodrigues
Departamento de Engenharia Química, Universidade Federal do Ceará, Campus do Pici, Bloco 709, CEP 60455-760, Fortaleza, CE, Brazil

Maria Valderez Ponte Rocha
Departamento de Engenharia Química, Universidade Federal do Ceará, Campus do Pici, Bloco 709, CEP 60455-760, Fortaleza, CE, Brazil

Luciana Rocha Barros Gonc, alves
Departamento de Engenharia Química, Universidade Federal do Ceará, Campus do Pici, Bloco 709, CEP 60455-760, Fortaleza, CE, Brazil

Manuela P. Klein[1]
Chemical and Food Engineering Department, Federal University of Santa Catarina, ZC 88040-900 Florianópolis, SC, Brazil
Enzymology Laboratory, Food Science and Technology Institute, Federal University of Rio Grande do Sul, Av. Bento Gonc, alves 9500, P.O. Box 15090, ZC 91501-970 Porto Alegre, RS, Brazil

Lucas P. Fallavena
Enzymology Laboratory, Food Science and Technology Institute, Federal University of Rio Grande do Sul, Av. Bento Gonc, alves 9500, P.O. Box 15090, ZC 91501-970 Porto Alegre, RS, Brazil

Jéssie da N. Schöffer
Enzymology Laboratory, Food Science and Technology Institute, Federal University of Rio Grande do Sul, Av. Bento Gonc, alves 9500, P.O. Box 15090, ZC 91501-970 Porto Alegre, RS, Brazil

Marco A.Z. Ayub
Enzymology Laboratory, Food Science and Technology Institute, Federal University of Rio Grande do Sul, Av. Bento Gonc, alves 9500, P.O. Box 15090, ZC 91501-970 Porto Alegre, RS, Brazil

Rafael C. Rodrigues
Enzymology Laboratory, Food Science and Technology Institute, Federal University of Rio Grande do Sul, Av. Bento Gonc, alves 9500, P.O. Box 15090, ZC 91501-970 Porto Alegre, RS, Brazil

Jorge L. Ninow
Chemical and Food Engineering Department, Federal University of Santa Catarina, ZC 88040-900 Florianópolis, SC, Brazil

Plinho F. Hertz
Enzymology Laboratory, Food Science and Technology Institute, Federal University of Rio Grande do Sul, Av. Bento Gonc, alves 9500, P.O. Box 15090, ZC 91501-970 Porto Alegre, RS, Brazil

Sarah Erich
University of Hohenheim, Institute of Food Science and Biotechnology, Department of Biotechnology and Enzyme Science, Garbenstr. 25, 70599 Stuttgart, Germany

Beatrice Kuschel
University of Hohenheim, Institute of Food Science and Biotechnology, Department of Biotechnology and Enzyme Science, Garbenstr. 25, 70599 Stuttgart, Germany

Thilo Schwarz
University of Hohenheim, Institute of Food Science and Biotechnology, Department of Biotechnology and Enzyme Science, Garbenstr. 25, 70599 Stuttgart, Germany

Jacob Ewert
University of Hohenheim, Institute of Food Science and Biotechnology, Department of Biotechnology and Enzyme Science, Garbenstr. 25, 70599 Stuttgart, Germany

Nico Böhmer
University of Hohenheim, Institute of Food Science and Biotechnology, Department of Biotechnology and Enzyme Science, Garbenstr. 25, 70599 Stuttgart, Germany

Frank Niehaus
B.R.A.I.N. Biotechnology Research and Information Network AG, Darmstaedter Str. 34-36, 64673, Zwingenberg, Germany

Jürgen Eck
B.R.A.I.N. Biotechnology Research and Information Network AG, Darmstaedter Str. 34-36, 64673, Zwingenberg, Germany

Sabine Lutz-Wahl
University of Hohenheim, Institute of Food Science and Biotechnology, Department of Biotechnology and Enzyme Science, Garbenstr. 25, 70599 Stuttgart, Germany

Timo Stressler
University of Hohenheim, Institute of Food Science and Biotechnology, Department of Biotechnology and Enzyme Science, Garbenstr. 25, 70599 Stuttgart, Germany

Lutz Fischer
University of Hohenheim, Institute of Food Science and Biotechnology, Department of Biotechnology and Enzyme Science, Garbenstr. 25, 70599 Stuttgart, Germany

J.C. Moreno-Piraja`n
Facultad de Ciencias, Departamento de Quı´mica, Grupo de Investigacio´n en So´lidos Porosos y Calorimetrı´a, Universidad de los andes, Carrera 1 No 18 A 10, Edifcio Q Oficina 837, Bogota´, Colombia

Liliana Giraldo
Facultad de Ciencias, Departamento de Quı´mica, Universidad Nacional de Colombia, Universidad de los andes, Carrera 30 No 45-00, Ciudad Universitaria, Bogota´, Colombia

Preface

The book is a comprehensive review of immobilized enzyme which is to guide the engineer and scientist along the path toward their industrial applications. The first section of this book aims to clarify the principle of immobilized enzymes. Several contributing factor in an immobilized enzyme system will be analyzed in an effort to accomplish this objective. Those factors that contribute to the performance of an immobilized enzyme system include the carrier, the immobilization technique, cofactor requirements, the reactor, and carrier regeneration. In earlier days, immobilized enzymes were carried out using the biotechnology methods, which were classified as Reverse enzyme immobilization and irreverse enzyme immobilization methods. Reverse enzyme immobilization includes absorption, ionic binding, affinity binding and metal binding. Irreverse enzyme immobilizations includes covalent binding and entrapment such as encapsulation, entrapping. The technique and ideal materials selected depend on enzyme process. Several studies compared the biochemical characterizations of soluble enzyme and immobilized enyzme. Because of rapid growth of nanotechnology, the synergistic interaction of biotechnology methods and nanotechnology methods, various enzymes are immobilized using conventional methods and variety nanomaterial such as carbon nanotubes, chitosan nanoparticles and polyaniline–carbon nanotubes. This book also reports that the development of techniques and support material/matrix improve significant enzyme activity and increasing their self life.

The remaining content of this book focuses on the variety applications of enzyme immobilization in industrial processing such as wood processing, paper manufacturing, sucrose hydrolysis, lactose hydrolysis and production of glucose syrup which is further fermented to bioethanol production and fermentation of food processing.The immobilized enzymes are always more stable than soluble enzymes, thereby allowing their applications even under harsh conditions such as pH, high temperature, non-aqueous solvents and organic solvents, then enable following their application in extra temperature and humidity biosensor for detecting pesticide level in food, water and formaldehyde in pollutant as well as biofuel cell. Beside that, their applications in biodiesel is also discussed.

Editor
Nguyen Minh Nhut

CHAPTER 1

ENZYME IMMOBILIZATION: AN OVERVIEW ON TECHNIQUES AND SUPPORT MATERIALS

Sumitra Datta, L. Rene Christena, and Yamuna Rani Sriramulu Rajaram

School of Chemical and Biotechnology, Shanmuga Arts, Science, Technology and Research Academy (SASTRA) University, Tirumalaisamudram, Thanjavur 613401, Tamilnadu, India

ABSTRACT

The current demands of the world's biotechnological industries are enhancement in enzyme productivity and development of novel techniques for increasing their shelf life. These requirements are inevitable to facilitate large-scale and economic formulation. Enzyme immobilization provides an excellent base for increasing availability of enzyme to the substrate with greater turnover over a considerable period of time. Several natural and synthetic supports have been assessed for their efficiency for enzyme immobilization. Nowadays, immobilized enzymes are preferred over their free counterpart due to their prolonged availability that curtails redundant downstream and purification processes. Future investigations should endeavor at adopting logistic and sensible entrapment techniques along with innovatively modified supports to improve the state of enzyme immobilization and provide new perspectives to the industrial sector.

INTRODUCTION

Enzymes or 'biocatalysts' are remarkable discovery in the field of bioprocess technology. Biocatalysis has been widely accepted in diverse sectors owing to their ease of production, substrate specificity and green chemistry. However, for large extent commercialization of these bio-derived catalysts, their reusability factor becomes mandatory, failing which they would no longer be economic. Maintenance of their structural stability during any biochemical reaction is highly challenging. Consequently, immobilized enzymes with functional efficiency and enhanced reproducibility are used as alternatives in spite of their expensiveness. Immobilized biocatalysts can either be enzymes or whole cells (Kawaguti et al. 2006). Enzyme immobilization is confinement of enzyme to a phase (matrix/support) different from the one for substrates and products. Inert polymers and inorganic materials are usually used as carrier matrices. Apart from being affordable, an ideal matrix must encompass characteristics like inertness, physical strength, stability, regenerability, ability to increase enzyme specificity/activity and reduce product inhibition, nonspecific adsorption and microbial contamination (Singh 2009). Immobilization generates continuous economic operations, automation, high investment/capacity ratio and recovery of product with greater purity (D'Souza1998). Several methods

Citation: Datta S, Christena LR, Rajaram YRS. Enzyme immobilization: an overview on techniques and support materials. 3 Biotech. 2013;3(1):1-9. doi:10.1007/s13205-012-0071-7.

are used for immobilization and various factors influence the performance of immobilized enzymes (Table 1). Adsorption/carrier-binding method uses water-insoluble carriers such as polysaccharide derivatives, synthetic polymers and glass (Al-Adhami et al. 2002; Rosa et al. 2002; Wu and Lia 2008; Cordeiro et al. 2011). In cross-linking/covalent method, bi/multifunctional reagents such as glutaraldehyde, bisdiazobenzidine and hexamethylene diisocyanate are used (Lee et al. 2006; Singh 2009). Polymers like collagen, cellulose and κ-carrageenan are employed by entrapment method, while the membrane confinement method includes formulation of liposomes and microcapsules (Katwa et al. 1981; Wang and Hettwer 1982; Mislovicová et al. 2004; Hilal et al. 2006; Tümtürk et al. 2007; Rochefort et al.2008; Jegannathan et al. 2010; Chen et al. 2011a, b; Klein et al. 2011).

Table 1: Factors influencing performance of immobilized enzymes (Cao 2006)

Factors	Implications of immobilization
Hydrophobic partition	Enhancement of reaction rate of hydrophobic substrate
Microenvironment of carrier	Hydrophobic nature stabilizes enzyme
Multipoint attachment of carrier	Enhancement of enzyme thermal stability
Spacer or arm of various types of immobilized enzymes	Prevents enzyme deactivation
Diffusion constraints	Enzyme activity decreases and stability increases
Presence of substrates or inhibitors	Higher activity retention
Physical post-treatments	Improvement of enzyme performance
Different binding mode	Activity and stability can be affected
Physical structure of the carrier such as pore size	Activity retention was often pore-size dependent
Physical nature of the carrier	Carriers with large pore size mitigate diffusion limitation, leading to higher activity retention

This article reviews the existing techniques used for immobilization along with providing insights into the recent developments for each of them. We have tried to throw light on significant modifications with respect to the techniques and innovative support materials employed for immobilization of biocatalysts that have potential implication on future enzyme market.

DIFFERENT TECHNIQUES USED FOR IMMOBILIZATION

Adsorption

Enzyme adsorption results from hydrophobic interactions and salt linkages where either the support is bathed in enzyme for physical adsorption or the enzyme is dried on electrode surfaces. Adsorbed enzymes are shielded from aggregation, proteolysis and interaction with hydrophobic interfaces (Spahn and Minteer2008). Researchers have used eco-friendly supports like coconut fibers having good water-holding capacity and high cation exchange property; microcrystalline cellulose with irreversible binding capacity; kaolin with high enzyme retainability by chemical acetylation; and micro/mesoporous materials having thiol functionalized, large surface area ideally suited for reduction and oxidation reactions (Dey et al. 2002; Hernández et al. 2007; Karagulyan et al. 2008; Brígida et al. 2010; Mitchell and Ramírez 2011; Huang et al.2011). Silanized molecular sieves have also been successfully used as supports for enzyme adsorption owing to the presence of silanols on pore walls that facilitate enzyme immobilization by hydrogen bonding (Diaz and Balkus 1996). Various chemical modifications of the currently used supports would definitely help in better immobilization. Water activity profiles of lipase adsorbed using polypropylene-based hydrophobic granules/Accurel EP-100 has been reported (Persson et al. 2000). It would be important to note that Accurel with smaller particle sizes increases reaction rates and enantiomeric ratios during biocatalyzation (Sabbani et al. 2006).

For better process control and economic production, *Yarrowia lipolytica* lipase was immobilized on octyl-agarose and octadecyl-sepabeads supports by physical adsorption that resulted in higher yields and greater (tenfold) stability than that of free lipase. This was accounted by the hydrophobicity of octadecyl-sepabeads that enhances affinity between the enzyme and support (Cunha et al. 2008). *Candida rugosa* lipase adsorbed on biodegradable poly (3-hydroxybutyrate-co-hydroxyvalerate) showed 94 % residual activity after 4 h at 50 °C and reusability till 12 cycles (Cabrera-Padilla et al. 2011). These supports were preferred because they are less tough and crystalline than polyhydroxybutyrate. 1, 4-Butenediol diglycidyl ether-activated byssus threads have been suitable basement for urease that increased pH stability and retained 50 % enzyme activity under dried conditions (Mishra et al. 2011). Eco-friendly supports of biological origin not only prevent cropping up of ethical issues, but also cut down the production costs. Of late, biocompatible mesoporous silica nanoparticles (MSNs) supports have been used for biocatalysis in energy applications owing to their long-term durability and efficiency (Popat et al. 2011).

Covalent binding

Covalent association of enzymes to supports occurs owing to their side chain amino acids like arginine, aspartic acid, histidine and degree of reactivity based on different functional groups like imidazole, indolyl, phenolic hydroxyl, etc. (D'Souza 1998; Singh 2009). Peptide-modified surfaces when used for enzyme linkage results in higher specific activity and stability with controlled protein orientation (Fu et al. 2011). Cyanogen bromide (CNBr)-agarose and CNBr-activated-Sepharose containing carbohydrate moiety and glutaraldehyde as a spacer arm have imparted thermal stability to covalently bound enzymes (Hsieh et al.2000; Cunha et al. 2008). Highly stable and hyperactive biocatalysts have been reported by covalent binding of enzymes to silica gel carriers modified by silanization with elimination of unreacted aldehyde groups and to SBA-15 supports containing cage-like pores lined by Si–F moieties (Lee et al. 2006; Szymańska et al.2009). Increase in half-life and thermal stability of enzymes has been achieved by covalent coupling with different supports like mesoporous silica, chitosan, etc. (Hsieh et al. 2000; Ispas et al. 2009). Cross-linking of enzymes to electrospun nanofibers has shown greater residual activity due to increased surface area and porosity. Use of such nanodiametric supports have brought a turning point in the field of biocatalyst immobilization (Wu et al. 2005; Kim et al. 2006; Ren et al. 2006; Li et al. 2007; Huang et al. 2008; Sakai et al. 2010). Covalent binding of alcohol dehydrogenase on attapulgite nanofibers (hydrated magnesium silicate) has been opted owing to its thermal endurance and variable nano sizes (Zhao et al. 2010). Biocatalytic membranes have been useful in unraveling effective covalent interactions with silicon-coated enzymes (Hilal et al. 2006). Cross-linked enzyme aggregates produced by precipitation of enzyme from aqueous solution by addition of organic solvents or ionic polymers have been reported (Sheldon 2011). Different orientations of immobilized enzyme on magnetic nanoclusters obtained by covalent binding have found their applications in pharmaceutical industries owing to their enhanced longevity, operational stability and reusability (Yusdy et al. 2009). Maintaining the structural and functional property of enzymes during immobilization is one of the major roles played by a cross-linking agent. One such agent is glutaraldehyde, popularly used as bifunctional cross-linker, because they are soluble in aqueous solvents and can form stable inter- and intra-subunit covalent bonds.

Affinity immobilization

Affinity immobilization exploits specificity of enzyme to its support under different physiological conditions. It is achieved by two ways: either the matrix is precoupled to an affinity ligand for target enzyme or the enzyme is conjugated to an entity that develops affinity toward the matrix (Sardar et al. 2000). Affinity adsorbents have also been used for simultaneous purification of enzymes (Ho et al. 2004). Complex affinity supports like alkali stable chitosan-coated porous silica beads and agarose-linked multilayered concanavalin A harbor higher amounts of enzymes which lead to increased stability and efficiency (Shi et al. 2003; Sardar and Gupta 2005). Bioaffinity layering is an improvisation of this technique that exponentially increases enzyme-binding capacity and reusability due to the presence of non-covalent forces such as coulombic, hydrogen bonding, van der Waals forces, etc. (Sardar and Gupta 2005; Haider and Husain 2008).

Entrapment

Entrapment is caging of enzymes by covalent or non-covalent bonds within gels or fibers (Singh 2009). Efficient encapsulation has been achieved with alginate–gelatin–calcium hybrid carriers that prevented enzyme leakage and provided increased mechanical stability (Shen et al. 2011). Entrapment by nanostructured supports like electrospun nanofibers and pristine materials have revolutionalized the world of enzyme immobilization with their wide-ranging applications in the field of fine chemistry, biomedicine biosensors and biofuels (Dai and Xia 2006; Kim et al. 2006; Wang et al. 2009; Wen et al. 2011). Prevention of friability and leaching and augmentation of entrapment efficiency and enzyme activity by *Candida rugosa*lipase entrapped in chitosan have been reported. This support has also been reported to be non-toxic, biocompatible and amenable to chemical modification and highly affinitive to protein due to its hydrophilic nature (Betigeri and Neau 2002). Entrapment by mesoporous silica is attributed to its high surface area, uniform pore distribution, tunable pore size and high adsorption capacity (Ispas et al. 2009). Simultaneous entrapment of lipase and magnetite nanoparticles with biomimetic silica enhanced its activity in varying silane additives (Chen et al. 2011a). Sol–gel matrices with supramolecular calixarene polymers have been used for entrapment of *C. rugosa* lipase keeping in view their selective binding and carrying capacities (Erdemir and Yilmaz 2011). Lipases entrapped κ-carrageenan has been reported to be highly thermostable and organic solvent tolerant (Tümtürk et al. 2007; Jegannathan et al. 2010).

MATERIALS USED FOR FABRICATION OF IMMOBILIZATION SUPPORTS

Natural polymers as supports

Alginate

Alginate derived from cell walls of brown algae are calcium, magnesium and sodium salts of alginic acid and have been extensively used for immobilization as xanthan–alginate beads, alginate–polyacrylamide gels and calcium alginate beads with enhanced enzyme activity and reusability. Cross-linking of alginate with divalent ions (like Ca^{2+}) and glutaraldehyde improves the stability of enzymes (Elçin 1995; Flores-Maltos et al. 2011).

Chitosan and chitin

Natural polymers like chitin and chitosan have been used as supports for immobilization (Vaillant et al. 2000; Kapoor and Kuhad 2007). The protein or carbohydrate moieties of enzymes are used for binding them to chitosan (Hsieh et al. 2000). Chitosan has been used in combination with alginate where chitosan-coated enzymes had less leaching effect compared to alginate owing to the physical and ionic interactions between the enzyme and support (Betigeri and Neau 2002). Similarly, a wet composite of chitosan and clay proved to be more reliable for enzyme trapping, because it has hydroxyl and amino groups, which easily link with enzymes, together with good hydrophilicity and high porosity. Chitosan in the form of beads can entrap twice as much of the enzymes (Chang and Juang 2007). According to Chern and Chao (2005), the chitin-binding domain of chitinase A1 from *Bacillus circulans* has a high affinity to chitin; so, this property has been exploited to retain D-hydantoinase.

Collagen

Being a natural polymer, collagen has been used for immobilization of tannase employing glutaraldehyde as cross-linking agent (Katwa et al. 1981). Fe^{3+}-collagen fibers proved to be excellent supporting matrix for catalase immobilization by retaining significant activity even after 26 reuses (Chen et al.2011b).

Carrageenan

Carrageenan, a linear sulfated polysaccharide, has been consistently used for immobilizing a variety of enzymes, like lipase for improving stability (Tümtürk et al. 2007). This support is pseudoplastic in nature, which helps it to thin under shear stress and recover its viscosity once the stress is removed. Jegannathan et al. (2010) could achieve an encapsulation efficiency of 42.6 % by the co-extrusion method using the same support for biodiesel production. Carrageenan has been reported as a cheap and durable support with better entrapment for lactic acid and α-galactosidase enzyme (Rao et al. 2008; Girigowda and Mulimani 2006).

Gelatin

Gelatin is a hydrocolloid material, high in amino acids, and can adsorb up to ten times its weight in water. Its indefinite shelf life has attracted attention for enzyme immobilization. Gelatin has been utilized in mixed carrier system with polyacrylamide where cross-linking with chromium (III) acetate proved better than chromium (III) sulfate and potassium chromium (III) sulfate (Emregul et al. 2006). Calcium alginate with gelatin forms a good template for calcium phosphate deposition for enzyme immobilization, and gelatin in combination with polyester films promoted 75 % loading efficiency, compared to previous studies which had 50 % loading efficiency (Shen et al. 2011 Ateş and Doğan 2010).

Cellulose

This most abundant natural polymer has been widely used to immobilize fungi laccase, penicillin G acylase, glucoamylase, α-amylase, tyrosinase, lipase and β-galactosidase (Al-Adhami et al. 2002; Mislovicová et al. 2004; Bryjak et al. 2007; Namdeo and Bajpai 2009; Labus et al. 2011; Huang et al. 2011; Klein et al. 2011). Diethylaminoethyl (DEAE)-modified cellulosic supports have longer storage capacity (Al-Adhami et al. 2002). Cellulose-coated magnetite nanoparticles have been used for starch degradation where the attachment of α-amylase to cellulose dialdehyde-coated magnetite nanoparticles resulted in the formation of a novel starch degrading system (Namdeo and Bajpai 2009). Immobilization with ionic liquid-cellulose film activated by glutaraldehyde gave better formability and flexibility (Klein et al. 2011).

Starch

Made of linear amylase and branched amylopectin units, starch has been used as enzyme immobilizer. Calcium alginate–starch hybrid supports were applied for surface immobilization and entrapment of bitter gourd peroxidase. Entrapped enzyme was more stable in the presence of denaturants like urea due to internal carbohydrate moieties, while surface-immobilized enzyme had superior activity (Matto and Husain 2009). Radiation grafting of substances like acrylamide and dimethylaminoethyl methacrylate onto starch are among the widely used industrial techniques for a high product yield (Dung et al. 1995; Raafat et al. 2011).

Pectin

This structural heteropolysaccharide along with 0.2–0.7 % glycerol acts as plasticizer to reduce brittleness of support and has been used to immobilize papain and for development of new materials for skin injury treatment (Ceniceros et al. 2003). Pectin–chitin and pectin–calcium alginate support have enhanced thermal and denaturant resistance and catalytic properties of entrapped enzymes due to the formation of high stable polyelectrolyte complexes between the enzyme and the pectin-coated support (Gómez et al. 2006; Satar et al. 2008).

Sepharose

CNBr-activated Sepharose-4B has been used to immobilize amylase and glucoamylase owing to its porosity and easy adsorption of macromolecules. Further matrix modifications like alkyl substituted Sepharose with multipoint attachment between hydrophobic clusters of the enzyme and alkyl residues of the support play a major role in retaining the catalytic properties at extremes of pH, high salt concentrations and elevated temperatures (Hosseinkhani et al. 2003). Another example of modified Sepharose matrix is concanavalin A (Con A)–Sepharose 4B where biospecific interaction between the glycosyl chains of the enzyme and Con A plays a pivotal role in fabrication of various biosensors (Mirouliaei et al. 2007).

Synthetic polymers as supports

Ion exchange resins/polymers are insoluble supports with porous surface for enzyme trapping. Amberlite and DEAE cellulose, renewable matrices with large surface area, have been used for immobilization of α-amylase (Kumari and Kayastha 2011). During white radish peroxidase immobilization, glutaraldehyde and polyethylene glycol act as an additive and protective layer around the active center of the enzyme to prevent the attack of free radicals (Ashraf and Husain 2010). Some synthetic polymers used as enzyme supports are stated as follows: polyvinyl chloride that prevents enzyme, cyclodextrin glucosyltransferase from thermal inactivation; polyurethane microparticles derived from polyvinyl alcohol and hexamethyl diisocyanate in the ratio of 1:3 with high enzyme loading and efficiency; UV-curable methacrylated/fumaric acid-modified epoxy that is proposed to be useful for industrial applications; polyaniline in two different forms, viz. emeraldine salt and emeraldine base powder used for covalent binding of α- amylase; glutaraldehyde-activated nylon for immobilizing lipase and UV-activated polyethylene glycol having high porosity employed for wastewater treatment (Abdel-Naby 1999; Kahraman et al. 2007; Pahujani et al. 2008; Romaskevic et al.2010; Xiangli et al. 2010; Ashly et al. 2011).

Inorganic materials as supports

Zeolites

Zeolites or 'molecular sieves' are microporous crystalline solids with well-defined structures and shape-selective properties and are widely used in molecular adsorption. Microporous zeolites were found to be a better support for α-chymotrypsin immobilization than microporous dealuminized ones because of the presence of more hydroxyl groups that form strong hydrogen bonds with the enzyme (Xing et al. 2000). Likewise, Na Y zeolite was used to immobilize lysozyme because it had higher activity compared to other supports as reported by Chang and Chu (2007). The heterogeneous surface of zeolites with multiple adsorption sites are considered to be suitable for modulating the enzyme and support interactions (Serralha et al. 1998).

Ceramics

Immobilization of *Candida antarctica* lipase on ceramic membrane showed that this inert support could be exploited for carrying out hydrolytic and synthetic reactions by limiting feedback inhibition (Magnan et al. 2004). Ceramic foams containing both macro (77 nm) and micropores (45 μm) was found to be efficient in lowering diffusion rate and increasing the specific surface area (Huang and Cheng 2008). Another example of ceramics is toyonite whose variable pore structure can be modified using different organic coatings (Kamori et al. 2000).

Celite

Celite is highly porous diatomaceous, bioaffinity material and has been used for immobilization of lipase, polyphenol oxidases and β-galactosidase, because it is an inexpensive support having low polarity and large adhesion area (Khan et al. 2006; Liu et al. 2009; Ansari and Husain 2011). It provides resistance against high pH or temperature, urea, detergents and organic solvents (Khan et al. 2006). Celite acts as an additive in sol–gel matrix for ω-transaminases immobilization. It has been preferred due to its chemical inertness and interconnected pore structure (Koszelewski et al. 2010).

Silica

Enzymes like lignin peroxidase and horseradish peroxidase (HRP) immobilized on activated silica have been effectively used for the removal of chlorolignins from eucalyptus kraft effluent (Dezott et al.1995). α-Amylase immobilized on silica nanoparticles improves cleaning performance of detergents. They have been used because of their nano-sized structures with high surface area, ordered arrangement and high stability to chemical and mechanical forces (Soleimani et al. 2011). Surface modifications of silica by amination of hydroxyl and reactive siloxane groups and addition of methyl or polyvinyl alcohol groups strengthen enzyme and support bonds (Rao et al. 2000; Shioji et al. 2003; Pogorilyi et al. 2007).

Glass

Glass is a highly viscous liquid and has been employed in immobilizing α-amylase; phthaloyl chloride containing amino group functionalized glass beads was found to be robust and renewable for the process (Kahraman et al. 2007). Another enzyme nitrite reductase was immobilized on controlled pore glass beads, which served as a biosensing device for continuous monitoring (Rosa et al. 2002). Urease immobilized on glass pH-electrodes has provided a stable biosensor for monitoring as low as 52 μg/ml urea in blood samples (Sahney et al. 2005).

Activated carbon

Both natural and hydrochloric acid-modified activated carbon has provided valuable support for enzymé adsorption (Alkan et al. 2009). Lately, mesoporous-activated carbon particles containing large contact sites for enzyme immobilization have been used for immobilizing acid protease and acidic lipases where catalytic efficiency has been significantly maintained after 21 cycles of reuse (Kumar et al. 2010; Ramani et al. 2011). It was also found that activated carbon with a high surface area (600–1,000 $m^2 g^{-1}$) and a significant fraction of its pore volume in the 300–1,000 Å range was suitable for enzyme immobilization (Daoud et al. 2010).

Charcoal

Chemical modification of charcoal by adsorbing papain with sulfhydryl groups increased the number of active sites and has been utilized for recovery of mercury from aqueous solution and efficiently employed for industrial wastewater treatment (Dutta et al. 2009). Charcoal supports have been also used in food industries for immobilizing amyloglucosidase for starch hydrolysis without any cross-linking agent and has 90 % catalytic activity (Rani et al. 2000). As reported earlier by Kibarer and Akovali (1996), charcoal is an excellent adsorbent with high adsorptive capacity and minimum fine particulate matter release.

APPLICATIONS AND SCOPE

Biocatalysts are the key players in various industrial processes. Constant efforts are being made to improve the enzyme's activity, efficiency, reproducibility and stability during industrial processes (Wang et al. 2010). Production of regioselective and enantioselective compounds for biomedical application has been possible by immobilized enzymes (Ren et al. 2006; Lee et al. 2009). Glucose biosensors have been developed using electrospun PVA and surface-modified carbon nanotubes (Wen et al. 2011). Hydrogen peroxide biosensors have been devised using γ-aluminum trioxide nanoparticles/chitosan film-modified electrode (Liu et al.2010). Agarose–guar has been successfully utilized for designing phenol biosensors (Bagal and Karve 2006). Currently, keen efforts are being taken for increasing the stability of biosensors. Immobilization of biosensing enzymes into nanocavities showed significant results (Vamvakaki and Chaniotakis 2007). Biosynthesis of polyester has been facilitated by immobilized *C. antarctica* lipase B, a greener alternative to petroleum-based conventional catalysts (Idris and Bukhari 2011). With the advent of nanotechnology, silica nanoparticles with immobilized laccase have been applied for elimination of micropollutants from wastewater (Zimmermann et al. 2011). Increasing environmental concerns have led to the use of immobilized biocatalysts for biodiesel production (Jegannathan et al. 2010).

The different factors influencing enzyme immobilization and the possible modifications for their enhancement in activity have been chalked out in Fig. 1.

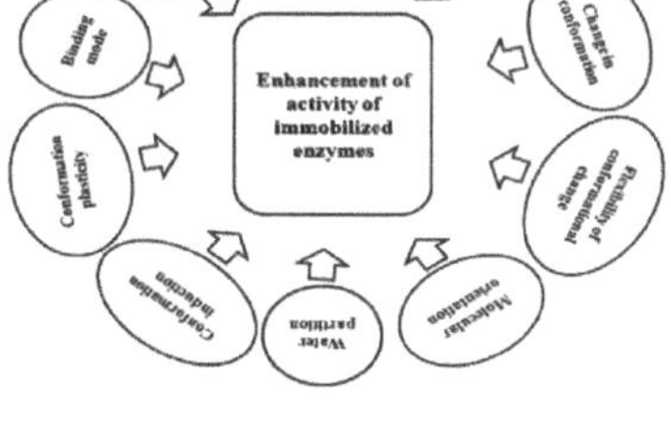

Figure 1: Determinants of enzyme immobilization and activity

CONCLUSION

With the vast array of research on enzyme immobilization, we can conclude that it is one of the most promising techniques for highly efficient and economically competent biotechnological processes in the field of environmental monitoring, biotransformation, diagnostics, pharmaceutical and food industries. Enzyme-based strategies are increasingly replacing conventional chemical methods in both laboratories and industries with attributes like efficiency, quicker performance and multifarious use. However, commercialization of immobilized enzymes is still at a lower pace because of their costs and storage problems. Research should be focused to overcome the current limitations related to immobilization techniques, so as to expand the horizon for all-round application.

ACKNOWLEDGMENTS

The authors wish to express their sincere gratitude to Dr. S. Swaminathan (Director, Centre for Nanotechnology and Advanced Biomaterials and Dean, Sponsored Research, SASTRA, Thanjavur) for his constant support and encouragement.

REFERENCES

1. Abdel-Naby MA (1999) Immobilization of Paenibacillus macerans NRRL B-3186 cyclodextrin glucosyltransferase and properties of the immobilized enzyme. Process Biochem 34:399–405
2. Al-Adhami AJH, Bryjak J, Greb-Markiewicz B, Peczynska-Czoch W (2002) Immobilization of wood-rotting fungi laccases on modified cellulose and acrylic carriers. Process Biochem 37:1387– 1394
3. Alkan S, Gu¨r A, Ertan M, Savran A, Gu¨r T, Genel Y (2009) Immobilization of catalase via adsorption into natural and modified active carbon obtained from walnut in various methods. Afr J Biotechnol 8:2631–2635
4. Ansari SA, Husain Q (2011) Lactose hydrolysis from milk/whey in batch and continuous processes by concanavalin A-celite 545 immobilized Aspergillus oryzae b-galactosidase. Food Bioprod Process. doi:10.1016/j.fbp.2011.07.003
5. Ashly PC, Joseph MJ, Mohanan PV (2011) Activity of diastase a- amylase immobilized on polyanilines (PANIs). Food Chem127:1808–1813
6. Ashraf H, Husain Q (2010) Use of DEAE cellulose adsorbed and crosslinked white radish (Raphanus sativus) peroxidase for the removal of a-naphthol in batch and continuous process. Int Biodeter Biodegr 64:27–31
7. Ates¸ S, Dog˘an NS (2010) Properties of immobilized phenylalanine ammonia lyase and investigation of its use for the prediagnosis of phenylketonuria. Turk J Biochem35:58–62
8. Bagal D, Karve MS (2006) Entrapment of plant invertase within novel composite of agarose–guar gum biopolymer membrane. Anal Chim Acta 555:316–321
9. Betigeri SS, Neau SH (2002) Immobilization of lipase using hydrophilic polymers in the form of hydrogel beads. Biomate- rials 23:3627–3636
10. Brı´gida AIS, Calado VMA, Gonc¸alves LRB, Coelho MAZ (2010) Effect of chemical treatments on properties of green coconut fiber. Carbohydr Polym 79:832–838
11. Bryjak J, Aniulyte J, Liesiene J (2007) Evaluation of man-tailoredcellulose-based carriers in glucoamylase immobilization. Car- bohyd Res 342:1105–1109
12. Cabrera-Padilla RY, Lisboa MC, Fricks AT, Franceschi E, Lima AS, Silva DP, Soares CM (2011) Immobilization of Candida rugosa lipase on poly (3-hydroxybutyrate-co-hydroxyvalerate) a neweco-friendly support. J Ind Microbiol Biotechnol. doi:10.1007/s10295-011-1027-3
13. Cao L (2006) Immobilized enzymes: past, present and prospects. In:Carrier-bound immobilized enzymes: principles, application and design. Wiley-VCH Verlag GmbH & Co. KGaA, Weinheim. doi:10.1002/3527607668.ch1
14. Ceniceros EPS, Ilyina A, Esquivel JCC, Menchaca DR, Espinoza JCF, Rodriguez OEM (2003) Entrapment of enzymes in natural polymer extracted from residue of food industry: preparation methods, partial characterisation and possible application. Becth Mock 44:84–87
15. Chang YK, Chu L (2007) A simple method for cell disruption by immobilization of lysozyme on the extrudate-shaped Na Y zeolite. Biochem Eng J 35:37–47
16. Chang MY, Juang RS (2007) Use of chitosan-clay composite as immobilization support for improved activity and stability of b- glucosidase. Biochem Eng J 35:93–98
17. Chen GC, Kuan IC, Hong JR, Tsai BH, Lee SL, Yu CY (2011a) Activity enhancement and stabilization of lipase from Pseudo- monas cepacia in polyallylamine-mediated biomimetic silica. Biotechnol Lett 33:525–529
18. Chen S, Song N, Liao X, Shi B (2011b) Immobilization of catalase on ferric modified collagen fibers. Chin J Biotechnol 27:1076–1081Chern JT, Chao YP (2005) Chitin-binding domain based immobili-zation of D-hydantoinase. J Biotechnol 117:267–275
19. Cordeiro AL, Lenk T, Werner C (2011) Immobilization of Bacillus licheniformis a-amylase onto reactive polymer films. J Biotech- nol 154:216–221
20. Cunha AG, Ferna´ndez-Lorente G, Bevilaqua JV, Destain J, Paiva LM, Freire DM, Ferna´ndez-Lafuente R, Guisa´n JM (2008) Immobi- lization of Yarrowia lipolytica lipase—a comparison of stability of physical adsorption and covalent attachment techniques. Appl Biochem Biotechnol 146:49–56
21. D'Souza SF (1998) Immobilized enzymes in bioprocess. Curr Sci77:69–79
22. Dai D, Xia L (2006) Effect of lipase immobilization on resolution of (R, S)-2-octanol in non aqueous media using modified ultrast-able-Y molecular sieve as support. Appl Biochem Biotechnol134:39–49
23. Daoud FBO, Kaddour S, Sadoun T (2010) Adsorption of cellulase Aspergillus niger on a commercial activated carbon: kinetics and equilibrium studies. Colloid Surface B Biointerfaces 75:93–99
24. Dey G, Nagpal V, Banerjee R (2002) Immobilization of alpha- amylase from Bacillus circulans GRS 313 on coconut fiber. Appl Biochem Biotechnol 102–103:303–313
25. Dezott M, Innocentini-Mei LH, Dura´n N (1995) Silica immobilized enzyme catalyzed removal of chlorolignins from eucalyptus kraft effluent. J Biotechnol 43:161–167
26. Diaz FJ Jr, Balkus KJ (1996) Enzyme immobilization in MCM-41molecular sieve. J Mol Catal B-Enzym 2:115–126

27. Dung NA, Huyen ND, Hang ND, Canh TT (1995) Immobilization of urease on grafted starch by radiation method. Radiat Phys Chem 46:1037–1042

28. Dutta S, Bhattacharyya A, De P, Ray P, Basu S (2009) Removal of mercury from its aqueous solution using charcoal-immobilizedpapain (CIP). J Hazard Mater 172:888–896

29. Elc¸in YM (1995) Encapsulation of urease enzyme in xanthan– alginate spheres. Biomaterials 16:1157–1161

30. Emregul E, Sungur S, Akbulut U (2006) Polyacrylamide–gelatincarrier system used for invertase immobilization. Process Biochem 38:27–30

31. Erdemir S, Yilmaz M (2011) Catalytic effect of calix[n] arene basedsol–gel encapsulate or covalent immobilized lipases on enantio- selective hydrolysis of (R/S)-naproxen methyl ester. J Incl Phenom Macrocycl Chem. doi:10.1007/s10847-011-9962-1

32. Flores-Maltos A, Rodrı´guez-Dura´n LV, Renovato J, Contreras JC, Rodrı´guez R, Aguilar CN (2011) Catalytical properties of free and immobilized Aspergillus niger tannase. Enzyme Res. doi:10.4061/2011/768183

33. Fu J, Reinhold J, Woodbury NW (2011) Peptide-modified surfaces for enzyme immobilization. PLoS ONE. doi:10.1371/journal.pone. 0018692

34. Girigowda K, Mulimani VH (2006) Hydrolysis of galacto-oligosac-charides in soymilk by j-carrageenan-entrapped a-galactosidase from Aspergillus oryzae. World J Microbiol Biotechnol22:437–442

35. Go´mez L, Ramı´rez HL, Neira-Carrillo A, Villalonga R (2006) Polyelectrolyte complex formation mediated immobilization ofchitosan–invertase neoglycoconjugate on pectin-coated chitin. Bioproc Biosyst Eng 28:387–395

36. HaiderT, HusainQ(2008)ConcanavalinAlayeredcalcium alginate– starch beads immobilized b-galactosidase as a therapeutic agent for lactose intolerant patients. Int J Pharm 359:1–6

37. Herna´ndez MR, Kispert L, Ramı´rez ET, Rosales DR, Ulloa RZ, Ferrara JT (2007) Electron paramagnetic resonance analyses of biotransformation reactions with cytochrome P-450 immobilized on mesoporous molecular sieves. Biotechnol Lett 29:919–924

38. Hilal N, Kochkodan V, Nigmatullin R, Goncharuk V, Al-Khatib L (2006) Lipase-immobilized biocatalytic membranes for enzy- matic esterification: comparison of various approaches to membrane preparation. J Membr Sci 268:198–207

39. Ho LF, Li SY, Lin SC, Hsu WH (2004) Integrated enzyme purification and immobilization processes with immobilized metal affinity adsorbents. Process Biochem 39:1573–1581

40. Hosseinkhani S, Szittner R, Nemat-Gorgani M, Meighen EA (2003) Adsorptive immobilization of bacterial luciferases on alkyl- substituted Sepharose 4B. Enzyme Microb Tech 32:186–193

41. Hsieh HJ, Liu PC, Liao WJ (2000) Immobilization of invertase via carbohydrate moiety on chitosan to enhance its thermal stability. Biotcchnol Lett 22:1459–1464

42. Huang L, Cheng ZM (2008) Immobilization of lipase on chemically modified bimodal ceramic foams for olive oil hydrolysis. Chem Eng J 144:103–109

43. Huang XJ, Yu AG, Xu ZK (2008) Covalent immobilization of lipase from Candida rugosa onto poly (acrylonitrile-co-2-hydroxyethyl methacrylate) electrospun fibrous membranes for potential bioreactor application. Bioresource Technol 99:5459–5465

44. Huang XJ, Chen PC, Huang F, Ou Y, Chen MR, Xu ZK (2011) Immobilization of Candida rugosa lipase on electrospun cellu- lose nanofiber membrane. J Mol Catal B-Enzym 70:95–100

45. Idris A, Bukhari A (2011) Immobilized Candida antarctica lipase B: hydration, stripping off and application in ring opening polyester synthesis. Biotechnol Adv. doi:10.1016/j.biotechadv.2011.10. 002

46. Ispas C, Sokolov I, Andreescu S (2009) Enzyme-functionalizedmesoporous silica for bioanalytical applications. Anal Bioanal Chem 393:543–554

47. Jegannathan KR, Jun-Yee L, Chan ES, Ravindra P (2010) Production of biodiesel from palm oil using liquid core lipase encapsulated in j-carrageenan. Fuel 89:2272–2277

48. Kahraman MV, Bayramoglu G, Kayaman-Apohan N, Gu¨ngo¨r A (2007) UV-curable methacrylated/fumaric acid modified epoxy as a potential support for enzyme immobilization. React Funct Polym 67:97–103

49. Kamori M, Hori T, Yamashita Y, Hirose Y, Naoshima Y (2000) Immobilization of lipase on a new inorganic ceramics support, toyonite, and the reactivity and enantioselectivity of the immo- bilized lipase. J Mol Catal B-Enzym 9:269–274

50. Kapoor M, Kuhad RC (2007) Immobilization of xylanase from Bacillus pumilus strain MK001 and its application in production of xylo-oligosaccharides. Appl Biochem Biotechnol 142:125– 138

51. Karagulyan HK, Gasparyan VK, Decker SR (2008) Immobilization of fungal beta-glucosidase on silica gel and kaolin carriers. Appl Biochem Biotechnol 146:39–47

52. Katwa LC, Ramakrishna M, Rao MRR (1981) Spectrophotometric assay of immobilized tannase. J Biosci 3:135–142

53. Kawaguti HY, Manrich E, Sato HH (2006) Production of isomaltu- lose using Erwinia sp. D12 cells: culture medium optimization and cell immobilization in alginate. Biochem Eng J29:270–277

54. Khan AA, Akhtar S, Husain Q (2006) Direct immobilization of polyphenol oxidases on celite 545 from ammonium sulphate fractionated proteins of potato (Solanum tuberosum). J Mol Catal B-Enzym 40:58–63

55. Kibarer GD, Akovali G (1996) Optimization studies on the features of an activated charcoal-supported urease system. Biomaterials17:1473–1479

56. Kim J, Jia H, Wang P (2006) Challenges in biocatalysis for enzyme- based biofuel cells. Biotechnol Adv 24:296–308

57. Klein MP, Scheeren CW, Lorenzoni ASG, Dupont J,

Frazzon, Hertz PF (2011) Ionic liquid-cellulose film for enzyme immobilization. Process Biochem 46:1375–1379

58. Koszelewski D, Mu¨ller N, Schrittwieser JH, Faber K, Kroutil W (2010) Immobilization of x-transaminases by encapsulation in a sol–gel/celite matrix. J Mol Catal B-Enzym 63:39–44
59. Kumar AG, Perinbam K, Kamatchi P, Nagesh N, Sekaran G (2010) In situ immobilization of acid protease on mesoporous activated carbon packed column for the production of protein hydroly- sates. Bioresour Technol 101:1377–1379
60. Kumari A, Kayastha AM (2011) Immobilization of soybean (Glycine max) a-amylase onto chitosan and amberlite MB-150beads: optimization and characterization. J Mol Catal B-Enzym69:8–14
61. Labus K, Turek A, Liesiene J, Bryjak J (2011) Efficient Agaricus bisporus tyrosinase immobilization on cellulose-based carriers. Biochem Eng J 56:232–240
62. Lee DH, Park CH, Yeo JM, Kim SW (2006) Lipase immobilization on silica gel using a cross-linking method. J Ind Eng Chem12:777–782
63. Lee CH, Lin TS, Mou CY (2009) Mesoporous materials for encapsulating enzymes. NANO 4:165–179
64. Li SF, Chen JP, Wu WT (2007) Electrospun polyacrylonitrile nanofibrous membranes for lipase immobilization. J Mol CatalB-Enzym 47:117–124
65. Liu CH, Lin YH, Chen CY, Chang JS (2009) Characterization of Burkholderia lipase immobilized on celite carriers. J Taiwan Inst Chem E 40:359–363
66. Liu X, Luo L, Ding Y, Xu Y, Li F (2010) Hydrogen peroxide biosensor based on the immobilization of horseradish peroxidase on c-Al2O3 nanoparticles/ chitosan film-modifiedelectrode. J Solid State Electrochem 15:447–453
67. Magnan E, Catarino I, Paolucci-Jeanjean D, Preziozi-Belloy L, Belleville MP (2004) Immobilization of lipase on a ceramic membrane: activity and stability. J Membr Sci 241:161–166
68. Matto M, Husain Q (2009) Calcium alginate–starch hybrid support for both surface immobilization and entrapment of bitter gourd (Momordica charantia) peroxidase. J Mol Catal B-Enzym 57:164–170
69. Mirouliaei M, Nayeri H, Shariat SZS, Atar AM (2007) Biospecific immobilization of lactoperoxidase on Con A–sepharose 4B. Sci Iran 14:303–307
70. Mishra N, Pithawala K, Bahadur A (2011) Byssus Thread: a novel support material for urease immobilization. Appl Biochem Biotechnol 165:1568–1576
71. Mislovicova´ D, Masarova J, Vikartovska A, Germeiner P, Michalk- ova E (2004) Biospecific immobilization of mannan–penicillinG acylase neoglycoenzyme on Concanavalin A-bead cellulose. J Biotechnol 110:11–19
72. Mitchell S, Ramı´rez JP (2011) Mesoporous zeolites as enzyme carriers: synthesis, characterization, and application in biocatal- ysis. Catal Today 168:28–37
73. Namdeo M, Bajpai SK (2009) Immobilization of a-amylase ontocellulose-coated magnetite (CCM) nanoparticles and preliminary starch degradation study. J Mol Catal B-Enzym 59:134–139
74. Pahujani S, Kanwar SS, Chauhan G, Gupta R (2008) Glutaraldehyde activation of polymer nylon-6 for lipase immobilization: enzyme characteristics and stability. Bioresour Technol 99:2566–2570
75. Persson M, Wehtje E, Adlercreut P (2000) Immobilisation of lipases by adsorption and deposition: high protein loading gives lower water activity optimum. Biotechnol Lett 22:1571–1575
76. Pogorilyi RP, Siletskaya EY, Goncharik VP, Kozhara LI, Zub YL (2007) Immobilization of urease on the silica gel surface by sol– gel method. Russ J Appl Chem 80:330–334
77. Popat A, Hartono SB, Stahr F, Liu J, Qiao SZ, Lu GQ (2011) Mesoporous silica nanoparticles for bioadsorption, enzyme immobilisation and delivery carriers. Nanoscale 3:2801–2818
78. Raafat AI, Araby E, Lotfy S (2011) Enhancement of fibrinolytic enzyme production from Bacillus subtilis via immobilization process onto radiation synthesized starch/dimethylaminoethyl methacrylate hydrogel. Carbohyd Polym 87:1369–1374
79. Ramani K, Karthikeyan S, Boopathy R, Kennedy LJ, Mandal AB, Sekaran G (2011) Surface functionalized mesoporous activated carbon for the immobilization of acidic lipase and their application to hydrolysis of waste cooked oil: isotherm and kinetic studies. Process Biochem. doi:10.1016/j.procbio.2011.11.025
80. Rani AS, Das MLM, Satyanarayana S (2000) Preparation and characterization of amyloglucosidase adsorbed on activated charcoal. J Mol Catal B-Enzym 10:471–476
81. Rao MN, Kembhavi AA, Pant A (2000) Immobilization of endo- polygalacturonase from Aspergillus ustus on silica gel. Biotech- nol Lett 22:1557–1559
82. Rao CS, Prakasham RS, Rao AB, Yadav JS (2008) Functionalized alginate as immobilization matrix in enantioselective L (?) lactic acid production by Lactobacillus delbrucekii. Appl Biochem Biotechnol 149:219–228
83. Ren G, Xu X, Liu Q et al (2006) Electrospun poly(vinyl alcohol)/ glucose oxidase biocomposite membranes for biosensor appli- cations. React Funct Polym 66:1559–1564
84. Rochefort D, Kouisni L, Gendron K (2008) Physical immobilization of laccase on an electrode by means of poly (ethyleneimine) microcapsules. J Electroanal Chem 617:53–63
85. Romaskevic T, Vikantiene E, Budriene S, Ramanaviciene A, Dienys G (2010) Immobilization of maltogenase onto polyurethane microparticles from poly (vinyl alcohol) and hexamethylene diisocyanate. J Mol Catal B-Enzym 64:172–176
86. Rosa CC, Cruz HJ, Vidal M, Oliva AG (2002) Optical biosensor based on nitrite reductase immobilised in controlled pore glass. Biosens Bioelectron 17:45–52
87. Sabbani S, Hedenstro¨m E, Nordin O (2006) The enantioselectivity of Candida rugosa is influenced by

the particle size of the immobilizing support material Accurel. J Mol Catal B-Enzym42:1–9

88. Sahney R, Puri BK, Anand S (2005) Enzyme coated glass pH- electrode: its fabrication and applications in the determination of urea in blood samples. Anal Chim Acta 542:157–161

89. Sakai S, Liu Y, Yamaguchi T, Watanabe R, Kawabe M, Kawakami K (2010) Immobilization of Pseudomonas cepacia lipase onto electrospun polyacrylonitrile fibers through physical adsorption and application to transesterification in nonaqueous solvent. Biotechnol Lett 32:1059–1062

90. Sardar M, Gupta MN (2005) Immobilization of tomato pectinase on Con A–Seralose 4B by bioaffinity layering. Enzyme Microb Tech 37:355–359

91. Sardar M, Roy I, Gupta MN (2000) Simultaneous purification and immobilization of Aspergillus niger xylanase on the reversibly soluble polymer Eudragit(TM) L-100. Enzyme Microb Tech27:672–679

92. Satar R, Matto M, Husain Q (2008) Studies on calcium alginate– pectin gel entrapped concanavalin A–bitter gourd (Momordica charantia) peroxidase complex. J Sci Ind Res India 67:609–615Serralha FN, Lopes JM, Lemos F, Prazeres DMF, Aires-Barros MR, Cabral JMS, Ribeiro FR (1998) Zeolites as supports for an enzymatic alcoholysis reaction. J Mol Catal B-Enzym 4:303–311

93. Sheldon RA (2011) Characteristic features and biotechnological applications of cross-linked enzyme aggregates (CLEAs). Appl Microbiol Biotechnol 92:467–477

94. Shen Q, Yang R, Hua X, Ye F, Zhang W, Zhao W (2011) Gelatin- templated biomimetic calcification for b-galactosidase immobi- lization. Process Biochem 46:1565–1571

95. Shi Q-H, Tian Y, Dong X-Y, Bai S, Sun Y (2003) Chitosan-coatedsilica beads as immobilized metal affinity support for protein adsorption. Biochem Eng J 16:317–322

96. Shioji S, Hanada M, Hayashi Y, Tokami K, Yamamoto H (2003) Continuous surface modification of silica particles for enzyme immobilization. Adv Powder Technol 14:231–245

97. Singh BD (2009) Biotechnology expanding horizons. Kalyani, India Soleimani M, Khani A, Najafzadeh K (2011) a-Amylase immobili- zation on the silica nanoparticles for cleaning performance towards starch soils in laundry detergents. J Mol Catal B-Enzym 74:1–5

98. Spahn C, Minteer SD (2008) Enzyme immobilization in biotechnol- ogy. Recent Pat Eng 2:195–200

99. Szyman′ska K, Bryjak J, Jarze˛bski AB (2009) Immobilization of invertase on mesoporous silica to obtain hyper active biocata- lysts. Top Catal 52:1030–1036

100. Tu¨mtu¨rk H, Karaca N, Demirel G, S¸ ahin F (2007) Preparation and application of poly(N, N-dimethylacrylamide-co-acrylamide) and poly(N-isopropylacrylamide-co-acrylamide)/j-Carrageenanhydrogels for immobilization of lipase. Int J Biol Macromol40:281–285

101. Vaillant F, Millan A, Millan P, Dornier M, Decloux M, Reynees M (2000) Co-immobilized pectinlyase and endocellulase on chitin and nylon supports. Process Biochem 35:989–996

102. Vamvakaki V, Chaniotakis NA (2007) Immobilization of enzymes into nanocavities for the improvement of biosensor stability. Biosens Bioelectron 22:2650–2655

103. Wang HY, Hettwer DJ (1982) Cell immobilization in j-carrageenanwith tricalcium phosphate. Biotechnol Bioeng 14:1827–1838

104. Wang ZG, Wan LS, Liu ZM, Huang XJ, Xu ZK (2009) Enzyme immobilization on electrospun polymer nanofibers: an overview. J Mol Catal B-Enzym 56:189–195

105. Wang L, Wei L, Chen Y, Jiang R (2010) Specific and reversible immobilization of NADH oxidase on functionalized carbon nanotubes. J Biotechnol 150:57–63

106. Wen H, Nallathambi V, Chakraborty D, Barton SC (2011) Carbon fiber microelectrodes modified with carbon nanotubes as a new support for immobilization of glucose oxidase. Microchim Acta175:283–289

107. Wu SC, Lia YK (2008) Application of bacterial cellulose pellets in enzyme immobilization. J Mol Catal B-Enzym 54:103–108

108. Wu L, Yuan X, Sheng J (2005) Immobilization of cellulase in nanofibrous PVA membranes by electrospinning. J Membr Sci250:167–173

109. Xiangli Q, Zhe L, Yinglin Z, Zhengjia Z (2010) Immobilization of activated sludge in poly (ethylene glycol) by UV technology and its application in micro-polluted wastewater. Biochem Eng J 50:71–76Xing GW, Li XW, Tian GL, Ye YH (2000) Enzymatic peptide synthesis in organic solvent with different zeolites as immobi-lization matrices. Tetrahedron 56:3517–3522

110. Yusdy, Patel SR, Yap MGS, Wang DIC (2009) Immobilization ofL-lactate dehydrogenase on magnetic nanoclusters for chiral synthesis of pharmaceutical compounds. Biochem Eng J 48:13–21Zhao Q, Hou Y, Gong GH, Yu MA, Jiang L, Liao F (2010) Characterization of alcohol dehydrogenase from permeabilized brewer's yeast cells immobilized on the derived attapulgite nanofibers. Appl Biochem Biotechnol 160:2287–2299

111. Zimmermann YS, Shahgaldian P, Corvini PFX, Hommes G (2011) Sorption-assisted surface conjugation: a way to stabilize laccase enzyme. Appl Microbiol Biotechnol 92:169–178

CHAPTER 2

ENZYME IMMOBILIZATION BY MICROENCAPSULATION: METHODS, MATERIALS, AND TECHNOLOGICAL APPLICATIONS

Christina Rother[1] and Bernd Nidetzky[1,2]

[1]Graz University of Technology, Graz, Austria

[2]Austrian Centre of Industrial Biotechnology, Graz, Austria

INTRODUCTION

Enzyme immobilization is a key technology of biocatalytic process development. Generally, enzymes are immobilized with the aim of making continuous process operation possible. Less often, immobilization is applied with the main goal of enhancing the enzyme stability. In the best case, both effects are combined to generate robust biocatalysts suitable for establishment of a continuous transformation. Immobilized enzymes are in most cases readily separated from product and are therefore reusable in a straightforward manner. Transition from simple batch to (partly) continuous processing involves reactor configurations adapted to the basic type of enzyme immobilizate used. The majority of biocatalytic processes at industrial manufacturing scale employ immobilized enzymes, emphasizing the high importance of the technology in the field (1-5).

Methods of enzyme immobilization are large in number and diverse in their underlying principles (2). However, they can be categorized broadly according to whether the immobilization involves attachment of the enzyme onto a solid material, which is often referred to as the *enzyme carrier*(Fig. 1a). Carrier-free immobilization of enzymes constitutes the alternative option, which comprises solid aggregates of cross-linked enzyme molecules (3, 4) (Fig. 1b). Microencapsulation presents a distinct method of enzyme immobilization by inclusion, which is primarily based on the principle of immobilization without attachment to the carrier material (1) (Fig. 1c). However, microencapsulation also integrates features of carrier-based immobilization, often to a degree that the major categories of enzyme immobilization are practically united within one single method. Both soluble and insoluble enzyme preparations are encapsulated in practice (2-6).

Citation: Christina Rother and Bernd Nidetzky, Enzyme Immobilization by Microencapsulation: Methods, Materials, and Technological Applications, DOI: 10.1002/9780470054581.eib275

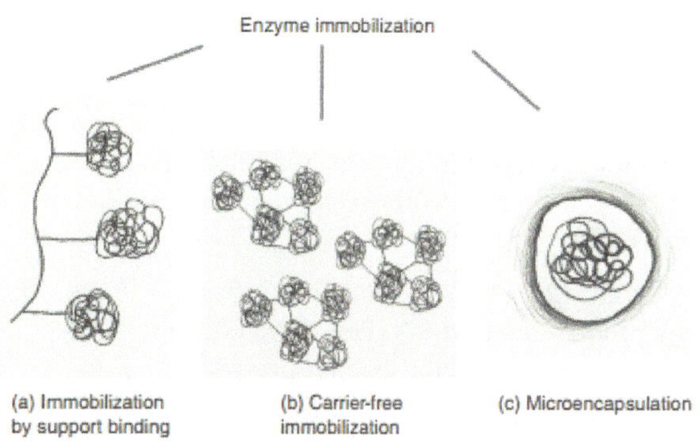

Figure 1: Categorization of enzyme immobilization methods. (a) Enzyme immobilization by attachment to a solid carrier, (b) carrier-free

immobilization by cross-linking of enzymes, (c) microencapsulation within a solid material without binding of the enzyme.

For enzyme immobilization to be effective, two general criteria must be fulfilled. Firstly, physical separation of the enzyme from the continuously or periodically flowing product stream under conditions of process operation needs to be efficient. Secondly, biological activity originally present in the free enzyme has to be retained in adequate amount in the immobilized enzyme preparation. Note that "activity" in the sense used here refers to the total turnover number of the enzyme (mol product formed/mol enzyme consumed). This parameter not only reflects the true activity of the enzyme preparation used (e.g. specific activity or apparent turnover frequency) but also includes effects of the enzyme›s kinetic stability under the operational conditions. A third criterion, which is vital for carrier-bound immobilized enzymes but also for microencapsulated enzymes, is that the enzyme mass or activity loading on the unit mass of noncatalytic material should be as high as possible (6). The percentage amount of the initially supplied enzyme activity or mass that was transferred from the solution onto or into the noncatalytic material is an important feature of immobilization methods. If the immobilization is carried out by encapsulation, this percentage is called *encapsulation efficiency*, sometimes also referred to as the *EE value of a method* (7). The abovementioned major categories of enzyme immobilization differ in the extent to which they satisfy the criteria of an effective immobilization procedure. Immobilization on carriers often provides excellent separation characteristics, but experiences difficulties regarding retention of activity (1-3). By contrast, carrier-free immobilization generally preserves well the enzyme activity, but often results in immobilizates showing insufficient performance in the separation and in enzyme reuse (4). As an immobilization method, therefore, microencapsulation holds promise to combine the advantages of carrier-free and carrier-bound immobilized enzymes while avoiding largely their respective drawbacks. Disadvantage of microencapsulated enzyme immobilizates is however that preparation of the microcapsules often requires relatively complex procedures (1). Microcapsule stability against mechanical forces and pressure presents another important issue requiring special attention. Mass transport into and out of the microcapsule is potential factor limiting the apparent catalytic effectiveness of thus immobilized enzymes.

A rather unique opportunity of microencapsulation lies in the immobilization of two or more enzymes catalyzing multistep biotransformations. Microcapsules constitute interesting small (nano- to microscale) bioreactors in which several enzymes are spatially confined to carry out the overall conversion intended. Advanced microencapsulation techniques even provide compartmentalization within the microcapsule, sometimes to a degree that is already reminiscent of subcellular organizations seen in biology (5, 8, 9).

Another important advantage of enzyme immobilization by microencapsulation is effective protection of the enzymes from protein-denaturing factors prevailing in the external environment (bulk). For example, enzymes are prevented from being directly contacted with gas–liquid or liquid–liquid interfaces, which is known to result often in protein unfolding and aggregation. Generally, exposure of enzymes to nonconventional (organic) solvents is minimized. Effect of the combined stresses of mixing operations on enzyme stability is largely eliminated upon encapsulation in stable microcapsules (5, 9). Confinement of the enzymes in the interior of a microcapsule where a substantial portion of the available volume is often occupied by other soluble macromolecules (e.g. polymers), or gel structures formed by them, can result in considerable enzyme stabilization, presumably owing to macromolecular crowding-like effects (6). Unfolding movements of the enzymes and subsequent aggregation processes are probably inhibited under these conditions, resembling a little bit the natural environment of the enzyme located in the interior of the cell (9-11).

Finally, microencapsulation is useful for immobilization of insoluble enzyme preparations. Enzyme aggregates such as CLEAs (cross-linked enzyme aggregates), but also enzymes entrapped in cells can be encapsulated effectively. Microencapsulation provides enzyme immobilizates that exhibit improved robustness and are easier to handle during process operation than the original preparations.

Microencapsulation of enzymes has attracted considerable interest for development of different applications in biotechnology and medicine, as will be reviewed herein. Industrial sectors using immobilized enzymes for bioproduction are generically interested in advanced techniques of enzyme immobilization, hence also in microencapsulation. Major (potential) applications of microencapsulated enzymes have been described for food and feed, in the pharmaceutical and health-care industries, and also in the fine-chemical industries (9, 12). New methods of microencapsulation are continuously being developed, and the existing protocols are optimized for improved performance in the industrial applications (9).

Herein we review the state of the art in enzyme microencapsulation. The paper is structured into two main parts according to the major methodological principles of microencapsulation used, namely encapsulation in semipermeable membrane systems and entrapment in network matrices. Each part is subdivided according to the different microencapsulation methods developed from the basic principle. Each method is described comprehensively including a brief evaluation of its pros and cons. Materials used and procedures of microcapsule preparation are discussed. Representative examples for application of the respective method for enzyme microencpasulation are presented. Emerging trends and new approaches in enzyme microencapsulation are also considered. New materials and innovative methods of microcapsule processing are discussed.

MICROENCAPSULATION IN SEMIPERMEABLE MEMBRANES

The membrane systems referred to herein are built from amphiphilic molecules such as natural lipids, lipid-like compounds, or block copolymers, which share the property of forming self-enclosed bilayer structures (vesicles) in aqueous solution (7). Vesicular structures derived from lipids or lipid-like compounds are called *liposomes*; however, those derived from block copolymers are called *polymersomes*.

Liposomes

Liposomes are bilayer vesicles that mainly consist of self-assembled natural or synthetic lipid molecules (13). Liposomes are effective systems for enzyme microencapsulation owing to their ability to contain water-soluble enzymes within their aqueous confinement (7). Typical liposome structures are shown in Fig. 2.

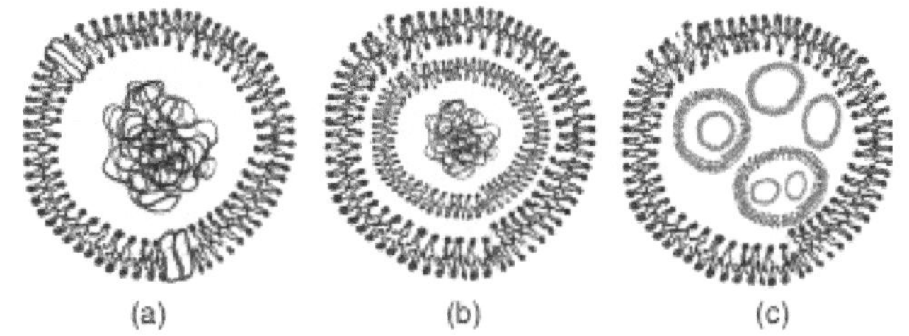

Figure 2: Different forms of liposomes. (a) Unilamellar vesicle, consisting of a bilayer-shell built from amphiphilic molecules that capture the enzyme in an aqueous confinement. The vesicle›s diameter is typically in the range 100–300 nm. Membrane channels can be inserted for increased permeability. (b) Multilamellar vesicle with a typical diameter of 1 µm, here containing the enzyme in innermost vesicle. (c) Multivesicular vesicle (not containing enzyme as shown) with typical diameters in the range 10–20 µm.

Phosphatidylcholines are the (phospho)lipids most commonly used for liposome preparation (7). They are often applied as natural mixtures from soybeans or egg yolk. Defined compounds such as dipalmitoyl- or palmitoyl-oleoyl-glycerophosphocholine (POPC) are also used alone or in combination. In addition to phosphatidylcholines as main constituents of liposomes, other components are included to fortify liposome structure and function. For example, cholesterol is often used to enhance liposome stability (14). Anionic or cationic amphiphiles are applied to prevent vesicle fusion (10).

Depending on the preparation method used, the vesicles formed can contain one or more bilayers. They are called *unilamellar* or *multilamellar vesicles*, respectively. Liposomes can also consist of multiple bilayer shells within one vesicle, resulting in multivesicular vesicles (7). Figure 2schematically shows these vesicular structures.

The methods used for liposome synthesis can be divided into four basic types, as depicted in Fig. 3. First, there is a variety of methods of liposome preparation that have in common that they start from a *dried lipid film* (Fig. 3a). Use of dried lipid films for enzyme microencapsulation often provides the advantage that direct contact of the enzyme with organic solvents is avoided and therefore protein denaturation is prevented effectively (7, 15). For dried lipid film preparation, the amphiphilic molecules that later constitute the liposome are first dissolved in an organic solvent. In the second step, the solvent is removed completely, typically at elevated temperature or reduced pressure. Aqueous enzyme solution is then added, and under extensive shaking to desorb the dried lipid film multilayers from the solid surface of the vessel used, formation of multilamellar vesicles occurs. The resulting vesicles typically exhibit heterogeneous size distribution with average diameters in the low micrometer range. Moreover, the number of lipid layers can vary considerably. The EE value for this method is often in the range of no more than 5–15%. The dried lipid film-derived multilamellar vesicles can be further modified in suspension using different procedures. For example, repeated freezing and thawing is applied to increase the relative abundance of multivesicular vesicle structures. Sonication by contrast was shown to result in the formation of unilamellar vesicles with sizes typically from about 30 to 100 nm. In addition, extrusion that can be performed as repeated passage of the vesicle suspension through polycarbonate filters is useful to achieve controlled reduction of size and lamellarity of the liposomes (7). In an effort to enhance the EE value of enzyme encapsulation in liposomes derived from dried lipid films, unilamellar vesicles were first prepared without the enzyme. The vesicles were then mixed with enzyme solution, and the resulting mixture was subjected to dehydration under reduced pressure, followed by rehydration with distilled water. The dehydration-rehydration procedure was shown to result in improved encapsulation efficiencies, depending of course on the enzyme and the combination of amphiphiles used (7, 16). Typical EE values were between 15% and 45%. A particularly successful example for the application of dehydration-rehydration procedure for microencapsulation is asparaginase. This enzyme was encapsulated with EE values of 68–70% within liposomes derived from egg phosphatidylcholine-cholesterole-diacetylphosphate in a ratio of 7 : 2 : 1 (7).

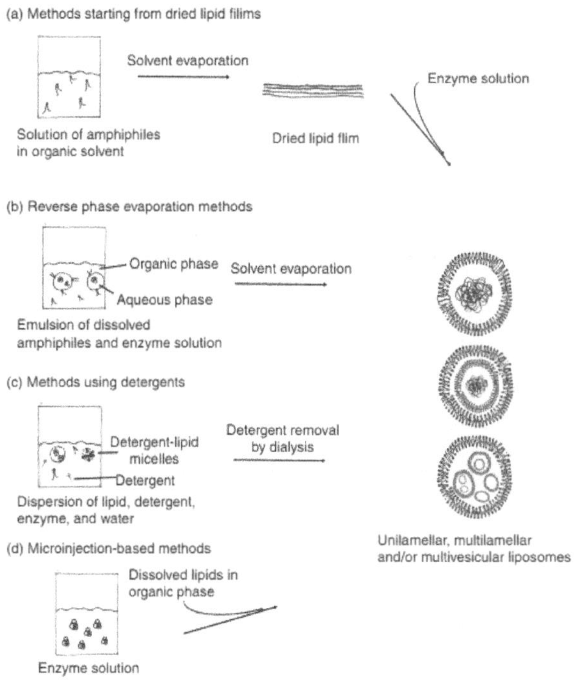

Figure 3: Schematic depiction of the four basic types of liposome preparation methods. (a) Liposome preparation from dried lipid film by the addition of enzyme solution and shaking. The dried lipid film is prepared by solvent evaporation of an organic amphiphile solution. (b) Reverse-phase evaporation methods starting from emulsions from which the organic phase is removed by evaporation. (c) Liposome preparation using micelle-forming detergents that are removed from the lipid-containing solution by dialysis. (d) Methods based on injection of dissolved lipids into enzyme solution.

A second type of liposome preparation methods starts from *emulsions from which the organic solvent is removed by evaporation* (Fig. 3b). Typically, the amphiphilic compounds making up the liposomes are dissolved in a relatively highly volatile organic solvent, and the aqueous solution of the enzyme is then admixed to the organic phase. Amphiphiles are thought to form interfacial layers around the enzyme-containing water droplets and turn into an aqueous vesicle suspension once the organic solvent is removed by evaporation. The principle is referred to as*reverse-phase evaporation method* (7). The liposomes formed are relatively heterogeneous in size, and encapsulation efficiencies in the range 10–30% have been reported (7). In a modification of the procedure, the initially prepared emulsion is additionally mixed with highly concentrated sugar (glucose and sucrose) or lysine solution before evaporating the organic solvent. A water-in-oil-in-water double emulsion can be obtained that proved useful in the preparation multivesicular liposomes with average diameters of around 10 µm (17) or 20 µm (18).

A third method type for liposome preparation involves the use of *micelle-forming detergents* for which sodium cholate is a representative example (Fig. 3c). Detergent is mixed with dispersed amphiphiles and afterward removed by controlled and continuous dialysis. During detergent removal, the detergent-lipid micelles initially formed in the solution are transformed more and more into mixed detergent-lipid vesicles. When dialysis is completed, the resulting vesicles are nearly detergent free and contain enzymes that were present in the aqueous solution (7). The method proved especially suitable for microencapsulation of membrane proteins and membrane-bound enzymes because micelle-forming detergents are already involved in the protein isolation. The proteins are incorporated inside the bilayer vesicles, forming the so-called proteoliposomes (19). Finally, a fourth type of liposome preparation employs *mixing of a lipid-containing organic phase* (e.g. ether or ethanol) with the aqueous enzyme solution (Fig. 3d). Injection of lipids in ether into the enzyme solution leads to rapid vesicle formation, and enzymes are encapsulated in liposomes with diameters of 300 nm or smaller. Despite considerable promise that detergent micelle- or microinjection-based methods of liposome preparation hold for enzyme microencapsulation, there is currently insufficient evidence for detailed comparative evaluation with established procedures such as dried lipid films and emulsion evaporation (7).

The membrane thickness of liposomes consisting of natural lipids is estimated to lie between 3 and 4 nm, being surprisingly independent of the lipid used (7, 20). Preparation of liposomes using phospholipids with mono-unsaturated fatty acids of 14–24 carbon atoms resulted in membranes with 2.00–3.75 nm thickness of the hydrophobic bilayer region (21).

Essentially, all methods of liposome preparation involve steps that have potentially adverse effects on enzyme stability: sonication, cycles of freezing and thawing, extrusion, or contact with organic solvents. Inactivation of the encapsulated enzymes therefore results to a greatly variable extent. For example, β-galactosidase was inactivated to 13.5 and 28.0%, respectively, when subjected to microencapsulation using dehydration-rehydration and reverse-phase evaporation method. Encapsulation efficiencies were however higher with the more strongly inactivating reverse-phase evaporation method. An EE value of 27.6% was obtained that can be compared to an EE value of 18.3% achieved by rehydration-dehydration (22). When using the "ether-injection method" for microencapsulation of δ-aminolaevulinate dehydratase, precipitation and thus inactivation of 40–50% of the enzyme were observed (23). Clearly, these examples provide a note of caution to consider enzyme stability during microencapsulation.

In addition to the amount of enzyme activity lost during microencapsulation, the observable activity of a liposome-based enzymatic microcapsule also depends heavily on how effective is the mass transfer across the membrane(s) of the liposomal vesicle. It is not uncommon that the mass transfer is completely rate limiting for the achievable transformation rate. A number of studies also show that vesicles provide a somewhat selective barrier for influx of certain substrates. For instance, glucose oxidase encapsulated in POPC vesicles exhibited just 4.2–5.9% of the activity measured in the free enzyme after liposome solubilization because of low glucose permeability (24). α-Chymotrypsin encapsulated in POPC vesicles exhibited enhanced selectivity for small substrates, which was due to permeability restraints for large compounds. The comparably small benzyltyrosinenitroanilide (405 Da) was evidently converted by the enzyme, while the large R-casein substrate (24 kDa) was not converted at all (25). Recently, altered substrate selectivity of microencapsulated d-amino acid oxidase as compared to free enzyme was reported, explained by permselectivity of the liposome membrane. The encapsulated enzyme was found to prefer hydrophobic d-Phe as a substrate over the relatively hydrophilic amino acids d-Ala and d-Ser because of permeation resistance of the liposome membrane. In contrast, the free enzyme showed quite broad substrate specificity for various d-amino acids (26).

The permeability and permselectivity of liposomes can be influenced, to some extent at least, using different physical or chemical methods (7). For this purpose, liposomes built from stearylamine, dipalmitoyl-phosphatidyl-choline, and cholesterol were exposed to low frequency, low amplitude magnetic fields. The procedure led to increased permeability of the liposomes for *p*-nitrophenyl acetate, which served as a substrate for the encapsulated carboanhydrase. Enhanced permeability resulted in an increase in enzyme activity from 17% to 80%, compared to carboanhydrase activity after liposome disruption (27). Modification of POPC liposomes with detergents such as Triton X-100 or cholate was also tested for improvement of liposome permeability. Liposome modification with cholate enabled up to four times higher activity of encapsulated glucose oxidase, compared to the initial encapsulated enzyme activity without detergent. Using Triton X-100 for this approach, even five times higher enzyme activity was measured. The extent of induced permeability was typically dependent on the detergent concentration, but authors also noted the possibility that activity from leaked enzymes could have been measured (24). In another approach, insertion of (bacterial) membrane porines such as OmpF was applied to enhance microcapsule permeability. The porines serve as size-selective filters that allow for permeation of molecules smaller than 400 Da (10, 28) (Fig. 1a). The assembly of OmpF porines into the liposomes was achieved by various strategies, among them rehydration of the dried lipid film with porine solution containing 1% detergent, or consecutive freeze and thaw cycles (28,29). β-Lactamase encapsulated in POPC liposomes showed no conversion of ampicillin owing to inhibited permeation before insertion of OmpF into the liposomal membrane. After insertion of two OmpF trimers per liposome in average, substrate diffusion was dependent on the bulk concentration of ampicillin. Below 0.5 μM ampicillin in the bulk solution, diffusion was still limiting, resulting in 86% β-lactamase activity with 0.4 μM ampicillin, compared to the nonencapsulated enzyme. In contrast, there was no difference in activity between the encapsulated and nonencapsulated enzymes above an ampicillin concentration of 0.5 μM (28).

Process stability of liposomes is often low. Liposomes exhibit only low durability and resistance to stresses such as osmotic pressure and mechanical forces. The noncovalent interactions between the amphiphiles that make up the liposomes are rather easily disturbed and this often leads to the release of the encapsulated enzymes into the bulk solution. Therefore, enhancement of liposome stability is often necessary. Methods commonly used are temperature- (30) or UV-induced (28) cross-linking polymerization (e.g. of cross-linker ethylene glycol methacrylate) in the hydrophobic membrane-interior (28, 31). Cross-linking polymerization of methacrylate inside the capsule walls enabled liposomes to tolerate detergent concentrations at least eight times higher than the concentration that otherwise led to lysis of untreated liposomes. Furthermore, the internally polymerized liposomes tolerated lyophilization and returned to the same size upon resolvatization (28). In another approach, coating with a polyelectrolyte shell or adding surface active polymers was used for liposome stabilization (31).

The major application of liposomes is in controlled drug delivery (10, 13, 31), but they have also been successfully used for encapsulating enzymes in biosensing applications. For example, acetylcholesterinesterase (10) and tyrosinase (29) are used in environmental analysis for measurement of various pesticides and insecticides. Moreover, enzymes encapsulated in liposomes are also used in the food industry. For example, encapsulated proteases and lipases are applied for acceleration of cheese ripening (32).

Pseudomonas sp. formaldehyde dehydrogenase (FaDH) is another interesting example of enzyme encapsulation in liposomes. Potential application of the enzyme is in reaction-based biosensing, where FaDH enables quantification of the toxic formaldehyde. The enzyme is also of interest for bioprocess development, where the reverse enzymatic reaction plays a role in a consecutive reduction cascade for CO_2 conversion(33, 34). The activity of FaDH in the NAD^+-dependent oxidation of formaldehyde depends strongly on the reaction temperature: it is 27 times higher at 60°C than at 25°C. However, stability of the free enzyme is not sufficient for use at elevated temperature, and it is therefore useful that enzyme inactivation can be slowed down by microencapsulation in liposomes. However, only co-encapsulation of FaDH with reduced coenzyme NADH resulted in significant improvement of thermal stability of the enzyme. After co-encapsulation with NADH in POPC liposomes, FaDH showed 49% remaining enzyme activity after 60 min-long incubation at 60°C, compared to only 20% when the free FaDH was used. Interestingly, activity loss in liposome-encapsulated FaDH mainly occurred in the first 20 min of incubation, while activity stayed almost unchanged for the remaining 40 min (35).

Co-encapsulation of yeast alcohol dehydrogenase (YADH) and NAD^+ has also been tried in an effort to improve enzyme stability. Frcc YADII incubated in the presence of the stabilizing NAD^+ at 50°C retained only 35% of its initial activity after 1 h. By contrast, liposomal YADH co-encapsulated with NAD^+ retained over 80% of its initial activity under these conditions (36). Therefore, these results provide examples supporting the notion that enzymes potentially receive substantial extra stability upon microencapsulation in liposomes. The underlying mechanism of stabilization is complex; however, it is likely to involve stabilizing interactions of the protein with the liposomal membranes (36, 37).

Polymersomes

Clear drawbacks of liposomes are enzyme leakage, complicated formulation, and liposome instability. In an effort to overcome these problems, vesicles made of other amphiphiles, namely block copolymers, were developed, and they are generally referred to as *polymersomes*. Block copolymers are repeated sequences of different polymerized monomers that mimic the amphiphilic character of lipids (38). AB diblock copolymers such as polystyrene-poly(acrylic acid) (PS-PAA), polystyrene-poly(3-(isocyano-l-alanylaminoethyl)thiophene) (PS-PIAT), orpoly(ethylene oxide)-poly(butadiene) (PEO-PBD) can be used for polymersome preparation. Other building blocks comprise ABA triblocks such as poly(2-methyloxazoline)-poly(dimethylsiloxane)-poly(2-methyloxazoline) (PMOXA-PDMS-PMOXA) and ABC triblock copolymers such as poly(ethylene oxide)-poly(□-caprolactone)-poly(acrylic acid) (PEO-PCL-PAA) (39, 40). While the amphiphilic AB diblock copolymers form bilayered vesicles in the way phospholipids do, ABA and ABC triblock copolymers tend to form monolayered vesicles to expose their hydrophilic A and C polymers to the aqueous bulk and capsule interior, respectively. The hydrophobic B polymer is typically hidden in the interior of the 'membrane' (39). In addition, multilamellar vesicles have been observed with all kinds of copolymers (41).

For polymersome preparation, self-assembly of the amphiphilic block copolymers is initiated with techniques similar to the ones used for liposome preparation. A very useful method for preparation of well-defined polymersomes is the solvent-switch procedure, the principle of which resembles the ether-injection method of liposome-preparation. The block copolymers are dissolved in an organic solvent and then hydrated with the enzyme dissolved in aqueous medium. This can be done by injection of the organic phase into the hydrophilic phase or by slowly adding the hydrophilic phase to the organic phase (42). Another method commonly used is rehydration of a dried copolymer film with protein solution, followed by extrusion to complete the encapsulation (39, 43, 44). In addition, water/oil/water emulsions have been used successfully for copolymer self-assembly (39).

Because of the higher molecular mass of the block copolymers in comparison to phospholipids (e.g. 23.4 kDa for $PMOXA_{65}$-$PDMS_{165}$-$PMOXA_{65}$ (45) vs 760 Da for POPC), the resulting nano- or microscale vesicles are often larger aggregates than liposomes and exhibit membranes of increased thickness. The thickness in turn depends on the molecular mass of the hydrophobic part of the polymers (46). Capsule thickness is around 24 nm for $PMOXA_{65}$-$PDMS_{165}$-$PMOXA_{65}$ capsules (45) and 20–40 nm for various PS-PAA polymersomes (20). Liposome capsules typically show nearly 10-fold lower capsule thickness. As compared to liposomes, enhanced thickness of polymersomes results in lower structural dynamics of the capsules and thus to increased mechanical stability and better preservation of the encapsulated enzyme (43,47). PMOXA-PDMS-PMOXA polymersomes containing superoxide dismutase, for example, were reported to be fully stable for over a month at 5°C (45). PEO-PBS capsules also were extraordinarily stable: 15 min of sterilization at 121°C were endured without damage and enzyme release by about 10% of the capsules (44).

Size, which can range from 50 nm to 10 µm (48), and morphology of the polymersomes depend strongly on the methods and conditions used for preparation. For instance, the size of PMOXA-PDMS-PMOXA polymersomes, formulated using the solvent-switch method, ranges from 100 nm to 10 µm before extrusion, whereby the average size and size distribution are both dependent on the polymer concentration used during preparation (49). Using extrusion, size can be reduced to 50–250 nm (49) or 100–300 nm. Final capsule size was independent of the length of hydrophilic and hydrophobic parts of the constituent polymers, but varied depending on pore size of the filter applied to extrusion (45, 48).

Encapsulation efficiencies of film hydration methods were relatively low when applied to encapsulation of proteins such as hemoglobin (4.5%) and fluor-albumin (5.0%) into PEO-PBD microcapsules (44). Using the same preparation method for encapsulation of superoxide dismutase in different PMOXA-PDMS-PMOXA polymersomes, encapsulation efficiencies of 33–38% were obtained (45). The preparation of PEO-poly(propylenesulfide) polymersomes by solvent switch enabled EE values of 37% for ovalbumin and 19% for bovine serum albumin (50). Because of the large amount of organic solvent involved in this method, it is less often applied to encapsulation of enzymes (51).

Coblock polymers have been modified and functionalized by reaction with fluorescent proteins, gold nanoparticles, or biotin (40, 52). Generally, diversity of copolymers available for polymersome preparation is large. Chemical, physical, and biological parameters of resulting microcapsules can be well tailored by varying block lengths or chemical structure of the polymer building blocks (52). As already described for liposome microcapsules, increase in polymersome microcapsule stability can be achieved through conjugation and internal cross-linking (47, 52). However, the improved stability often includes the downside of a further loss of membrane permeability. Some polymersomes made of specific block copolymers such as PS-PIAT (Fig. 4) are inherently porous because of their specific way of bilayer packaging. Other polymersomes require introduction of additional features to permit transport of substrate and product across the capsule wall (40, 43). Following the principles applied to liposome preparation, incorporation of bacterial porines such as OmpF, LamB (47), or aquaporin Z (54) into the capsule wall is often carried out to improve permeability. Incorporation can be initiated by applying electric potential to block copolymer membranes and a solution of porines (47) or by adding the porines during the formation of the polymer vesicles (54). The channel proteins can improve permeability for molecules beyond certain size, such as OmpF does, or for molecules of certain properties. Among the latter are aquaporin Z, which is selective for water (54), and LamB, which is selective for maltodextrins (47). Another approach is to use sacrificial block copolymers such as poly(styrenylboronic acid)-PEO that can be solubilized by increasing pH and sugar concentration. They are thus removed after assembly of the polymersome and allow generation of membranes with tuned porosity (40).

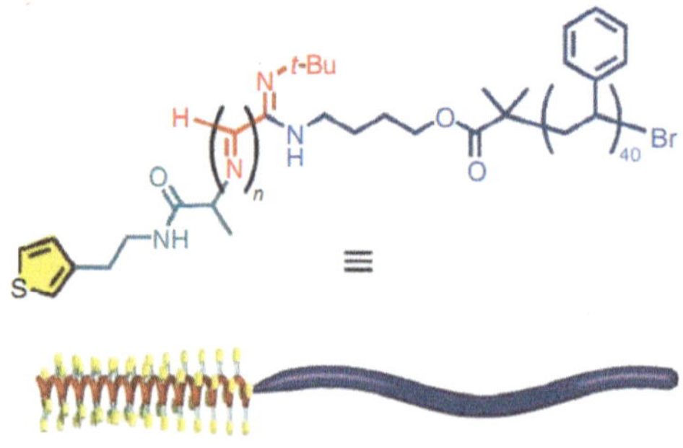

Figure 4: Chemical structure and schematic model of PS-PIAT, block copolymer for synthesis of inherently porous polymersomes

Besides medical applications for controlled drug delivery (39, 43), polymersome microcapsules can also be used as nanometer-sized enzyme reactors. The Baeyer–Villiger monooxygenase phenylacetone monooxygenase (PAMO) was encapsulated as a fusion-protein named "CRE-PAMO" with a protein capable of recycling the cofactor (NADPH) used in the monooxygenase reaction. PS_{40}-$PIAT_{50}$ polymersomes were prepared by solvent-switch method, and CRE-PAMO was encapsulated with EE values of 18–20%. The polymersomes were sufficiently permeable for external NADPH to diffuse into the capsule's interior and thus allowed conversion of phenylacetone into phenylacetate by the enzyme. Complete conversion was accomplished within 15 h, which indicated a significant decrease in the overall reaction rate of the encapsulated system, compared to free CRE-PAMO that converted a four times higher amount of phenylacetone within 5 h (55). This difference was assumed to be a result of diffusion limitations across polymersome membrane (56).

For ring-opening polymerization of lactones, CalB (lipase B from *Candida antarctica*) was encapsulated either in the aqueous core or in the bilayer of polymersomes made of PS-PIAT block copolymers. Encapsulation of the enzyme into the polymersome bilayer was achieved by lyophilization of a mixture containing dissolved copolymer and enzyme. After resuspension in organic solvent, the resulting mixture was added dropwise to water. It was shown that CalB activity in the interior of microcapsules was five times higher than that in the bilayer of the polymersomes. Activity was measured via hydrolytic conversion of a reporter molecule (DiFMU-octanoate) into a fluorescent product (DiFMU). CalB activity inside the capsule was nearly as high as the activity of the free enzyme. In addition, CalB in the interior of the capsule showed similar product specificity as the enzyme in aqueous solution. Distribution of oligomerization products derived from CalB-catalyzed conversion of 8-octanolactone and dodecalactone were similar for free and encapsulated enzymes. Octanolactone oligomers of molecular masses of 266–2701 Da were obtained using CalB encapsulated in the polymersomes' interior, whereas the molecular masses of the corresponding oligomers produced by free CalB were around 247–3303 Da. Molecular mass distributions of the dodecalactone oligomers (490–1143 Da) formed were practically identical using free and encapsulated CalBs. The enzyme immobilized in the polymersome bilayer, however, produced shorter oligomers from 8-octanolactone (256–917 Da) (57).

As already mentioned, encapsulation of superoxide dismutase is another example for the application of polymersomes (45, 48). The enzyme is applied in biomedicine for detoxification of superoxide radicals.

Furthermore, research has recently been focused on the encapsulation of not only one, but a number of enzymes to perform cascade reactions(40). One of the first reports was the synthesis of ATP by combined actions of bacteriorhodopsin (creating a light-driven proton-gradient across the polymersome membrane) and F_oF_1-ATP-synthase (using that gradient as energy source) that were both inserted in the membrane of the copolymer vesicle (39, 40).

Even a three-enzyme cascade has already been realized, using all three available domains of the polymersome available for positioning of the individual enzyme catalysts: CalB was placed inside the membrane, glucose oxidase (GOx) was in the interior of the polymersome, andhorseradish peroxidase (HRP) was attached to the periphery of the microcapsule (Fig. 5). The initial substrate, glucose monoacetate, was converted into glucose by CalB and then further oxidized by GOx, producing H_2O_2, which in turn was used by HRP to oxidize the assay molecule, creating a colored product (58, 59). As encapsulation efficiencies were rather low (17.2% for CalB, 25% for GOx, and 25.5% for HRP), this system may currently have no catalytic advantage over a mixture of the free enzymes, even though it clearly holds promise to potentially achieve reaction intensification in the future. Moreover, it provides a clear example of the possibility of enzyme positioning in polymersomes (58).

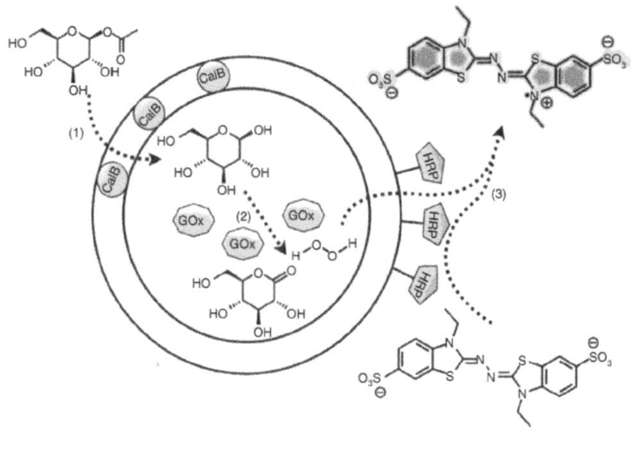

Figure 5: Enzyme positioning in polymersomes. Monoacetylated glucose (1) is deprotected by CalB, enabling GOx to oxidize the product (2), producing H_2O_2, which is further used by HRP for turning the assay molecule ABTS into a colored product (3).

Another three-enzyme cascade was realized in multicompartmentalized polymersomes: CalB and NAD^+-dependent ADH were each encapsulated in PS-PIAT polymersomes with 21 and 51% encapsulation efficiency, respectively. These enzyme-loaded polymersomes were then mixed with "cytosolic enzymes" (NADPH-dependent PAMO) and substrates, and the whole mixture was encapsulated in PB-PEO vesicles. In the cascade biocatalytic transformation, followed by nonenzymatic, spontaneous beta-elimination, a profluorescent substrate was converted into the fluorescent resorufin. The overall velocity of this reaction was 14% lower than the velocity of the reaction catalyzed by the free enzymes. In both cases, conversions of 10–15% were obtained (60).

A variation of polymersomes is peptosomes, which are vesicles formed by self-assembly of polypeptide-containing copolymers (39). The polypeptide can comprise the hydrophilic part (like with PBD-poly(glutamate) or PS-poly(isocyano dipeptide)), the hydrophobic part (like with PEO-PBD-poly(cystein phenylalanine) (61)), or the copolymer consists completely of peptides, such as poly(l-lysine)-poly(glycine) (62) or poly(glutamic acid)-poly(lysine) diblock compounds (63). Applications of peptosomes are in biomedicine where they are used as drug or DNA delivery vehicles (39) or for tumor imaging (64). Peptosomes might be also useful for enzyme encapsulation.

ENTRAPMENT IN NETWORK MATRICES

The entrapment of enzymes in polymer networks can be realized with various organic or inorganic polymers, resulting in well-defined matrices such as hydrogels, sol–gels, and layer-by-layer (LbL) structures that shall be described in the following sections.

Hydrogels

Hydrogels are networks that consist of cross-linked polymers that contain highly hydrophilic components. Therefore, the gels have a strong affinity to absorb water and consequently show swelling behavior in hydrophilic solutions (65, 66). Among the polymers used, there are natural polymers such as chitosan or alginate and synthetic polymers such as poly(vinylalcohol) (PVA) or poly(ethyleneimine) (PEI) (65) (Fig. 6). Synthetic polymers usually display increased chemical stability and a decreased proneness to abrasion as compared to the natural polymers(67).

Figure 6: Polymers used for hydrogel synthesis.

The diversity of polymers that can be used for hydrogel preparation has necessitated development of suitable methods of cross-linking. These cross-linking methods can be basically divided according to whether chemical or physical induction of cross-linking between the polymers is applied. While chemical induction of cross-linking can involve UV-irradiation, use of cross-linking agents, or an electron beam, physical induction can be carried out by freeze-thaw processing or controlled dehydration. For some applications, both chemical and physical strategies are used in combination. Preparation of hydrogel nanoparticles based on PVA presents a relevant example (65-

67). For synthesis of PVA hydrogels under mild conditions, a particular formulation of PVA and polyethyleneglycol (PEG), named LentiKats®, is often used. In this procedure, the enzyme solution is added to LentiKats liquid, which is afterward partially dried at room temperature and controlled humidity. After incubation with stabilizer solution and washing, stable and easily separable lense-shaped hydrogel particles are obtained (1, 67). The resulting LentiKats exhibit diameters of 3–4 mm and about 400 μm thickness (68, 69). The LentiKats-enzyme solution can also be extruded into PEG, thus presenting a method that skips some steps compared to the original procedure and is also easily reproducible and perfectly suitable for scale-up. However, the resulting PVA/PEG particles are hemispheric instead of lense-shaped (67), but show the same average diameters as the LentiKats (68). Applying both methods to the encapsulation of inulinase in a comparative evaluation, average EE values of 27% were obtained for the LentiKats, which can be compared to an EE value of 21% for the hemispheric particles. A 24 h-long bioconversion of inulin into fructose showed that LentiKats (93%) and hemispheric particles (96%) were similarly active as the free enzyme that was used as 100% reference. However, it was also shown that K_m was nearly twofold increased, whereas V_{max} was decreased roughly fourfold in the encapsulated inulinase preparations, most probably owing to diffusional limitations. Another problem was that, because of the high optimal temperature of the free inulinase (55°C), LentiKats already tended to melt under conditions of use. The hemispheric hydrogel particles by contrast were still stable up to 60°C (68). Generally and somewhat in contrast to the PVA-derived hydrogels just discussed, diffusion plays a relatively unimportant role in most hydrogels. Because of their hydrophilicity, hydrogels often allow for nearly free diffusion of hydrophilic substrates. Relatively unrestricted diffusion in hydrogels presents a major difference to liposomes, where sometimes substantial activity loss of the encapsulated enzymes due to limited diffusion is observed, as discussed earlier.

Using natural polymers, gelation of alginate and chitosan is often induced by adding cross-linking reagents or by adding certain counter-ions. For chitosan, these counter-ions need to be anionic, whereas alginate requires cationic counterions for induction of cross-linking (65, 70, 71). Hydrogels derived from chitosan, alginate, and agarose were evaluated for encapsulation of lipase as a model enzyme. It was found that chitosan particles were superior to agarose and alginate in terms of enzyme leakage and lipase activity. In fresh and freeze-dried chitosan beads prepared using 2.0 and 2.5% polymer concentration, an enzyme release of about 8% within 8 h of shaking at 25°C was observed. Further shaking for 40 h at the same conditions did not result in any more enzyme release. Alginate beads, however, were found to be more prone to enzyme leakage, presumably because of their higher hydrophilicity. Within 20 h, 15% of the encapsulated enzymes were released from fresh alginate beads. Within the same time span, enzyme leakage of even 20–25% was observed from freeze-dried alginate gels. Concerning the lipase activity that was retained upon encapsulation, chitosan beads also showed better results. Enzyme activity within the chitosan beads was five times higher than in alginate beads. Still, only a maximum of 42% activity compared to the activity of the free enzyme could be measured after encapsulation. Encapsulation efficiencies, however, were all around 43–50% when different chitosan and alginate concentrations were used for hydrogel capsule preparation (70).

Alginate and chitosan have also been combined for the purpose of enzyme encapsulation, thus overcoming individual drawbacks of each alginate (enzyme leakage) and chitosan, which is only soluble at acidic pH. An alginate core matrix was first synthesized to encapsulate and protect the enzyme. This was then surrounded by a chitosan shell. The chitosan shell was also applied for regulating the permeability of the capsule. $Ca^{2+}$and Ba^{2+} were tested for induction of cross-linking of the alginate polymers, whereas phosphate ions served as counterions for chitosan cross-linking. The alginate core prepared by Ca^{2+}-cross-linking was liquefied by the phosphate ions, whereas cores prepared using Ba^{2+}-induced cross-linking were not (72).

When free enzymes are encapsulated, the resulting hydrogels are only suitable for use in organic media. In aqueous medium, by contrast, the gel-pores get too large owing to swelling so that leakage of enzymes can no longer be prevented. However, hydrogels can be used for encapsulating enzymes in the presence of water, provided that the molecular weight of the enzyme is increased by cross-linking. Prominent approaches involve encapsulating CLEAs (73), or forming enzyme complexes with polyelectrolytes (1) where a combination of glutaraldehyde and chitosan presents a notable example (69). Another interesting approach is to use an additional chemical immobilization of the encapsulated enzyme by covalent binding to the hydrogel matrix. For this to happen, alginate carboxyl groups were activated by treatment of the enzyme-containing alginate beads with *N*-(3-dimethylaminopropyl)-*N'*-ethylcarbodiimide hydrochloride (EDC) and *N*-hydroxysulfosuccinimide sodium salt(NHSS). The activated carboxyl groups reacted with the protein's amino groups to form a covalent bond. Using this method, enzyme leakage from alginate-encapsulated glucose oxidase could be reduced during storage at room temperature for 1 week from 50% without conjugation to only 16.9% after conjugation (74).

The major application of hydrogels is in encapsulating drugs for controlled delivery (65, 71), but also in encapsulation of enzymes for biotransformation. A relatively new approach in this field is ferrogels, which are hydrogels with magnetic components. Their advantage is that the particles can be separated relatively easily and product-film formation can be prevented through application of an altered magnetic field (75, 76). Glucoamylase for dextrin hydrolysis is one example of enzyme immobilization in ferrogels, even though immobilization was done by adsorption and/or cross-linking instead of using true encapsulation (75).

For stereoselective synthesis of chiral alcohols by asymmetric reduction of prochiral ketones, alcohol dehydrogenase from *Lactobacillus kefir*, along with its cofactor NADPH, was encapsulated in PVA hydrogels. The hydrophilic gel protected the enzyme in the organic solvent, but still allowed diffusion of the hydrophobic substrate by partitioning between the organic and aqueous phases. The synthesis was carried out using a coupled-substrate method where oxidation of isopropanol into acetone is used for NADH recycling. Overall conversion and enantiomeric excess in product were evaluated as shown in Table 1. In the reduction of acetophenone to (*R*)-phenylethanol as model reaction, 60% of the substrate was converted with an enantiomeric excess of 98% for (*R*)-phenylethanol. Using 1-phenoxyacetone as a substrate, 100% conversion could be obtained with an enantiomeric excess of 96%. Reduction of substrates such as 3-octanone or 1-octen-3-one was catalyzed with good enantioselectivity of 90–98%, but with comparably low conversion of only 30–42%. For storage at 4°C, Mg^{2+} ions were added, resulting in 6% slower enzyme inactivation as compared to conditions of enzyme storage without supplementation (77).

Table 1: Asymmetric Reduction of Prochiral Ketones by *Lactobacillus kefir* **Alcohol Dehydrogenase, Co-encapsulated in PVA Hydrogels with NADPH**

Substrate	Product	Conversion (%)	Enantiomeric Excess (%)
Acetophenone	Phenylethanol	60	98
1-Phenoxyacetone	1-Phenoxy-2-propanol	100	96
3-Octanone	3-Octanol	30	98
1-Octen-3-one	1-Octen-3-ol	42	90

Data are taken from Ref. 77.

Alginate microcapsules have been used for encapsulation and controlled release of proteins for medical application. IgG, myoglobin, and gamma globulin were encapsulated in that way (71). Moreover, carbonic anhydrase was immobilized through encapsulation in alginate beads. The enzyme was used for CO_2 sequestration through conversion of CO_2 to bicarbonate followed by calcite formation in the presence of Ca^{2+}. Enzyme stability under storage conditions at 4°C was significantly improved by encapsulation. In addition, the encapsulated enzyme could be washed and reused more than six times, retaining approximately 67% of its initial activity (78).

SOL–GELS

Sol–gels are inorganic gels that are chemically, thermally, and photochemically stable, transparent, and can be shaped to films, fibers, or nanospheres (11, 67). Materials that are most commonly used for sol–gel preparation are oxide mixtures of silica, alumina, alumnosilica, titanium, or zirconium (79, 80). However, for encapsulation of biomaterials including enzymes, silica sol–gels are often the material of choice, as they can be synthesized under relatively mild conditions (67, 80, 81). Moreover, porosity, network structures, and surface functionalities can be easily tailored by careful selection of the precursor and gelation conditions for silica sol–gels (67, 80, 82). Structures of most precursor molecules that are mentioned in this chapter can be seen in Fig. 7.

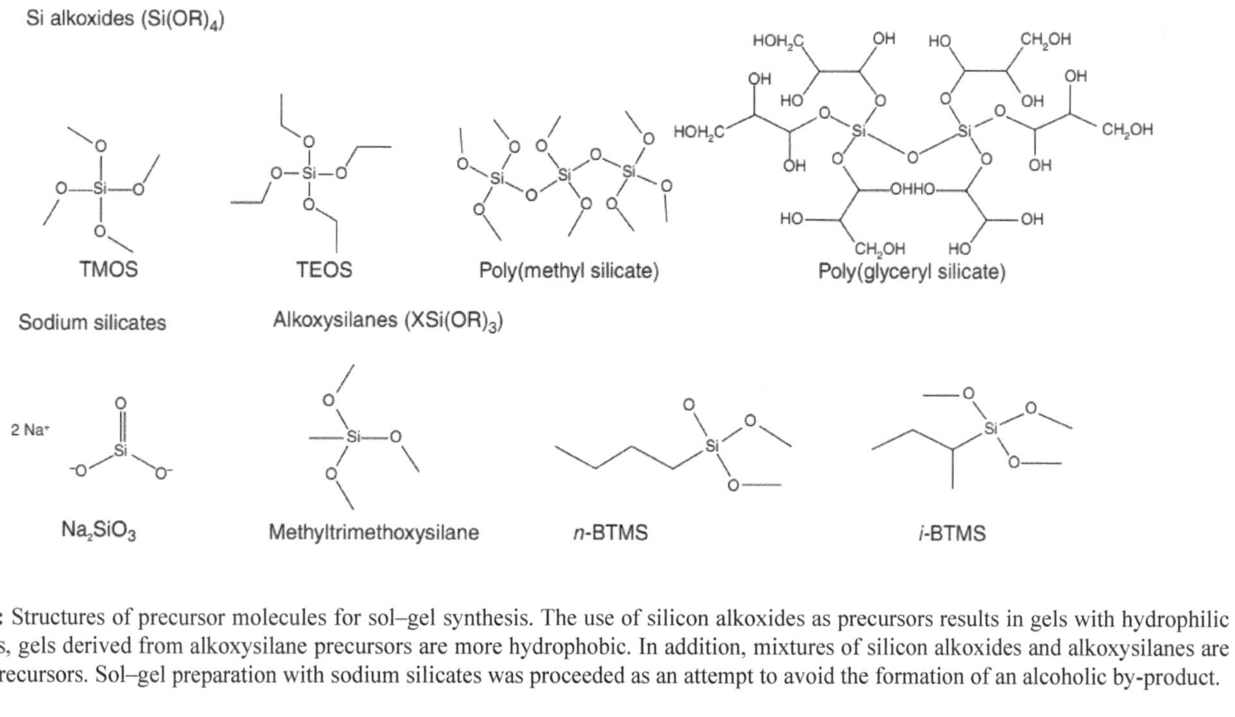

Figure 7: Structures of precursor molecules for sol–gel synthesis. The use of silicon alkoxides as precursors results in gels with hydrophilic properties, gels derived from alkoxysilane precursors are more hydrophobic. In addition, mixtures of silicon alkoxides and alkoxysilanes are used as precursors. Sol–gel preparation with sodium silicates was proceeded as an attempt to avoid the formation of an alcoholic by-product.

The precursors that have been most frequently used are Si alkoxides of the general form $Si(OR)_4$ where TMOS (tetramethyl orthosilicate) andTEOS (tetraethyl orthosilicate) are important in particular. Both TMOS and TEOS are converted into a silica sol–gel in a two-step reaction within minutes (83). In step 1, hydrolysis of the precursor in acidic conditions (83) or in the presence of a catalyst such as NaF (84) takes places to create a silicic-acid sol (Eq. **1**). In step 2, condensation at neutral pH results in the formation of the silica network, consisting of corner-sharing $[SiO_4]$ tetrahedra (Eq. **2**).

$$Si(OR)_4 + 4\,H_2O \rightarrow Si(OH)_4 + 4\,ROH \quad (1)$$

$$Si(OH)_4 \rightarrow SiO_2 + 2\,H_2O \quad (2)$$

The alcoholic by-product that arises in step 1 can be methanol, when TMOS is used as precursor, or ethanol, when the precursor is TEOS. As low pH and high concentrations of alcohol represent adverse conditions for most enzymes, the enzymatic solution is often only added in the second step after evaporation of the alcohol to avoid enzyme inactivation (11, 80, 83). Often, however, contamination with the alcohol by-product cannot be prevented completely. Therefore, an alternative approach of silica sol–gel synthesis uses conventional sodium silicate as precursor in the two-step method. This method is absolutely free of alcoholic by-product and the sodium ions that arise instead usually do not affect the encapsulated enzymes (83). An alternative precursor to $Si(OR)_4$ for sol–gel synthesis is an alkoxysilane of the general form $XSi(OR)_3$ where X stands for an organic croup linked to the silicon atom in an Si–C bond. In addition, mixtures of both precursors are used (1, 80). The organic groups of the alkoxysilane precursors result in hydrophobic properties of the prepared sol–gels, which makes them a good environment for lipophilic enzymes (10, 85).

Mild silica sol–gel synthesis can also be realized by biosilification. This method uses cationic amine-rich scaffold molecules for induction of rapid gelation that leads to entrapment of any material contained in the reaction mixture. Polyethyleneimine, protamine, and the synthetic peptide R5 have been used as scaffold molecules. R5 is a derivative of the marine diatom silaffin protein and shows the most versatile encapsulation abilities among these molecules (82).

The obtained sol–gels can be dried before use, resulting in gels with different pore sizes called *xerogels*, *aerogels*, or *ambigels*. *Xerogels* are obtained by drying hydrophilic sol–gels at room temperature, which results in significant shrinkage (typically 70% (80)) of the nano-cages and pores. This network of hydrated amorphous silica contains pores of 1–10 nm in diameter (11) (Fig. 8a). Wet gels in contrast usually display pore diameters with a large size distribution of 4–130 nm (83, 86), dependent on the pH of the gelation solution. The lower the pH is, the smaller are the pores obtained (83). For further condensation and a stronger network creation in the wet gel, aging for a few days is recommended before use(87).

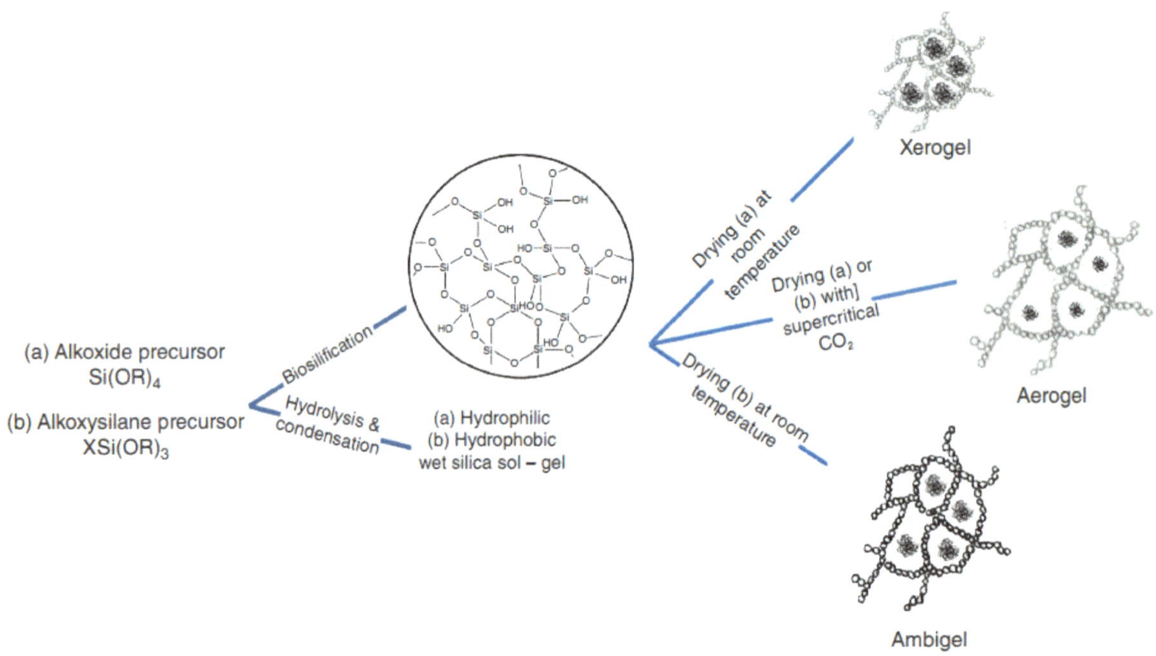

Figure 8: (a,b) Synthesis of silica sol–gels. Biosilification or two-step reaction (hydrolysis and condensation of the precursors) results in the formation of wet sol–gels, encapsulating enzymes into a porous network (enzymes not shown in the wet gel). Drying *hydrophilic* gels by evaporation at room temperature leads to xerogels with significantly shrunk pores; drying the *hydrophobic* gels at the same conditions results in ambigels with unaltered pore size. Pore size is also not changed when the wet silica gels are dried using supercritical CO_2 (aerogels).

Aerogels, which show nearly unaltered porosity compared to wet gels, can be obtained by drying the sol–gel using supercritical CO_2 (1, 80, 83)(Fig. 8b). Using dialysis, the buffer of the wet gel is first exchanged to an intermediate liquid such as acetone, amyl acetate, or ethanol. This enables a further solvent change to liquid CO_2, which is evaporated afterward (80). Gelation at pH7 and drying with supercritical CO_2, for example, leads to aerogels with large distribution in pore diameters (2–80 nm), mainly lying between 12 and 40 nm (83). By drying hydrophobic gels at room temperature, *ambigels* are obtained, which also remain in the same size and porosity as the wet gels (80) (Fig. 8c). The hydrophobic gels can be dried with supercritical CO_2 as well, resulting in aerogels with hydrophobic properties (88, 89).

Encapsulation efficiencies for entrapment of enzymes such as α-chymotrypsin, superoxide dismutase, or HRP into silica sol–

gels have been shown to range from 41% to 99%. Poly(methyl silicate) (condensed TMOS) was used as precursor. After encapsulation, 11–76% activity of the free enzymes remained. Using more biocompatible precursors such as poly(glyceryl sililcate), EE values of 84–96% were achieved, while at the same time, 83–93% of the enzyme activity was retained upon encapsulation. After storage of these gels at neutral pH for 10 months, leakage of only 1–13% of the encapsulated enzymes was reported, whereas 86–99% of the initial activity was still measurable (90). Complete enzyme encapsulation with EE values of 100% was reported for the encapsulation of HRP and glucose oxidase into sol–gels derived from sodium silicate precursors. However, HRP lost 27% of its initial activity owing to true inactivation or diffusion limitations during the encapsulation process. For glucose oxidase, even larger loss of activity (64–69%) was observed (83). In general, K_m values of sol–gel encapsulated enzymes were higher than K_m values of their free counterparts, probably because of the effect of limited diffusion. K_m of alkaline phosphatase increased 89 times upon encapsulation, whereas for alcohol dehydrogenase and bovine carboanhydrase II, K_m increased 14 and 1.4 times compared to free enzyme (87).

Shrinkage of the xerogels, which is often not desirable as it causes capillary stress for their protein carriage, can be reduced by adding organic polymers such as hydroxyethyl-carboxymethyl-cellulose to the gelation solution. The polymer acts as pore-fillers and thus preserves the porous gel structure (81). Other polymers such as PVA were found to have positive impact on the activity of lipophilic enzymes owing to their hydrophobic groups. For example, addition of PVA before the gelation step was extremely efficient in preserving the activity of "Lipase PS" in a hydrophobic sol–gel made from methyltrimethoxysilan. The enzyme showed up to 130-fold increased specific activity (μmol product/min/g protein) compared to enzyme activity in the sol–gel prepared without supplement (91, 92). Biomolecules such as the highly hydrophilic dextran and chitosan or proteins such as BSA can also create a more biocompatible environment that helps stabilizing enzymes in sol–gels (93).

Despite the stress that evaporation and shrinkage can potentially cause in the encapsulated protein cargo, xerogels have been used for the encapsulation of enzymes since 1990, when alkaline phosphatase, chitinase, aspartase, and β-glucosidase were first successfully encapsulated in a stable and catalytically active way in TMOS- and TEOS-derived xerogels. In this study, 30% of initial phosphatase activity was recovered after encapsulation. This activity was then retained for at least two months of storage (94). Because of their transparency, xerogels are perfectly suited for encapsulation of active agents for use as biosensor components (11, 67). HRP was encapsulated for hydrogen peroxide sensing; urease and acetylcholinesterase were used for sensing metal ions (95). As with hydrogels, it is possible to simultaneously entrap enzymes and magnetic particles for easier isolation of the xerogel microparticles (67, 96).

Aerogels have recently been used for encapsulation of various enzymes for biocatalysis, including different lipases for biodiesel synthesis (91,97); tyrosinase for phenol removal from aqueous solution (98); or cellulase for hydrolysis of carboxymethylcellulose (99). However, the chemicals used as intermediate liquids for supercritical drying can be harmful to enzymes (80). For enzymes utilizing hydrophobic substrates, the silica particle-derived colloidosomes are often used. Colloidosomes allow for encapsulation of aqueous enzyme solutions but are particularly useful for biocatalytic synthesis in organic media. They are prepared from water-in-oil emulsions of the enzyme solution with silica nanoparticles, which are cross-linked at the water/oil interface with silica precursor polymers. Laccase, which catalyzes oxidation of aromatic substrates, was encapsulated as a model enzyme in these particles (100).

Hydrophobic sol–gels using organic silane derivates have been investigated for immobilizing lipases (84). After encapsulation of lipase from *C. antarctica* in TMOS/PTMS (*n*-propyl trimethoxysilane) gels, an increase in enzyme activity by fivefold was reported in comparison to the commercial lipase powder. A model esterification of lauric acid with *n*-octanol to octyl laureate in isooctane was studied. Encapsulated lipase from*Aspergillus niger* showed 17% increased activity. Lipase SP 523 (NOVO) exhibited even 88% higher activity than the free lipase preparation(101). Various lipases (from 10 different organisms) were encapsulated in sol–gels derived from the precursors TMOS and *n*-butylsilane (*n*BTMS)/isobutylsilane (iBTMS) in combination with isopropyl alcohol and PVA as additives. Apart from isopropyl alcohol and PVA, which were used in each preparation, additional additives were evaluated for effects on EE values and activity in the esterification of lauric acid by *n*-octanol. Crown ether (18-crown-6) and the surfactant Tween 80® were identified as being particularly useful. Encapsulated lipase B from *C. antarctica*, for example, showed about 30-fold increase in specific activity compared to lipase powder under conditions where celite and 18-crown-6 were added for encapsulation in iBTMS/TMOS networks. *Thermomyces lanuginosa* lipase encapsulated in *n*BTMS/TMOS sol–gels under use of celite and Tween 80® showed a nearly 1400 times higher specific activity than the commercial lipase (102).

Enzyme recycling was also improved by encapsulation in TMOS sol–gels. After 20 cycles of reaction, separation, washing, and reuse, 70% activity of encapsulated CalB was still measurable, whereas only 40% activity was retained by commercial CalB under the same reaction conditions. Three lipases were also tested for enzyme enantioselectivity in the acylating kinetic resolution of *rac*-2-octanol, but only one of them, CalB, showed enhanced properties after encapsulation. The enantiomeric excess was shifted from 70 to >100 upon encapsulation in iBTMS/TMOS (102).

Layer-by-Layer (LbL) Structures

These are polymeric structures that consist of two organic polyelectrolytes of opposite charge that are held together by electrostatic forces. Usually, a combination of weak and strong polyelectrolytes is chosen (103, 104). A polymer combination often used is the weakly negatively charged polystyrene sulfonate (PSS) and the strongly positively charged polyallylamine hydrochloride (PAH). Alternatively, PEI and protamine are used as polycations, and dextran sulfate and silica are frequently used polyanions (104-106). LbL-structures assembled from nonpolyelectrolyte polymers exploit hydrophobic interactions between polymers, hydrogen bonding (104), or even DNA base pair hybridization(103).

Depending on the preparation conditions, spherical microcapsules with different diameters or enzyme-containing films can be obtained. The thickness of the walls or films generated is estimated to increase approximately 2 nm per polyelectrolyte layer (105).

The assembly of alternatingly charged polyelectrolyte layers requires a core or template around which the LbL construction can be built. The assembly is then achieved by resuspending the core alternately in aqueous solutions containing positively or negatively charged polyelectrolytes(104). Another possibility for layer assembly is to spray the dissolved polyelectrolytes onto the core, which can be even done simultaneously for both polyelectrolytes, saving time compared to the conventional (sequential) method (107, 108). Depending on the nature of the core, LbL structure preparation can be subdivided into direct and indirect encapsulation methods, as shown in Fig. 9. Direct methods involve incorporation of the enzyme into a solid porous structure such as mesoporous silica (109) or $CaCO_3$ (110, 111), by either adsorption or co-precipitation (111). The cores are then surrounded by the polyelectrolyte layers. The $CaCO_3$ microparticles within the formed capsules, however, can be easily decomposed by precipitation with EDTA at neutral pH (110).

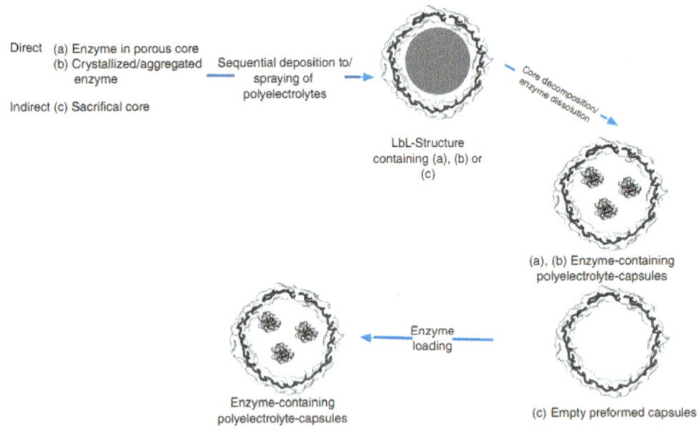

Figure 9: Schematic description of LbL-capsule synthesis by direct (a and b) or indirect methods (c). Alternate layers of oppositely charged polyelectrolytes (gray and black layers) are assembled on a core by sequential deposition or spraying. Enzyme-containing cores (a) and sacrificial cores (c) are decomposed afterward, followed by enzyme loading for (c). Crystallized enzymes are dissolved (b). The final result in all preparation methods is a polyelectrolyte-shell encapsulating enzymes in its confinement.

Size of the final microcapsules critically depends on the diameter of the core. For example, $CaCO_3$/α-lactalbumin particles with diameters of 4–6 μm were encapsulated within PAH/PSS bilayers, and the resulting microcapsules exhibited average diameters of approximately 7 μm (110). Encapsulation of catalase in protamine/silica capsules by $CaCO_3$ coprecipitation was possible with excellent encapsulation efficiency of over 98%. Nevertheless, only 24% of the enzyme activity compared to the free enzyme could be measured after the encapsulation, which can be ascribed to limited diffusion across the capsule wall, assuming that there was no enzyme inactivation owing to use of a mild encapsulation method and biocompatible materials (112).

The other possibility of direct enzyme encapsulation within LbL structures is the use of a colloidal core, which consists of the enzyme directly, either in crystalline (113) or in aggregated (114) form. The difficulty with these methods is to achieve crystals or aggregates of the normally highly soluble enzymes that are large enough to serve as cores for LbL assembly. The aggregation is normally induced in conditions with high salt concentrations (113, 114). At acidic pH, the enzyme crystals inside the capsules can be dissolved, and spherical capsules with diameters of approximately 12 μm are often obtained. Capsule size is typically increased upon air-drying to about 20 μm. Using catalase, 97% of the free enzyme›s was activity retained after encapsulation. Enzyme activity was still present after a 24 h-long exposure to protease, probably because of effective protection of the catalase by the four polyelectrolyte layers surrounding it (105).

In addition to formation of enzyme aggregates by precipitation crystallization, there are alternative ways to prepare enzyme aggregates to be later used as colloidal protein cores for LbL structure generation. Fusion of target protein to the cellulose binding domain (CBD) of *Clostridium cellulovorans* is an interesting possibility. This CBD is only poorly soluble and therefore CBD-protein fusions are often produced as aggregates(115). d-Amino acid oxidase from *Trigonopsis variabilis* harboring an *N*-terminal CBD accumulated to a great extent as insoluble aggregates during protein production in *Escherichia coli*. The fusion protein exhibited about 40% of the specific activity of the soluble enzyme lacking the CBD (116).

In indirect methods of LbL structure formation, the enzyme is not present at the time of microcapsule preparation. Instead, a sacrificial core is used that can be dissolved after the assembly. Materials serving as sacrificial cores are melamine formaldehyde (117-119), polystyrene, but also silica (104, 120) and $CaCO_3$ (121). In contrast to the other materials just mentioned, polystyrene- and melamine formaldehyde-derived sacrificial cores cannot be removed completely (104, 110). After dissolution of the sacrificial core, hollow capsules are obtained, whereas partial dissolution leads to microspheres, which can also take up enzymes upon incubation (106).

For loading of the preformed capsules with the desired enzymes, its pores have to be opened in order to allow diffusion of enzymes into them. This can be achieved by reversible expansion by pH increase (117, 118), resulting in swelling of the capsule and its pores. For capsules containing light-activatable molecules such as azobenzene that changes between *cis* and *trans* configurations, reversible expansion can also be induced by light exposure (104). The resulting microcapsules, though, are mechanically rather weak (104) and their permeability strongly depends on the environmental conditions (122), typically pH and ionic strength. Often, permeability increases at high salt concentration (104,123). Mechanically stronger microcapsules are obtained by initially synthesizing highly permeable capsules that allow the diffusion of enzymes, and then stabilizing them after loading. Stabilization is achieved through irreversible shrinkage induced by change in temperature or salt concentration (104), or by cross-linking of the polyelectrolytes in the shell. This can be done, for example, by exposure to UV light in the case of diazoresin-based microcapsules, where irradiation causes cross-linking between the multilayer walls (122). Cross-linking of PAA/PAH films can be induced either thermally by heating the film to 160°C in an inert atmosphere or chemically using

1-ethyl-3-(3-dimethylaminopropyl) (EDC) as an activator. The cross-linking reaction using EDC is also appropriate for dispersed capsules (120). Encapsulation efficiencies into preformed microcapsules are sometimes low, depending on the concentrations of microcapsules and enzymes during the loading procedure. At optimal concentrations of both components, around 95% encapsulation efficiency of α-chymotrypsin (used as model enzyme) was achieved. However, the enzyme lost 70% of its initial activity in the procedure. The activity loss could be attenuated to 40–50% at lowered microcapsule concentrations, however, with the drawback that the EE value was also decreased substantially (6–40%) (124). Another downside of encapsulation into preformed microcapsules is the harsh conditions used for permeabilization and cross-linking that may not be appropriate for many enzymes (103).

A totally different approach of enzyme encapsulation in LbL-structures is to use the enzyme itself as a polyanion for microcapsule assembly. This is possible with enzymes having acidic isoelectric point, such as cholesterol oxidase or cholesterol esterase. Being negatively charged at neutral pH, these enzymes were used at pH 7.5 as polyanions for microcapsule construction along with the polycation PEI. The microcapsule was assembled from the PEI-enzyme combination after assembling an initial polyanion-polycation layer with PSS and PEI (106). With this method, enzyme-containing coating films on colloidal particles or planar surfaces can be designed (106).

By varying the nature of the polyelectrolytes and the number of layers, it is possible to develop microcapsules with accurately tailored properties(103, 108). Permeability can be enhanced or restricted by chemically modifying the polyelectrolytes, for example, by esterification. In that way, hydrophobicity and charge density can be adjusted with consequent effects on permeability (125). In addition, ability to respond to various physiochemical stimuli, such as light or altered pH, can be introduced. Thus, triggered change in microcapsule permeability enables simple control of enzymatic processes (103, 108).

The ability of the LbL-assembled microcapsules to respond to a number of biological, chemical, and physical stimuli makes them also of great interest for controlled drug delivery and release (103, 104).

Other interesting applications are the use of LbL-encapsulated enzymes in biosensors and as biocatalysts. Encapsulated forms of cholesterin oxidase and glucose oxidase were used in biosensing applications. Glucose isomerase and various peroxidases were applied in encapsulated form to different biotransformation applications. In addition, multiple enzymes were encapsulated to allow for enzymatic cascade transformations. The challenging task of multienzyme LbL-microcapsule development was recently addressed through preparation of enzyme layers separated by polyelectrolyte layers. Coupling of HRP and glucose oxidase was achieved by encapsulating the peroxidase through coprecipitation with $CaCO_3$, which was followed by assembly of PSS/PAH polyelectrolyte layers around this capsule. In a second coprecipitation step, the resulting construct was again surrounded by an enzyme layer containing the glucose oxidase. Finally, outer polyelectrolyte layers were prepared and the $CaCO_3$cores were then dissolved. An enzymatic cascade of glucoamylase and glucose oxidase was also realized using the enzymes as polycations and fixing them on a film, again separated by PEI/PSS layers. In both cases, the substrates were able to diffuse through the polyelectrolyte layers(106).

The LbL-microcapsules generally exhibit good storage stability and enzyme retention over time. α-Chymotrypsin encapsulated in LbL-structures was found to retain 73% of its initial activity after 6 days storage at pH3 and 4°C, whereas only 43% activity remained when the free enzyme was stored at the same conditions (105, 106).

Like described earlier, much of the original biological activity is often lost in the encapsulated enzyme preparation. This occurs not only because of mass transfer limitations resulting in apparent enzyme inactivation but also because of unfavorable interactions of the polyelectrolyte layers with the enzyme. These interactions can result in decreased intrinsic enzyme activity (e.g. creation of steric hindrance around the active site) or true inactivation (126, 127). Benzaldehyde lyase from *Pseudomonas fluorescens* and alcohol dehydrogenase from *Lactobacillus brevis* showed almost no activity after encapsulation into PAH/PSS microcapsules. In *Bacillus coagulans* esterase, the interaction of the polyelectrolyte layers with the enzyme not only resulted in 86% decrease of original activity, but was also responsible for decreasing the enzyme's enantioselectivity. In the kinetic resolution of 1,2-*O*-isopropylidenglycerolesters, enantioselectivity dropped from 95% to 34% upon encapsulation. In contrast, the use of the same encapsulation protocol resulted in 1.32-fold increase in the activity of CalB (127).

To overcome the drawbacks of limited mass transfer across the polyelectrolyte layers, catalase was coencapsulated with hydrophilic silica nanospheres. Enzymes and silica nanospheres were incorporated within a $CaCO_3$ core using coprecipitation, and this was followed by assembly of the polyelectrolyte layers In this particular case, protamine/silica microcapsules were prepared, in which the positive protamine layer induced hydrolysis and condensation of sodium silicate, which in turn formed the negative silica layer. After dissolution of $CaCO_3$, the silica nanospheres caused enhanced flow into the encapsulated solution and thus improved the mass transfer through the capsule wall. Catalase activities of 41.7% compared to the enzyme in solution were observed within these microcapsules. In addition, high encapsulation efficiencies of 97.5% were reported (112).

OTHER METHODS

Recent trend in enzyme encapsulation is that not only one, but several methods of encapsulation are combined in one application to capitalize on the particular strengths of each single method to achieve optimal performance of the created system. Furthermore, new materials for enzyme encapsulation are being developed to satisfy increasing demands for broad application.

Encapsulating Capsules

One approach is to encapsulate an enzyme more than once with two (or even more) different methods, creating the so-called core-shell capsules. Alginate hydrogels are often used in combination with other methods to prevent leakage during long-term usage or storage (128). For example, polyelectrolyte layers are assembled around a negatively charged microcapsule, so that retention of encapsulated enzymes eventually depends on the polycationic layer used (129). Various different polyelectrolyte layers were applied in coating of alginate capsules. For example, diazoresin/PSS (130), PAH/PSS (74, 128), and PAH/PAA were used (128). A porous outer membrane was produced by incubating alginate beads in the presence of the polycation poly(methylene co-guanidine), which resulted in polyelectrolyte condensation (116, 131). Leaching of alginate-encapsulated glucose oxidase could be decreased from 45% in noncoated microcapsules to 25% and 15% after with PAH/PSS and cross-linked PAH/PAA, respectively. In addition, positive effects of the polyelectrolyte layers on enzyme activity were found. After three months of storage at room temperature, the glucose oxidase in uncoated alginate microcapsules only retained less than 20% of its initial activity. However, enzymes encapsulated in coated microspheres retained 84% (PAH/PSS) and 60% (cross-linked PAH/PAA) of their initial activities. It has to be noted though that the activity of the glucose oxidase after encapsulation into alginate microspheres only represented 24% of the free enzyme's activity (128).

Other enzymes encapsulated in gelatin particles (132), liposomes (133, 134), alginate particles (135), or in Ca-alginate/protamine layers (136-138) have been additionally encapsulated within silica sol–gel for better stability and to introduce the ability of the microcapsules to respond to altered pH. For preparation of Ca-alginate/protamine layers, enzyme-containing Ca-alginate microcapsules were dispersed in protamine solution, resulting in capsular structures with loosened or even liquefied core. Adding sodium silicate solution caused biosilification by the protamine, creating an additional outer silica shell. Encapsulation of β-glucuronidase with this method was possible with 69% efficiency and resulted in 125% relative activity compared to the free enzyme. Moreover, the enzyme activity remained unaltered during 10 cycles of reaction and recycling that included filtering and washing (138).

In another approach to prevent enzyme leakage by combining LbL- and alginate hydrogel techniques, the microcapsules were designed the other way round, using alginate as the outermost layer. Enzyme encapsulation was carried out by $CaCO_3$ coprecipitation and the resulting particles were then covered by polyelectrolyte layers (PAH/PSS). Exposing these capsules to alginate and glucono-δ-lactone caused internal ionotropic gelation of the alginate where glucono-δ-lactone (or the gluconic acid produced from it by spontaneous hydrolysis) dissolved the $CaCO_3$ core, which in turn promoted alginate gelation. The outcome was a highly porous structure with enzymes being confined to the microcavities and thoroughly retained by the polyelectrolyte layers. With this preparation, approximately 50 times higher retention of glucose oxidase was possible than with uncoated alginate microspheres (139).

Silica sol–gels containing encapsulated enzymes were themselves additionally encapsulated by polyelectrolyte-layers or polyelectrolyte-nanoparticle-layers. The example of catalase showed that activity loss after 25 successive batch reactions was reduced from 65% to 30% comparing LbL-coated and untreated sol–gel microspheres. Furthermore, encapsulated enzyme was protected effectively from attack by proteases, enabling 98% activity retention after proteolysis treatment with protease Sigma P6911 (140).

ORGANIC/INORGANIC HYBRID BEADS

Hybrid materials consisting of organic polymers and inorganic components, such as the sol–gels derived from biosilification described earlier, can be synthesized with a number of other materials and shall be distinguished here from the core-shell capsules by their way of synthesis, which does not involve more than one separate encapsulation step. The organic/inorganic hybrid beads do also not exhibit core-shell structure, but rather have one microcapsule wall with homogeneous structure.

One example is the preparation of calcium carbonate-mineralized alginate/gelatin microcapsules, made by adding droplets of sodium alginate solution, containing sodium carbonate and YADH, to Ca^{2+}-containing gelatin solution. Formation of an alginate/gelatin membrane and precipitation of calcium carbonate are promoted, thus creating a calcificated shell around the enzymes, which were encapsulated with 89% efficiency. The resulting capsular shell enabled preservation of 55% of the initial enzyme activity after six reaction/recycling cycles. In contrast, only 10% activity was retained in nonmineralized alginate/gelatin capsules prepared according to the same procedure and used identically. Furthermore, the calcificated capsules showed 60% lower enzyme leakage than nonmineralized capsules (141).

Another promising material for encapsulation is offered by boehmite/alginate hybrid-gels, also synthesized in a one-pot method, where boehmite-enzyme solution mixed with sodium alginate is added dropwise into Ca^{2+} solution. In the resulting beads, boehmite (γ-AlOOH) particles were homogeneously embedded in a Ca-alginate network that also encapsulated enzymes (YADH). Remarkable encapsulation efficiencies of nearly 100% could be reached, and 86.6% of the initial YADH activity was still measurable after 12 recycling steps. Almost no enzyme leakage from the beads was observed within 67 h incubation at 4°C (142).

In addition to hybrid networks, also organic/inorganic hybrid vesicles were made. For their preparation, an alxosysilane-lipid was dispersed in water, followed by incubation to allow for sol–gel reaction of the alkoxysilane headgroups to take place. The resulting vesicles enable transport of small hydrophilic molecules across the membrane without the requirement of introducing any structural features (pores and channels) for transport across the vesicle wall. They can stably encapsulate water-soluble enzymes. α-Chymotrypsin was encapsulated without loss of activity compared to free enzyme. The encapsulated enzyme was protected against the effect of bovine pancreas trypsin inhibitor, whereas in contrast, the free enzyme totally lost its activity (143).

In order to overcome diffusional limitations, hybrid-microcapsules with several mesopores and a single macropore per capsule wall were produced. For this task, vaterite-type $CaCO_3$ microspheres were prepared by precipitation in a solution containing PAH. While PAH was entrapped in the near-surface region of the spheres, further stirring led to the formation of a calcite projection on the microsphere surface that was responsible for the creation of the capsule's macropore. The resulting templates were subsequently incubated in

Ti(IV)-bis-(ammonium lactate) dihydroxide and a solution of oligodopa, which resulted in mineralization and metal-organic coordination inside the capsule wall. After dissolution of the $CaCO_3$ core, the microspheres were incubated in enzyme solution, which resulted in physical adsorption and covalent binding of the enzymes to the microcapsule. For comparison, also enclosed hybrid microspheres encapsulating enzymes were prepared. After adding enzymes to the $CaCl_2$ solution in the first step, coprecipitation occurred, resulting in enzyme-containing templates. Stirring was shortened in the next step to prevent projection formation, and the microspheres were further prepared as described earlier. Catalase was immobilized in both systems as a model enzyme and evaluated in the conversion of H_2O_2. Conversion of H_2O_2 was nearly completed after 300 s in the reaction mixture containing catalase-adsorbed macroporous capsules. In contrast, enzyme-encapsulated enclosed microspheres only reached 51.2% conversion within 600 s owing to mass transfer resistance (144).

NEW MATERIALS

Apart from efforts to improve conventional and well-established systems for enzyme encapsulation, also new materials and methods of synthesis are examined for their applicability to microcapsule preparation.

Cryogels are derived from cryotopic gelation of any gel-forming agents, for example, PVA. A solution of these agents is frozen just a few degrees below the crystallization point of the solvent. Thus, unfrozen liquid microphases remain, where the gel-forming agents are concentrated and gelation occurs around the already existing crystals. As the crystals act as pore-forming agents, the gels are highly porous after thawing and thus potentially allow for the processing of macromolecular substrates (145). In reality though, cryogels have shown fast leakage (145) and partly reduced activity (146) of the encapsulated enzymes. Ionogels, consisting of an ionic liquid (organic salts with low melting temperature), an organic gelator, and organic silane derivates, show good results in biosensing with encapsulated enzymes in regard to response time and biocompatibility (147). Conventional polyacrylamide has seen a revival in combination with other materials, such as poly(*N*-isopropylacrylamide) for better performance of biocatalysis in nonpolar solvents (148). Polyacrylamide has also been combined with organic materials, such as alginate (149), or gelatin and pectin (150), creating further organic/inorganic hybrid materials. Poly(*N*-isopropylacrylamide) itself has also been used for preparing gels for enzyme encapsulation (151, 152).

In addition, new materials for capsules rather than gels are investigated, one of them being silk fibroin, assembled on enzyme-containing liposomes, which were afterward dissolved by methanol or sodium chloride. This procedure resulted in silk fibroin microcapsules that were used for drug delivery and release (153). Another rather novel material is a polymer-polyphenol complex made of a polyphenol derived from epigallocatechin gallate and water-soluble cellulose ethers. The polymer-polyphenol beads were prepared by one-pot method, where solution containing the enzyme and cellulose ether is dropped into phenol solution. Even though retention of enzyme activity (α-amylase) was promising, the capsules were decomposed within five cycles of reaction and washing (154).

In addition, different polydopamine (PDA) microcapsules were synthesized. First, $CaCO_3$ cores were prepared to encapsulate enzymes via coprecipitation, some of them being doped with PAH or PSS. When exposing the enzyme-$CaCO_3$-cores to dopamine hydrochloride solution, self-polymerization of the dopamine and therefore synthesis of an outer PDA layer occurred. Afterward, the $CaCO_3$ core was dissolved. PAH-PDA capsules proved to be most suitable for encapsulation of catalase, where EE values of 95% and unaltered activity compared to the free enzyme were observed. No leakage was observed after seven cycles of washing and almost no loss in specific activity (about 4%) occurred after 10 cycles of washing and reuse, but activity decreased about 20% during storage for 30 days (155). In addition, a multienzyme system was established using PDA microcapsules. $CaCO_3$ cores containing glucosidase and PSS were prepared by coprecipitation and resuspended in PDA-hydrochloride solution, which also contained β-amylase in the experiment reported (Fig. 10). After assembly, the microcapsules were incubated in α-amylase solution and the $CaCO_3$ core dissolved afterward. The resulting microspheres contained glucosidase in their interior, β-amylase entrapped within the PDA wall, and α-amylase chemically attached to the outer surface. Conversion of starch to isomaltooligosaccharide was catalyzed with a maximum yield of 53%, compared to 39% maximum yield with the free enzyme cascade (156).

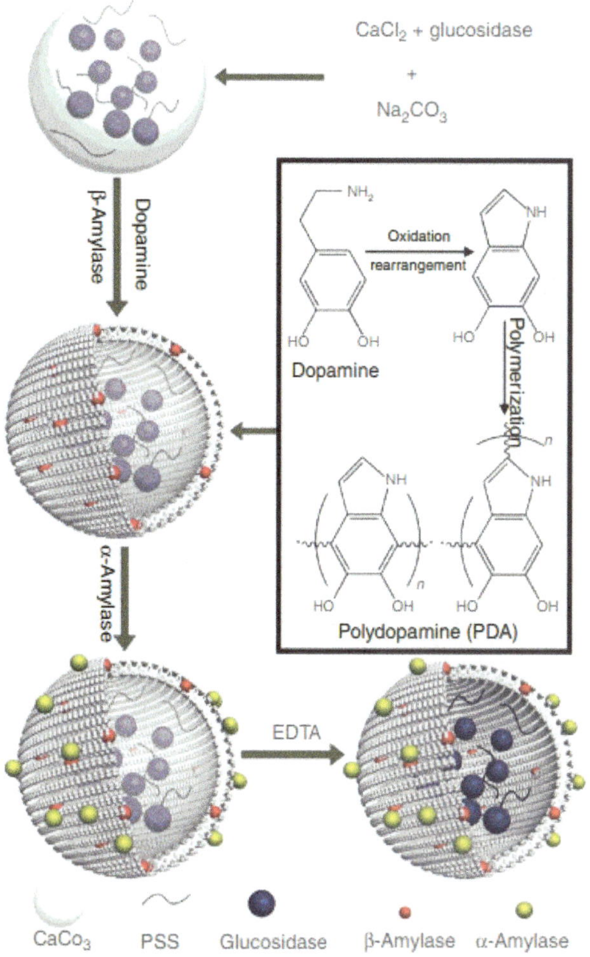

Figure 10. Preparation process of polydopamine capsules enabling multienzyme system. Glucosidase, β-amylase and α-amylase were immobilized for concerted conversion of starch to isomaltooligosaccharide.

Another silicon-containing matrix suitable for encapsulating enzymes can be produced by hydrosilylation reaction. This method, called *coating*, is not an encapsulation method in the closest meaning, as it uses the solid biocatalyst that is noncovalently immobilized to a carrier for encapsulation. In hydrosilylation reaction, the carrier-bound biocatalyst is disposed to divinyl-terminated polydimethylsiloxane (a siloxane that contains Si–H groups) and Pt-based catalyst (157, 158). This results in the formation of silicone and its penetration into the porous carrier, which in turn leads to the entrapment of the enzyme in the silicone matrix. The conformational freedom of the biocatalyst is thereby fully retained (158). Unlike initially predicted, no core-shell particles but composite particles with a silicone and a core-material phase were obtained (157). This method was at first used with lipophilic enzymes such as the lipase preparation NZ435. The enzyme preparation thus obtained showed improved performance in industrial applications in respect to catalyst stability and enzyme leakage as compared to the uncoated enzyme preparation (159,160). Synthesis of myristyl myristate was carried out as a model reaction at lab-scale in a stirred flask reactor, achieving a 23-fold increase of the total turnover number (mass of product produced per mass of enzyme used) compared to the native NZ435. This clearly broadens the range of applications of this lipase, which could be only used efficiently in fixed-bed reactors until then (159).

The physiochemical properties of the silicone polymer can be tailored by varying the degree of cross-linking, the number of modifications, and the chain length of the polymers (157). The matrix-material has thus been made suitable for encapsulation of hydrophilic enzymes such as carbonyl reductase (161).

CONCLUSIONS

Microencapsulation presents a diverse and powerful methodology for the development of heterogeneous biocatalysts. In its optimum case, microencapsulation succeeds in combining the most important advantages of carrier-bound and carrier-free methods of enzyme immobilization and therefore holds considerable promise for implementation in biocatalytic processes and in other enzyme applications. Problems of enzyme microencapsulation are sometimes laborious procedures of microcapsule preparation, relatively low encapsulation efficiencies, control of intracapsule transport, and microcapsule stability under operational conditions. Unique possibilities of enzyme microencapsulation are in multienzyme immobilization where controlled compartimentalization of individual enzymatic transformations provides a new paradigm for multistep cascade biotransformations.

REFERENCES

1. Sheldon RA. Enzyme immobilization: the quest for optimum performance. *Adv Synth Catal* 2007;349:1289–1307.
2. Cao L. *Carrier-bound immobilized enzymes: principles, application and design*. Weinheim: *Wiley-VCH*; 2005.
3. Sheldon RA. Characteristic features and biotechnological applications of cross-linked enzyme aggregates (CLEAs). *Appl Microbiol Biotechnol* 2011;92:467–477.
4. Roessl U, Nahálka J, Nidetzky B. Carrier-free immobilized enzymes for biocatalysis. *Biotechnol Lett* 2010;32:341–350.
5. Brady D, Jordaan J. Advances in enzyme immobilization. *Biotechnol Lett* 2009;31:1639–1650.
6. Cao L. Immobilised enzymes: science or art? *Curr Opin Chem Biol* 2005;9:217–226.
7. Walde P, Ichikawa S. Enzymes inside lipid vesicles: preparation, reactivity and applications. *Biomol Eng* 2001;18:143–177.
8. van Oers MCM, Rutjes FPJT, van Hest JCM. Cascade reactions in nanoreactors. *Curr Opin Biotech* 2014;28:10–16.
9. Fonseca L, Meesters G. Enzyme encapsulation. *Bioencaps Innovations* 2013;13:1–2.
10. Park BW, Yoon DY, Kim DS. Recent progress in bio-sensing techniques with encapsulated enzymes. *Biosens Bioelectron* 2010;26:1–10.
11. Livage J, Coradin T, Roux C. Encapsulation of biomolecules in silica gels. *J Phys Condens Matter* 2001;13:673–691.
12. Gibbs BF, Kermasha S, Alli I, *et al*. Encapsulation in the food industry: a review. *Int J Food Sci Nutr* 1999;50:213–224.
13. Discher DE, Eisenberg A. Polymer vesicles. *Science* 2002;297:967–973.
14. Silvius JR. Role of cholesterol in lipid raft formation: lessons from lipid model systems. *Biochim Biophys Acta* 2003;1610:174–183.
15. Mozafari MR, Johnson C, Hatziantoniou S, *et al*. Nanoliposomes and their applications in food nanotechnology. *J Liposome Res*2008;18:309–327.
16. Kirby C, Gregoriadis G. Dehydration-rehydration vesicles: a simple method for high yield drug entrapment in liposomes. *Nat Biotechnol*1984;2:979–984.
17. Spector MS, Zasadzinski JA. Topology of multivesicular liposomes, a model biliquid foam. *Langmuir* 1996;12:4704–4708.
18. Katre NV, Asherman J, Schaefer H, *et al*. Multivesicular liposome (DepoFoam) technology for the sustained delivery of insulin-like growth factor-I (IGF-I). *J Pharm Sci* 1998;87:1341–1346.
19. Rigaud JL, Pitard B, Levy D. Reconstitution of membrane proteins into liposomes: application to energy-transducing membrane proteins.*Biochim Biophys Acta* 1995;1231:223–246.
20. Discher BM, Hammer DA, Bates FS, *et al*. Polymer vesicles in various media. *Curr Opin Colloid Interf Sci* 2000;5:125–131.
21. Paula S, Volkov AG, Van Hoek AN, *et al*. Permeation of protons, potassium ions and small polar molecules through phospholipid bilayers as a function of membrane thickness. *Biophys J* 1996;70:339–348.
22. Rao DR, Chawan CB, Veeramachaneni R. Liposomal encapsulation of β-galactosidase: comparison of two methods of encapsulation and in vitro lactose digestibility. *J Food Biochem* 1994;18:239–251.
23. Espinola LG, Wider EA, Stella AM, *et al*. Enzyme

replacement therapy in porphyrias-II: entrapment of δ-aminolaevulinate dehydratase in liposomes. *Int J Biochem* 1983;15:439–445.

24. Yoshimoto M, Wang S, Fukunaga K, *et al.* Preparation and characterization of reactive and stable glucose oxidase-containing liposomes modulated with detergent. *Biotechnol Bioeng* 2003;81:695–704.
25. Walde P, Marzetta B. Bilayer permeability-based substrate selectivity of an enzyme in liposomes. *Biotechnol Bioeng* 1998;57:216–219.
26. Yoshimoto M, Okamoto M, Ujihashi K, *et al.* Selective oxidation of d-amino acids catalyzed by oligolamellar liposomes intercalated with d-amino acid oxidase. *Langmuir* 2014;30:6180–6186.
27. Ramundo-Orlando A, Morbiducci U, Mossa G, *et al.* Effect of low frequency, low amplitude magnetic fields on the permeability of cationic liposomes entrapping carbonic anhydrase I. *Bioelectromagnetics* 2000;21:491–498.
28. Graff A, Winterhalter M, Meier W. Nanoreactors from polymer-stabilized liposomes. *Langmuir* 2001;17:919–923.
29. Guan H, Liu X, Wang W. Encapsulation of tyrosinase within liposome bioreactors for developing an amperometric phenolic compounds biosensor. *J Solid State Electrochem* 2013;17:2887–2893.
30. Hotz J, Meier W. Vesicle-templated polymer hollow spheres. *Langmuir* 1998;14:1031–1036.
31. Ruysschaert T, Germain M, Pereira de Silva Gomes JF, *et al.* Liposome-based nanocapsules. *IEEE Trans Nanobiosci* 2004;3:49–55.
32. Kheadr EE, Vuillemard JC, El-Deeb SA. Impact of liposome-encapsulated enzyme cocktails on cheddar cheese ripening. *Food Res Int*2003;36:241–252.
33. Baskaya FS, Zhao X, Flickinger MC, *et al.* Thermodynamic feasibility of enzymatic reduction of carbon dioxide to methanol. *Appl Biochem Biotechnol* 2010;162:391–398.
34. Yoshimoto M, Yamashita T, Yamashiro T. Stability and reactivity of liposome-encapsulated formate dehydrogenase and cofactor system in carbon dioxide gas–liquid flow. *Biotechnol Prog* 2010;26:1047–1053.
35. Yoshimoto M, Yamashita T, Kinoshita S. Thermal stabilization of formaldehyde dehydrogenase by encapsulation in liposomes with nicotinamide adenine dinucleotide. *Enzyme Microb Technol* 2011;49:209–214.
36. Yoshimoto M, Sato M, Yoshimoto N, *et al.* Liposomal encapsulation of yeast alcohol dehydrogenase with cofactor for stabilization of the enzyme structure and activity. *Biotechnol Prog* 2008;24:576–582.
37. Yoshimoto M. Stabilization of enzymes through encapsulation in liposomes. In: Minteer SD, editor. Volume 679, *Enzyme stabilization and immobilization: methods and protocols, methods in molecular biology*. New York: © Springer Science+Business Media; 2011. pp. 9–18.
38. Bates FS, Fredrickson GH. Block copolymer thermodynamics: theory and experiment. *Annu Rev Phys Chem* 1990;41:525–557.
39. Blanzas A, Armes SP, Ryan AJ. Self-assembled block copolymer aggregates: from micelles to vesicles and their biological applications.*Macromol Rapid Commun* 2009;30:267–277.
40. Peters RJRW, Louzao I, van Hest JCM. From polymeric nanoreactors to artificial organelles. *Chem Sci* 2012;3:335–342.
41. Kita-Tokarczyk K, Grumelard J, Haefele T, *et al.* Block copolymer vesicles—using concepts from polymer chemistry to mimic biomembranes. *Polymer* 2005;46:3540–3563.
42. Lee JS, Feijen J. Polymersomes for drug delivery: design, formation and characterization. *J Control Release* 2012;161:473–483.
43. Tanner P, Baumann P, Enea R, *et al.* Polymeric vesicles: from drug carriers to nanoreactors and artificial organelles. *Acc Chem Res*2011;44:1039–1049.
44. Lee JC, Bermudez H, Discher BM, *et al.* Preparation, stability, and in vitro performance of vesicles made with diblock copolymers.*Biotechnol Bioeng* 2001;73:135–145.
45. Onaca O, Hughes DW, Balasubramanian V, *et al.* SOD antioxidant nanoreactors: influence of block copolymer composition on the nanoreactor efficiency. *Macromol Biosci* 2010;10:531–538.
46. Rameez S, Alosta H, Palmer AF. Biocompatible and biodegradable polymersome encapsulated hemoglobin: a potential oxygen carrier.*Bioconjugate Chem* 2008;19:1025–1032.
47. Meier W, Nardin C, Winterhalter M. Reconstitution of channel proteins in (polymerized) ABA triblock copolymer membranes. *Angew Chem Int Ed* 2000;39:4599–4602.
48. Axthelm F, Casse O, Koppenol W, *et al.* Antioxidant nanoreactor based on superoxide dismutase encapsulated in superoxide-permeable vesicles. *J Phys Chem B* 2008;112:8211–8217.
49. Nardin C, Hirt T, Leukel J, *et al.* Polymerized ABA triblock copolymer vesicles. *Langmuir* 2000;16:1035–1041.
50. O'Neil CP, Suzuki T, Demurtas D, *et al.* A novel method for the encapsulation of biomolecules into polymersomes via direct hydration.*Langmuir* 2009;25:9025–9029.
51. Brinkhuis RP, Rutjes FPJT, van Hest JCM. Polymeric vesicles in biomedical applications. *Polym Chem* 2011;2:1449–1462.
52. Antonietti M, Förster S. Vesicles and liposomes: a self-assembly principle beyond lipids. *Adv Mater* 2003;15:1323–1333.
53. Kuiper SM, Nallani M, Vriezema DM, *et al.* Enzymes containing porous polymersomes as nano reaction vessels for cascade reactions.*Org Biomol Chem* 2008;6:4315–4318.
54. Kumar M, Grzelakowski M, Zilles J, *et al.* Highly permeable polymeric membranes based on the incorporation of the functional water channel protein Aquaporin Z. *Proc Natl Acad Sci U S A* 2007;104:20719–20724.

55. Meeuwissen SA, Rioz-Martinez A, de Gonzalo G, *et al.* Cofactor regeneration in polymersome nanoreactors: enzymatically catalyzed Baeyer-Villiger reactions. *J Mater Chem* 2011;21:18923–1826.

56. Meeuwissen SA, Rioz-Martinez A, de Gonzalo G, *et al.* Cofactor regeneration in polymersome nanoreactors: enzymatically catalysed Baeyer-Villiger reactions. *J Mater Chem* 2011;21:18923–18926.

57. Nallani M, de Hoog HPM, Cornelissen JJLM, *et al.* Polymersome nanoreactors for enzymatic ring-opening polymerization. *Biomacromolecules* 2007;8:3723–3728.

58. van Dongen SFM, Nallani M, Cornelissen JJCM, *et al.* A three enzyme cascade reaction through positional assembly of enzymes in a polymersome nanoreactor. *Chem Eur J* 2009;15:1107–1114.

59. Vriezema DM, Garcia PMI, Oltra NS, *et al.* Positional assembly of enzymes in polymersome nanoreactors for cascade reactions. *Angew Chem Int Ed* 2007;46:7378–7382.

60. Peters RJRW, Marguet M, Marais S, *et al.* Cascade reactions in multicompartmentalized polymersomes. *Angew Chem Int Ed* 2014;53:146–150.

61. Geng Y, Discher DE, Justynska J, *et al.* Grafting short peptides onto polybutadiene-block-poly(ethylene oxide): a platform for self-assembling hybrid amphiphiles. *Angew Chem Int Ed* 2006;118:7740–7743.

62. Gaspard J, Silas JA, Shantz DF, *et al.* Supramolecular assembly of lysine-b-glycine block copolypeptides at different solution conditions. *Supramol Chem* 2010;22:178–185.

63. Rodríguez-Hernández J, Lecommandoux S. Reversible inside-out micellization of pH-responsive and water-soluble vesicles based on polypeptide diblock copolymers. *J Am Chem Soc* 2005;127:2026–2027.

64. Makino A, Kimura S. Preparation of peptide- and protein-based molecular assemblies and their utilizations as nanocarriers for tumor imaging. *React Funct Polym* 2011;71:272–279.

65. Hamidi M, Azadi A, Rafiei P. Hydrogel nanoparticles in drug delivery. *Adv Drug Deliv Rev* 2008;60:1638–1649.

66. Peppas NA, Hilt JZ, Khademhosseini A, *et al.* Hydrogels in biology and medicine: from molecular principles to bionanotechnology. *Adv Mater* 2006;18:1345–1360.

67. Fernandes P. Synthetic gels for enzyme encapsulation: PVA and sol–gel matrices. *Bioencaps Innovations* 2013;13:9–11.

68. Cattorini S, Marques MPC, Carvalho F, *et al.* Lentikat®-based biocatalysts: effective tools for inulin hydrolysis. *Chem Biochem Eng Q* 2009;23:429–434.

69. Gröger H, Capan E, Barthuber A, *et al.* Asymmetric synthesis of an (R)-cyanohydrin using enzymes entrapped in lens-shaped gels. *Org Lett* 2001;3:1969–1972.

70. Betigeri SS, Neau SH. Immobilization of lipase using hydrophilic polymers in the form of hydrogel beads. *Biomaterials* 2002;23:3627–3636.

71. Gombotz WR, Wee SF. Protein release from alginate matrices. *Adv Drug Deliv Rev* 1998;31:267–285.

72. Taqieddin E, Amiji M. Enzyme immobilization in novel alginate–chitosan core-shell microcapsules. *Biomaterials* 2004;25:1937–1945.

73. Wilson L, Illanes A, Pessela BCC, *et al.* Encapsulation of crosslinked penicillin G acylase aggregates in lentikats: evaluation of a novel biocatalyst in organic media. *Biotechnol Bioeng* 2004;86:558–562.

74. Zhu H, Srivastava R, Brown JQ, *et al.* Combined physical and chemical immobilization of glucose oxidase in alginate microspheres improves stability of encapsulation and activity. *Bioconjugate Chem* 2005;16:1451–1458.

75. Bayramoglu G, Altintas B, Arica MY. Immobilization of glucoamylase onto polyaniline-grafted magnetic hydrogel via adsorption and adsorption/crosslinking. *Appl Microbiol Biotechnol* 2013;97:1149–1159.

76. Li Y, Huang G, Zhang X, *et al.* Magnetic hydrogels and their potential biomedical applications. *Adv Funct Mater* 2013;23:660–672.

77. Metrangolo-Ruiz De Temiño D, Hartmeier W, Ansorge-Schumacher MB. Entrapment of the alcohol dehydrogenase from *Lactobacillus kefir* in polyvinyl alcohol for the synthesis of chiral hydrophobic alcohols in organic solvents. *Enzyme Microb Technol* 2005;36:3–9.

78. Yadav RR, Mudliar SN, Sekh AY, *et al.* Immobilization of carbonic anhydrase in alginate and its influence on transformation of CO_2 to calcite. *Process Biochem* 2012;47:585–590.

79. Hench LL, West JK. The sol–gel process. *Chem Rev* 1990;90:33–72.

80. Pierre AC. The sol–gel encapsulation of enzymes. *Biocatal Biotransform* 2004;22:145–170.

81. Avnir D, Coradin T, Lev O, *et al.* Recent bio-applications of sol–gel materials. *J Mater Chem* 2006;16:1013–1030.

82. Betancor L, Luckarift HR. Bioinspired enzyme encapsulation for biocatalysis. *Trends Biotechnol* 2008;26:566–572.

83. Bhatia RB, Brinker CJ, Gupta AK, *et al.* Aqueous sol–gel process for protein encapsulation. *Chem Mater* 2000;12:2434–2441.

84. Adlercreutz P. Immobilisation and application of lipases in organic media. *Chem Soc Rev* 2013;46:6406–6436.

85. Reetz MT, Wenkel R, Avnir D. Entrapment of lipases in hydrophobic sol–gel materials: efficient heterogeneous biocatalysts in aqueous medium. *Synthesis* 2000;6:781–783.

86. Gill I, Ballesteros A. Bioencapsulation with synthetic polymers (part 1): sol–gel encapsulated biologicals. *Trends Biotechnol* 2000;18:282–296.

87. Jin W, Brennan JD. Properties and applications of proteins encapsulated within sol–gel derived materials. *Anal Chim Acta* 2002;461:1–36.

88. El Rassy H, Maury S, Buisson P, *et al.* Hydrophobic silica aerogel–lipase biocatalysts - Possible interactions between the enzyme and the gel. *J Non-Cryst Solids* 2004;350:23–30.

89. Hüsing N, Schubert U, Misof K, *et al.* Formation and

structure of porous gel networks from $Si(OMe)_4$ in the presence of $A(CH_2)_nSi(OR)_3$(A=functional group). *Chem Mater* 1998;10:3024–3032.

90. Gill I, Ballesteros A. Encapsulation of biologicals within silicate, siloxane, and hybrid sol–gel polymers: an efficient and generic approach.*J Am Chem Soc* 1998;120:8587–8598.
91. Buisson P, Hernandez C, Pierre M, *et al.* Encapsulation of lipase in aerogels. *J Non-Cryst Solids* 2001;285:295–302.
92. Reetz MT, Zonta A, Simpelkamp J. Efficient immobilization of lipase by entrapment in hydrophobic sol–gel materials. *Biotechnol Bioeng*1996;49:527–534.
93. Satvekar RK, Rohiwal SS, Raut AV, *et al.* A silica-dextran nanocomposite as a novel matrix for immobilization of horseradish peroxidase, and its application to sensing hydrogen peroxide. *Microchim Acta* 2014;181:71–77.
94. a Braun S, Rappoport S, Zusman R, *et al.* Biochemically active sol–gel glasses: the trapping of enzymes. *Materials Lett*2007;61:2843–2846;
95. b Braun S, Rappoport S, Zusman R, *et al.* Biochemically active sol–gel glasses: the trapping of enzymes. *Materials Lett* 1990;10:1–5.
96. Cheung MC, Yung KY, Xu H, *et al.* Porous nanostructured encapsulation and immobilization materials for optical biosensors. *IEEE J Sel Top Quantum Electron* 2012;18:1147–1159.
97. Bernardino S, Estrela N, Ochoa-Mendes V, *et al.* Optimization in the immobilization of penicillin G acylase by entrapment in xerogel particles with magnetic properties. *J Sol–Gel Sci Technol* 2011;58:545–556.
98. Orcaire O, Buisson P, Pierre AC. Application of silica aerogel encapsulated lipases in the synthesis of biodiesel by transesterification reactions. *J Mol Catal B: Enzym* 2006;42:106–113.
99. Sani S, Muhid MNM, Hamdan H. Design, synthesis and activity study of tyrosinase encapsulated silica aerogel (TESA) biosensor for phenol removal in aqueous solution. *J Sol–Gel Sci Technol* 2011;59:7–18.
100. Paljevac M, Primožič M, Habulin M, *et al.* Hydrolysis of carboxymethyl cellulose catalyzed by cellulase immobilized on silica gels at low and high pressures. *J Supercrit Fluids* 2007;43:74–80.
101. Zhang C, Hu C, Zhao Y, *et al.* Encapsulation of laccase in silica colloidosomes for catalysis in organic media. *Langmuir*2013;29:15457–15462.
102. Reetz MT, Zonta A, Simpelkamp J. Efficient heterogeneous biocatalysts by entrapment of lipases in hydrophobic sol–gel materials.*Angew Chem Int Ed* 1995;34:301–303.
103. Reetz MT, Tielmann P, Wiesenhöfer W, *et al.* Second generation sol–gel encapsulated lipases: robust heterogeneous biocatalysts. *Adv Synth Catal* 2003;345:717–728.
104. Johnston APR, Cortez C, Angelatos AS, *et al.* Layer-by-layer engineered capsules and their applications. *Curr Opin Colloid Interf Sci*2006;11:203–209.
105. Skirtach AG, Yashchenok AM, Möhwald H. Encapsulation, release and applications of LbL polyelectrolyte multilayer capsules. *Chem Commun* 2011;47:12736–12746.
106. Caruso F, Trau D, Möhwald H, *et al.* Enzyme encapsulation in layer-by-layer engineered polymer multilayer capsules. *Langmuir*2000;16:1485–1488.
107. Sakr OS, Borchard G. Encapsulation of enzymes in layer-by-layer (LbL) structures: latest advances and applications.*Biomacromolecules* 2013;14:2117–2135.
108. Lefort M, Popa G, Seyrek E, *et al.* Spray-on organic/inorganic films: a general method for the formation of functional nano- to microscale coatings. *Angew Chem Int Ed* 2010;49:10110–10113.
109. Dierendonck M, De Kooker S, De Rycke R, *et al.* Just spray it—LbL assembly enters a new age. *Soft Matter* 2014;10:804–807.
110. Yu A, Wang Y, Barlow E, *et al.* Mesoporous silica particles as templates for preparing enzyme-loaded biocompatible microcapsules. *Adv Mater* 2005;17:1737–1741.
111. Volodkin DV, Larionova NI, Sukhorukov GB. Protein encapsulation via porous $CaCO_3$ microparticles templating. *Biomacromolecules*1962–1972;2004:5.
112. Petrov AI, Volodkin DV, Sukhorukov GB. Protein-calcium carbonate coprecipitation: a tool for protein encapsulation. *Biotechnol Prog*2005;21:918–925.
113. Shi J, Zhang X, Zhang S, *et al.* Incorporating mobile nanospheres in the lumen of hybrid microcapsules for enhanced enzymatic activity.*Appl Mater Interf* 2013;5:10433–10436.
114. Trau D, Renneberg R. Encapsulation of glucose oxidase microparticles within a nanoscale layer-by-layer film: immobilization and biosensor applications. *Biosens Bioelectron* 2003;18:1491–1499.
115. Balabushevitch NG, Sukhorukov GB, Moroz NA, *et al.* Encapsulation of proteins by layer-by-layer adsorption of polyelectrolytes onto protein aggregates: factors regulating the protein release. *Biotechnol Bioeng* 2001;76:207–213.
116. Nahalka J, Nidetzky B. Fusion to a pull-down domain: a novel approach of producing *Trigonopsis variabilis* d-amino acid oxidase as insoluble enzyme aggregates. *Biotechnol Bioeng* 2007;97:454–461.
117. Nahalka J, Dib I, Nidetzky B. Encapsulation of *Trigonopsis variabilis* d-amino acid oxidase and fast comparison of the operational stabilities of free and immobilized preparations of the enzyme. *Biotechnol Bioeng* 2008;99:251–260.
118. Ghan R, Shutava T, Patel A, *et al.* Enzyme-catalyzed polymerization of phenols within polyelectrolyte microcapsules. *Macromolecules*2004;37:4519–4524.
119. Tiourina OP, Sukhorukov GB. Multilayer alginate/protamine microsized capsules: encapsulation of α-chymotrypsin and controlled release study. *Int J Pharm* 2002;242:155–161.
120. Balabushevich NG, Tiourina OP, Volodkin DV, *et al.* Loading the multilayer dextran sulfate/protamine microsized capsules with peroxidase. *Biomacromolecules* 2003;4:1191–1197.

121. Schuetz P, Caruso F. Copper-assisted weak polyelectrolyte multilayer formation on microspheres and subsequent film crosslinking. *Adv Funct Mater* 2003;13:929–937.

122. Volodkin DV, Petrov AI, Prevot M, *et al.* Matrix polyelectrolyte microcapsules: new system for macromolecule encapsulation. *Langmuir*2004;20:3398–3406.

123. Zhu H, McShane M. Macromolecule encapsulation in diazoresin-based hollow polyelectrolyte microcapsules. *Langmuir* 2005;21:424–430.

124. Antipov A, Sukhorukov G, Möhwald H. Influence of the ionic strength on the polyelectrolyte multilayer's permeability. *Langmuir*2003;19:2444–2448.

125. Tiourina OP, Antipov AA, Sukhorukov GB, *et al.* Entrapment of α-chymotrypsin into hollow polyelectrolyte microcapsules. *Macromol Biosci* 2001;1:209–214.

126. Dai J, Jensen AW, Mohanty DK, *et al.* Controlling the permeability of multilayered polyelectrolyte films through derivatization, cross-linking, and hydrolysis. *Langmuir* 2001;17:931–937.

127. Garbers E, Mitlöhner R, Georgieva R, *et al.* Activity of immobilized trypsin in the layer structure of polyelectrolyte microcapsules (PEMC).*Macromol Biosci* 2007;7:1243–1249.

128. Wiemann LO, Buthe A, Klein M, *et al.* Encapsulation of synthetically valuable biocatalysts into polyelectrolyte multilayer systems.*Langmuir* 2009;25:618–623.

129. Srivastava R, Brown JQ, Zhu H, *et al.* Stable encapsulation of active enzyme by application of multilayer nanofilm coatings to alginate microspheres. *Macromol Biosci* 2005;5:717–727.

130. Srivastava R, McShane MJ. Application of self-assembled ultra-thin film coatings to stabilize macromolecule encapsulation in alginate microspheres. *J Microencapsulation* 2005;22:397–411.

131. Srivastava R, Brown JQ, Zhu H, *et al.* Stabilization of glucose oxidase in alginate microspheres with photoreactive diazoresin nanofilm coatings. *Biotechnol Bioeng* 2005;91:124–131.

132. Hearn E, Neufeld RJ. Poly(methylene co-guanidine) coated alginate as an encapsulation matrix for urease. *Process Biochem*2000;35:1253–1260.

133. Allouche J, Boissière M, Hélary C, *et al.* Biomimetic core–shell gelatine/silica nanoparticles: a new example of biopolymer-based nanocomposites. *J Mater Chem* 2006;16:3120–3125.

134. Besanger TR, Chen Y, Deisingh AK, *et al.* Screening of inhibitors using enzymes entrapped in sol–gel-derived materials. *Anal Chem*2003;75:2382–2391.

135. Macario A, Verri F, Diaz U, *et al.* Pure silica nanoparticles for liposome/lipase system encapsulation: application in biodiesel production.*Catal Today* 2013;204:148–155.

136. Mei L, Xie R, Yang C, *et al.* pH-responsive Ca-alginate-based capsule membranes with grafted poly(methacrylic acid) brushes for controllable enzyme reaction. *Chem Eng J* 2013;232:573–581.

137. Mei L, Xie R, Yang C, *et al.* Bio-inspired mini-eggs with pH-responsive membrane for enzyme immobilization. *J Membr Sci*2013;429:313–322.

138. Wang J, Yu HR, Xie R, *et al.* Alginate/protamine/silica hybrid capsules with ultrathin membranes for laccase immobilization. *AIChE J*2013;59:380–389.

139. Zhang Y, Wu H, Li J, *et al.* Protamine-templated biomimetic hybrid capsules: efficient and stable carrier for enzyme encapsulation. *Chem Mater* 2008;20:1041–1048.

140. Roberts JR, Ritter DW, McShane MJ. A design full of holes: functional nanofilm-coated microdomains in alginate hydrogels. *J Mater Chem B* 2013;1:3195–3201.

141. Wang Y, Caruso F. Mesoporous silica spheres as supports for enzyme immobilization and encapsulation. *Chem Mater* 2005;17:953–961.

142. Zhang L, Jiang Y, Shi J, *et al.* Biomimetic polymer-inorganic hybrid microcapsules for yeast alcohol dehydrogenase encapsulation. *React Funct Polym* 2008;68:1507–1515.

143. Ai Q, Yang D, Zhu Y, *et al.* Fabrication of boehmite/alginate hybrid beads for efficient enzyme immobilization. *Ind Eng Chem Res*2013;52:14898–14905.

144. Yasuhara K, Kawataki T, Okuda S, *et al.* Spontaneously formed semipermeable organic–inorganic hybrid vesicles permitting molecular organic–inorganic hybrid vesicles permitting molecular. *Chem Commun* 2013;49:665–667.

145. Shi J, Zhang S, Wang X, *et al.* Open-mouthed hybrid microcapsules with elevated enzyme loading and enhanced catalytic activity. *Chem Commun* 2014;50:12500–12503.

146. Lozinsky VI, Galaev IY, Plieva FM, *et al.* Polymeric cryogels as promising materials of biotechnological interest. *Trends Biotechnol*2003;21:445–451.

147. Jiang PJ, Wynn-Jones G, Grover LM. A calcium phosphate cryogel for alkaline phosphatase encapsulation. *J Mater Sci*2010;45:5257–5263.

148. Le Bideau J, Viau L, Vioux A. Ionogels, ionic liquid based hybrid materials. *Chem Soc Rev* 2011;40:907–925.

149. Du M, Lu D, Liu Z. Design and synthesis of lipase nanogel with interpenetrating polymer networks for enhanced catalysis: molecular simulation and experimental validation. *J Mol Catal B* 2013;88:60–68.

150. Pachariyanon P, Agar DW. Selectivity enhancement using permselective microcapsule. *Chem Eng J* 2012;207–208:481–489.

151. Gassara-Chatti F, Brar SK, Ajila CM, *et al.* Encapsulation of ligninolytic enzymes and its application in clarification of juice. *Food Chem*2013;137:18–24.

152. Schachschal S, Adler HJ, Pich A, *et al.* Encapsulation of enzymes in microgels by polymerization/cross-linking in aqueous droplets.*Colloid Polym Sci* 2011;289:693–698.

153. Milašinović N, Milosavljević N, Filipović J, *et al.* Efficient immobilization of lipase from *Candida rugosa* by entrapment into poly(N-isopropylacrylamide-co-itaconic acid) hydrogels under mild conditions,'. *Polym Bull* 2012;69:347–361.

154. Wang X, Wenk E, Matsumoto A, *et al.* Silk microspheres for encapsulation and controlled release. *J Controlled Release*2007;117:360–370.

155. Patel AR, Nijsse J, Velikov KP. Novel polymer–polyphenol beads for encapsulation and microreactor applications. *Soft Matter*2011;7:4294–4301.

156. Shi J, Yang C, Zhang S, *et al.* Polydopamine microcapsules with different wall structures prepared by a template-mediated method for enzyme immobilization. *Appl Mater Interf* 2013;5:9991–9997.

157. Zhang L, Shi J, Jiang Z, *et al.* Bioinspired preparation of polydopamine microcapsule for multienzyme system construction'. *Green Chem*2011;13:300–306.

158. Wiemann LO, Nieguth R, Eckstein M, *et al.* Composite particles of novozyme 435 and silicone: advancing technical applicability of macroporous enzyme carriers. *ChemCatChem* 2009;1:455–462.

159. Wiemann LO, Weisshaupt P, Nieguth R, *et al.* Enzyme stabilization by deposition of silicone coatings,'. *Org Process Res Dev*2009;13:617–620.

160. Nieguth R, Eckstein M, Wiemann LO, *et al.* Enabling industrial biocatalytic processes by application of silicone-coated enzyme preparations. *Adv Synth Catal* 2011;353:2522–2528.

161. Ansorge-Schumacher MB, Thum O. Immobilised lipases in the cosmetics industry. *Chem Soc Rev* 2013;42:6475–6490.

162. Scholz A, Eckstein M, Ansorge-Schumacher MB. Hydrophilized silicone matrix for the preparation of stable carbonyl reductase immobilizates. *ChemCatChem* 2013;5:815–821.

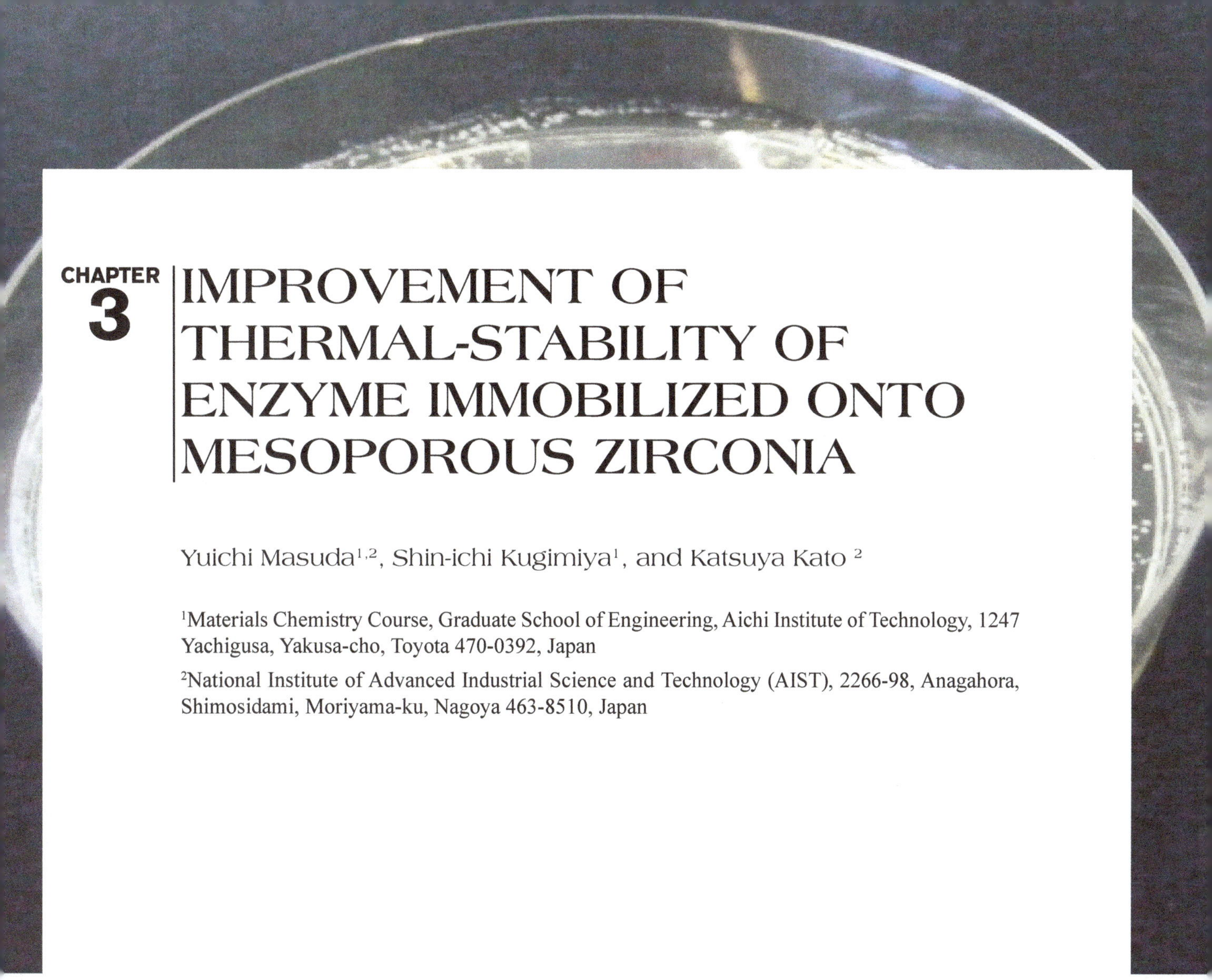

CHAPTER 3

IMPROVEMENT OF THERMAL-STABILITY OF ENZYME IMMOBILIZED ONTO MESOPOROUS ZIRCONIA

Yuichi Masuda[1,2], Shin-ichi Kugimiya[1], and Katsuya Kato [2]

[1]Materials Chemistry Course, Graduate School of Engineering, Aichi Institute of Technology, 1247 Yachigusa, Yakusa-cho, Toyota 470-0392, Japan

[2]National Institute of Advanced Industrial Science and Technology (AIST), 2266-98, Anagahora, Shimosidami, Moriyama-ku, Nagoya 463-8510, Japan

ABSTRACT

In our previous research, the immobilization of formaldehyde dehydrogenase (FDH) onto mesoporous silica (MPS: pore size = 12.3 nm) was investigated. However, this method could not obtain the thermal stability of the enzyme. To solve this problem, FDH was immobilized on mesoporous zirconia material (MPZ) with a lower thermal conductivity. MPZ was synthesized using [poly(ethylene glycol)–poly(propylene glycol)–poly(ethylene glycol) (Pluronic P123, $EO_{20}PO_{70}EO_{20}$)], zirconium(IV) *n*-propoxide (ca. 75% in 1-propanol), acetylacetone, 1, 3, 5-trimethylbenzene, and ethanol. The material retained high surface area (113.6 m^2/g) and pore volume (0.27 cm^3/g). It was characterized by Brunauer–Emmett–Teller surface area, Barrett–Joyner–Halenda pore size distribution, X-ray diffraction, field-emission scanning electron microscopy, transmission electron microscopy, and energy-dispersive X-ray spectroscopy.

Thereafter, FDH immobilized on MPZ showed higher catalytic activity than that on MPS. Enhancement of catalytic activity was obtained by improving the substrate affinity derived from interparticle voids of MPZ. In addition, the FDH immobilized on MPZ had a very great higher thermal stability. Further investigation using transmittance Infrared spectroscopy indicated that the high-order structure of the FDH immobilized on MPZ did not get altered after the heat-treatment.

INTRODUCTION

Among these pollutants, volatile organic compounds (VOCs) have drawn significant interest since they have been reported to induce sick house syndrome and to have chronic and acute health effects [1] and [2]. In particular, formaldehyde, one of the harmful VOCs, is of great significance. Even low concentration of formaldehyde, such as 10 μM, adversely affects the human body, for example, anger and cancer [2], [3],[4] and [5]. Therefore, development of a sensor is highly required for measuring VOCs, such as formaldehyde, with high sensitivity and selectivity from the environmental and medical viewpoints.

Currently, standard methods for the detection of formaldehyde include visible absorption, HPLC, gas chromatography, and fluorometry [6], [7] and [8]. Even though

Citation: Yuichi Masuda, Shin-ichi Kugimiya, Katsuya Kato, Improvement of thermal-stability of enzyme immobilized onto mesoporous zirconia, Journal of Asian Ceramic Societies, Volume 2, Issue 1, March 2014, Pages 11-19, ISSN 2187-0764, http://dx.doi.org/10.1016/j.jascer.2013.12.003.

these formaldehyde sensors can be measured at a very low concentration of <0.1 nM, the size of equipments is very large, and it takes a long time for measurements. Smaller size of sensors unfortunately results in lower sensitivity.

From the above reasons, we pay attention to high sensitivity and selectivity of enzymes. Campanella et al. [9] and [10] have developed an enzyme electrode for phenol determination which is capable of being used also in non-aqueous solvents and a recently published paper investigated the sensitivity of biosensors to phenol or *p*-cresol, working in*n*-hexane, chloroform or mixtures thereof. Formaldehyde dehydrogenase (FDH) is known to react to formaldehyde with high selectively and sensitivity. FDH derived from*Pseudomonas putida* (molecular size: 8.6 nm × 8.6 nm × 19.0 nm) is a dimeric molecule with a molecular weight of 85,757.2 Da, which has a Zn[II] ion within its active site. Therefore, we are working on the development of formaldehyde biosensor using FDH.

While enzymes can be used as biosensors, enzymes are generally known to be unstable under high temperature and to organic solvents besides having a higher cost, thus limiting their practical application. In order to find a solution to stability issue, we have worked on the enzyme immobilization onto mesoporous silica (MPS), with the aim of enhancing the enzyme activity and stability under a wide range of practical conditions [11]. In recent years, MPS was verified to be one of the most effective solid supports to immobilize a wide variety of enzymes owing to its large specific surface area and uniform pore size [12], [13], [14], [15], [16], [17] and [18]. In a previous research, we immobilized an enzyme in MPS to be used in various environments [11]. However, thermal stability of the enzyme immobilized on MPS did not improve. It was considered that the thermal energy was transmitted to the enzyme by thermal conductance of silica [11]. From the above results, we turned our focus on the zirconia with a lower thermal conductivity as an enzyme immobilization support.

Zirconia has high boiling point, strength, toughness, hardness, wear resistance, and good frictional resistance, thus making it an excellent refractory material. It acts as a non-magnetic electrical insulator with low thermal conductivity, and has good corrosion resistance in acidic and alkaline environments [19]. Due to its high oxygen ion conductivity it is also used in structural ceramics, soled catalysts, oxygen sensors, electric materials for high temperature fuel cells [19], [20], [21], [22], [23], [24], [25], [26], [27] and [28]. The most important application of the zirconia is that it can be used as a catalyst. It has a high melting point and high thermal stability under reducing or oxidizing environments, marking it an excellent catalyst support even under harsh conditions [29]. It is chemically more stable than the traditional support materials such as γ-alumina, zeolite, and MPS[30]. Additionally, it also possesses unique bifunctional characteristics of weak basic and acidic properties, and can function as a catalyst support or catalyst by itself [19],[31] and [32].

The effective use of zirconia as a catalyst and catalyst support requires a high specific surface area with suitable pore structure. Thus, synthesis of mesoporous zirconia (MPZ) by using liquid crystal template technique was demonstrated by Sudhakar Reddy and Sayari [33] in 1996 for the first time. Most of the synthesized pure MPZ can be pore size controlled only a small size. However, Rezaei et al. [34] succeeded in the synthesis of MPZ with large pore size by using ethylene diamine along with P123 block copolymer as a template. For the materials calcined at 600 °C, the surface area, pore volume, and pore size ranged from 128.5 to 138.2 m^2/g, 0.1 to 0.4 cm^3/g, 4.00 to 11.3 nm, respectively. In recent years, those studies are progressing for a wide range of applications such as thin film of MPZ on a substrate.

Next, we presented the improved activity and cost performance of enzyme immobilized on MPS with an interparticle mesoporous structure (IPMPS), in order to solve the problem of high cost [35]. IPMPS materials have a uniform pore size with very low volume compared to conventional MPS. Using such a material, the aggregation of the enzymes immobilized on this support is spatially suppressed. Simultaneously, the substrate affinity is greater in the case of enzymes immobilized on conventional supports. Thus, the enzymes immobilized on IPMPS exhibited a high relative activity, despite the small amount of adsorbed enzyme. We achieved improvement of enzyme activity and cost reduction.

To the best of our knowledge, there was no report on improvement in the thermal stability and structural change of enzyme in the immobilized FDH on zirconia. We synthesized MPZ with interparticle pore structure using Pluronic P123, zirconium(IV) *n*-propoxide (ca. 75% in 1-propanol), acetylacetone, trimethylbenzene (TMB), and ethanol. In this study, we present that FDH raised substrate affinity and thermal stability though immobilization on the MPZ with interparticle pore structure.

EXPERIMENTAL PROCEDURE

Materials

FDH from *P. putida* (molecular size: 8.6 nm × 8.6 nm × 19.0 nm) obtained from Toyobo. Co., Japan. Triblock co-polymer of poly(ethylene glycol)–poly(propylene glycol)–poly(ethylene glycol) (Pluronic P123, $EO_{20}PO_{70}EO_{20}$, Mn ca. 5800 g/mol) was purchased from Sigma–Aldrich Co., St. Louis, MO. Phenyltriethoxysilane was purchased from Shin-Etsu Chemical Co., Tokyo, Japan. Zirconium(IV) *n*-propoxide (ca. 75% in 1-propanol), acetylacetone, 25% ammonia solution, and 99.5% ethanol were purchased from Wako Co., Tokyo, Japan. All the reagents are special-grade chemicals.

Synthesis of MPZ

The following cocktails C-1 were employed: C-1: 0.03 M of zirconium(IV) *n*-propoxide, 0.017 M of acetylacetone, and 0.17 M of ethanol were mixed.

MPZ was prepared as follows: 1 g of triblock copolymer (Pluronic P123) was dissolved in 26 mL of deionized water and added to 0.75 g of 28% aqueous ammonia solution. The solution was stirred for 2 h at room temperature, and then C-1 was added to the mixture. The suspension was stirred for 3 h at RT and for another 48 h at 90 °C in temperature-controlled bath. After cooling, solid material was collected by the centrifugation, and washed with de-ionized water and acetone. The final product was dried at 100 °C for 24 h and calcined at 500 °C for 4 h.

Synthesis of MPS

The MPS materials were prepared by using the triblock copolymer Pluronic P123, TEOS, and *n*-decane as reagents for controlling the morphology. In a typical procedure, TEOS (2.13 g) was added drop wise to a solution of P123 (1 g) in 10 M HCl (3.7 mL) and H_2O (31.3 mL). This mixture was stirred at 40 °C for 20 h and then transferred to an autoclave for further reaction at 100 °C for 24 h. The product was filtered, dried at 80 °C for 10 h, and then calcined at 550 °C for 4 h in air (heating rate of 1 °C/min). In some cases, *n*-decane (7.5 g) was added to the HCl solution containing P123 before addition of the TEOS.

Synthesis of MPS with IPMPS

The IPMPS materials were prepared as follows: a 28% ammonium solution (400 μL) was added to deionized water (2.6 mL). After gently stirring this solution, TEOS (3 mL) was added to the mixed solution. The resulting mixture was stirred for 1 day at room temperature and then frozen at −30 °C for one day. The frozen sample was then freeze-dried to obtain the IPMPS.

Characterization of porous materials

The particle morphology of the MPS materials was observed via field-emission scanning electron microscopy (FE-SEM: S-4300, Shimadzu Co., Kyoto, Japan) with an accelerating voltage of 10.0 kV. Transmission electron microscopy (TEM) and scanning TEM (STEM) images were taken using a JEOL JEM 2010 (JOEL Ltd., Tokyo, Japan) operated at 200 kV. The surface area, pore diameter, and pore volume were determined from nitrogen adsorption/desorption measurements using a Shimadzu TriStar 3000 system. Pore diameter distributions were calculated from the desorption branches using the Barrett–Joyner–Halenda (BJH) method. The specific surface area was calculated using the Brunauer–Emmett–Teller (BET) method based on the desorption isotherms. Small-angle X-ray diffraction (SAXRD) spectra were recorded on an X-ray diffractometer (RINT-2550VB3L, Rigaku Co., Tokyo, Japan) with CuKα, radiation at 40 kV, and 200 mA. The particle diameter and surface potential in the buffer solution were measured using dynamic light scattering (DLS, Otsuka Electronics Co., Tokyo, Japan) and zeta-potential (ζ-potential, Otsuka Electronics Co., Tokyo, Japan) analyses, respectively.

Immobilization of FDH onto porous materials

Facile preparation of immobilized FDH was performed as follows: 100 μL of FDH/50 mM–pH 7.0–phosphate buffer solution (0.5 mg/100 μL) was added to 700 μL of the same buffer. Three milligrams of porous materials were added in FDH solution and the mixture was stirred at 4 °C for overnight. The precipitate was separated by centrifuge at 4 °C with 12,000 rpm for 10 min. Then, FDH immobilized on MPZ was washed with cold deionized water, and stored at 4 °C. FDH immobilized MPZ are named, for example, MPZ–FDH.

Assay of FDH activities

Three milligrams of immobilized FDH were added to a mixture containing 100 μL of 7.68 mM formaldehyde solution as a substrate, 100 μL of nicotinamide adenine dinucleotide (NAD) solution (4 mg/mL), and 800 μL of 50 mM–pH 7.0–phosphate buffer solution. The mixture was stirred at room temperature for 8 min. After the enzyme reaction, immobilized enzyme was separated using centrifuge at 4 °C with 12,000 rpm for 2 min. Amount of NADH was determined with UV–vis spectrophotometer at fixed wavelength of 340 nm (BioSpec-1600, Shimadzu, Kyoto, Japan), and the enzyme reaction is as under Eq. (I):

$$HCHO + NAD+ + H_2O \rightarrow HCOOH + NADH + H^+ \quad (I)$$

The activity of non-immobilized FDH (the native) was measured as a positive control. The enzyme activity (Unit) of FDH was determined by using Eq. (II):

$$\text{Enzyme (Unit)} = (\text{Absorbance value})/((\text{Molecular weight of NADH}) \times (\text{Amount of enzyme adsorption}) \times (\text{reaction time})) \quad (II)$$

STABILITY OF FDH UNDER VARIOUS CONDITIONS

Thermal stability of immobilized FDH was investigated. Three hundred micrograms of the native or 3.0 mg of immobilized FDH was added to 800 μL of 50 mM–pH 7.0–phosphate buffer solution. The suspension was incubated at 40 °C for at a constant interval (0, 10, 20, 30, 40, and 50 min), and then cooled to room temperature. After the treatment, immobilized FDH was washed using cool, deionized water, and then remaining activity of enzyme was measured. The thermal-treated suspension was added to the mixture of 100 μL of NAD solution (4 mg/mL) and 100 μL of 7.68 mM HCHO solution, and then remaining activity of FDH was measured. The activity of native FDH was measured as a positive control. Denaturation temperature of the native, immobilized FDH on MPZ, and MPS were measured by DSC (DSC-120U, Seiko Instrument Inc., Tokyo, Japan).

Analyses on structural orderliness of immobilized FDHs

The secondary structures of the native and immobilized FDH were analyzed using FT-IR. The FT-IR spectra of native FDH and immobilized FDH were recorded at room temperature on a Perkin-Elmer Spectrum GX FT-IR Spectrometer System using a mercury–cadmium–tellurium detector (MA, resolution, 4 cm^{-1}; number of scans, 64).

The secondary structure of native and immobilized FDH was determined by circular dichroism (CD) spectropolarimeter (J-820K, JASCO Co., Tokyo, Japan) in the wavelength range 190–260 nm using a quartz cell (optical path length 0.1 cm) and an integration number of 16. CD spectra were measured when concentrations of the native, FDH immobilized on MPZ, MPS, and IPMPS were prepared at 6.0×10^{-7}, 4.1×10^{-6}, 4.1×10^{-6}, and 4.1×10^{-6} M, respectively.

RESULTS AND DISCUSSION

Properties of mesoporous materials

The structures of the MPZ, MPS, and IPMPS samples were characterized based on their BET surface areas, BJH pore-size distributions, TEM, FE-SEM images, and ζ-potentials. The structural properties of the MPZ, MPS, and IPMPS materials as determined based on the BET, SAXRD, ζ-potential measurements, and average particle diameter analyses are summarized in Table 1.

Table 1: Structural properties of mesoporous materials

Sample name	Pore size[a][nm]	Surface area[b] [m^2/g]	Pore volume[a][cm^3/g]	Pore structures[c]	Zeta potential[d][mV]	Average particle diameter[d] [nm]
MPZ	12.2	70.55	0.2	Disordered	19.5	274.7 ± 146.0
MPS	13.2	723.7	1.8	2D-hexagonal	−13.8	480.6 ± 281.4
IPMPS	14.4	291.8	0.9	Disordered	−30.9	288.5 ± 168.0

[a]Calculated according to BJH analysis (adsorption branch between 1.7 and 300 nm diameters).

[b]Specific surface area according to BET.

[c]Pore structure was obtained from small-angle X-ray diffraction (SAXRD) peaks.

[d]Calculated value was obtained from zeta potential and particle size distribution measurements.

In Fig. 1, the pore-size distribution curves and the nitrogen adsorption–desorption isotherms obtained using the BJH and the BET methods, respectively, are plotted for the MPZ [circular], MPS [square], and IPMPS [triangle] samples. The MPS material exhibited type IV [H1] curves, indicating the presence of a cylindrical mesoporous structure on its surface [35]. On the other hand, MPZ and IPMPS did not have a type IV [H1] structure. The surface areas of synthesized MPZ, MPS, and IPMPS materials exhibited 70.55, 723.7, and 291.8 m^2/g and pore volumes of 0.2, 1.8, and 0.9 cm^3/g, respectively. The average pore diameters of the MPZ, MPS, and IPMPS samples were controlled at approximately 13 nm, respectively. The suitability of this MPS pore size for the molecular size of FDH has been demonstrated previously [11], and thus all samples had an optimum pore size for FDH immobilization. In Fig. 1, it can be seen that the MPZ and IPMPS materials had a significantly reduced pore volume because of its noncylindrical pore structure. FE-SEM images of the MPZ, MPS, and IPMPS materials are shown in Fig. 2(a)–(c). The MPS sample was composed of large particles (~500 nm), while the MPZ and IPMPS materials were composed of aggregates of differently sized particles (100–400 nm).

This result indicates that a heterotypic sol–gel reaction occurred because of the presence of an excessive amount of ammonia in the case of the MPZ and IPMPS materials. The TEM images of the MPS and IPMPS materials are presented inFig. 2(A)–(C), respectively. As can be seen in these images, the MPS material was formed as uniform and small domains of cylindrical small pores (13.2 nm), while the mesopores of MPZ and IPMPS were not regularly arranged. Thus, the porous structure of the MPZ and IPMPS sample was determined from the TEM images. Next, the pore structures were investigated using SAXRD analysis, and the results are shown in Fig. S1. The MPS material exhibited diffraction peaks (1 0 0, 1 1 0, and 2 0 0) corresponding to an ordered two-dimensional hexagonal mesostructure (*p6mm*) [36].

In contrast, the respective mesoporous structure of the MPZ and IPMPS materials was not well-ordered (Fig. S1). It was thus concluded that the shape of the pore structure of the MPZ and IPMPS materials is regular, but with varying pore spaces. Therefore, the supports have interparticle pore structure appears to have a "disordered pore structure" based on the low-angle X-ray diffraction analysis.

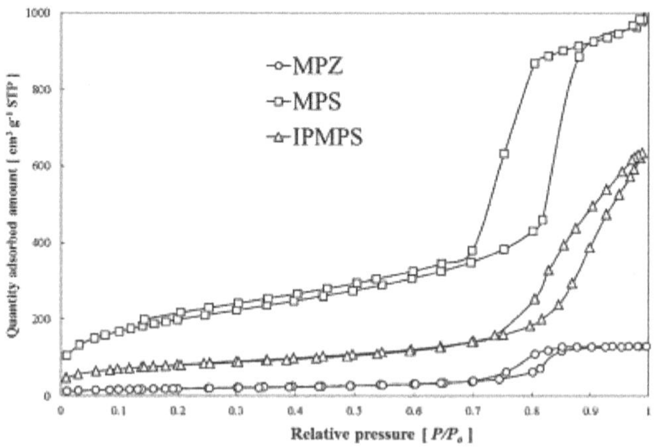

Figure 1: Nitrogen adsorption–desorption isotherms and BJH analysis [inset] for MPZ [circular], MPS [square] and IPMPS [triangle].

Figure 2: FE-SEM (a–c) and TEM (A–C) images of the synthesized mesoporous materials. (A, a): MPS, (B, b): IPMPS, (C, c): MPZ.

We investigated the surface composition and elemental composite by wide-angle XRD and EDX, respectively. The XRD patterns of MPZ material are shown in Fig. S2. The sample exhibited the diffraction peak (1 0 1, 1 1 0, 1 1 2, and 2 1 1) corresponding to crystal structure of tetragonal phase with high thermal, mechanical, and structural stability[37]. Fig. 3 shows an EDX, scanning TEM (STEM) image, and several elemental mapping images. Not only the zirconium mapping images but also the oxygen ones conform to the STEM image.

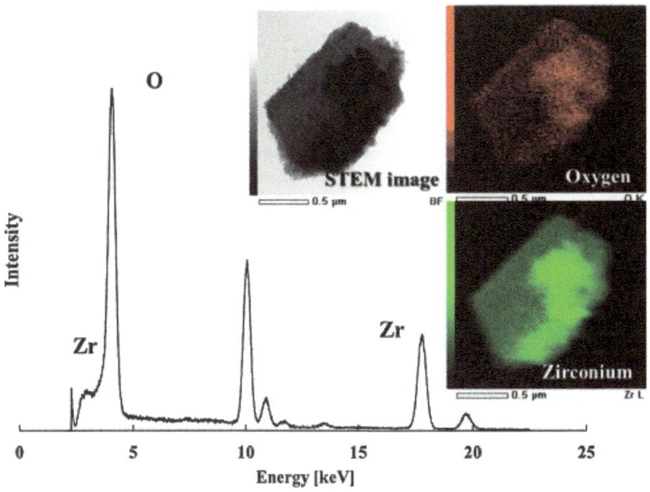

Figure 3: EDX profiles and EDX mapping images of MPZ. The zirconia and oxygen mapping images all agree well the STEM image.

Activity evaluation of FDH immobilized on porous materials

We compared the activity of the FDH immobilized on those supports. FDH was immobilized on MPZ, IPMPS, and MPS with the highest activity in previous study [11].Fig. 4 shows the relative activity and adsorbed amount of FDH immobilized on MPZ, MPS, and IPMPS. Adsorbed amount of enzyme on MPZ and IPMPS was less than that on MPS, because MPZ (surface area: 70.55 m^2/g, pore volume: 0.20 cm^3/g) and IPMPS (surface area: 291.8 m^2/g, pore volume: 0.90 cm^3/g) have lower surface area and pore volume than MPS (surface area: 723.7 m^2/g, pore volume: 1.8 cm^3/g). Activity of native FDH was measured as a positive control. Relative activity to the native was lower than 100%, but activity had a maximum when FDH was immobilized on MPZ and IPMPS. In the previous study, we showed that substrate affinity of FDH is improved by immobilizing on a support having an interparticle pore structure [34]. Therefore, we performed a kinetic analysis of FDH, which was immobilized on the MPZ.

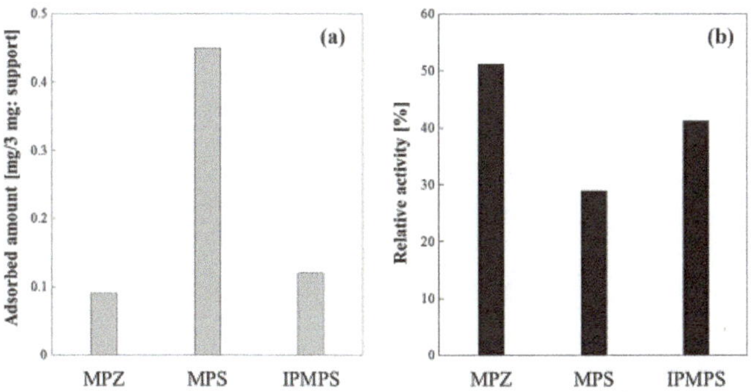

Figure 4: (a) FDH adsorption on different supports. Immobilized method was shown in Section 2.6. (b) Relative activities of FDH immobilized on each mesoporous materials.

Kinetic studies of native FDH and immobilized FDH

The kinetic properties of FDH immobilized on MPZ and each MPSs, including the Michealis constant, K_m; the maximum rate, V_{max}; and the turn over number, K_{cat}; were determined for the enzyme reaction (Table 2). The K_{cat}/K_m value of FDH immobilized on MPZ was comparable to that of the native. These results show that FDH immobilized on MPZ has better effect on enzyme activity than that on MPS. Particularly, K_m value of MPZ–FDH was conspicuously higher than that of others. This is consistent with our hypothesis that interparticle voids can facilitate substrate diffusion and, therefore, enhance the apparent enzymatic activity. However, the K_{cat} value of MPZ was lower than that of IPMPS. From these results, we suggested that FDH immobilized on MPZ led to denaturation.

Table 2: Apparent K_m, K_{cat}, and K_{cat}/K_m values at three mesoporous materials, calculated using the Michaelis–Menten equation

Sample name	K_m [mM]	K_{cat} [min^{-1}]	K_{cat}/K_m
Native	0.249	12.1	48.6
MPZ	0.117	5.23	44.7
MPS	0.509	3.62	7.11
IPMPS	0.285	6.18	21.7

Structural changes in the immobilized FDH

Next, we analyzed the structure of FDH in order to study the influence of the electrostatic interactions between the enzyme and the silica materials. Changes in the secondary structure of FDH immobilized onto the three porous materials were analyzed using FT-IR and CD spectra.

According to the literature [38], [39], [40], [41], [42] and [43], the contribution of the β-sheet, unordered, and α-helix content can be identified based on the amide I peaks at approximately 1670, 1660, and 1650–1640 cm^{-1}, respectively, while peaks in the 1680–1690 cm^{-1} region reflect the contribution of the antiparallel β-sheet. Fig. 5(a) shows the amide I region of FT-IR spectra measured over the wavelengths range 1700–1580 cm^{-1}for native and immobilized FDH samples. The α-helix and β-sheet conformation of FDH are presented in Scheme S1. The bands at 1650, 1690, and 1630 cm^{-1} suggest the presence of the α-helix, antiparallel β-sheet, and β-sheet, respectively [43] and [44]. The band indicating the presence of a random coil arises near 1540 cm^{-1}. The maximum peak for the native FDH [solid], FDH immobilized on MPZ [dashed line], MPS [dotted line], and IPMPS [dot–dot dashed line] occur at 1655, 1648, 1655, and 1655 cm^{-1}. This result suggested that FDH immobilized on MPZ increased content of the β-sheet. These results indicate that the enzyme did not alter any structural changes when immobilized on the two pore structures, but enzyme led to structural changes by immobilization on zirconia.

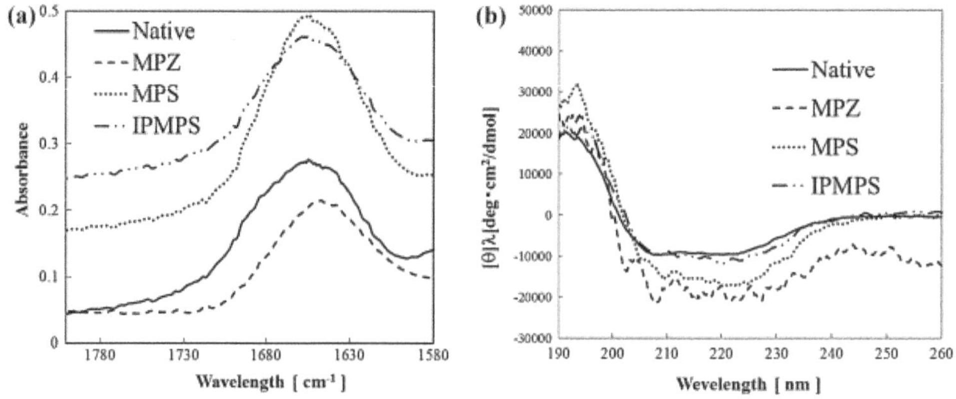

Figure 5: (a) FT-IR spectra of the native and immobilized FDHs on various MPS materials. The native FDH is denoted by the solid line. The FDHs on MPZ, MPS and IPMPS are denoted by the dashed, dotted and dot–dot dashed lines, respectively. (b) Circular dichroism spectra of the native and immobilized FDHs on various absorbents. Native FDH is denoted by solid line. FDHs on MPZ, MPS and IPMPS are denoted by dashed, dotted and dot–dot dashed lines, respectively.

CD spectra measurements were carried out to analyze the changes in secondary structure of the enzyme. Fig. 5(b) shows the CD spectra measured at wavelengths of 190–260 nm for native and immobilized FDH suspended in 50 mM–pH 7.0–phosphate buffer solution. The spectrum of the native [solid line] shows broad negative peaks from 208 to 222 nm, suggesting a combination of two different structures: α-helix with the inverted peaks 222 and 208 nm, and β-sheets with the inverted peak at 217 nm. Similar to the results of the FT-IR analysis, FDH immobilized on MPZ supports undergo a structural change.

Enzyme immobilization by physical adsorption is mainly affected by the electrostatic interaction [44]. Table S1 shows the ζ-potential of FDH, MPZ, MPS, and IPMPS. The value is −17.4, 19.5, −13.8, and −30.9, respectively. As a result, the ζ-potential of only MPZ will have a positive charge. The interaction between the FDH and MPZ is very strong, because surface potential of enzyme has a negative charge. Therefore, FDH immobilized on MPZ caused the structural change.

Assays of FDH stability after thermal treatment

Numerous papers have been published to point out that the stability of enzymes can be enhanced by immobilizing onto inorganic materials [45]. Therefore, we also evaluated the stability of immobilized FDH after thermal treatment.

Fig. 6(a) illustrates the residual activities of native and immobilized FDH after thermal treatments at 40 °C for different time periods. FDH immobilized on MPZ showed improved thermal stability, but with FDH immobilized on MPS and IPMPS thermal stability was not improved. FDH immobilized on the MPS had a denaturalization owing to thermal energy transmitted from silica wall (thermal conductivity of silica and zirconia: 8 and 3 W/m K), respectively.

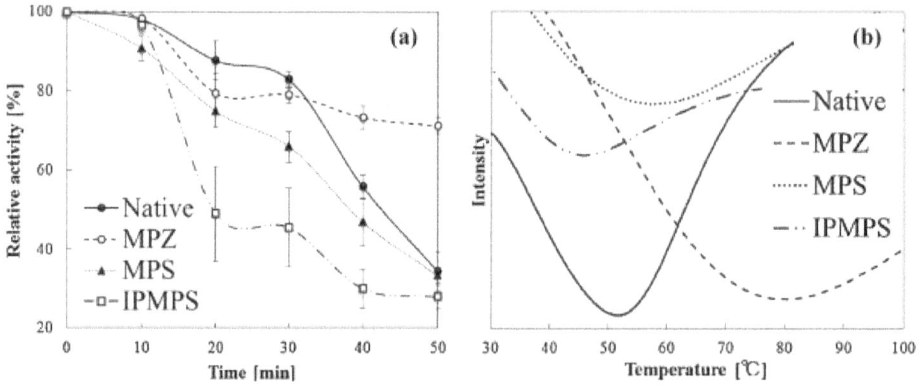

Figure 6: (a) Thermal stability as a function of time for the native [circle] and immobilized FDHs on MPZ [white circle], MPS [triangle] and IPMPS [square] at 40 °C in 50 mM–pH 7.0–phosphate buffer. Initial activities are considered as 100%. (b) Differential scanning calorimetry curves for the Native [solid line] and immobilized, FDH on MPZ [dashed line], MPS [dotted line] and MPZ [dot–dot dashed line].

The starting temperature of denaturation of the native and FDH immobilized on MPZ, MPS, and IPMPS was measured using DSC (Fig. 6(b)). The denaturation temperature of the native was 51.8 °C. In contrast, those temperatures of FDH immobilized on MPZ, MPS, and IPMPS were 78.9, 57.7, and 46.8 °C, respectively. This also indicated that thermal stability of FDH was improved by encapsulated on MPZ. We believe that structural change of the enzyme was less likely to occur by the strong interaction with the zirconia and the zirconia prevented the heat transfer from external to immobilized FDH.

Structural changes in the immobilized FDH after thermal process

In the next step, we incubated the native and immobilized FDH at 40 °C for 50 min, and the thermal treated enzymes were carried out to analyze the changes in secondary structure of the enzyme. Fig. 7(a)–(c) demonstrated IR spectrum of immobilized FDH before and after thermal treatment. The maximum peak for FDH immobilized on MPZ, MPS, and IPMPS occur at 1649, 1648, and 1646 cm^{-1}. By comparing the data obtained before and after thermal treatment, it can be observed that FDH immobilized on MPSs induced an increase in the β-sheet content. On the other hand, FDH immobilized on MPZ did not alter any structural changes after thermal treatment. Fig. 7(A)–(C) shows the CD spectra measured of native FDH and immobilized FDH before and after thermal treatment. From this result, native FDH increase the β-sheet at the same time occur the thermal denaturation. FDH immobilized on MPSs was also increased β-sheet by a thermal treatment. Therefore, we suggested that enzyme activity decreased by enzyme structural changes due to thermal treatment.

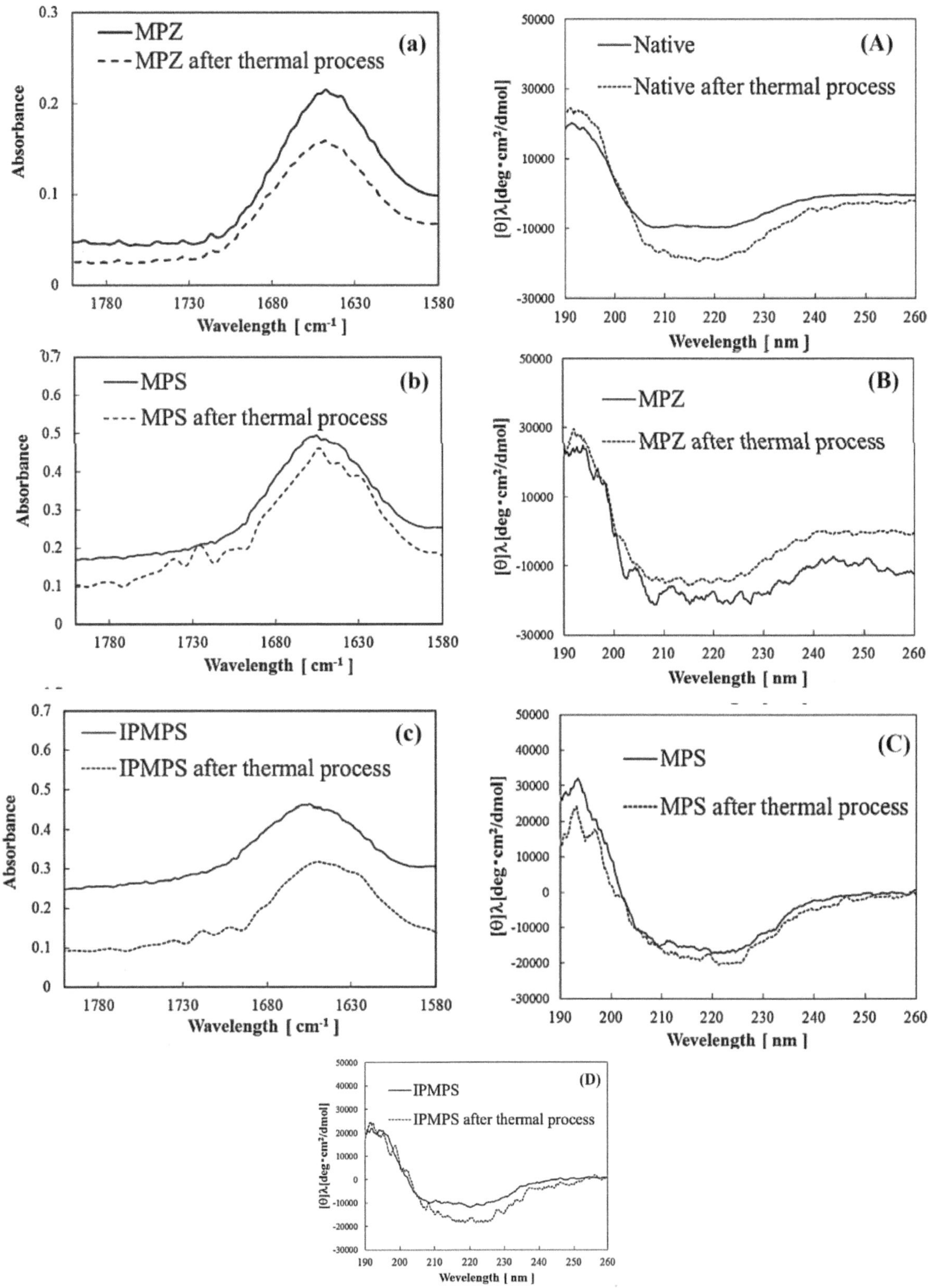

Figure 7: Analysis of changes in the secondary structure of native and immobilized FDHs on various supports after thermal process. FT-IR: (a)–(c), CD spectra: (A)–(D).

Cycle test

The cycle performance of MPZ, MPS, and IPMPS was compared. Fig. S3 illustrates the cycle tests of FDH immobilized on MPZ, MPS, and IPMPS. After each cycle, the remaining activity was recorded. FDH immobilized on MPZ was only retained ~80% of its initial activity even after seven repeated reactions. These results indicate that FDH immobilized on MPZ had higher cycle performance than that on MPSs. FDH immobilized on MPZ decreased enzyme release because of the strong electrostatic interaction with the zirconia.

CONCLUSIONS

FDH was immobilized through physical adsorption into the carefully controlled, size-optimized pores of three absorbents, and the performance of these catalysts was evaluated based on their enzyme activity, cycling characteristics, kinetic behavior, and enzyme structure analyses in order to determine the optimal structure as the carrier.

FDH immobilized on the MPZ support exhibited a higher activity than that on the MPS material because of an increase in the substrate affinity resulting from the interparticle pore space. However, FDH immobilized on MPZ was occurred a structural change by a strong interaction of the zirconia has positive charge.

In addition, we also evaluated the stability of immobilized FDH after thermal treatment. As a result, we suggested that structural change of the enzyme was less likely to occur by the strong interaction with the zirconia and the zirconia prevented the heat transfer from external to FDH.

Finally, we examined structural changes in the immobilized FDH after thermal process. It was found that the enzyme immobilized on MPZ is not affected even thermal treatment by the above experiments. It should also be noted that the potential of new MPZ immobilization carriers was also demonstrated.

ACKNOWLEDGMENT

This study is supported by A-STEP (Adaptable & Seamless Technology Transfer Program through Target-driven R&D) Project No. AS242Z00150M, Japan Science and technology Agency (JST).

REFERENCES

1. National Air Quality and Emission Trend Report: Air Toxics, US Environment Protection Agency (1996) (Chapter 5).
2. World Health Organization, Environmental Health Criteria 89, Formaldehyde, WHO, Geneva (1989).
3. M.A. Flyvholm and P. Andersen, Am. J. Ind. Med., 24, 533–552 (1993).
4. I.J. Chasnoff, J.W. Ellis and Z.S. Fainman, Family Health and Medical Guide, Publications International, Lincoln-Wood, IL (1989).
5. H. Kim, Y.D. Kim and S.H. Cho, Arch. Environ. Health, 54, (2) 115–118 (1999).
6. G.R. Mohlmann, Appl. Spectrosc., 39, 98–101 (1985).
7. T. Dumas, J. Chromatogr., 247, 289–295 (1982).
8. B. Mann and J. Grajeski, J. Chromatogr., 386, 149–158 (1987).
9. L. Campanella, M.P. Sammartino and M. Tomassetti, Sens. Actuators B, 7, 383–388 (1992).
10. L. Campanella, G. Favero, M.P. Sammartino and M. Tomassetti, Talanta, 6, 1015–1023 (1994).
11. Y. Masuda, S. Kugimiya, K. Murai, A. Hayashi and K. Kato, Colloids Surf. B, 101, 26–33 (2013).
12. R.M. Blanco, P. Terreros, M. Fernández-Pérez, C. Otero and G. Díaz-González, J. Mol. Catal. B: Enzym., 30, 83–93 (2004).
13. S. Shen, P.S. Chow, S. Kim, K. Zhu and R.B.H. Tan, J. Colloid Interface Sci., 321, 365–372 (2008).
14. K. Kato, R. Irimescu, T. Saito, Y. Yokogawa and H. Takahashi, Biosci. Biotechnol. Biochem., 67, 203–206 (2003).
15. K. Kato, S. Seelan and T. Saito, J. Biosci. Bioeng., 108, 310–313 (2009).
16. R. Atluri, Y. Sakamoto and A.E. Garcia-Bennett, Langmuir, 25, 3189–3191 (2009).
17. J.Y. Lee, J.Y. Lim and H.K. Lim, Abdom. Imaging, 30, 744–747 (2005).
18. J. Huang, Y. Liu and X. Wang, J. Mol. Catal. B: Enzym., 57, 10–15 (2009).
19. S. Jaenicke, G.K. Chuah, V. Raju and Y.T. Nie, Catal. Surv. Asia, 12, 153–169 (2008).
20. M.J. Mayo, J.R. Seidensticker, D.C. Hague and A.H. Carim, NanoStruct. Mater., 11, 271–282 (1999).
21. X. Yang, R.E. Jentoft and F.C. Jentoft, Catal. Lett., 106, 196–202 (2006).
22. Y. Sun, S. Walspurger, B. Louis and J. Sommer, Appl. Catal. A, 292, 200–207 (2005).
23. M. Faticanti, N. Cioffi, S. De Rossi, N. Ditaranto, P. Porta, L. Sabbatini and T. Bleve-Zacheo, Appl. Catal. B, 60, 73–82 (2005).
24. O.K. Tana, W. Caob, Y. Hua and W. Zhu, Solid State Ionics, 172, 309–316 (2004).
25. G. Liu and Y. Lin, Anal. Chem., 77, 5894–5901 (2005).
26. Q. Zhu and B. Fan, Solid State Ionics, 176, 889–894 (2005).
27. S.T. Aruna and K.S. Rajam, Scr. Mater., 48, 507–512 (2003).
28. V.G. Deshmane and Y.G. Adewuyi, Microporous Mesoporous Mater., 148, 88–100 (2012).
29. R. Gopalan, C.H. Chang and Y.S. Lin, J. Mater. Sci., 30, 3075–3081 (1995).
30. B.Q. Xu, T. Yamaguchi and K. Tanabe, Chem. Lett., 17, 1663–1666 (1988).
31. K. Tanabe and T. Yamaguchi, Catal. Today, 20, 185–197 (1994).
32. J. Sudhakar Reddy and A. Sayari, Catal. Lett., 38, 219–223 (1996).
33. M. Rezaei, S.M. Alavi, S. Sahebdelfar, L. Xinmei and Z.F. Yan, J. Mater. Sci., 42, 7086–7092 (2007).
34. Y. Masuda, S. Kugimiya, Y. Kawachi and K. Kato,

Chem. Lett., 42, 1252–1254 (2013).

35. R. Yadav, K.A. Singh and A. Sakthivel, Chem. Lett., 42, 1160–1162 (2013).
36. M. Rezaei, S.M. Alavi, S. Sahebdelfar and Z.-F. Yan, J. Porous Mater., 15, 171–179 (2008).
37. S.F. Parker, Proteins and Polypeptides, Plenum Press, New York (1971)(Chapter 10).
38. T.A. Tu,RamanSpectroscopy inBiology: Principles andApplications,JohnWiley and Sons, Ltd., Chichester (1982).
39. M.D. Byler, Biopolymers, 25, 469–487 (1986).
40. H. Torii and M. Tasumi, J. Chem. Phys., 96, 3379–3387 (1992).
41. S. Spassov, M. Beekes and D. Naumann, Biochim. Biophys. Acta, 1760, 1138–1149 (2006).
42. T. Miyazawa and R.E. Blout, J. Am. Chem. Soc., 83, 712–719 (1961).
43. M.J. Winningham and D.Y. Sogah, Macromolecules, 30, 862–876 (1997).
44. Y. Chang-Cheng, A.S. Sarit, D. Mrinmoy, K.J. Michael and R.M. Vincent, J. Am. Chem. Soc., 128, 14612–14618 (2006).
45. T. Orita, K. Kato and M. Tomita, J. Ceram. Soc. Jpn., 119, 238–245 (2011).

CHAPTER 4

BIODECOLORIZATION OF TEXTILE DYES BY IMMOBILIZED ENZYMES IN A VERTICAL BIOREACTOR SYSTEM

Dede Heri Yuli Yanto[1,2], Sanro Tachibana[3], and Kazutaka Itoh[3]

[1]The United Graduate School of Agricultural Sciences, Ehime University. 3-5-7 Tarumi, Matsuyama, Ehime 790-8566, Japan

[2]Research and Development Unit for Biomaterials, Indonesian Institute of Science. Jl. Raya Bogor Km. 46, Cibinong, Bogor, 16911 Indonesia.

[3]Department of Applied Bioscience, Faculty of Agriculture, Ehime University. 3-5-7 Tarumi, Matsuyama, Ehime, 790-8566,

ABSTRACT

A capacity of immobilized enzymes from Trametes versicolor U97 or Pestalotiopsis sp. NG007 to decolorize three textile dyes was investigated in a vertical bioreactor system. Immobilization was conducted using a double layer of alginate bead (1.5% w/w) and crude enzymes. The effect of mediators: Tween 80, 1-hydroxybenzotriazol (HBT), or 2,2,6,6-tetramethylpiperidin-1-yl)oxyl (TEMPO) was investigated on decolorization rate. The results showed that application of immobilized enzymes in a vertical bioreactor enhanced the decolorization efficiency of dyes in the order LB16>RRV9>RRN4, respectively. The decolorization rate was accelerated when a mediator TEMPO was used in immobilized enzyme of U97 or a Tween 80 was used in NG007. Reaction with glutaraldehyde (0.6%) for 4 h maintained the longevity and reusability of the beads. The study suggests that double layer immobilized enzyme in a bioreactor system has a terrific potential strategy for treating the textile dye effluents.

INTRODUCTION

Water pollution by waste dye stuff released from textile industries and dye houses represents a major environmental concern. It is estimated around 10-15% of textile dyes are lost during the dyeing process and 2-20% are directly discharged as aqueous effluents in different environmental components [1]. Unavoidably, these textile effluents would be a significant threat to public and environmental health[2]. The conventional technology based on physical or chemical treatments is usually effective only if the effluent volume is small [3]. In addition, a biological treatment in liquid state fermentation is less effective of removing dyes from continuous effluents [3]. Recently, enzymatic process has attracted much attention in the treatment of textile dyes in wastewater due to their ecofriendly and offers a rapid decolorization process [4, 5]. However, few studies have investigated the use of bioreactor system to decolorize these effluents. The purposes of this study are to investigate the potential use of a bioreactor system developed with double layer immobilized enzymes to decolorize textile dyes, to investigate the effect of enzymatic mediator addition in the second layer of the bead on decolorization rate of dyes, and finally, to investigate the longevity of the bead in sequential batch cycles decolorization in order to establish a bioreactor system that allows decolorization of reactive dyes over an extended period without addition of new enzymes.

Citation: Dede Heri Yuli Yanto , Sanro Tachibana, and Kazutaka Itoh, Biodecolorization of textile dyes by immobilized enzymes in a vertical bioreactor system, doi:10.1016/j.proenv.2014.03.030

MATERIALS AND METHOD

Chemicals

Textile dyes Lefavix Blue 16 (LB16), Reactive Remazol Violet 9 (RRV9), and Reactive Remazol Navy 4 (RRN4) were used as substrates. Trametes versicolor U97 and Pestalotiopsis sp. NG007 were isolated from Ehime Prefecture, Japan. Tween 80, manganese (II) sulfate, hydrogen peroxide, 1-hydroxybenzotriazole (HBT), (2,2,6,6- tetramethylpiperidin-1-yl)oxyl (TEMPO), and glutaraldehyde were purchased from Wako Pure Chemical Industries, Ltd. (Japan).

Crude enzyme extraction

Isolate Trametes versicolor U97 or Pestalotiopsis sp. NG007 was pre-grown in wood meals (1 kg) containing glucose (100 g) and shitake nutrient (150 g) with final water content 60% for 1 month. Crude enzyme was extracted from the pre-grown fungus using homogenizer 10000 rpm for 10 minutes with malonate buffer pH 4.5. After filtration, filtrate was subjected to centrifuge at 8000 rpm, 4o C for 20 minutes. Supernatant was collected and placed in container. Ammonium sulfate 75% was added into container, stirred for 1 hour, then centrifuged at 8000 rpm, 4o C for 20 minutes. Pellet was collected and dilute in the buffer while stirred. The powder of crude enzyme was obtained after freeze-drying for 3 days.

Immobilization of crude enzymes

Immobilization of crude enzyme was conducted using double layer of alginate bead (1.5% w/w) and crude enzyme [0.36 U/mL] for T. versicolor or Pestalotiopsis sp. Mixture of Mn^{2+} (1 mM), H_2O_2 (1 mM) and Tween 80 (1%) was performed in the second layer. The effect of ligninolytic redox mediator such as HBT (1 mM) and/or TEMPO (1 mM) was investigated on decolorization rate. The chemical structure of Tween 80, HBT and TEMPO are shown in Fig 1.

Decolorization of textile dyes using bioreactor

Experiment was conducted using a bioreactor system MasterFlex®L/S® (Cole-Parmer Instrument Company) with two easy-loaders (model 7518-10) and tubing size no. 16. The immobilized enzyme was placed into bioreactor column (Ø = 2.5 cm, h = 10 cm, v = 49 cm^3) and the dye simulating textile effluent (100 mgL^{-1}) was flowed from Erlenmeyer flask (100 mL) into the beads with flow rate 1.5 mL/min.

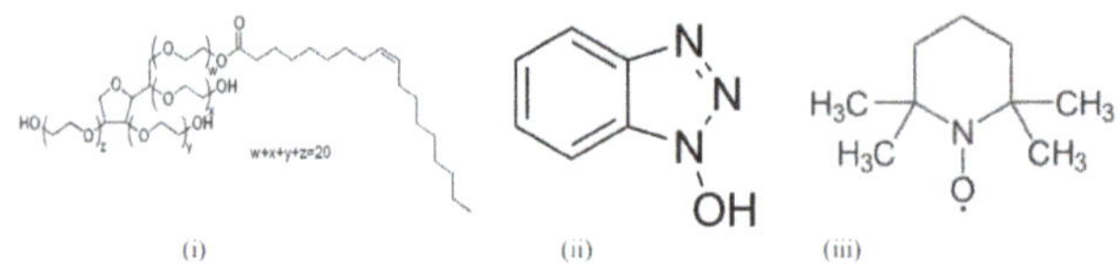

Figure 1: Chemical structure of enzyme mediators used in this study. Tween 80 (i), HBT (ii), and TEMPO (iii)

Dye decolorization was determined by monitoring the decrease in the absorbance at the wavelength of the maximum absorbance of each dye: LB16 (λmax = 610 nm), RRV9 (λmax = 555.5 nm), RRN4 (λmax = 606 nm). The absorbance of the dye solution was monitored for 1, 2, 3, 6, 12, 24, and 48 h reaction. The decolorization efficiency (R, %) was calculated as the following equation (1).

$$R = (1 - (A\ \text{observed})/(A\ initial))x\ 100\% \quad (1)$$

where, $A_{initial}$ is the initial absorbance, and $A_{observed}$ is observed absorbance. Experimental design of vertical bioreactor is shown in Fig. 2.

Sequential dye decolorization

Sequential batch decolorization was carried out to evaluate the longevity of the beads. For this purpose, the effects of glutaraldehyde addition on five sequential cycles of decolorization were evaluated. The first cycle was run using bioreactor system with dye concentration of 100 mg/L. At the end of each cycle (interval of 360 min), the dye simulating textile effluent was substituted and the beads was washed with 50 mM malonate buffer (pH 4.5) before starting for the next cycle.

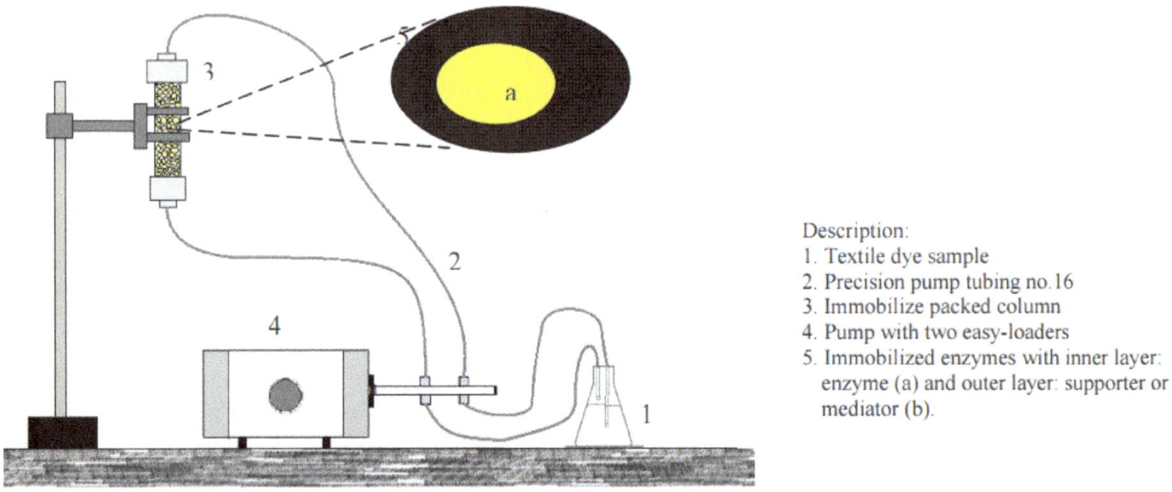

Figure 2: Experimental design of vertical bioreactor system for decolorization of textile dyes by immobilized enzymes.

RESULTS AND DISCUSSION

Decolorization by immobilized enzyme

Dye decolorization of LB16, RRV9, and RRN4 by immobilized enzyme from T. versicolor U97 or Pestalotiopsis sp. NG007 was first evaluated. Data of the decolorization for 24 h was shown in Table 1. Strong decolorization of LB16 was obtained when immobilized enzyme from U97 was employed. However, no significantly different of decolorization efficiency was observed when the two immobilized enzymes were employed to RRV9 or RRN4. In general, capacity of enzyme to decolorize one dye may be different for other dyes. Therefore, chemical structure of dyes as well as type of dyes influenced the removal efficiency by the enzymes. In this study, decolorization efficiency was decreased in the order LB16 > RRV9 and > RRN4.

Initial pH condition significantly affected the decolorization by the immobilized enzyme, specifically by U97. Decolorization efficiency generally decreased at pH 8.2, except in LB16 by NG007. The decreasing of the decolorization might be due to the decreasing of the enzymatic activities of beads. Under pH 4.5 the ligninolytic activities may involve more optimum to decolorize dyes than at pH 8.2 which is basic condition. It was reported that most basidiomycetes and their enzyme worked significantly in decolorization of textile dyes at ligninolytic condition [6, 7]. In addition, the pH of the environment may affect the mass transfer into the bead caused by the influence of the transport of bulk H+ ion; therefore, it affects the decolorization rate [8].

Efficiency decolorization of textile dyes can also be attributed to the enzyme activities involved in the reaction. This study showed that the efficiency of immobilized enzyme from NG007 to decolorize textile dyes was lower than that of U97 due to this ligninolytic activity characteristic. T. versicolor U97 is a basidiomycete fungus mainly produced manganese peroxidase (MnP), laccase, and lignin peroxidase (LiP) that important for removing pollutant such as Remazol Brilliant Blue R (RBBR) and 1,1,1-trichloro-2,2-bis(4-chlorophenyl)-ethane (DDT) in the environment [9]. However, NG007 is an ascomycete fungus with small activities of MnP, laccase, and LiP [10,11]. Although the ligninolytic activities of NG007 were low, the remarkable decolorization was observed by the immobilized enzymes. Some ascomycetes fungi decolorized dyes by an alternative mechanisms such as participation of P450-monooxygenases [12].

Table 1: Decolorization of textile dyes by immobilized enzymes over 24 h

Immobilized enzyme	Dyes	Decolorization (%) at pH 4.5	at pH 8.2
Pestalotiopsis sp. NG007	LB16	53	56
	RRV9	48	44
	RRN4	44	45
T. versicolor U97	LB16	84	69
	RRV9	51	47
	RRN4	46	45

Flow rate dependent

Interaction of dyes with the bead is the important factor considered to the success of decolorization. In this study, the flow rate of reaction influenced the decolorization efficiency by immobilized enzyme (Fig. 3). Generally, the faster the flow rate, the lower the biodecolorization rate. For all dyes investigated in this study, when 10 ml/min of flow rate was employed, the biodecolorization was the lowest. In contrast, when the flow rate was 1.0 or 1.5 ml/min, the biodecolorization was optimum. In LB16, the optimum was obtained at the flow rate 1.5 mL/min. However, in RRV9 or RRN4 no significant result was obtained when the flow rate 1.0 or 1.5 mL/min was employed. Therefore, flow rate 1.5 mL/min was used for further investigation.

Biodecolorization of textile dyes using bioreactor had a higher efficiency than in a liquid medium. In liquid medium, the immobilized enzyme decolorized the dyes maximum 36% of the dyes. However, when it was applied in bioreactor, the biodecolorization enhanced to 49 – 100% by flow rate 1.5 mL/min. We suggest that decreasing efficiency in liquid medium is caused by the over-load capacity of the beads when directly applied to dyes.

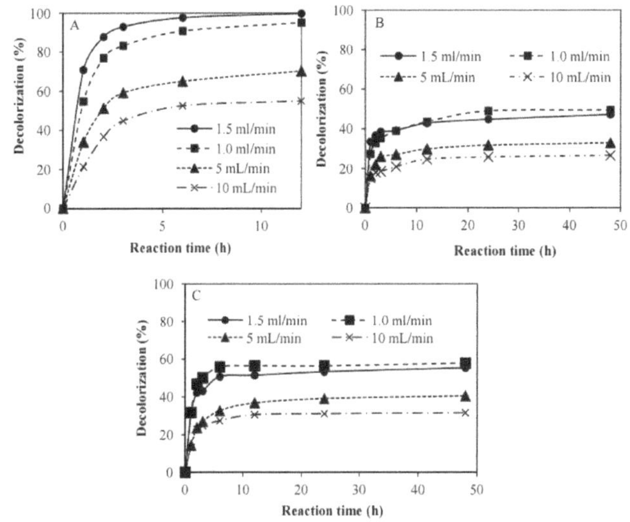

Figure 3: Effect of flow rate on biodecolorization efficiency of LB16 (A), RRV9 (B), and RRN4 (C) by immobilized enzyme of Trametes versicolor U97.

However, these constraints can be overcome by periodically dropped of the dye effluent to the bead with small volume in bioreactor. When the flow rate was fit with the time of reaction, the decolorization reached the optimum condition. In this study, the optimum flow rate was 1.5 mL/min. When the drop of dyes fell on the beads, the reaction was started and enough to be decolorized before another drop came. However, the optimum flow rate may be different for various dyes and immobilized enzymes, and critically influenced by a combination of mass transfer and the capacity of immobilized enzyme to degrade dyes.

Effect of enzyme mediators

Addition of ligninolytic enzyme mediators such as Tween 80 (1%), HBT (1 mM) or TEMPO (1 mM) generally enhanced the decolorization efficiency for the three textile dyes by two immobilized enzymes either at pH 4.5 or 8.2 (Fig. 4 - 6). Addition of Tween 80 in the presence of Mn2+ promoted the oxidation by the enzyme. Tween 80 provides an unsaturated fatty acid chain that may be readily oxidized to form aperoxide which was subsequently turned into a peroxyl radical. This radical was the actual oxidant for the oxidation of the dyes [13]. In many research, addition of HBT has enhanced the decolorization of textile dyes by fungal and enzymes treatment [14, 15, 16, 17]. On the other hand, TEMPO is one of laccase mediators with `stable` N-oxyl radical that involves the most effective oxidation of some non-phenolic substrates [18].

Immobilized enzyme from T. versicolor U97 had a greater effect of enhancement when 1 mM TEMPO was put in second layer of the beads. Conversely, the most efficient decolorization of textile dyes by immobilized enzyme from Pestalotiopsis sp. NG007 was obtained when 1% Tween 80 was employed. Different behavior of immobilized enzyme can be due to the enzyme activities involved in the beads.

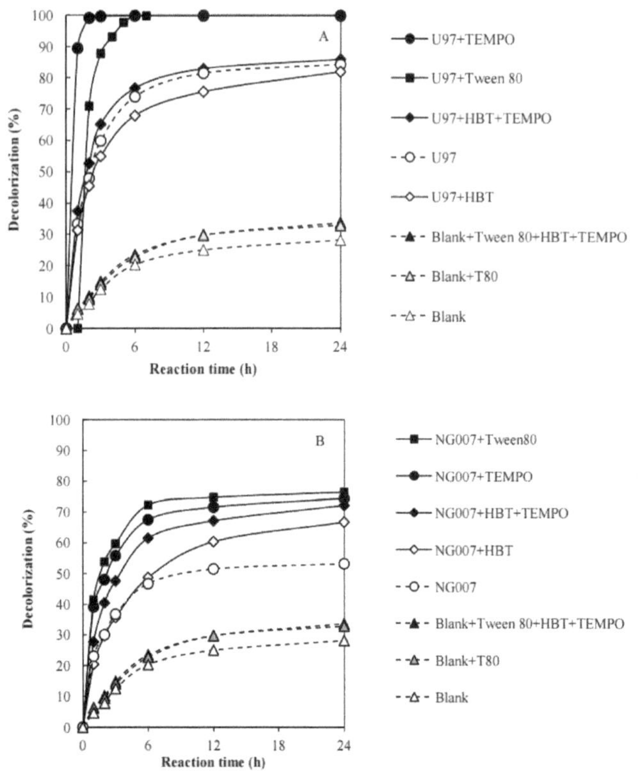

Figure 4: Effect of various mediators on biodecolorization of LB16 by immobilized enzymes from Trametes versicolor U97 (A) and Pestalotiopsis sp. NG007 (B) in bioreactor system.

In LB16, the decolorization significantly improved when mediators were used (Fig. 4). In immobilized enzyme of U97, the accelerated decolorization was obtained by the addition of TEMPO (100% decolorization for 3 h reaction) followed by Tween 80 (100% for 6 hours). Unfortunately, addition of HBT could not improve the decolorization rate since it was decreased by 4% compared to the control (without addition of HBT). In addition, when the mixture of HBT and TEMPO was employed to the beads, no significant improvement was observed due to the negative effect of HBT. Approximately 28-32% of dyes were decolorized due to the absorption processes by the blank alginate, by means the decolorization of dyes was due to the enzymatic degradation process. In contrast with U97, immobilized enzyme of NG007 accelerated the decolorization of LB16 by addition of Tween 80 (72% for 6 hours) followed by TEMPO (67% for 6 hours). The addition of HBT did not show a negative effect to decolorization but the efficiency was lower than the addition of Tween 80 or TEMPO. In addition, when the mixture of HBT and TEMPO was added to the beads, the decolorization was still lower than the addition of TEMPO alone. It can be suggested that the biodecolorization of dyes is due to the enzymatic degradation process because of the low absorption by the blank alginate.

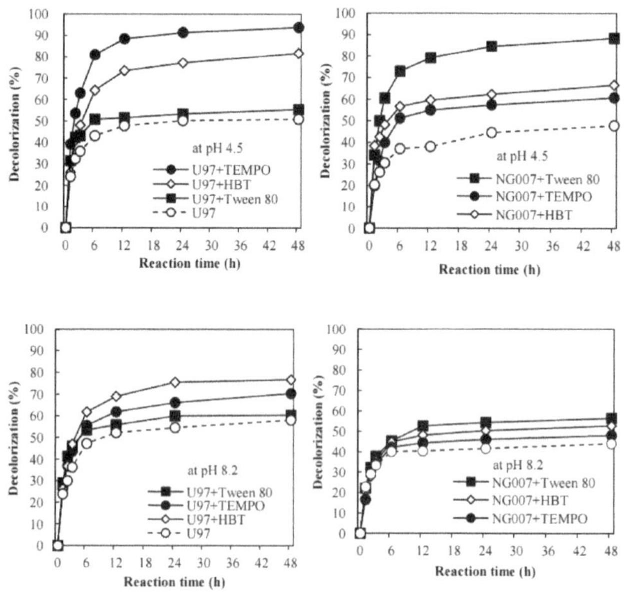

Figure 5: Effect of various mediators on biodecolorization of RRV9 by immobilized enzymes of Trametes versicolor U97 and Pestalotiopsis sp. NG007 in bioreactor system at pH 4.5 or at pH 8.2.

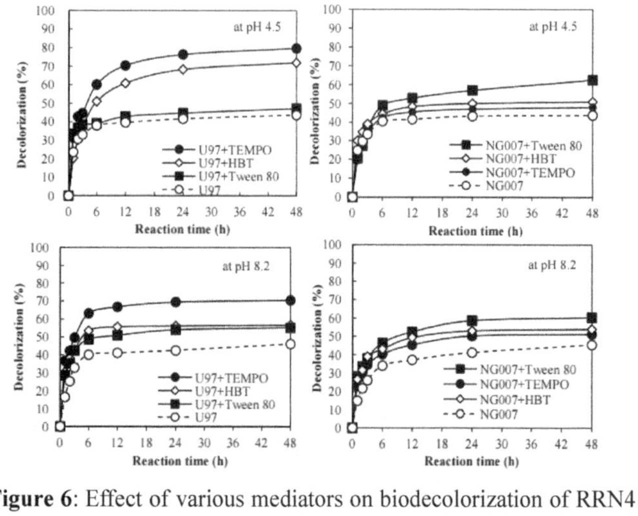

Figure 6: Effect of various mediators on biodecolorization of RRN4

by immobilized enzymes of Trametes versicolor U97 and Pestalotiopsis sp. NG007 in bioreactor system at pH 4.5 or at pH 8.2.

The same effect of mediator was also observed in RRV9 and RRN4. When the beads was applied to dyes at pH 4.5, addition of TEMPO in U97 beads, enhanced the decolorization by 84% and 82% for RRV9 and RRN4, respectively. The enhancement, however, decreased when the beads was applied to the pH of dyes 8.2 (21% for RRV9 and 52% for RRN4, respectively). In NG007, addition of Tween 80, enhanced the decolorization by 85% of RRV9 and 43% of RRN4, respectively, at pH 4.5. The enhancement was decreased when applied to the pH 8.2 to become 28% of RRV9 and 32% of RRN4, respectively. These results confirmed that the characteristic and chemical structure of dyes affected the effectiveness of mediator for enhancing decolorization by the beads. The effectiveness of mediators on decolorizing the dyes was decreased in the order LB16 > RRV9 > RRN4, respectively.

Sequential dye decolorization

Immobilization allows both their reuse for continuous process and the stability of enzyme upon wide pH, temperature range and several of dyes [19]. Therefore, the longevity of the beads is important for the efficient strategy of textile dyes decolorization. In this research, the longevity of the beads was performed by reaction with 0.6% glutaraldehyde for 4 h in ice bath reactor. The result showed that glutaraldehyde increased the longevity of the beads although the efficiency decolorization was decreased in the first cycle when compared with the control (without reaction with glutaraldehyde) (Fig. 7).

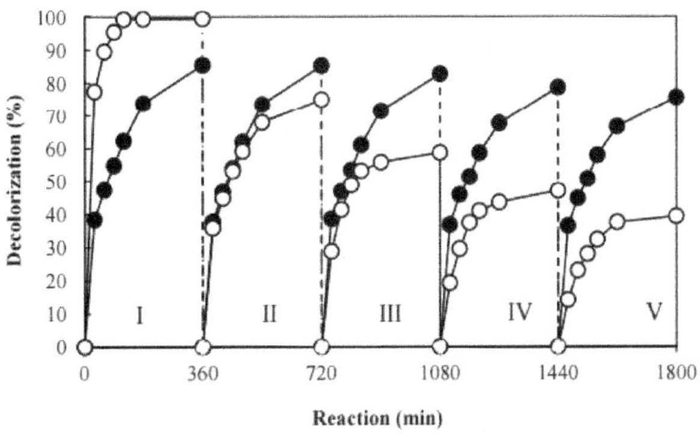

Figure 7: Sequencing batch decolorization of 100 mgL-1 LB16 by immobilized enzyme of T. versicolor U97 in the presence of 1 mM TEMPO using bioreactor system and treated by glutaraldehyde (filled cycle) or without addition of glutaraldehyde (opened cycle).

When sequential batch analysis was performed in bioreactor system, the efficiency decolorization of the beads was drastically decreased in the second cycle and attained only near 40% decolorization in the last cycle. However, reaction of the beads with glutaraldehyde maintained the decolorization in the last cycle at 75%. The results suggest that double layer immobilized enzyme decolorization using bioreactor system is a terrific potential strategy for treating the textile dyes effluents.

CONCLUSION

The immobilized enzymes from fungi showed high decolorization efficiency when they were applied in a vertical bioreactor system. The addition of specific mediators for enzymatic process enhanced the decolorization of dyes by particular immobilized enzymes. The longevity of the beads can be improved by the reaction with glutaraldehyde 0.6% which is important for the reusability of the beads. This study suggests that application of immobilized enzyme in a vertical bioreactor system is a potential tool to improve the biodecolorization of textile dyes from industrial effluents.

ACKNOWLEDGEMENT

This research was partly supported by JUTAKUKENKYUHI from Ehime Prefecture, Japan.

REFERENCES

1. Jin XC, Liu GQ, Xu ZH, Tao WY. Decolorization of a dye industry effluent by Aspergillus fumigatus XC6. Appl Microbiol Biotechnol. 2007;74:239-43.
2. Walsh GE, Bahner LH, Horning WB. Toxicity of textile mill effluents to freshwater and estuarine algae, crustaceans and fishes. Environ. Poll. 1980;21(3):169-79.
3. Robinson T, McMullan G, Marchan R, Nigam P. Remediation of dyes in textile effluent: a critical review on current treatment technologies with a proposed alternative. Bioresour. Technol. 2001;77:247-55.
4. Iqbal HMN, Asgher M. Decolorization applicability of sol-gel matrix immobilized manganese peroxidase produced from an indigenous white rot fungal strain Genoderma lucidum. BMC Biotechnol. 2013;13:1-7.
5. Sun QY, Hong YZ, Xiao YZ, Fang W, Fang J. Decolorization of textile reactive dyes by the crude laccase produced from solid-state fermentation of agro-byproducts. World J. Microbiol. Biotechnol.2009; 25:1153-60.
6. Kandelbauer A, Maute O, Kessler RW, Erlacher A, Gübitz GM. Study of dye decolorization in an immobilized laccase enzyme-reactor using online spectroscopy. Biotechnol.Bioeng. 2004; 87(4):552-63.
7. Zhan FM, Knapp JS, Tapley KN. Development of bioreactor systems for decolorization of Orange II using white rot fungus. Enzyme Microb. Technol. 1999; 24:48-53.
8. Chen JP, Lin YS. Decolorization of azo dye by immobilized Pseudomonas luteola entrapped in alginate-silicate sol-gel beads. Process Biochem. 2007; 42(6):934-42.
9. Sari AA, Tachibana S, Muryanto. Correlation of ligninolytic enzymes from the newly-found species Trametes versicolor U97 with RBBR decolorization and DDT degradation. Water Air Soil Pollut. 2012; 223:5781-92.
10. Yanto DHY, Tachibana S. Biodegradation of petroleum hydrocarbons by a newly isolated Pestalotiopsis sp. NG007. Int. Biodeter. Biodegradation. 2013; 85:438–450.
11. Yanto DHY, Tachibana S. Enhanced biodegradation of asphalt in the presence of Tween surfactants, Mn2+ and H2O2 by Pestalotiopsis sp. in liquid medium and

soil. Chemosphere. 2013; http://dx.doi.org/10.1016/j.chemosphere.2013.11.044.

12. Ambrósio ST, Campos-Takaki GM. Decolorization of reactive azo dyes by Cunninghamella elegans UCP 542 under co-metabolic condition. Bioresour. Technol. 2004; 91:69-75.
13. Bao W, Fukushima Y, Jensen JrKA, Moen MA, Hammel KE. Oxidative degradation of non-phenolic lignin during lipid peroxidation by fungal manganese peroxidase. FEBS Lett.1994; 354:297-300.
14. Reyes P, Pickard MA, Duhalt RV. Hyderoxybenzotriazole increase the range of textile dyes decolorized by immbolized laccase. Biotechnol. Lett. 1999; 21:875-80.
15. Domínguez A, Couto SR, Sanromá MÁ. Dye decolorization by Trametes hirsuta immobilized into alginate beads. World J. Microbiol. Biotechnol.2005; 21:405-09.
16. Clause H, Faber G, König H. Redox-mediated decolorization of synthetic dyes by fungal laccases. Appl Microbiol. Biotechnol.2002; 59:672- 78.
17. Khlifi R, Sayadi S, Belbahri L, Woodward S, Mechichi T. Effect of HBT on the stability of laccase during the decolourization of textile wastewaters. J.Chem. Technol. Biotechnol. 2009; 84(12):1828-1833.
18. Fabbrini M, Galli C, Gentili P.Comparing the catalytic efficiency of some mediators of laccase. J. Mol. Cat.2002; 16:231-40.
19. Leonowicz A, Sarkar JM, Bollag JM. Improvement in stability of an immobilized fungal laccase. Appl. Microbiol. Biotechnol. 1988; 29:129- 35.

CHAPTER 5

ENZYMATIC EXTRACT CONTAINING LIGNIN PEROXIDASE IMMOBILIZED ON CARBON NANOTUBES: POTENTIAL BIOCATALYST IN DYE DECOLOURIZATION

Sabrina Feliciano Oliveira [1], Jose´ Maria Rodrigues da Luz [2], Maria Catarina Megumi Kasuya[2], Luiz Orlando Ladeira[1] , and Ary Correa Junior [1]

[1]Department of Microbiology, Universidade Federal de Minas Gerais, Av. Anto^nio Carlos, 6627 – Pampulha, 31270-901 Belo Horizonte, Minas Gerais, Brazil

[2]Department of Microbiology, Universidade Federal de Vic¸ osa, Av. Peter Henry Rolfs, s/n° – Campus Universita´rio, Vic¸ osa, Minas Gerais, Brazil

ABSTRACT

The majority of the textile dyes are harmful to the environment and potentially carcinogenic. Among strategies for their exclusion, the treatment of dye contaminated wastewater with fungal extract, containing lignin peroxidase (LiP), may be useful. Two fungi isolates, *Pleurotus ostreatus* (PLO9) and *Ganoderma lucidum* (GRM117), produced the enzymatic extract by fermentation in the lignocellulosic residue, *Jatropha curcas* seed cake. The extracts from PLO9 and GRM117 were immobilized on carbon nanotubes and showed an increase of 18 and 27-fold of LiP specific activity compared to the free enzyme. Also, LiP from both fungi extracts showed higher Vmax and lower Km values. Only the immobilized extracts could be efficiently reused in the dye decolourization, contrary, the carbon nanotubes became saturated and they should be discarded over time. This device may offer a final biocatalyst with higher catalytic efficiency and capability to be reused in the dye decolourization process.

INTRODUCTION

Dyes are an important class of macromolecules which are an integral part of our lives and used by, textile, paint and plastic, solar cells, optics, metal extraction and sensor industries, among others (Rauf and Ashraf, 2012). Due to the extensive use, they have become a part of industrial effluent. It is estimated that up to 50% of dye contents are lost after dyeing textiles and about 10–15% are discarded in effluents (Papadoulou et al., 2013). Further to esthetic pollution, the colored effluent promotes the inhibition of light penetration along the depth of water bodies, affecting biological cycles, particularly the photosynthesis processes. Most of these dyes are harmful and potentially carcinogenic in nature and their exclusion from wastewaters is a major environmental challenge (Chequer et al., 2011).

The use of enzymes for environmental remediation purposes has increased due to peculiar properties of this class of proteins. These molecules operate in a wide range of contaminant concentrations, pH, temperature and salinity (Rao et al., 2014). The ligninolytic enzyme system (lignin peroxidase, manganese peroxidase and laccase) produced by the white rot fungi is of particular interest because it has a low substrate specificity, no steric selectivity and strong oxidative abilities (Ashger et al., 2014). Besides, due to the

Citation: Sabrina Feliciano Oliveira, José Maria Rodrigues da Luz, Maria Catarina Megumi Kasuya, Luiz Orlando Ladeira, Ary Correa Junior, Enzymatic extract containing lignin peroxidase immobilized on carbon nanotubes: Potential biocatalyst in dye decolourization, Saudi Journal of Biological Sciences, Available online 10 March 2016, ISSN 1319-562X, http://dx.doi.org/10.1016/j.sjbs.2016.02.018.

high cost of biotechnology processes, the current challenge is to increase the production of these enzymes, using low-cost substrates. Therefore the use of lignocellulosic residues, such as the *Jatropha curcas* seed cake, as substrate for the synthesis of enzymes is an alternative for cost reduction and at the same time solves an environmental problem regarding the disposal of these agro-industrial wastes. Among those enzymes, lignin peroxidase (LiP) stands out as a relatively nonspecific enzyme, able to mineralize various recalcitrant aromatic and halogenated phenolic compounds, with high redox potential compared with the classical peroxidases, therefore being a stronger oxidant (Wong, 2009). Regarding the biotechnological potential, this enzyme has been successfully applied in food and pharmaceutical industries, wastewater treatment of pulp and paper and textile industry, bioremediation and biomass delignification (Wong, 2009 and Ashger et al., 2014).

Although the natural catalysts are sustainable, selective and efficient, they are often not perfectly adapted for industrial applications. To promote the use of enzymes in industrial processes, its stability and reuse ability should be considered (Gardossi et al., 2010). Hence enzyme immobilization emerges as a key enabling technology (Liese and Hilterhaus, 2013). Enzyme immobilization on various solid supports offers a number of advantages such as easy handling and a possible increase in the thermal stability and pH range supported by the protein. An additional benefit is greater stability under storage and operating conditions, for example, to heat denaturation, organic solvents or autolysis (Gardossi et al., 2010).

Recent interest in nanotechnology has provided a wide variety of materials that may potentially be used as supports for immobilization. Among the various nanostructures, carbon nanotubes (CNTs) have played a key role in the nanotechnology revolution. Due to the large specific surface area of CNTs, these carbon nanoparticles have been preconized as a good biomolecules carrier, being promising scaffolds for peptides, proteins and enzymes (Feng and Ji, 2011 and Marchesan and Prato, 2015). Furthermore, carbon nanotubes exhibit extraordinary mechanical properties, electrical, thermal and chemical stability (Zhang and Henthorn, 2010); hence this nanoparticle has been applied in diverse fields such as materials, electronics and nanomedicine (Oliveira et al., 2015). One of the primary fields to benefit from nanoparticle-enzyme conjugates is biocatalysis, which is becoming one of the most powerful tools in biotechnology, having a profound impact on environmental protection (Illanes et al., 2012).

The main objective of this study is the production of fungal enzymatic extract, containing LiP, and its immobilization on carbon nanotubes. We also evaluated the LiP catalytic efficiency, stability and the capability of reuse in dye decolourization. The results obtained from this study can contribute to developing a biocatalyst that can be applied in diverse environmental applications in the future, especially in the treatment of textile effluents.

MATERIALS AND METHODS

Strain and chemicals

The fungi, *Pleurotus ostreatus* (PLO9) and *Ganoderma lucidum* (GRM117) were obtained from Mycorrhizal Associations Laboratory at Universidade Federal de Viçosa (Minas Gerais, Brazil). They were stored on potato dextrose agar (PDA) slants at 4 °C and multiplied, as needed, in the same medium.

All analyses were made with analytical reagent grade. Veratryl alcohol (VA) and Remazol brilliant blue R (RBBR) dye were purchased from Sigma–Aldrich (St. Louis, MO – USA) and the commercial lignin peroxidase (LiPc) was purchased from Santa Cruz Biotechnology, Inc (Dallas, TX – USA).

Enzymatic extract production: solid state fermentation (SSF)

Fungal cultures were multiplied in PDA plates at 25 °C for up to 7 days for inoculum production. Four mycelial disks, containing 1 cm agar-disk, were grown in substrate containing 20 g of *J. curcas L.* seed cake, previously humidified (70% of water) and autoclaved for one hour at 121 °C. Substrates were supplemented with 2 mL of Kirk's medium (Tien and Kirk, 1983) and 1 mL of exogenous inductor Tween 80 (0.3 mmol l^{-1}), and incubated at 25 °C (room temperature). After 7 days of inoculation, the crude enzymatic extract (CEE) was obtained by adding 100 mL sodium tartrate buffer (100 mmol l^{-1}, pH 3.5) supplemented with EDTA (5 mmol l^{-1}), then centrifuged at 4000×*g*for 30 min and filtrated (Whatman N°. 1 filter paper).

This procedure was done for SSF of both fungal isolates in triplicates, including the control (without inoculum) and the one flask inoculated but without exogenous inductor.

Partial purification of enzymatic extract

The crude enzymatic extract was centrifuged at 4000×*g* for 15 min. The supernatant was firstly brought to 40% saturation by the gradual addition of solid crystals of ammonium sulfate and kept overnight at 4 °C. The precipitate was collected by centrifugation (4000×g) for 15 min at 4 °C and to the supernatant more crystals of ammonium sulfate were added to achieve 80% saturation. It was again kept overnight at 4 °C and centrifuged as described previously. After centrifugation, the sediments were dissolved in sodium tartrate buffer (100 mM l^{-1}, pH 3.5) and dialyzed against the same buffer. The dialysate was applied to DEAE cellulose (Sigma–Aldrich, Brazil) ion-exchange column (3 × 14 cm) which had been equilibrated with sodium tartrate buffer (100 mmol l^{-1}, pH 5.5). The column was washed stepwise with 50, 100, 150, 200, 300 and 500 mmol l^{-1} de sodium chloride solution (1 mol l^{-1}). The fractions were collected with a flow rate of 1 mL min^{-1} and the LiP activity and protein contents were determined for each fraction. The active fractions were then pooled and utilized as the partial purified enzymatic extract (PPE) after lyophilization.

LiP activity and protein content

The lignin peroxidase activity was determined by spectrophotometry based on change in absorbance at 310 nm at 30 °C (Multiskan™ FC Microplate Photometer). The enzyme assay contained 100 µL of sodium tartrate buffer (100 mmol l^{-1}, pH 3.5), 100 µL of veratryl alcohol (4 mmol l^{-1}), 50 µL hydrogen peroxide (0.2 mol l^{-1}) and 10 µL of PPE.

LiP activity was expressed in units (U) per mL. One unit (U) LiP was defined as the amount of enzyme required to oxidize 1 µmol of veratryl alcohol in 1 min, at 30 °C and pH 3.5. The activity of reaction mixture was measured against reagent blank at 310 nm (310 = 9300).

To determine the protein contents of the crude and purified enzyme extracts, the Bradford micro assay (Bradford, 1976) using bovine serum albumin as standard was used.

PPE immobilization on carbon nanotubes

Carboxylated carbon nanotubes with HNO_3 and H_2SO_4 were used. Carbon nanotubes were produced by the chemical vapor deposition method using iron as a catalyst. The CNT produced had a diameter of 20–90 nm with a purity of 83% and a length between 10 and 30 µm.

Carbon nanotubes, 5 mg, were suspended in 1.5 mL of sodium phosphate buffer (0.2 mol l^{-1}, pH 7.7) and stirred for 30 min in a sonicator (Thornton – INPEC Electronic S/A). Different amounts of the lyophilized enzymatic extract (5, 25 and 50 mg) were suspended in 2 mL of the same buffer. The whole volume was added to nanotube solution and this mixture was sonicated for 3 h. After that, enzymes not linked to the nanotubes were removed by repeated washing (up to 4×) with 2 mL of phosphate buffer (0.2 mol l^{-1}, pH 7.7).

Immobilized LiP activity was determined, as described in the previous section for the free enzyme.

Supernatant, just after immobilization, and from washing solutions was analyzed to quantify the protein by the Bradford method.

The immobilized protein concentration ([LiP]immob) was determined by the difference between the initial protein concentration ([LiP]i) and the concentration of proteins present in the supernatant after immobilization ([LiP]s). In turn, the protein concentration that remained bound to the carbon nanotube ([LiP]b) was determined by the difference between ([LiP]immob) and the sum of the protein concentration present in washing solutions (Σ [LiP]w). The immobilization efficiency was determined by the equation (i).

Immob. Efficiency = {([LiP]b)/([LiP]i)} × 100. (i)

where: [LiP]b = Enzyme bound to the Carbon nanotube

[LiP]i = Initial protein concentration

Effect of pH and temperature on LiP activity

The effect of varying pH (pH 3–9) on the activities of free and immobilized LiP was investigated using sodium tartrate buffers in pH 3.0, 5.0, 7.0 or 9.0. The free and immobilized LiPs were incubated at room temperature (±25 °C) for 15 min.

Effect of varying temperatures on LiP activity was also studied. Both the free and immobilized LiPs were incubated with sodium tartrate buffer pH 3.5 for 15 min, at temperatures of 15, 30, 45 or 60 °C.

Effect of substrate concentration: determination of Km and Vmax

The Michalis–Menten kinetic constants (Km) and maximum velocity (Vmax) for free and immobilized LiP were determined using varying concentrations of veratryl alcohol (25, 50, 75, 100, 150, 200, 250, 300, 500, 700 and 900 µmol l^{-1}) as substrate. To determine the kinetic constants free and immobilized LiPs were incubated for 15 min at room temperature (25 °C) in the same conditions previously described for enzyme assay.

The Km and Vmax for free and immobilized LiPs were calculated with at least seven initial substrate concentrations using the program SigmaPlot V. 12.0 (Systat Software, San Jose, CA).

Dye decolourization

Decolourization experiment of the liquid RBBR dye by free PPEs (100 mg mL^{-1}) and immobilized (25 mg ml^{-1}) was carried out in tubes containing sodium tartrate buffer (100 mM L^{-1}, pH 3.5), hydrogen peroxide (0.2 mol l^{-1}) and RBBR concentrations of 10 and 30 mg l^{-}. The decolourization was monitored spectrophotometrically at 595 nm after 24 h of incubation at 25 °C.

The free and immobilized PPEs were reused once in RBBR decolourization of 10 mg $l^{-1}$and again in RBBR decolourization at the concentration of 30 mg l^{-1}.

The decolourization rate was determined by the difference between the initial absorbance and absorption observed divided by the initial absorbance. In the calculation, the effect of time of incubation as well as the effect of the sample composition was discounted through the control sample which contained the carbon nanotubes and PPE from the solid fermentation control sample.

In order to evaluate the capability of the free and immobilized commercial lignin peroxidase (LiPc) and just carbon nanotubes (without enzymes) in decolorizing the RBBR dye, a second decolourization experiment was performed.

The LiPc was immobilized according to the methodology previously described. In this assay, the absorbance decrease caused by dyes adsorbed on carbon nanotubes was not negligible. The assay was carried out in tubes containing sodium tartrate buffer (100 mmol l^{-1}, pH 3.5), hydrogen peroxide (0.2 mol l^{-1}) and RBBR concentrations of 10, 30 and 50 mg l^{-1}.

The decolourization was also monitored spectrophotometrically at 595 nm after 1, 3, 11 and 14 days of incubation at room temperature. Just the immobilized LiPc was reused 3 times subsequently, 2, 8, and 3 days after the first, second and third measurement, respectively. In this case, the free LiPc was diluted in the solution, so they could not be reused.

RESULTS AND DISCUSSION

Enzymatic extract production: solid state fermentation (SSF)

Both fungi tested in solid fermentation condition in *J. curcas* seed cake plus exogenous inductor Tween 80 showed a high potential for LiP production (Table 1). We have obtained similar volumes of crude enzymatic extract from both isolates, but the LiP specific activity from GRM117 was higher than from PLO9 extract (Table 1).

Table 1: Characteristics of enzymatic extract containing lignin peroxidase (LiP) from *Pleurotus ostreatus* PLO9 and *Ganoderma lucidum* GRM117.

Purification steps	Total volume (mL)	Total LiP activity (U mL^{-1})	Total protein content (mg mL^{-1})	LiP specific activity (U mg^{-1})	Purification fold	Yield (%)
Pleurotus ostreatus PLO9						
Crude enzymatic extract	123	10602.68	0.34	30623.93	1	100
$(NH_4)_2SO_4$precipitation	37	10556.27	0.25	41283.92	1.3	100
DEAE-cellulose	25	4247.3118	0.063	66686.76	1.6	40
Ganoderma lucidum GRM117						
Crude enzymatic extract	120	9419.35	0.13	69814.04	1	100
$(NH_4)_2SO_4$precipitation	21	7715.05	0.13	56836.99	0.8	82
DEAE-cellulose	25	1442.65	0.019	75424.97	1.3	15

The high potential of *P. ostreatus* PLO9 and *G. lucidum* GRM-117 in producing ligninolytic enzyme was expected, since both are basidiomycetes, belonging to the white rot fungus group, known to include generally robust microorganisms that produce extracellular enzymes such as lignin peroxidase, manganese peroxidase, and laccase (Da Luz et al., 2014).

Further to the fungal species used, the growth medium also favored the production of ligninolytic enzymes, in particular the lignin peroxidase, once *J. curcas* seed cake has considerable lignin contents (Da Luz et al., 2014). In Brazil, *J. curcas* has been widely used for biodiesel production, after extracting oil, a solid residue, called seed cake, is produced (Da Luz et al., 2014). The fermentation medium, in addition to the lignocellulosic residue, was supplemented with Tween 80. Several authors have demonstrated an increase in the secretion of enzymes in the presence of Tween 80 in cultures of the fungus *Phanerochaete chrysosporium*. It is believed that the surfactant transforms the structure of the cell membrane and promotes the permeability of lignin peroxidase from fungal cells to the extracellular medium (Asther et al., 1988). Moreover, supplementation of culture medium Kirk also favors for greater LiP production, since this medium is known to induce the production, in particular, of the lignin peroxidase enzyme (Tien and Kirk, 1983). In our experiments, we did not detect LiP activity in the extracts produced without Tween 80, suggesting that LiP is an inducible enzyme for our isolates.

Partial purification of enzymatic extract

We observed an increase in specific activity of PLO9 in the two steps of purification with 40% of yield after DEAE-cellulose (Table 1). In turn, for the GRM117 fungal extract, the yield after this later step of purification was 2.66 times smaller than the yield of PLO9 (Table 1).

The partially purified enzyme extract of these fungi presented high total and specific activities of LiP (Table 1), considering the values reported in the literature (Asgher et al., 2007, Asgher et al., 2012 and Iqbal et al., 2011). Although the purification steps present lower efficiency compared to that described in other studies (Table 1), the partial purity obtained in our extracts was adequate to our final purpose. Thus, once the purification is the most expensive step in the whole process of enzyme attainment, the crude or partially purified enzymatic extract can be used, as low-cost alternative, in some environmental applications, for example, in the dye decolourization.

PPEs immobilization on carbon nanotubes

After the immobilization, the lignin peroxidase present in both the partially purified enzymatic extract showed a higher specific activity (Table 2). The PPE from PLO9, at concentrations of 12.5 and 25 mg mL^{-1}, showed an increase of 18 and 9-fold in LiP specific activity, respectively, compared to the free enzyme (Table 2). For the PPE from GRM-117, only at the concentration of 25 mg mL^{-1}, it was possible to calculate the specific activity after the immobilization that was 27 times higher than the free enzyme (Table 2). These results are significant since, for example, Asgher et al. (2012) verified an increase in the specific activity of only 1.27 of LiP partially purified, after the immobilization by trapping in the xerogel.

Table 2: Immobilization of partial purified enzymatic extract (PPE) containing lignin peroxidase (LiP) from*Pleurotus ostreatus* PLO9 and *Ganoderma lucidum* GRM117 on carbon nanotubes (CNT)

CNT (mg)	PPE (mg)□	PPE (mg mL−1)	Free LiP		Immobilized LiP		Immobilization efficiency (%)
			Total activity (U mL^{-1})	Specific activity (U mg^{-1})	Total activity (U mL^{-1})	Specific activity (U mg^{-1})	
Pleurotus ostreatus PLO9							
5.0	0	0	0	0	0	0	0
5.0	5.0	2.5	ND	ND	586.02	ND	ND
5.0	25.0	12.5	39.42	4877	489.24	89522	67.6
5.0	50.0	25	170.25	13053	577.06	128,100	34.5
Ganoderma lucidum GRM117							
5.0	0	0	0	0	0	0	0
5.0	5.0	2.5	ND	ND	446.23	ND	ND
5.0	25.0	12.5	ND	ND	557.34	ND	ND
5.0	50.0	25	44.80	10,826	507.16	302,572	40.5

□Data referent to the enzymatic extract that was partially purified and lyophilized.

The high LiP specific activities, after the immobilization process, may be explained by the carrier used, once the catalytic behavior of immobilized enzymes strongly depends on the properties of their carriers, such as material types, structures, and compositions. Carbon nanotubes have been shown one of the best supports for immobilization as compared to traditional carriers such as zirconia, silica and epoxy (Saifuddin et al., 2013). Furthermore, they are more stable in extreme conditions, presenting the capacity to carry a larger amount of enzyme and increase up the catalytic activity by two times more than the other supports and by ten times more than the free enzyme (Feng and Ji, 2011 and Saifuddin et al., 2013). Xu et al. (2015) showed that the introduction of carbon nanotubes (MWNTs) to the membrane improved the properties of immobilized laccase, such as enzyme loading, activity retention, pH stability, thermal stability, operational stability, and storage stability. These improvements can be attributed to the enhanced electrical conductivity and biocompatible microenvironment provided by the carbon nanotubes.

Effect of pH and temperature on LiP activity

For the PPE from PLO9, the immobilization did not cause changes, maintaining optimum activity around pH 5 (Fig. 1), and the extract from GRM117, showed higher LiP activity at neutral to basic conditions after the immobilization (Fig. 1).

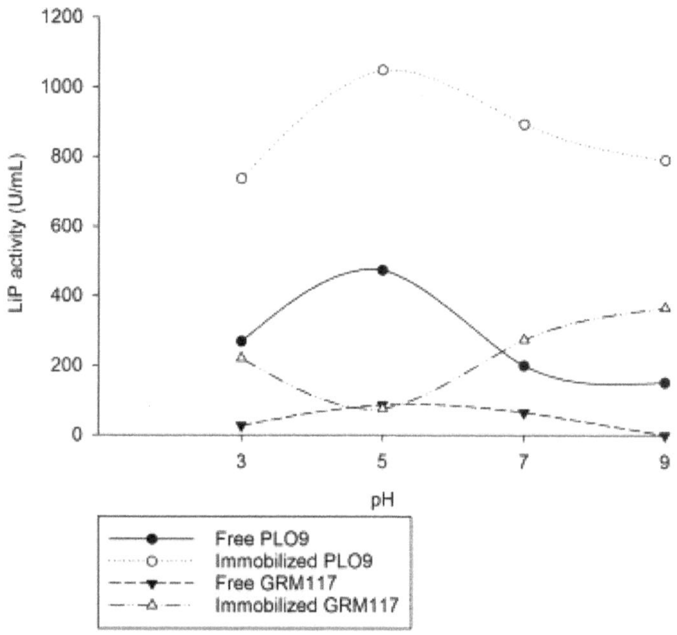

Figure 1: Effect of pH on LiP activity. LiP from partially purified enzymatic extract (PPE) of *Pleurotus ostreatus*PLO9 and *Ganoderma lucidum* GRM117. (Free PPE = 10 mg mL^{-1}; immobilized PPE = 25 mg mL^{-1}).

The LiP activities were higher in PPEs immobilized on carbon nanotubes than free PPEs. Despite being unknown mechanisms, the immobilization protects the enzyme of pH effects and this is important, for example, to apply the immobilized LiP in the treatment of textile effluents which present a wide range of pH.

Previous studies report that the higher activities for LiPs from white rot fungi are generally observed between pH 2 and 5 and it is usually verified, after the immobilization procedure, a slight change in pH of LiP toward more acidic conditions (Asgher et al., 2007 and Asgher et al., 2012). However, LiP of GRM117 has optimal pH in basic conditions after immobilization (Fig. 1). It is difficult to predict the enzyme behavior when interacting with the support, there are many possible conformations and thus it is variable how these respond to changing environmental conditions.

Free enzymes were thermostable between 15 and 60 °C (Fig. 2). Generally, the optimum temperature of different LiPs depends on fungal species. For example, LiP of *Loweporus lividus* MTCC-1178 showed optimum activity around 24 °C (Yadav et al., 2009a), in contrast, LiP of *P. chrysosporium* showed higher stability at 55 °C (Alam et al., 2009). After immobilization, the LiP from PLO9 showed the highest activity at 60 °C (Fig. 2), similar results were reported by Qiu et al. (2009), however, the immobilized LiP from GRM117 showed the highest activity at 15 °C. Generally the temperature profile of enzymes has been extended after attachment to material surfaces (Talbert and Goddard, 2012). This expansion of the temperature profile enables the enzyme to be used for applications that require the protein to be subjected to conditions not suitable for the free enzyme. In this case, these different enzymatic extracts can be used together or in sequence in order to extend the temperature range that these enzymes can be applied, keeping its activities.

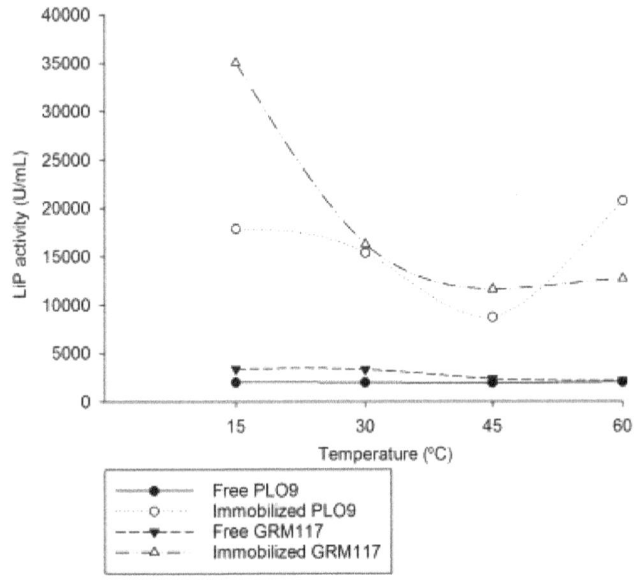

Figure 2: Effect of temperature on LiP activity. LiP from PPEs of *Pleurotus ostreatus* PLO9 and *Ganoderma lucidum* GRM117 (Free PPE from PLO9 = 100 mg mL^{-1}; free PPE from GRM117 = 200 mg mL^{-1}; immobilized PEE = 25 mg mL^{-1}).

Although the immobilized enzymes become more susceptible to temperature changes, their activities of immobilized LiP were higher compared to free enzymes (Fig. 2). Higher activities and a relative thermal stability are attractive and desirable characteristics for industrial applicability of enzymes.

Effect of substrate concentration: determination of Km and Vmax

After immobilization, LiP in the PPE from both fungi showed the Vmax higher than 400 U mL^{-1}and low values of Km, 12 μM for PLO9 and 53 μM for GRM117 (Figure 3 and Figure 4) which were lower values than those observed in free PPEs (Figure 3 and Figure 4). The lower Km and higher Vmax indicate that immobilization favored for higher affinity to the substrate and improved the enzyme catalytic efficiency.

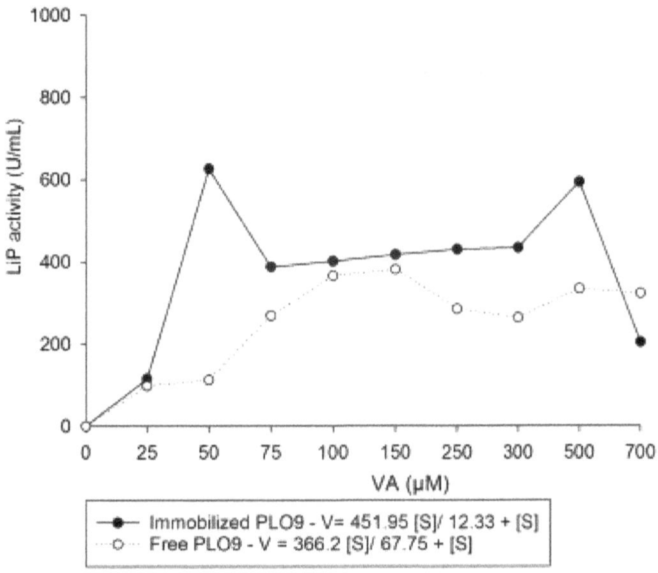

Figure 3: Michalis–Menten kinetic constants (Km) and maximum velocity determination of free and immobilized LiP from PPE of *Pleurotus ostreatus* PLO9. (Free PPE = 10 mg mL^{-1}, Vmax = 366.2, Km = 67.75 μM; immobilized PPE = 25 mg mL^{-1}, Vmax = 451.9, Km = 12.3 μM). VA = Veratryl alcohol.

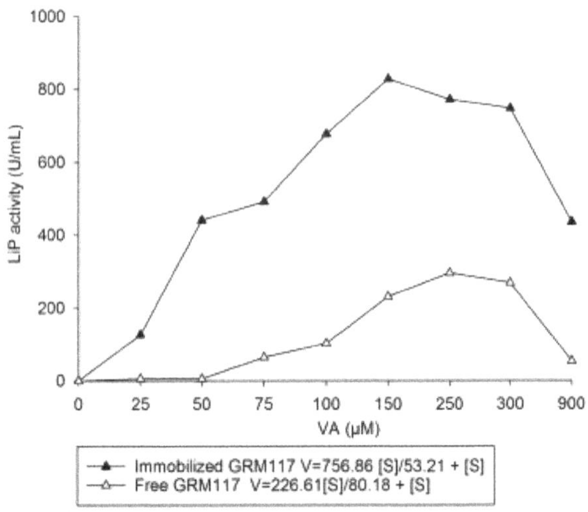

Figure 4: Michalis–Menten kinetic constants (Km) and maximum velocity determination of free and immobilized LiP from the PPE *Ganoderma lucidum* GRM117. (Free PPE = 10 mg mL^{-1}, Vmax = 226.61, Km = 80.18 μM; immobilized PEE = 25 mg mL^{-1}, Vmax = 756.86, Km = 53.21 μM).

We observed two points of higher activities in PPE from PLO9 that can be due to isoenzyme activities (Fig. 3). Ollikka et al. (1993) showed application of three LIP isoenzymes in decolourization of recalcitrant dyes. In relation to the PPE from GRM117 only one higher activity was observed between 150 and 300 µM (Fig. 4). These values were different compared to other studies. For LiP from *Loweporus lividus* MTCC-1178, the Km was 58 mM and 83 mM respectively for veratryl alcohol and hydrogen peroxide substrates (Yadav et al., 2009a and Yadav et al., 2009b). While *Trametes versicolor* LiP showed Km of 56 µM after immobilization in xerogel (Asgher et al., 2012). In turn, the Vmax values (Figure 3 and Figure 4) were higher compared to those found by Yadav et al., 2009a, Yadav et al., 2009b and Asgher et al., 2012. Immobilizing enzymes onto surfaces can limit their performance due to multiple factors including the distortion of native protein configuration, mass transfer limitations and enzyme orientation (Talbert and Goddard, 2012 and Ding et al., 2015). However, the immobilization of enzymatic extract on carbon nanotubes increases the LiP catalytic activity. The increase in activity could be attributed to changes in the collision frequency between the enzyme-immobilized nanoparticles and the substrate. According to the collision theory equations, the apparent Km can be reduced with a decreased size of the carrier, resulting in increased velocity of product formation (Wu et al., 2011). Furthermore, since carbon nanotubes maintain higher radii of curvature due to their smaller diameters, these materials can also promote increased activity of an immobilized enzyme due to decrease of the limiting unfavorable protein-to-protein interactions (Ding et al., 2015).

Dye decolourization

Both PPEs, free and immobilized, were able to decolorize the liquid dye RBBR at the concentrations of 10 mg l^{-1}, the first time that they were used (Fig. 5). The free extracts did not have the ability to decolorize the RBBR at concentrations of 10 mg l^{-1}, in the first reuse round, and 30 mg l^{-1} in second reuse (Fig. 5). Only the immobilized EEP continued to discoloring the dye on reuses. These reuse rounds should be considered since enzymes were kept active over time and hence could contribute to decreasing the costs of effluent treatment. In addition, the decolourization rate was higher than those observed for free enzymes (Fig. 5).

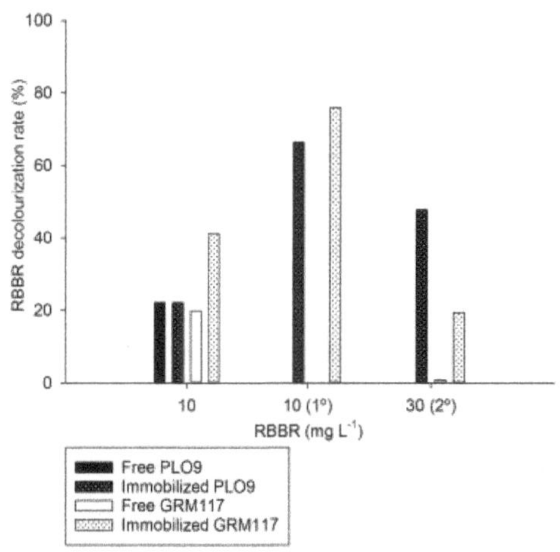

Figure 5: RBBR decolourization rate by free and immobilized PPEs from *Pleurotus ostreatus* PLO9 and*Genoderma lucidum* GRM117. (1°) = first reuse round. (2°) = second reuse round of the same PPE.

The improvement in catalytic efficiency of LiP reflected the ability of immobilized PPEs in decolorizing the RBBR dye (Fig. 5). The immobilization procedure favored, especially, the enzymatic extract from PLO9, regarding the effects of different pH values and temperatures and the catalytic efficiency (Figure 1, Figure 2, Figure 3 and Figure 4). This was evidenced by the decolourization in which the immobilized PPE from PLO9 had a higher percentage of decolourization (over 40%) in the highest concentration analyzed of the dye (Fig. 5). It is well established that fungi show the capacity to decolorize dyes (Zhuo et al., 2011), including the *P. ostreatus* species, which has already reported that its extracellular peroxidases may decolorize the RBBR dye and other groups structurally different, including azos and polymeric dyes (Kim et al., 1996). Furthermore, the immobilized LiP on different supports also showed a higher capability of decolorizing different dyes (Qiu et al., 2009 and Hu et al., 2012).

In the second decolourization experiment, just the commercial lignin peroxidase (LiPc) immobilized on carbon nanotubes could be reused (Fig. 6). In the first measurement, after 1 day of incubation, just the carbon nanotubes were responsible by the color removal (Fig. 6a), once the carbon nanotubes are able to adsorb dyes after determined time of incubation with the dyes (Kuo et al., 2008). In the first reused round, after 2 days of incubation, LiPc immobilized got started to participate in the decolourization at the concentration of 10 and 30 mg l^{-1} (Fig. 6b). In the second reused round, after 8 days of incubation, the immobilized LiPc at the dye concentration of 30 mg l^{-1} showed the similar contribution with carbon nanotubes in the color removal process (Fig. 6c). In the third reused round, after 14 days of incubation, the immobilized LiPc at the concentration of 30 mg l^{-1} showed the decolourization percentage higher than carbon nanotubes (Fig. 6d), indicating that the dye decolourization was mainly caused by the catalytic reaction of the enzyme. This sample evidences that the carbon nanotubes may be saturated by dyes over time, and it could not more be reused and should be discarded contrary, the enzyme keeps the capacity of dye decolourization, it not being spent in the reaction and besides showed a high stability after 14 days of the run experiment.

The free LiPc at the concentration of 30 mg l^{-1} showed higher decolourization percentage than the immobilized LiPc (data no shown). This was expected, once this sample of immobilized LiPc retained just 30% of the free LiP activity (data no shown). However, the free LiP cannot be reused. This case proves why it is promising to use the enzyme conjugated with the carbon nanotube than use each component separately, even if the enzyme has a decrease in its activity after the immobilization process.

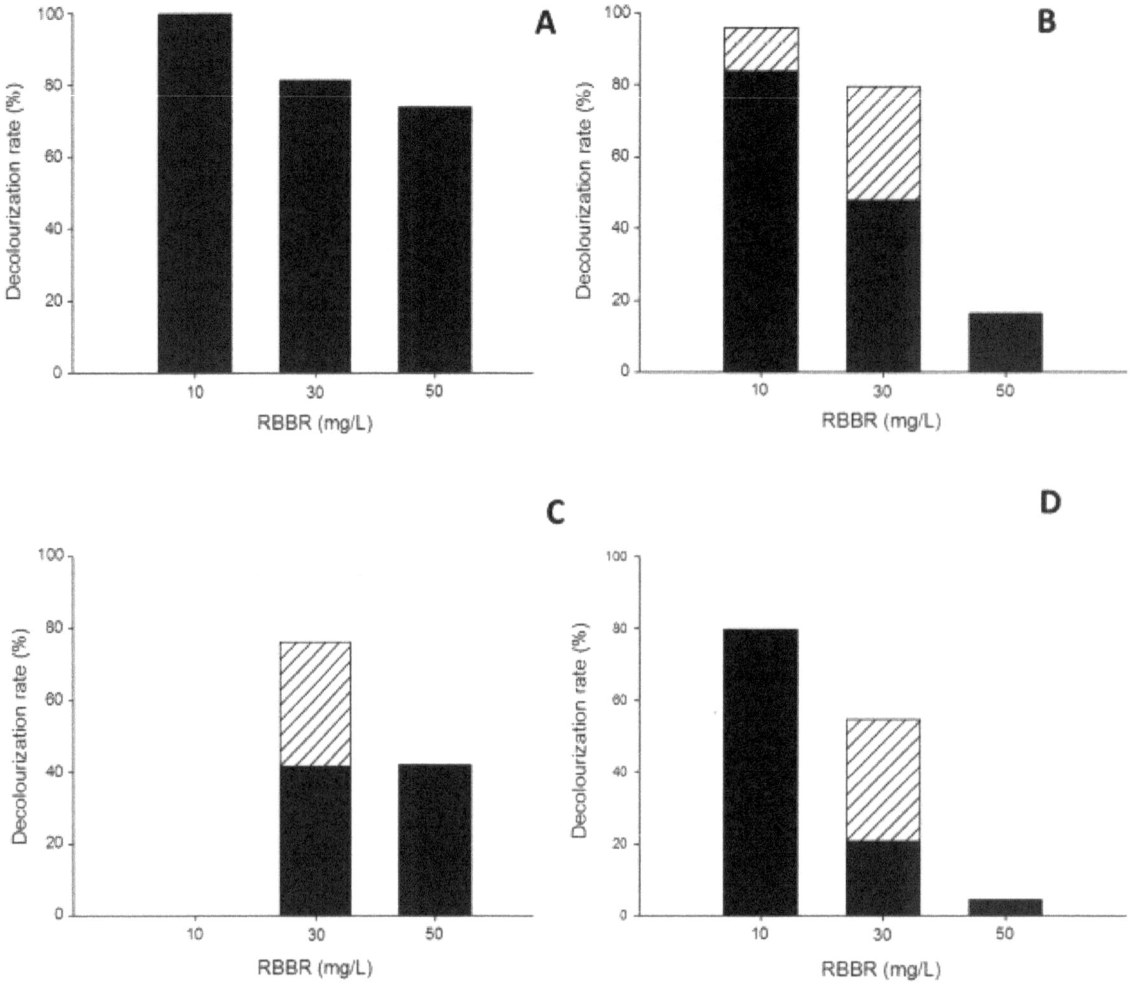

Figure 6: RBBR decolourization rate by commercial lignin peroxidase (LiPc) free and immobilized on carbon nanotubes (hatched columns) and by carbon nanotubes (black columns). (A) First measurement; (B) first reused round; (C) second reused round; (D) third reused round.

CONCLUSION

The fungal isolates and the fermentation conditions studied showed great potential for the ligninolytic enzyme production, especially LiP. Furthermore, the immobilization process of the partially purified enzymatic extracts on carbon nanotubes promoted an increase of LiP catalytic efficiency. Considering the environmental applications as the final goal, the enzymatic extract also can be used without the need for complete purification.

The immobilized enzymatic extract on carbon nanotubes also showed a great potential for the dye decolourization, with a higher catalytic efficiency, stability and capability to be reused in the dye decolourization process. Therefore, these preliminary results may contribute for developing a final biocatalyst that can be applied in diverse environmental applications in the future, especially in the treatment of textile effluents.

ACKNOWLEDGMENTS

The authors are grateful to the Coordenação de Aperfeiçoamento de Pessoal de Nível Superior (Capes), Conselho Nacional de Desenvolvimento Científico e Tecnológico (CNPq) and Fundação de Amparo à Pesquisa do estado de Minas Gerais (FAPEMIG) for financial support.

REFERENCES

1. Alam, M.Z., Mansor, M.F., Jalal, K.C.A., 2009. Optimization of lignin peroxidase production and stability by Phanerochaete chrysosporium using sewage treatment-plant sludge as substrate in a stirred-tank bioreactor. J. Ind. Microbiol. Biotechnol. 36, 757– 764.
2. Asgher, M., Asad, M.J., Bhatti, H.N., Legge, R.L., 2007. Hyperactivation and thermostablization of Phanerochaete chrysosporium lignin peroxidase by immobilization in xerogels. World J. Microbiol. Biotechnol. 23, 525–531.
3. Asgher, M., Iqbal, H.M.N., Irshad, M., 2012. Characterization of purified and Xerogel immobilized novel lignin peroxidase produced from Trametes versicolor IBL-04 using solid state medium of Corncobs. BMC Biotechnol. 12, 46.
4. Ashger, M., Shahid, M., Kamal, S., Iqbal, H.M.N., 2014. Recent trends and valorization of immobilization strategies and ligninolytic enzymes by industrial biotechnology. J. Mol. Catal. B: Enzym. 101, 56–66.
5. Asther, M., Lesage, L., Drapon, R., Corrieu, G., Odier, E., 1988. Phospholipid and fatty acid enrichment of Phanerochaete chrysosporium INA-12 in relation to ligninase production. Appl. Environ. Microbiol. 27, 393–398.
6. Bradford, M.M., 1976. A rapid and sensitive method

for the quantitation of microgram quantities of protein utilizing the principle of protein-dye binding. Anal. Biochem. 72, 248–254.

7. Chequer, F.M.D., Lizier, T.M., de Felı´cio, R., Zanon, M.V., Debonsi, H.M., Lopes, N.P., Marcos, R., de Oliveira, D.P., 2011. Analyses of the genotoxic and mutagenic potential of the products formed after the biotransformation of the azo dye Disperse Red 1. Toxicol. in Vitro 25, 2054–2063.

8. Da Luz, J.M.R., Nunes, M.D., Paes, S.A., Torres, D.P., Kasuya, M.C. M., 2014. Biodetoxification of Jatropha curcas seed cake by Pleurotus ostreatus. Afr. J. Microbiol. Res. 1148, 1156-1156.

9. Ding, S., Cargill, A.A., Medintz, I.L., Claussen, J.C., 2015. Increasing the activity of immobilized enzymes with nanoparticle conjugation. Curr. Opin. Biotechnol. 34, 242–250.

10. Gardossi, L., Poulsen, P.B., Ballesteros, A., Hult, K., Sˇvedas, V.K., Vasic´-Racˇki, D., Carrea, G., Magnusson, A., Schmid, A., Wohlgemuth, R., Halling, P.J., 2010. Guidelines for reporting of biocatalytic reactions. Trends Biotechnol. 28, 171–180.

11. Feng, W., Ji, P., 2011. Enzymes immobilized on carbon nanotubes. Biotechnol. Adv. 29, 889–895.

12. Hu, Z., Xu, L., Wen, X., 2012. Mesoporous silicas synthesis and application for lignin peroxidase immobilization by covalent binding method. J. Environ. Sci. 25, 181–187.

13. Illanes, A., Cauerhffb, A., Wilsona, L., Castro, G.R., 2012. Recent trends in biocatalysis engineering. Bioresour. Technol. 115, 48–57.

14. Iqbal, H.M.N., Asgher, M., Bhatti, H.N., 2011. Optimization of physical and nutritional factors for synthesis of lignin degrading enzymes by a novel strain of Trametes versicolor. Bioresources 6, 1273–1278.

15. Kim, B., Ryu, S., Shin, K., 1996. Effect of culture parameters on the decolorization of remazol brilliant blue R by Pleurotus ostreatus. J. Microbiol. 34, 101–104.

16. Kuo, C., Wu, C., Wu, J., 2008. Adsorption of direct dyes from aqueous solutions by carbon nanotubes: Determination of equilibrium, kinetics and thermodynamics parameters. J. Colloid Interface Sci. 328, 308–315.

17. Liese, A., Hilterhaus, L., 2013. Evaluation of immobilized enzymes for industrial applications. Chem. Soc. Rev. 42, 6236–6249.

18. Marchesan, S., Prato, M., 2015. Under the lens: carbon nanotube and protein interaction at the nanoscale. Chem. Commun. (Camb.) 51, 4347–4359.

19. Oliveira, S.F., Bisker, G., Bakh, N.A., Gibbs, S.L., Landry, M.P., Strano, M.S., 2015. Protein functionalized carbon nanomaterials for biomedical applications. Carbon 95, 767–779.

20. Ollikka, P., Alhonmaki, K., Leppanen, V.-M., Glumoff, T., Raijola, T., Suominen, I., 1993. Decolorization of azo, triphenyl methane, heterocyclic, and polymeric dyes by Lignin peroxidase isoenzymes from Phanerochaete chrysosporium. Appl. Environ. Microb. 59, 4010–4016.

21. Papadoulou, K., Kalagona, I.M., Philippoussis, A., Rigas, F., 2013. Optimization of fungal decolorization of azo and anthraquinone dyes via Box-Behnken design. Int. Biodeterior. Biodegrad. 77, 31– 38.

22. Qiu, H., Li, Y., Ji, G., Zhou, G., Huang, X., Qu, Y., Gao, P., 2009. Immobilization of lignin peroxidase on nanoporous gold: enzymatic properties and in situ release of H2O2 by co-immobilized glucose oxidase. Bioresour. Technol. 100, 3837–3842.

23. Rao, M.A., Scelza, R., Acevedo, F., Diez, M.C., Gianfreda, L., 2014. Enzymes as useful tools for environmental purposes. Chemosphere 107, 145–162.

24. Rauf, M.A., Ashraf, S.S., 2012. Survey of recent trends in biochemically assisted degradation of dyes. Chem. Eng. J. 209, 520–530.

25. Saifuddin, N., Raziah, A.Z., Junizah, A.R., 2013. Carbon nanotubes: a review on structure and their interaction with proteins. J. Chem. 2013, 1–18, ID 676815.

26. Talbert, J.N., Goddard, J.M., 2012. Enzymes on material surfaces. Colloids Surf. B Biointerfaces 93, 8–19.

27. Tien, M., Kirk, T.K., 1983. Lignin-degrading enzyme from the Hymenomycete Phanerochaete chrysosporium Burds. Science 221, 661–663.

28. Wong, D.W.S., 2009. Structure and action mechanism of ligninolytic enzymes. Appl. Biochem. Biotechnol. 157, 174–209.

29. Wu, C., Lee, C., Wu, C., Yang, Y., Ko, F., 2011. Size-modulated catalytic activity of enzyme-nanoparticle conjugates: a combined kinetic and theoretical study. Chem. Commun. 47, 7446–7448.

30. Xu, R., Tang, R., Zhou, Q., Li, F., Zhang, B., 2015. Enhancement of catalytic activity of immobilized laccase for diclofenac biodegradation by carbon nanotubes. Chem. Eng. J. 262, 88–95.

31. Yadav, M., Yadav, P., Yadav, K.D.S., 2009a. Purification and characterization of lignin peroxidase from Loweporus lividus MTCC-1178. Eng. Life Sci. 9, 124–129.

32. Yadav, M., Singh, S.K., Yadav, K.D.S., 2009b. Purification and characterisation of lignin peroxidase from Pleurotus sajor-caju MTCC-141. J. Wood Chem. Technol. 29, 1–15.

33. Zhang, P., Henthorn, D.B., 2010. Synthesis of PEGylated single wall carbon nanotubes by a photoinitiated graft from polymerization. AIChE J. 56, 1610–1615.

34. Zhuo, R., Ma, L., Fan, F., Gong, Y., Wan, X., Jiang, M., Zhang, X., Yang, Y., 2011. Decolorization of different dyes by a newly isolated white-rot fungi strain Ganoderma sp.En3 and cloning and functional analysis of its laccase gene. J. Hazard. Mater. 192, 855–873.

CHAPTER 6

PREPARATION AND CHARACTERIZATION OF K-CARRAGEENASE IMMOBILIZED ONTO MAGNETIC IRON OXIDE NANOPARTICLES

Anfeng Xiao[1,2,4], Caiyun Xu [1] Yan Lin[3] , Hui Ni[1,2,4], Yanbing Zhu[1,2,4], and Huinong Cai[1,2,4]

[1]Department of Microbiology, Universidade Federal de Minas Gerais, Av. Antônio Carlos, 6627 – Pampulha, 31270-901 Belo Horizonte, Minas Gerais, Brazil

[2]Department of Microbiology, Universidade Federal de Viçosa, Av. Peter Henry Rolfs, s/n° – Campus Universitário, Viçosa, Minas Gerais, Brazil

ABSTRACT

Background

Carboxyl-functionalized magnetic nanoparticles were synthesized via chemical co-precipitation method and modified with oleic acid which was oxidized by potassium permanganate, and κ-carrageenase from *Pseudoalteromonas* sp. ASY5 was subsequently immobilized onto them. The immobilization conditions were further optimized, and the characterizations of the immobilized κ-carrageenase were investigated.

Results

The κ-carrageenase was immobilized onto magnetic iron oxide nanoparticles, and the bonding was verified by Fourier transform infrared spectroscopy. The optimal conditions for κ-carrageenase immobilization were 2.5% (w/v) glutaraldehyde, 13.9 U κ-carrageenase for 20 mg of magnetic nanoparticles, a 2-h cross-linking time, and a 2-h immobilization time at 25°C. Under these conditions, the activity of the immobilized enzyme and the enzyme recovery rate were 326.0 U · g^{-1} carriers and 46.9%, respectively. The properties of the immobilized κ-carrageenase were compared with those of the free enzyme. The optimum temperatures of the free and immobilized κ-carrageenase were 60 and 55°C, respectively, and the optimum pH of κ-carrageenase did not change before and after immobilization (pH 7.5). After immobilization, κ-carrageenase exhibited lower thermal stability and improved pH stability, as well as better storage stability. The immobilized κ-carrageenase maintained 43.5% of the original activity after being used 4 times. The kinetic constant value (Km) of κ-carrageenase indicates that the immobilized enzyme had a lower binding affinity for the substrate.

Citation: Anfeng Xiao, Caiyun Xu, Yan Lin, Hui Ni, Yanbing Zhu, Huinong Cai, Preparation and characterization of κ-carrageenase immobilized onto magnetic iron oxide nanoparticles, Electronic Journal of Biotechnology, Volume 19, January 2016, Pages 1-7, ISSN 0717-3458, http://dx.doi.org/10.1016/j.ejbt.2015.10.001.

Conclusions

Under optimal conditions, the activity of the immobilized enzyme and enzyme recovery rate were 326.0 U · g^{-1}·κ-carrageenase-CMNPs and 46.9%, respectively. The thermal, pH, and storage stabilities of κ-carrageenase-CMNPs were relatively higher than those of free κ-carrageenase.

INTRODUCTION

Carrageenans are gel-forming, linear, and sulfated galactans extracted from certain marine red algae and consist of D-galactose residues with alternating α-1,3 and β-1,4 linkages [1]. Recent studies have shown that carrageenan oligosaccharides with small molecular weight possess certain biological activities, such as anti-oxidation [2], anti-viral[3], anti-angiogenic [4], anti-inflammatory [5], and immunomodulation [6], compared with other carrageenans. Therefore, carrageenans are often degraded into oligomers by using acid, active oxygen, microwave, sonication, or carrageenase to improve their bioactivity and application performance [7].

Compared with chemical and physical degradation, enzyme hydrolysis is more likely to produce carrageenan-derivatives with uniform molecular weights, which can be more advantageous because the observed physiological activities of oligo-carrageenans are associated with their molecular weights [8]. However, the industrial application of the enzyme as a biocatalyst is economically unattractive because of its high cost and inconvenience in separation, recycling, and reusing [9]. Enzyme immobilization presents interesting advantages over these problems, including highly concentrated enzymatic activity and high stability and reusability. Tang and Lee [10] immobilized β-glucosidase on κ-carrageenan hybrid matrix, and the immobilized enzyme tolerated broader range of pH values and higher reaction temperature up to 60°C compared with the free β-glucosidase. Bezerra et al. [11] immobilized laccase in green coconut fiber, and the immobilized enzyme retains up to 100% of the initial activity after the assay was reused 10 times. Among the various supports used in immobilization, the magnetic nanoparticles are receiving considerable attention because of their large specific surface area, low mass transfer resistance, less fouling, easy separation from reaction solution with magnetic field, less capital and operating costs, and surface modification with various active groups [12], [13] and [14].

To our knowledge, κ-carrageenase immobilization with magnetic nanoparticles has not been reported yet. In the present study, carboxyl-functionalized magnetic nanoparticles were synthesized via chemical co-precipitation method, and κ-carrageenase was subsequently immobilized onto carboxyl-functioned magnetic iron oxide nanoparticles (CMNPs). The immobilization conditions were further optimized, and the characterizations of the immobilized κ-carrageenase were investigated.

MATERIALS AND METHODS

Materials

$FeCl_3·6H_2O$, $FeCl_2·4H_2O$, NH_4OH, $KMnO_4$, oleic acid, 3,5-dinitrosalicylic acid, and potassium sodium tartrate were of analytical grade and purchased from Sinopharm Chemical Reagent Ltd., Corp. (China). κ-Carrageenan was provided by Greenfresh (Fujian) Foodstuff Co. Ltd.

Microorganism and κ-carrageenase preparation

Pseudoalteromonas carrageenovora CICC 23819, deposited at the China Center of Industrial Culture Collection, was used to produce κ-carrageenase. κ-Carrageenase production was performed by shaking flask fermentation using carrageenan as the sole carbon source. The flasks were incubated at 20°C for 60 h in a shaker at 180 rpm. Culture supernatant as crude enzyme was collected by centrifugation at 9800 × g and 4°C for 10 min.

Synthesis of oleic acid-coated magnetic nano-particles and CMNPs

Oleic acid-coated magnetic nanoparticles were prepared via chemical co-precipitation method as described by Chen et al. [13]. $FeCl_3·6H_2O$ (8.1 g) was dissolved in a flask with 142.5 mL of distilled water and then heated to 70°C. $FeCl_2·4H_2O$ (3.3 g) was dissolved in 7.5 mL of water, and the mixture was subsequently added into the flask. Under rapid stirring condition, 18 mL of ammonia (25%, w/v) was quickly added. After 1 min, 5.3 mL of oleic acid was added dropwise into the flask and maintained at 70°C for 1 h under continuous stirring. The black precipitate was separated using an external magnet and washed several times with alcohol and water to remove the excess oleic acid.

Thereafter, 160 mL of 10 g · L^{-1} $KMnO_4$ solution was added into the oleic acid-coated magnetic nanoparticles obtained earlier. The precipitate was separated using a magnet after 8 h of sonication treatment and washed several times with water to remove excess KMnO4. Powdered CMNPs were obtained after drying at -80°C under vacuum.

Immobilization of κ-carrageenase onto CMNPs

CMNPs (20 mg) were added into 5 mL glutaraldehyde solution (5%, w/v) and incubated at 4°C for 3 h for cross-linking. The magnetic nanoparticles were separated magnetically and washed five times with sodium phosphate buffer (50 mM, pH 7.5) to remove the excess glutaraldehyde. Subsequently, the mixture of the supports obtained above and 1 mL of κ-carrageenase solution were immobilized at 4°C for 3 h. CMNPs bound with κ-carrageenase were collected using an external magnetic field, washed with sodium phosphate buffer (50 mM, pH 7.5), dried at -80°C under vacuum using a freeze dryer, and stored at 4°C for activity assay.

Enzyme activity assay

The activities of free κ-carrageenase and κ-carrageenase-CMNPs were determined using the 3,5-dinitrosalicylic acid (DNS) method. The substrate used was 0.5% (w/v) κ-carrageenan dissolved in sodium phosphate buffer (50 mM, pH 7.5); the resulting solution was uniformly mixed with carrageenase and incubated at 55°C for 20 min. Subsequently, the DNS was added into the reaction solutions and incubated in boiling water for 10 min; the absorbance was measured at 520 nm thereafter. The amount of sugar was determined based on the D-galactose in standard curve. One unit of κ-carrageenase activity is defined as the amount of enzyme that produced 1 μmoL of D-galactose in 1 min at 55°C and pH 7.5.

Enzyme recovery rate was calculated according to the residual enzyme activity after immobilization. Free κ-carrageenase and κ-carrageenase-CMNP activities were assayed, and the enzyme activity recovery rate was calculated based on the following equation:

$$\text{Activity recovery rate } (\%) = ai/af \times 100\% \quad (1)$$

where ai is the activity of the immobilized κ-carrageenase, and af is the activity of the free κ-carrageenase added into immobilization system.

Determination of optimal conditions for κ-carrageenase immobilization

Glutaraldehyde concentration (0%–3.5%, w/v), enzyme dosage (6.95–83.4 U κ-carrageenase mL^{-1}), cross-linking time (0.5–6 h), immobilization time (0.5–6 h), and immobilization temperature (4 and 25°C) were investigated to determine the optimal immobilization conditions of κ-carrageenase. The immobilization efficiency during the immobilization process was evaluated using the activity recovery rate of κ-carrageenase.

Fourier transform infrared spectra (FTIR) of CMNPs and κ-carrageenase-CMNPs

FTIR spectra were obtained on a NEXUS 670 FTIR instrument using KBr discs in the 4000–400 cm^{-1} region at room temperature (25°C).

Properties of the free and immobilized κ-carrageenase

The effect of temperature on the κ-carrageenase activity was determined at various temperature levels from 45°C to 65°C, pH 7.5. The effect of pH on κ-carrageenase activity was assayed at 55°C with pH values that ranged from 4.0 to 10.0 in the following buffers: acetic acid-sodium acetate (50 mM, pH 4.0–5.0), Na_2HPO_4–NaH_2PO_4 (50 mM, pH 5.0–7.0), Tris–HCl (50 mM, pH 7.0–9.0), and glycine–sodium hydroxide (50 mM, pH 9.0–10.0). The maximum activity was considered to be 100% and used as the reference to determine the relative activity at different catalytic reaction conditions.

The thermal stability of κ-carrageenase was evaluated by measuring the residual activity of the free and immobilized enzyme after incubation from 45°C to 60°C without substrate. The residual activities of the free and immobilized enzymes were assayed after incubating κ-carrageenase at 4°C for 24 h in the aforementioned pH buffers to determine pH stability. The residual enzyme activity was defined as the proportion of the final value of enzyme activity to the initial activity (100%).

Reusability and storage stability assays

The reusability of the immobilized κ-carrageenase was evaluated by repeated utilization to catalyze the κ-carrageenan hydrolysis at 55°C and pH 7.5. The activity obtained in each round was compared with the initial activity (defined as 100%) to calculate the relative activity. The storage stability was assessed by storing the free and immobilized κ-carrageenase at 4°C for 30 d.

Kinetic parameters

Kinetic parameters (Km) of free and immobilized κ-carrageenase at optimum temperature were determined using different κ-carrageenan concentrations from 0.3% to 1.0% (w/v) κ-carrageenan at 55°C with pH 7.5.

RESULTS AND DISCUSSIONS

Immobilization of κ–carrageenase

The magnetic nanoparticles obtained using the co-precipitation method exhibited a strong magnetic response and superparamagnetic property. The prepared CMNPs were employed to immobilize κ-carrageenase using glutaraldehyde as the cross-linking agent. The immobilization efficiency and immobilization enzyme were 46.9% and 326.0 U · g^{-1}magnetic nanoparticles, respectively.

FTIR spectroscopy

FTIR determined the structures of CMNPs, κ-carrageenase-CMNPs, and free κ-carrageenase (Fig. 1). The vibrational peaks at 571 cm^{-1} on the CMNPs and κ-carrageenase-CMNPs were attributed to the Fe–O bond, and it was the characteristic absorption band for Fe3O4. Peaks around 2860 and 2925 cm^{-1} on the CMNPs and κ-carrageenase-CMNPs were due to the stretching of –CH2 and C–H in the oleic acid. A band at 1640 cm^{-1} was attributed to the bending vibrations of the –OH groups in water molecules or carboxyl groups [15]. Fig. 1 also shows a large band around 3420 cm^{-1}, probably due to the physical adsorption of water or the presence of NH_2 groups [16]. Peak around 1100 cm^{-1} on the κ-carrageenase-CMNPs and κ-carrageenase were due to the stretching of C–N from peptide bond in protein. However, this peak is not present in CMNPs. The results showed κ-carrageenase were successfully immobilized on CMNPs.

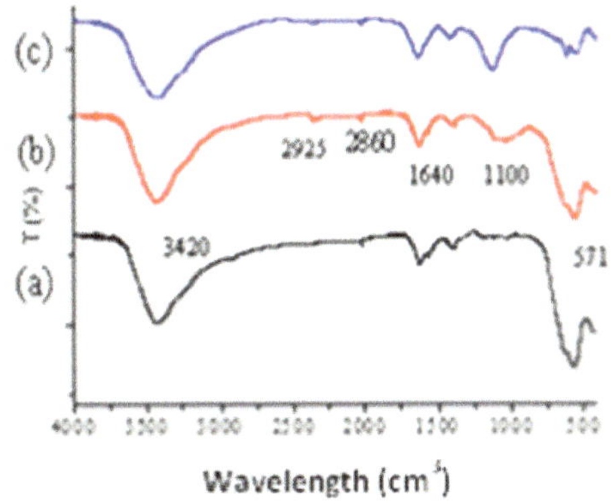

Figure 1: FTIR spectra of CMNPs (a), κ-carrageenase-CMNPs (b), and free κ-carrageenase (c).

Effects of immobilization parameters

Glutaraldehyde as a bifunctional cross-linking agent is generally used in inter- and intra-molecular cross-linking reactions [17]. However, excessive cross-linking may lead to aggregation, precipitation, loss of activity, and distortion of the 3D enzyme structure [18]. Therefore, glutaraldehyde concentration is an important factor for enzyme immobilization. The effects of glutaraldehyde concentration on the immobilization activity and efficiency were investigated at 0%–3.5% (Fig. 2). As a result of the bonding between the aldehyde and amine groups of κ-carrageenase, the highest enzyme activity of 278.6 U · g^{-1} carries was obtained when glutaraldehyde concentration reached 2.5%. A slight decrease in the activity of the immobilized κ-carrageenase was observed above 2.5% glutaraldehyde. This finding is probably due to the excessive glutaraldehyde that may cause unavoidable chemical modification and denaturation effects [19]. Another possible reason for this finding is that the active sites of the enzyme for substrate may be blocked by the carrier and other proteins [20].

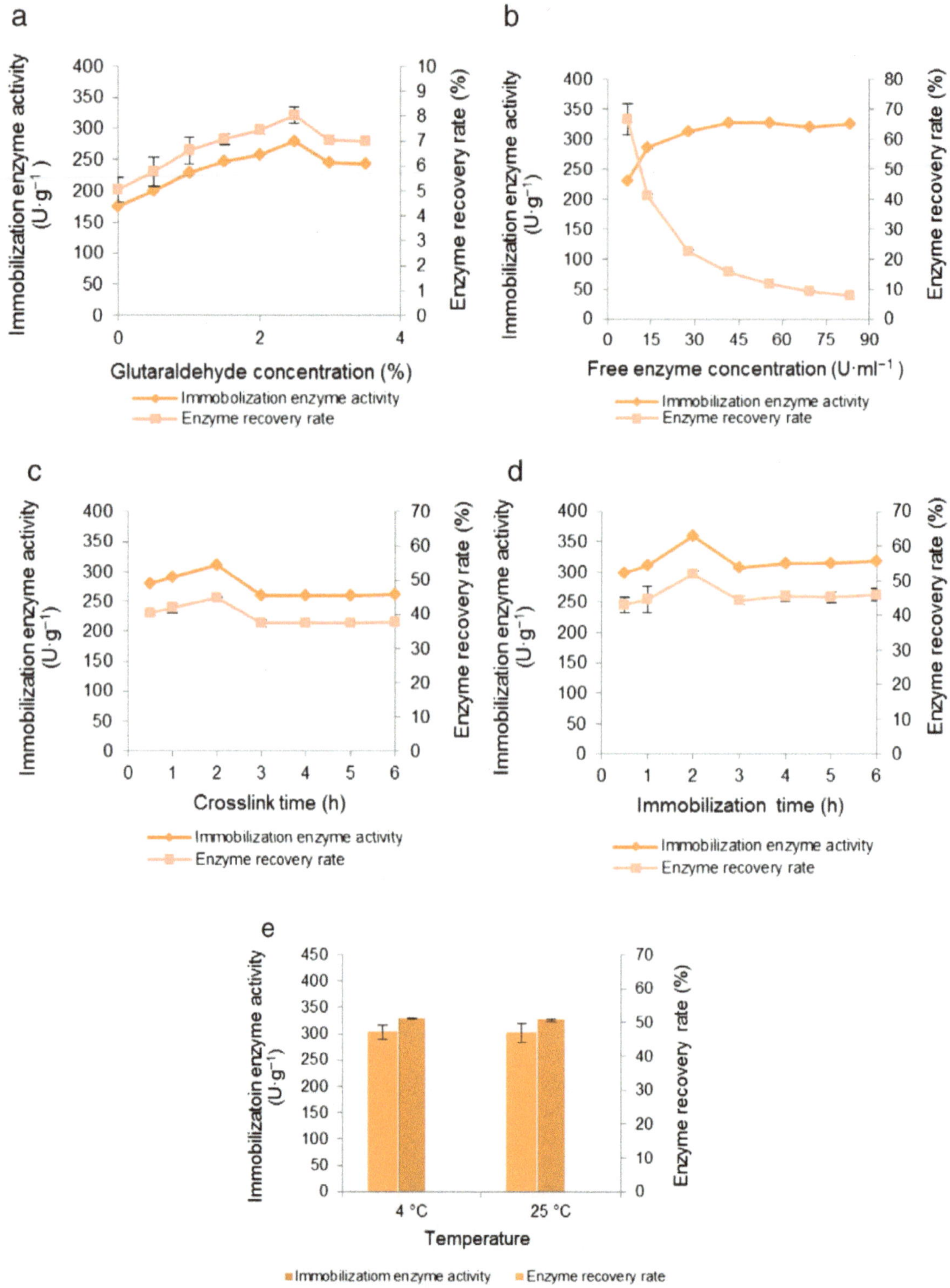

Figure 2: Effects of glutaraldehyde concentration (a), κ-carrageenase concentration (b), cross-linking time (c), immobilization time (d), and immobilization temperature (e) on the activity of the immobilized κ-carrageenase.

The enzyme concentration is an important factor that affects the enzyme recovery rate during the immobilization process. Enzyme solutions (6.95 U · mL^{-1} to 83.4 U · mL^{-1}) were used to analyze the effect of enzyme concentration on κ-carrageenase immobilization. Fig. 2 shows that the immobilized κ-carrageenase activity gradually increased with increasing initial enzyme concentration. When the magnetic nanoparticles reached its saturation adsorption capacity, the activity of the immobilized κ-carrageenase reached the highest value (326.9 U · g^{-1} carries) and still remained stable. In contrast to the trend of the enzyme activity, the enzyme recovery rate rapidly declined with the increasing concentration of κ-carrageenase. The κ-carrageenase concentration of 13.9 U · g^{-1} was selected for 20 mg of magnetic nanoparticles to ensure the enzyme activity and recovery rate.

The influence of cross-linking time on the activity of the immobilized enzyme is shown inFig. 2. The highest enzyme activity recovery rate observed was 44.8% (311.2 U · g^{-1}carries) for 2 h of cross-linking. Less cross-linking time may lead to rest of binding sites on the CNMPs, which can result in low enzyme activity. By contrast, prolonged cross-linking time may lead to a steric hindrance among the enzyme molecules, thereby blocking the active sites over CMNPs and making the binding sites for the substrate unavailable toward the enzyme [21]. Therefore, the cross-linking time was set for 2 h to obtain maximum activity recovery rate.

Immobilization time was also an important factor to determine the amount of enzyme immobilized on the magnetic nanoparticles and may finally affect the activity of the immobilized κ-carrageenase. Fig. 2 shows that the enzyme activity gradually increased to 359.5 U · g^{-1} carries during the first 2 h. However, the enzyme activity decreased to only 318.9 U · g^{-1} carries after 6 h of immobilization because of the denaturation effect. The recovery rate showed a similar tendency to the enzyme activity, reaching its maximum at 2 h.

Temperature is a significant factor that affects the catalytic activity and stability of the enzyme. The effect of temperature on enzyme immobilization was investigated (Fig. 2). The immobilization process at 25°C could obtain activity as high as that at 4°C.

Effects of temperature and pH on the activities of free and immobilized -carrageenase

The optimum temperatures of the activities of free and immobilized κ-carrageenase were investigated by conducting catalytic reactions from 45°C to 70°C. As shown in Fig. 3, the lower optimal temperature of the immobilized κ-carrageenase than the free enzyme was attributed to the changes in physico-chemical properties induced by immobilization. κ-Carrageenase activity did not improve after immobilization because the diffusional limitations increased after covalent attachment of the enzyme to the support matrix[22] and [23]. More than 80% activity of the immobilized enzyme was maintained at 50 to 65°C. However, the activity of the free enzyme was sharply changed within the same temperature range (Fig. 3). The immobilized κ-carrageenase exhibited significantly lower activity stability than free κ-carrageenase (Fig. 3). This phenomenon could be explained by the fact that the restricted conformational mobility of the molecules could be obtained after immobilization [18].

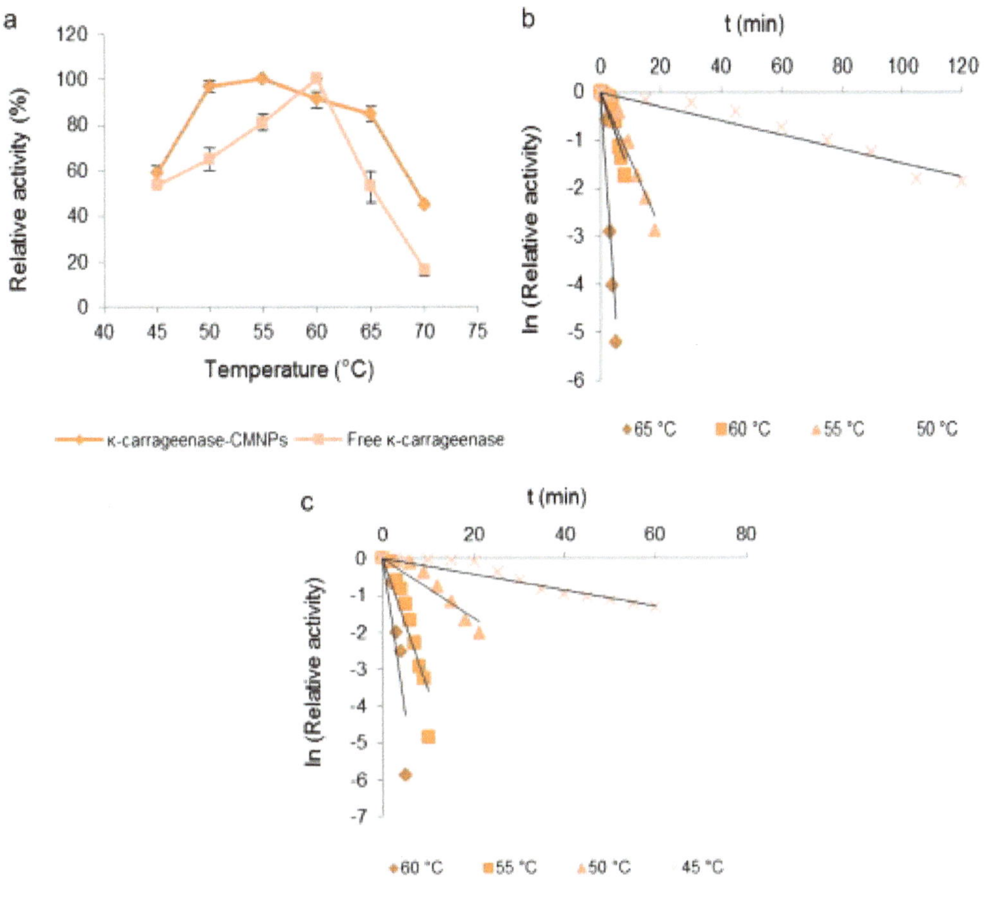

Figure 3: Optimum temperature (a) and temperature stability (b, c) of the free (hollow icon) and immobilized (solid icon) κ-carrageenase. (Note: the actual value of 100% relative activity for immobilized and free enzyme was 410 U · g^{-1} carries and 11.3 U · mL^{-1}, respectively).

The effect of pH on κ-carrageenase activity was determined under varied pH values (pH 4.0–10.0) in different buffers, and the results are presented in Fig. 4. The optimum pH value for the free and immobilized κ-carrageenase activities was 7.5. This phenomenon is also observed in another research [17]. The optimum pH of acetyl xylan esterase immobilization in the study of Saravanakumar et al. [16] is identical to that of the free enzyme. More than 80% activity was also observed in the pH range of 5.0–10.0 for the immobilized κ-carrageenase (Fig. 4), and the pH stability of the immobilized κ-carrageenase was apparently better than that of the free κ-carrageenase. This finding can be attributed to the enzyme protection by the microenvironment structure and properties formed by the CMNPs. Moreover, the microenvironment may have been buffered, thus the immobilized enzyme was less affected by the pH values [22].

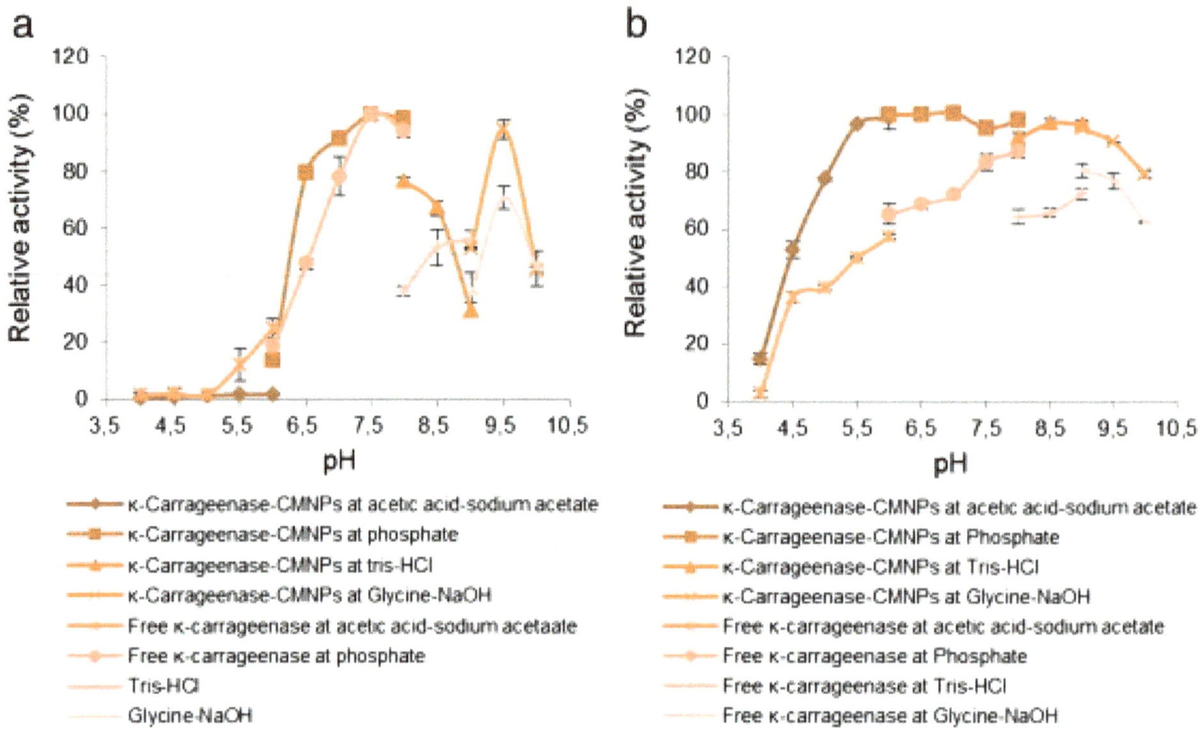

Figure 4: Optimum pH (a) and pH stability (b) of the free and immobilized κ-carrageenase. (Note: the actual value of 100% relative activity for immobilized and free enzyme was 410 U · g^{-1} carries and 11.3 U · mL^{-1}, respectively).

Reusability and storage stability

Reusability of the immobilized enzymes is an important parameter for industrial applications; immobilized enzymes lessen production cost because of their repeated, continuous, or batch uses. Baskar et al. [24] reported that α-amylase immobilized on a magnetic support exhibits 75% residual activity after five reactions. In the present study, the residual activities for κ-carrageenase immobilized on magnetic nanoparticles were 43.5% after four cycles, and the activity gradually decreased in the successive cycles (Fig. 5).

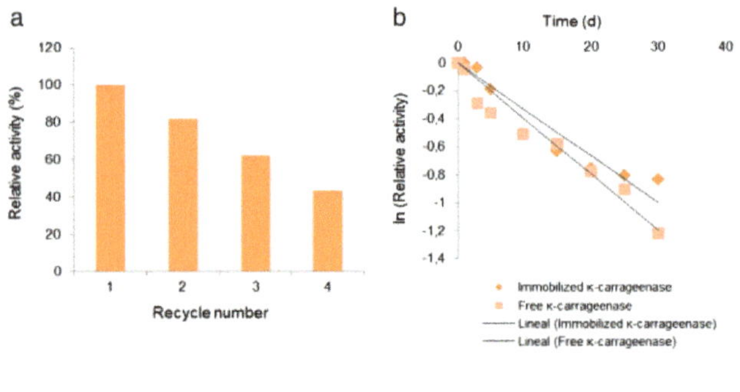

Figure 5: Reusability of the immobilized enzyme (a) and storage stability at 4°C. (Note: the actual value of 100% relative activity for immobilized and free enzyme was 410 U · g^{-1} carries and 11.3 U · mL^{-1}, respectively).

Enzyme activity is linked to the stability of its structure, and changes in this structure tend to decrease enzyme catalytic activity. As shown in Fig. 5, the half-life of free κ-carrageenase and κ-carrageenase-CMNPs at 4°C were 17.4 and 20.6 d, respectively. This result may be explained by the structure of immobilized κ-carrageenase that was stabilized as a result of the flexibility reduction of free κ-carrageenase. Therefore, multipoint covalent attachment prevents enzyme conformational changes induced by any distorting conditions [22].

Kinetic parameters

The Michaelis–Menten kinetics of the free and immobilized κ-carrageenase was investigated using a series of κ-carrageenan concentrations. The Michaelis constant (Km) was evaluated from double-reciprocal plots (Fig. 6). The calculated Km value of the immobilized κ-carrageenase was 22.5 mg · mL^{-1}, which was higher than that of free κ-carrageenase (2.28 mg · mL^{-1}). As previously reported [25] and [26], the increase in Km value indicated that the κ-carrageenase immobilized on CMNPs had a lower binding affinity for the substrate. This finding may be due to limitation on the substrate mass transfer through the CMNPs or the steric effect that arises from the structural rigidity of the entire enzyme structure, which was distorted after covalent immobilization. Decrease of affinity to the immobilized enzyme is caused by an effect of enzymatic orientation [20].

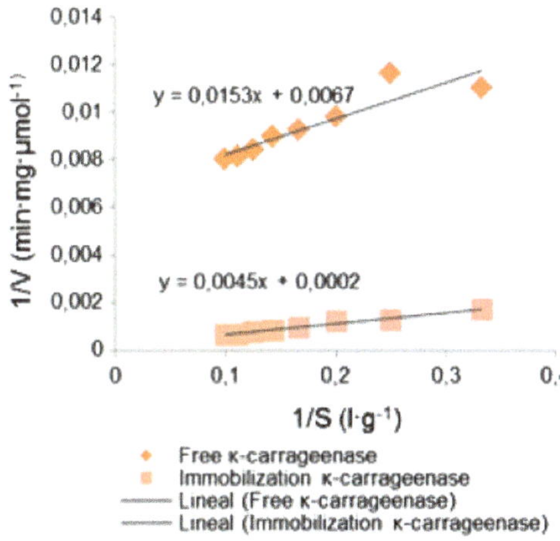

Figure 6: Lineweaver–Burk plot of free and immobilized κ-carrageenase.

CONCLUSIONS

In the present work, CMNPs were synthesized by chemical co-precipitation method and successfully immobilized with κ-carrageenase obtained from *P. carrageenovora* CICC 23819. FTIR analysis identified the bonding between κ-carrageenase and CMNPs. The optimal conditions for enzyme immobilization were 2.5% (w/v) glutaraldehyde, 13.9 U κ-carrageenase for 20 mg of magnetic nanoparticles, a 2-h cross-linking time, and a 2-h immobilization time at 25°C. Under these optimal conditions, the activity of the immobilized enzyme and enzyme recovery rate were 326.0 U · g^{-1} κ-carrageenase-CMNPs and 46.9%, respectively. Although the κ-carrageenase-CMNPs exhibited lower affinity after the immobilization process, the thermal, pH, and storage stabilities of κ-carrageenase-CMNPs were relatively higher than those of free κ-carrageenase.

CONFLICT OF INTEREST

The authors declare no conflict of interest.

FINANCIAL SUPPORT

Science and Technology Program of Xiamen City Foundation (3502Z20120005), Xiamen Southern Ocean Technology Center of China Foundation (13GZP004NF10).

REFERENCES

1. Michel G, Nyval-Collen P, Barbeyron T, Czjzek M, Helbert W. Bioconversion of red seaweed galactans: A focus on bacterial agarases and carrageenases. Appl Microbiol Biotechnol 2006;71:23–33. http://dx.doi.org/10.1007/s00253-006-0377-7.
2. Yuan H, Song J, Zhang W, Li X, Li N, Gao X. Antioxidant activity and cytoprotective effect of κ-carrageenan oligosaccharides and their different derivatives. Bioorg Med Chem Lett 2006;16:1329–34. http://dx.doi.org/10.1016/j.bmcl.2005.11.057.
3. Wang W, Zhang P, Yu GL, Li CX, Hao C, Qi X, et al. Preparation and anti-influenza a virus activity of κ-carrageenan oligosaccharide and its sulphated derivatives. Food Chem 2012;133:880–8. http://dx.doi.org/10.1016/j.foodchem.2012.01.108.
4. Chen H, Yan X, Lin J, Wang F, Xu W. Depolymerized products of λ-carrageenan as a potent angiogenesis inhibitor. J Agric Food Chem 2007;55:6910–7. http://dx.doi.org/10.1021/jf070183.
5. Kalitnik AA, Marcov PA, Anastyuk SD, Barabanova AOB, Glazunov VP, Popov SV, et al. Gelling polysaccharide from Chondrus armatus and its oligosaccharides: The structural peculiarities and anti-inflammatory activity. Carbohydr Polym 2015; 115:768–75. http://dx.doi.org/10.1016/j.carbpol.2014.04.070.
6. Xu L, Yao Z, Wu H, Wang F, Zhang S. The immune regulation of κ-carrageenan oligosaccharide and its desulfated derivatives on LPS-activated microglial cells. Neurochem Int 2012;61:689–96. http://dx.doi.org/10.1016/j.neuint.2012.06.019.
7. Yao Z, Wu H, Zhang S, Du Y. Enzymatic preparation of κ-carrageenan oligosaccharides and their anti-angiogenic activity. Carbohydr Polym 2014;101:359–67. http://dx.doi.org/10.1016/j.carbpol.2013.09.055.
8. Tayco CC, Tablizo FA, Regalia RS, Lluisma AO. Characterization of a κ-carrageenaseproducing marine bacterium, isolate ALAB-001. Philipp J Sci 2013;142:45–54.
9. Kanimozhi S, Perinbam K. Synthesis of amino-silane modified superparamagnetic Fe3O4 nanoparticles and its application in immobilization of lipase from Pseudomonas fluorescens Lp1. Mater Res Bull 2013;48:1830–6. http://dx.doi.org/10.1016/j.materresbull.2013.01.024.
10. Tang IS, Lee KT. Immobilization of β-glucosidase from Aspergillus niger on κ-carrageenan hybrid matrix and its application on the production of reducing sugar from macroalgae cellulosic residue. Bioresour Technol 2015;184:386–94. http://dx.doi.org/10.1016/j.biortech.2014.10.146.
11. Bezerra TMS, Bassan JC, Santos VTO, Ferraz A, Monti R. Covalent immobilization of laccase in green coconut fiber and use in clarification of apple juice. Process Biochem 2015;50:417–23. http://dx.doi.org/10.1016/j.procbio.2014.12.009.
12. Mosafa L, Moghadam M, Shahedi M. Papain enzyme supported on magnetic nanoparticles: Preparation, characterization and application in the fruit juice clarification. Chin J Catal 2013;34:1897–904. http://dx.doi.org/10.1016/S1872-2067(12)60663-9.
13. Chen G, Ma Y, Su P, Fang B. Direct binding glucoamylase onto carboxylfunctioned magnetic nanoparticles. Biochem Eng J 2012;67:120–5. http://dx.doi.org/10.1016/j.bej.2012.06.002.
14. Cui R, Bai C, Jiang Y, Hu M, Li S, Zhai Q.Well-defined bioarchitecture for immobilization of chloroperoxidase on magnetic nanoparticles and its application in dye decolorization. Chem Eng J 2015;259:640–6. http://dx.doi.org/10.1016/j.cej.2014.08.074.
15. Andjelković U, Milutinović-Nikolić A, Jović-Jovičić N, Banković P, Bajt T, Mojović Z, et al. Efficient stabilization of Saccharomyces cerevisiae external

invertase by immobilisation on modified beidellite nanoclays. Food Chem 2015;168:262–9. http://dx.doi.org/10.1016/j.foodchem.2014.07.055.

16. Saravanakumar T, Palvannan T, Kim DH, Park SM. Optimized immobilization of peracetic acid producing recombinant acetyl xylan esterase on chitosan coated-Fe3O4 magnetic nanoparticles. Process Biochem 2014;49:1920–8. http://dx.doi.org/10.1016/j.procbio.2014.08.008.

17. Tükel SS, Hürrem F, Yildirim D, Alptekin Ö. Preparation of crosslinked enzyme aggregates (CLEA) of catalase and its characterization. J Mol Catal B: Enzym 2013; 97:252–7. http://dx.doi.org/10.1016/j.molcatb.2013.09.007.

18. Wang B, Cheng F, Lu Y, Ge W, Zhang M, Yue B. Immobilization of pectinase from Penicillium oxalicum F67 onto magnetic cornstarch microspheres: Characterization and application in juice production. J Mol Catal B: Enzym 2013;97:137–43. http://dx.doi.org/10.1016/j.molcatb.2013.07.018.

19. Yavuz H, Bayramoğlu G, Kaçar Y, Denizli A, Yakup Arıca M. Congo Red attached monosize poly(HEMA-co-MMA) microspheres for use in reversible enzyme immobilisation. Biochem Eng J 2012;10:1–8. http://dx.doi.org/10.1016/S1369-703X(01)00146-2.

20. Matsushima K, Kaneda H, Harada K, Matsuura H, Hirata K. Immobilization of enzymatic extracts of Portulaca oleracea cv. roots for oxidizing aqueous bisphenol A. Biotechnol Lett 2015;37:1037–42. http://dx.doi.org/10.1007/s10529-014-1761-5.

21. Seenuvasan M, Malar CG, Preethi S, Balaji N, Iyyappan J, Kumar MA, et al. Fabrication, characterization and application of pectin degrading Fe3O4–SiO2 nanobiocatalyst. Mater Sci Eng C 2013;33:2273–9. http://dx.doi.org/10.1016/j.msec.2013.01.050.

22. Rodrigues RC, Ortiz C, Berenguer-Murcia Á, Torres R, Fernández-Lafuente R. Modifying enzyme activity and selectivity by immobilization. Chem Soc Rev 2013;42:6290–307. http://dx.doi.org/10.1039/c2cs35231a.

23. Guzik U, Hupert-Kocurek K, Marchlewicz A, Wojcieszynska D. Enhancement of biodegradation potential of catechol 1,2-dioxygenase through its immobilization in calcium alginate gel. Electron J Biomed 2014;17:83–8. http://dx.doi.org/10.1016/j.ejbt.2014.02.001.

24. Baskar G, Afrin Banu N, Helan Leuca G, Gayathri V, Jeyashree N. Magnetic immobilization and characterization of α-amylase as nanobiocatalyst for hydrolysis of sweet potato starch. Biochem Eng J 2015;102:18–23. http://dx.doi.org/10.1016/j.bej.2015.02.020.

25. Zhang L, Zhu X, Zheng S, Sun H. Photochemical preparation of magnetic chitosan beads for immobilization of pullulanase. Biochem Eng J 2009;46:83–7. http://dx.doi.org/10.1016/j.bej.2009.04.024.

26. Zhou Y, Pan S, Wei X, Wang L, Liu Y. Immobilization of β-glucosidase onto magnetic nanoparticles and evaluation of the enzymatic properties. BioResources 2013;8: 2605–19. http://dx.doi.org/10.15376/biores.8.2.2605-2619.

CHAPTER 7

HIGH OPERATIONAL STABILITY OF INVERTASE FROM SACCHAROMYCES CEREVISIAE IMMOBILIZED ON CHITOSAN NANOPARTICLES

Sheila G. Valerio[1], Joana S. Alves [1], Manuela P. Klein[1], Rafael C. Rodrigues [2], Plinho F. Hertz [1]

[1]Enzymology Lab, Institute of Food Science and Technology, Federal University of Rio Grande do Sul, Av. Bento Gonc¸ alves 9500, ZC 91501-970 Porto Alegre, RS, Brazil

[2]Biocatalysis and Enzyme Technology Lab, Institute of Food Science and Technology, Federal University of Rio Grande do Sul, Av. Bento Gonc¸ alves 9500, ZC 91501-970 Porto Alegre, RS, Brazil

ABSTRACT

Invertase (E.C.3.2.1.26) from *Saccharomyces cerevisiae* was covalently immobilized on chitosan nanoparticles prepared by the ionotropic gelation method and activated with glutaraldehyde. The support was characterized and it was studied its load capacity, the influence of the presence of substrate during immobilization, and determined the biocatalyst kinetic parameters and stabilities. The light scattering analysis (LSA) and transmission electron microscopy (TEM) techniques indicated a mixture of chitosan nano and aggregated nanoparticles, providing high superficial area for enzyme immobilization. The thermal and storage stabilities, the optimal pH and temperature of the enzyme were not altered. *Km* increased 3-fold, while V_{max} remained unaltered. The immobilized biocatalyst was reused for 59 batches with maximal invertase activity, the highest operational stability so far described in the literature. These results fulfill some important aspects for the enzyme immobilization: the simplicity of the protocols, the conservation of the enzyme activity, and the high operational stability.

INTRODUCTION

Invertase or β-d-fructofuranosidase (E.C. 3.2.1.26) from *Saccharomyces cerevisiae* is an enzyme that has major specificity for sucrose hydrolysis. This reaction results in an equimolar mixture of α-d-glucose and β-d-fructose, known as invert syrup. As a consequence, the crystallization phenomenon is avoided and, as fructose is sweeter than sucrose, the sweetness of the product is higher. The prevention of crystallization justifies the main food industrial application of the enzyme in the manufacture of fillings of sweets, keeping the softness. Also, invertase can be applied in fermentation process, when sucrose is the substrate, in the manufacture of artificial honey, humectant agent for candies production, besides other applications as in cosmetic, paper and drug industry (Kotwal & Shankar, 2009).

The solubility of free enzymes makes their uses for large-scale relatively costly since in a mixture containing the substrate, products and other components, their recoveries are difficult, being economically unattractive (Kotwal & Shankar, 2009). Nevertheless, biocatalysts are increasingly being

Citation: Sheila G. Valerio, Joana S. Alves, Manuela P. Klein, Rafael C. Rodrigues, Plinho F. Hertz, High operational stability of invertase from Saccharomyces cerevisiae immobilized on chitosan nanoparticles, Carbohydrate Polymers, Volume 92, Issue 1, 30 January 2013, Pages 462-468, ISSN 0144-8617, http://dx.doi.org/10.1016/j.carbpol.2012.09.001.

employed because of their high selectivity and potential as a greener alternative to chemical catalysts (Polizzi, Bommarius, Broering, & Chaparro-Riggers, 2007), which could result in the formation of undesirable color and flavoring agents.

The interest for enzymatic processes is over growing, which also reflects the great interest for biocatalysts immobilization. Taking into account distinct applications, there are diverse immobilization techniques with different methods of linkages among enzymes and supports. It is noteworthy that each protein and support has typical characteristics, so there is no universal ideal method of immobilization. For example, when the support for the immobilization is solid and non-porous, the size of the particles needs to be as small as possible, considering that nanoparticles provide a reasonable enzyme load capacity (Garcia-Galan, Berenguer-Murcia, Fernandez-Lafuente, & Rodrigues, 2011). In general, the enzyme immobilization aims at keeping or increasing storage, temperature, pH, and operational stabilities.

Chitosan has been used as a support for enzyme immobilization (Klein et al., 2012, Kuo et al., 2012, Li et al., 2012, Muzzarelli, 1980 and Orrego et al., 2010). Chitosan, isolated from chitin, is the linear and partly acetylated (1–4)-2-amino-2-deoxy-β-d-glucan (Muzzarelli, 1977, Muzzarelli, 2012 and Muzzarelli et al., 2012). Certain chitosan salts are soluble in water, thus they form gels and polyelectrolyte complexes, in particular with proteins (Krajewska, 2004). Invertase was encapsulated in chitosan microbeads (Siso et al., 1997), and immobilized covalently via carbohydrate moiety (Hsieh, Liu, & Liao, 2000). The enzyme was also modified with chitosan and immobilized in sodium alginate-coated chitin support, and in hyaluronic-acid-coated chitin support (Gomez et al., 2008 and Gómez et al., 2006). Chitosan was also prepared as films and utilized for the biocatalyst immobilization (Teodor, Radu, Dan, & Stanciu, 2006). Biró, Németh, Sisak, Feczkó, and Gyenis (2008) described protocols for the preparation of chitosan support suitable for biocatalysts immobilization, applying to β-galactosidase. The ionotropic gelation method proposed for nanoparticles production, modified from the one described by Berthold, Cremer, and Kreuter (1996), showed the highest enzyme activity, serving as model for the chitosan preparation described in the present study. Recently, it was studied the effect of support size for enzyme immobilization on chitosan (Klein et al., 2012). Macro and nanoparticles of chitosan were prepared and characterized for β-galactosidase immobilization. The authors found a high operational stability for lactose hydrolysis for both preparations.

Thus, the objective of this work was to immobilize the invertase from *S. cerevisiae* on chitosan nanoparticles. The evaluation of the enzyme properties and thermal and operational stability was also studied.

MATERIALS AND METHODS

Materials

Invertase from *S. cerevisiae* (Maxinvert L 10000, batch 409200451) was kindly donated by DSM Food Specialties (The Netherlands). Chitosan from shrimp shells (>75% deacetylated) was purchased from Sigma–Aldrich (St. Louis, USA). Sucrose and all other chemicals (acetic acid, dinitrosalicylic acid, ethylene glycol, glutaraldehyde, magnesium chloride, polysorbate 80, potassium phosphate, sodium and potassium tartrate, sodium chloride, sodium hydroxide and sodium sulfate) were of analytical grade and purchased either from Merck or Fluka (São Paulo, Brazil). All the following tests were performed in duplicate.

Preparation of chitosan particles

Chitosan support was prepared by ionotropic gelation according to a procedure already described (Berthold et al., 1996) with some modifications: 0.5 mL of sodium sulfate aqueous solution (1.4 M) were added dropwise into 9.5 mL of chitosan (0.25%, w/v) dissolved in 0.35 M acetic acid containing Tween 80 (1%, v/v) under sonication (30 min, 40 kHz, 25 °C). This suspension was magnetically stirred for 2 h (500 rpm) and the particles formed were collected by centrifugation (3500 × *g*, 15 min, 4 °C). The particles obtained were washed with distilled water and activated by suspending them in 10 mL of glutaraldehyde (1.25%, v/v) in 0.1 M phosphate–potassium buffer (pH 7.0) under agitation during 30 min. The glutaraldehyde excess was removed with successive washings using the same buffer.

Dry weight and support size determination

Support dry weight was performed by lyophilization, after activation with glutaraldehyde. The further immobilization results are expressed as units of enzyme activity by dry mass of support.

Two different techniques were used for support size determination, considering the procedures involved in the preparation of the samples for analysis and the sensitivity of each equipment. Results of the mean particle size before activation were performed using filtered sample (membrane of 45 μm of diameter) in light scattering analysis (LSA) on a Brookhaven Instruments standard setup (BI-200M goniometer, BI-9000AT digital correlator) with a He–Ne laser (k = 632.8 nm) as light source. Using a JEOL JEM 1200ExII transmission electron microscope (JEOL, Tokyo) operating at 120 kV, the transmission electron microscopy (TEM) was done for attainment morphological images of chitosan particles before and after activation. The support was suspended in distilled water and after homogenized with ultrasonic bath for 5 min, the samples were negatively stained with uranyl acetate solution (0.047 M). At room temperature the samples were air-dried.

Invertase covalent immobilization

Covalent immobilization of invertase on chitosan beads was carried out incubating 0.05 g (dry weight) of the chitosan activated support with 10 mL of each enzyme solution in acetate buffer (0.1 M; pH 4.5) overnight under gentle stirring, at room temperature. Then, the immobilized enzymes were centrifuged (15 min, 4 °C, 3000 × *g*) and washed with solutions of acetate buffer (0.1 M; pH 4.5), sodium chloride (1 M), ethylene glycol (5.38 M), and again with acetate buffer, until activity was no longer detected in the washing solutions. The immobilization yields were calculated using Eq. (1):

$$IY = \frac{U_i - (U_3 + U_w)}{U_i} 100 \quad (1)$$

where IY is the immobilization yield, *Ui* is the initial enzyme activity in the solution, *Us*the activity in the supernatant after immobilization, and *Uw* the activity in the washing solutions.

The efficiency of immobilization was defined as:

$$Ef = \frac{U_i}{IY} 100 \quad (2)$$

where Ef is the immobilization efficiency, *Ui* the activity in the immobilized enzyme, and IY the immobilization yield.

Except for free enzyme, all aliquots were submitted to the washing solutions.

Enzymatic activity assay

Based on a described method (Bryjak, Liesiene, & Stefuca, 2008), activities of free and immobilized enzyme were assayed by the addition of invertase suspension (1 mL of final volume) in acetate buffer (0.1 M; pH 4.5) to 2 mL of sucrose (8% in the same buffer). The samples were incubated during 5 min at 55 °C in a water bath (with agitation for the immobilized derivatives), and ice bath (5 min) was used to stop the reaction. An aliquot of 100 µL was withdrawn and analyzed with 1 mL of DNS reagent by DNS method (Miller, 1959), in order to quantify released reducing sugars by the sucrose hydrolysis. The absorbance was measured at 540 nm and the results were calculated using glucose as standard. The enzyme activity unit (U) was defined as the amount of enzyme liberating 1 µmol of reducing sugars per minute under the assay conditions. All samples were analyzed in duplicate.

Determination of optimal pH, temperature and kinetic parameters

For the determination of optimal pH and temperature of free and immobilized invertase, each parameter was individually changed (pH from 3.0 to 7.0; temperature from 40 °C to 70 °C). The buffers concentrations were 0.1 M, and except for pH 3 (citrate buffer) and for pH 7 (phosphate buffer), the others measurements were performed with acetate buffer.

The Michaelis–Menten constant (*Km*) and V_{max} were calculated under optimal conditions, using Lineweaver–Burk plot.

Thermal stability

The thermal stability of free and immobilized invertase was measured at 55 °C and 65 °C. Samples were collected periodically until the residual activity was 10%. Each sample was immersed in ice bath to stop the temperature damage and analyzed by previous enzymatic activity assay also being determined the half-life time ($t_{1/2}$).

Storage stability

Free and immobilized enzymes were stored at 8 °C in acetate buffer (0.1 M; pH 4.5), and their stability was measured once a week, during 25 weeks.

Operational stability

The operational stability of the immobilized enzyme was obtained by sucrose hydrolysis at 55 °C. The batches run up during 30 min, under gentle stirring, using 20 mL of substrate (8% in acetate buffer) and 0.05 g of support (200,000 U/g support). At the end of each batch, homogenous samples were taken, centrifuged (15 min, room temperature, 3000 × *g*), and the supernatant obtained utilized for reducing sugars determination. The precipitated pellet was recovered and used in a new batch (support washings among batches were not necessary).

RESULTS AND DISCUSSION

Support size determination

Light scattering analysis (LSA) indicated that the mean particle size of non-activated support was 410.9 nm, with a low polydispersity of 0.266 (parameter related to the particle size homogeneity, being higher as unequal is the sample). In the absence of stabilizing agents the nanoparticles form unstable systems with a tendency to aggregation. These events are addressed in a recent review (Zhang et al., 2011), in which are discussed various aspects of the production of nanoparticles providing particular reference to their influence on stability and size. Fig. 1 shows the transmission electron microscopy (TEM) pictures of the support not activated, indicating aggregated nanoparticles even before the utilization of glutaraldehyde. The TEM results after cross-linking (data not shown) indicate more particle aggregation phenomena, also showing the support with uneven surfaces and part of the samples reaching a micro size (Klein et al., 2012). The formation of larger support particles occurs because since the glutaraldehyde is a bifunctional agent, through its aldehyde groups it can crosslink two chitosan units from different polymeric chains (Monteiro & Airoldi, 1999). In a study for trypsin immobilization on linolenic acid modified chitosan, increasing the glutaraldehyde concentration for crosslinking, the particle size increased from 523 to 1372 nm (Liu, Desai, Chen, & Park, 2005).

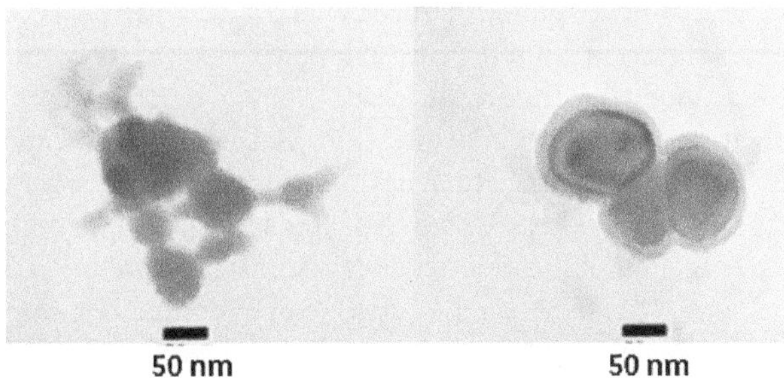

Figure 1: Transmission electron microscopy of non-activated chitosan particles.

LSA after activation (data not shown) without sample filtration presented higher values for polydispersity, showing a wide distribution of the aggregated particles, where it is not possible to determine the mean particle size formed. By other side, performing LSA after activation with the filtration step, considerable part of the support was retained. Thus, because of these factors this technique was used only before cross-linking.

Each additional unit operation for the support preparation represents an additional cost, being less attractive for use in large-scale process. In order to minimize the number of steps for support preparation and the possible material losses inherent in each step, in this work filtration step for particle size classification was not performed. So, it can be concluded that the invertase was immobilized on a chitosan support constituted by nano and aggregated nanoparticles, which have high superficial area for enzyme immobilization.

Support load capacity

Different enzyme solutions were prepared diluting Maxinvert L 10000, for the determination of the support load capacity. Immobilization yields and efficiencies for each concentration tests are presented in Table 1. Activity detected in the support increased with invertase concentrations, although not proportionally.

Table 1: Enzyme solutions for estimating support load capacity.

Sample	Offered free enzyme (U/g support)	Immobilized enzyme (U/g support)	Immobilization yield (%)	Immobilization efficiency (%)
1	10,400	9200	91.8	96.36
2	45,800	45,800	100	100
3	140,600	119,200	90.3	93.89
4	453,600	207,600	74.3	61.6
5	1,121,400	223,200	59.87	33.24

The aim of this study involves only covalent linkages, so through the washing solutions it was possible to remove the non-covalent attached enzymes (mainly adsorptions). One-point covalent immobilization in a non-porous support provides a system free of diffusion limitation (Amaya-Delgado, Hidalgo-Lara, & Montes-Horcasitas, 2006). However, when high enzyme concentrations are immobilized diffusion limitation phenomena can happen, as result of the blocking effect caused by neighboring proteins, in part because there are random enzyme orientations (Wong, Khan, & Micklefield, 2009), reducing the accessibility of substrate to the catalytic site (Giacomini, Villarino, Franco-Fraguas, & Batista-Viera, 1998). Besides the considerations about the support load and the enzyme concentration, it is important to regard the uneven surfaces from the aggregated nanoparticles that, in part, resemble a porous structure. This suggested 'porous nature' (Biró et al., 2008) a characteristic sustained by particles swelling ability (Shu & Zhu, 2002), could be considered as another factor for limitation diffusion effects. These indications may explain the lower efficiency values for samples 4 and 5, compared to the others. Considering the enzyme concentration offered to immobilization and the results of immobilization yield and efficiency, sample number 4 presented the better relationship among the tested conditions, since it was possible to immobilize a high load of enzyme on the support and to obtain interesting immobilizations yield and efficiency. Thus, sample number 4 was selected for further experiments.

Once defined the enzyme concentration for following studies, the immobilization kinetics was evaluated (Fig. 2a). It was demonstrated (Cadena et al., 2010), that immobilization of invertase on polyurethane in the presence of sucrose, to protect the catalytic site, improved the load of immobilization protein. Thus, it was performed experiments maintaining sucrose concentration at 8% in the immobilization solution, keeping the catalytic site saturated (Fig. 2b).

Immobilization occurred in 5 h, when around 50% of the offered enzyme was immobilized in the support. It was defined 5 h as the immobilization time, since activity in the supernatant and support kept unaltered after this time. The presence of substrate (sucrose) in the immobilization solution did not improve enzyme immobilization, in contrast to another study (Cadena et al., 2010), in which is shown the increased covalent immobilization of invertase to an organic polymer in a sucrose solution (10%).

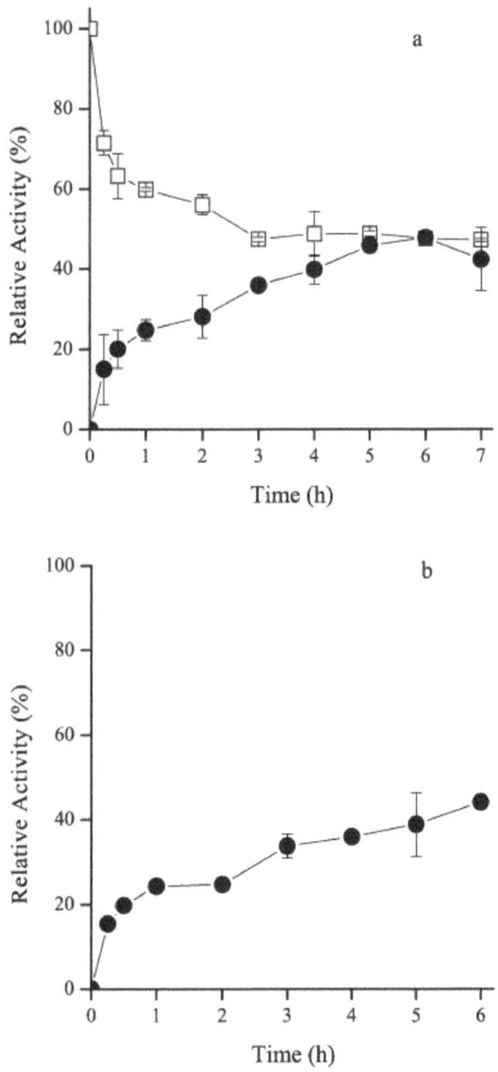

Figure 2: Supernatant of immobilization solution (□) and support activity (●) during invertase immobilization process. (a) Absence of sucrose. (b) In the presence of sucrose (8%).

pH, temperature and kinetic parameters

Results for the effects of pH and temperature on the activity of free and immobilized invertase are presented in Fig. 3 and Fig. 4, respectively. Activities are related to the highest value, which was considered as 100% activity.

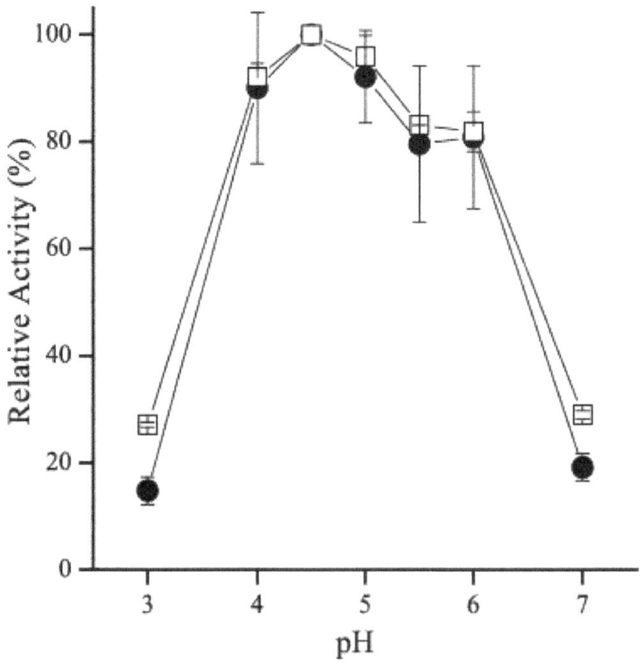

Figure 3: Effect of pH on free (□) and immobilized (•) invertase activity at 55 °C.

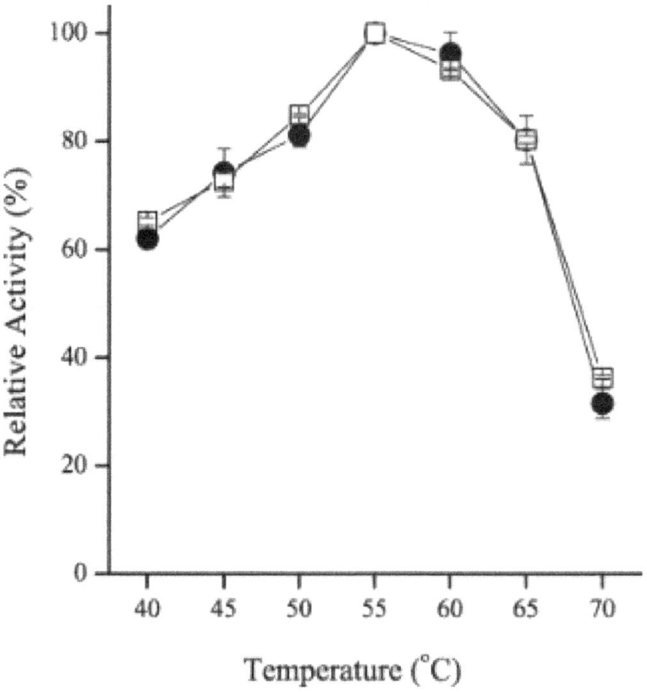

Figure 4: Effect of temperature on free (□) and immobilized (•) enzyme at pH 4.5.

Although it is possible to change the optimal pH, temperature and kinetic parameters for an immobilized derivative as a result of the immobilization method, support structure and conformation change of the enzyme after bounded, making the catalytic site more or less accessible to substrate (Bayramoglu et al., 2009), in our work no alterations in pH and temperature were detected.

Fig. 3 shows the same optimal pH (4.5) for both enzymatic preparations, indicating that there is no conformational alteration by immobilization process in the monomeric invertase. Other authors (Bahar and Tuncel, 2002 and Chen et al., 2000) also covalently immobilized the enzyme on polymeric supports and did not find shifts in optimal pH after immobilization. These authors considered this as a positive factor to the preservation of the biocatalyst during the immobilization process.

It can be observed in Fig. 4 that the behavior for free and immobilized invertase at different temperatures is similar, and 55 °C is the optimal temperature for both enzymatic forms. As stated in other studies (Mateo et al., 2007 and Rodrigues et al., 2009) the one-point covalent immobilization of the enzyme suggests that the immobilized biocatalyst presented the same rigidity than the soluble form. For this reason, the immobilization did not present rigidification and protective effects, as can be seen in multipoint or multisubunit covalent immobilization examples, where there are improvements in the stability of the enzymes (Garcia-Galan et al., 2011 and Mateo et al., 2007).

The determination of enzyme kinetic parameters is important because it is possible to measure the reaction rate and how it is influenced by physical and chemical conditions. The kinetic parameters for free and immobilized invertase (Table 2) were obtained by measuring the enzyme activities at sucrose concentration from 29.2 mM to 292.2 mM for free and immobilized invertase, in acetate buffer (0.1 M, pH 4.5) at 55 °C.

Table 2: Kinetic parameters for free and immobilized invertase

Enzyme form	*Km* (mM)	V_{max} (U/mL)	Catalytic efficiency[a]
Free	65.7	1670.0	25.4
Immobilized	205.7	1830.0	8.9

[a] V_{max}/Km.

Km for the biocatalyst after immobilization increased approximately 3-fold. The catalytic efficiency shows higher value for the free enzyme. Similar results were found for invertase *Km* (46 mM and 203 mM, for free and immobilized form, respectively) in a study using a polymeric membrane as support (Mazi, Emregul, Rzaev, & Kibarer, 2006). This considerable reduction in the affinity of the enzyme for the substrate, could be due to the uneven surface of the support, and also attributed to the high concentration of protein that was immobilized, generating diffusion effects. On the other hand, V_{max} did not change, showing the conservation of the maximal conversion of sucrose into its monosaccharides, since the invertase conversion capacity after immobilization was kept.

Thermal stability

One of the different targets of immobilization is the improvement of the enzyme stability (Garcia-Galan et al., 2011), which is quite dependent of the immobilization technique, besides the enzyme and support characteristics. Considering 55 °C the optimal temperature obtained for free enzyme and immobilized form, and the fact that many immobilization techniques can improve enzyme stability (Mateo et al., 2007), thermal stability of free and immobilized invertase were carried out at optimal conditions (Fig. 5) and at 65 °C. The enzymatic forms were incubated in the absence of sucrose.

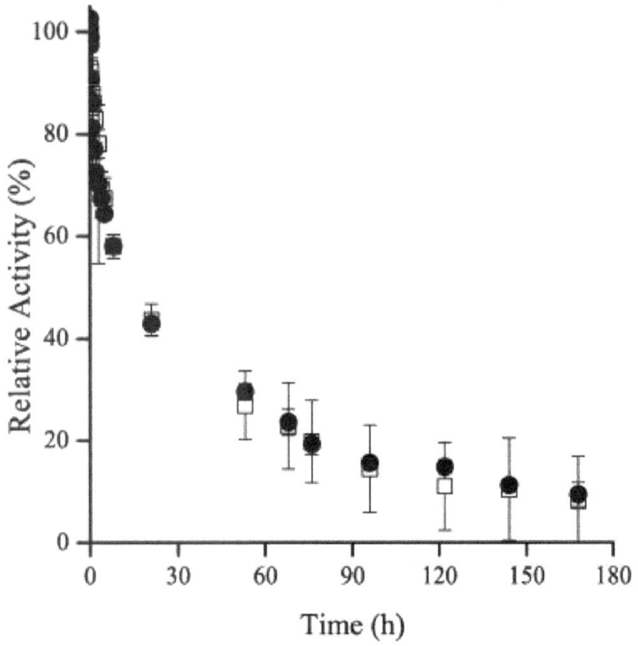

Figure 5: Influence of temperature on the stability of free (□) and immobilized (•) invertase at 55 °C and pH 4.5.

At 55 °C free biocatalyst and immobilized derivative showed very similar profile, with no difference between the half-life times before and after immobilization. At 65 °C (data not shown) the half-life times also did not change between free and immobilized invertase, but sharply decreased for few minutes of incubation in relation to the results obtained for optimal temperature. Kumar, Chauhan, and Nahar (2008) also observed significant decrease in immobilized derivative activity at temperatures higher than optimal, and obtained similar result for free invertase at similar conditions, since Gómez, Ramírez, and Villalonga (2000) increased this parameter for immobilized derivative, using an immobilization protocol that induced enzyme sugar moiety oxidation by sodium periodate. The possible explanation is that nonpolar surface area of proteins indirectly induces the enzyme inactivation in vitro (Turková, 1999), therefore oriented immobilization protocols, which couple the biocatalyst to the support through carbohydrate moiety, can contribute to derivative stabilization.

The results from the present work agreed with previous reports, which showed that one-point attached enzymes to the support, with no diffusion limitations, have similar profiles to free biocatalyst (Garcia-Galan et al., 2011). Although there was not an improvement in the invertase thermal stability, the conservation of whole enzyme activity observed after immobilization is important because of the possibility of reusing the immobilized biocatalyst is dependent upon the residual enzyme activity.

Storage stability

For practical applications, storage stability of immobilized derivatives is important, the more stable the better. Free and immobilized invertase did not lose their activities after 25 weeks of storage at 8 °C and optimal pH, showing that there were neither biocatalyst modifications nor stabilization by the immobilization process.

Operational stability

Due to the high cost of enzymes, they should be reused for several times to be competitive with chemical catalysts. Ideally, the immobilization of enzymes should provide, besides thermal and solvent stabilities, a high operational stability (Bayramoglu et al., 2009). The results for repeated uses of the immobilized invertase in the sucrose hydrolysis are presented in Fig. 6.

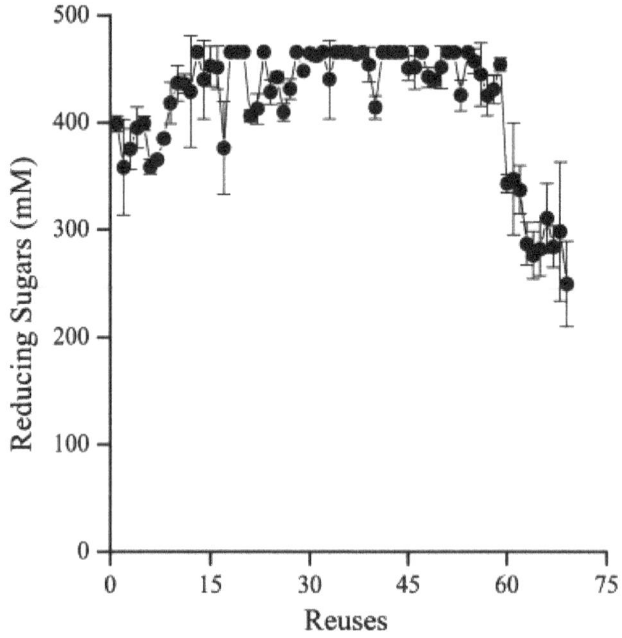

Figure 6: Residual enzyme activity using invertase immobilized in 0.05 g support in a sucrose solution (20 mL, 8%) during 69 batches (each batch of 30 min, 55 °C, gentle stirring).

Fig. 6 shows the expressive number of 69 reuses of the immobilized invertase, keeping the sucrose hydrolysis near to 100% (465.93 mM) until reuse 59. Immobilizing invertase onto nanogels, just remained 11.03% of initial activity at the ninth cycle (Raj, Chauhan, Azmi, Ahn, & Manuel, 2011). The immobilized derivative (hydrogel-clay) was reused for 26 successive batches retaining about 50% of residual activity (Oztop, Hepokur, & Saraydin, 2010). In other study the authors observed the enzyme hydrolysis during 30 cycles keeping at the last one 70% of activity using a mineral as support (Sanjay & Sugunan, 2008).

Probably the decrease in the activity shown at the final batches is a consequence to different factors as the progressive decay in mechanical resistance of the preparation (support-enzyme), inherent wastes by the material transfer, but mainly due to the natural denaturation of the biocatalyst. Nevertheless, since there was not found in literature any work for sucrose hydrolysis by immobilized invertase with similar number of reuses, the results obtained in this work represents the highest operational stability so far described.

CONCLUSIONS

The immobilization of invertase from *S. cerevisiae* on nano and aggregated chitosan nanoparticles presented high recoveries of activities, with good immobilization yields and efficiencies. After the immobilization the enzyme presented similar parameters of the soluble form (optimal pH, temperature, thermal and storage stabilities). V_{max} did not change, while *Km* presented a sharp increase (from 65.7 mM to 205.7 mM), showing the reduction of the affinity of the enzyme for sucrose, yet keeping the invertase maximal conversion capacity. In terms of operational stability, the immobilized preparation could be reused for 59 batches at maximal enzyme activity. These results fulfilled some important requirements for the invertase immobilization because the protocols for support preparation and enzyme immobilization are simple, the enzyme activity and optimal conditions were conserved, and it was obtained the highest operational stability for sucrose hydrolysis so far described in the literature.

Acknowledgments

The authors would like to thank Coordenação de Aperfeiçoamento de Pessoal de Nível Superior (CAPES), a Brazilian bureau for science and education, for the scholarships for Miss Valerio. This work was also supported by the Conselho Nacional de Desenvolvimento Científico e Tecnológico (CNPq) and Fundação de Amparo a Pesquisa do Rio Grande do Sul (FAPERGS).

REFERENCES

1. Amaya-Delgado, L., Hidalgo-Lara, M. E., & Montes-Horcasitas, M. C. (2006). Hydrolysis of sucrose by invertase immobilized on nylon-6 microbeads. Food Chemistry, 99, 299–304.
2. Bahar, T., & Tuncel, A. (2002). Immobilization of invertase onto crosslinked poly(pchloromethylstyrene) beads. Journal of Applied Polymer Science, 83, 1268–1279.
3. Bayramoglu, G.,Karakisla, M.,Altintas, B., Metin,A. U., Sac¸ ak, M., & Arica, M. Y.(2009). Polyaniline grafted polyacylonitrile conductive composite fibers for reversible immobilization of enzymes: Stability and catalytic properties of invertase. Process Biochemistry, 44, 880–885.
4. Berthold,A., Cremer, K., & Kreuter, J. (1996). Preparation and characterization of chitosan microspheres as drug carrier for prednisolone sodium phosphate as model for anti-inflammatory drugs. Journal of Controlled Release, 39, 17–25.
5. Biró, E., Németh, A. S., Sisak, C., Feczkó, T., & Gyenis, J. (2008). Preparation of chitosan particles suitable for enzyme immobilization. Journal of Biochemical and Biophysical Methods, 70, 1240–1246.
6. Bryjak, J., Liesiene, J., & Stefuca, V. (2008). Man-tailored cellulose-based carriers for invertase immobilization. Cellulose, 15, 631–640.
7. Cadena, P. G., Jeronimo, R. A. S., Melo, J. M., Silva, R. A., Filho, J. L., & Pimentel, M. C. B. (2010). Covalent immobilization of invertase on polyurethane, plast-film and ferromagnetic Dacron. Bioresource Technology, 101, 1595–1602.
8. Chen, Y., Kang, E. T., Neoh, K. G., & Tan, K. L. (2000). Covalent immobilization of invertase onto the surface-modified polyaniline from graft copolymerization with acrylic acid. European Polymer Journal, 36, 2095–2103.
9. Garcia-Galan, C., Berenguer-Murcia, A., Fernandez-Lafuente, R., & Rodrigues, R. C. (2011). Potential of different enzyme immobilization strategies to improve enzyme performance. Advanced Synthesis and Catalysis, 353, 2885–2904.
10. Giacomini, C., Villarino, A., Franco-Fraguas, L., & Batista-Viera, F. (1998). Immobilization of -galactosidase from Kluyveromyces lactis on silica and agarose: Comparison of different methods. Journal of Molecular Catalysis B: Enzymatic, 4, 313–327.
11. Gomez, L., Ramirez, H. L., Cabrera, G., Simpson, B. K., & Villalonga, R. (2008). Immobilization of invertase-chitosan conjugate on hyaluronic-acid-modified chitin. Journal of Food Biochemistry, 32, 264–277.
12. Gómez, L., Ramírez, H. L., & Villalonga, R. (2000). Stabilization of invertase by modi- fication of sugar chains with chitosan. Biotechnology Letters, 22, 347–350.
13. Gómez, L., Ramírez, H. L., Villalonga, M. L., Hernández, J., & Villalonga, R. (2006). Immobilization of chitosan-modified invertase on alginate-coated chitin support via polyelectrolyte complex formation. Enzyme and Microbial Technology, 38, 22–27.
14. Hsieh,H.J., Liu, P.C., & Liao,W.J.(2000). Immobilizationofinvertase via carbohydrate moiety on chitosan to enhance its thermal stability. Biotechnology Letters, 22, 1459–1464.
15. Klein, M. P., Nunes, M. R., Rodrigues, R. C., Benvenutti, E. V., Costa, T. M. H., Hertz, P. F., et al. (2012). Effect of the support size on the properties of -Galactosidase immobilized on chitosan: advantages and disadvantages of macro and nanoparticles. Biomacromolecules, 13, 2456–2464.
16. Kotwal, S. M., & Shankar, V. (2009). Immobilized invertase. Biotechnology Advances, 27, 311–322. Kotwal, S. M., & Shankar, V. (2009). Immobilized invertase. Biotechnology Advances, 27, 311–322.
17. Krajewska, B.(2004).Application of chitin- and chitosan-based materials for enzyme immobilizations: A review. Enzyme and Microbial Technology, 35, 126–139.
18. Kumar, S., Chauhan, V. S., & Nahar, P. (2008). Invertase embedded-PVC tubing as a flow-through reactor aimed at conversion of sucrose into inverted sugar. Enzyme and Microbial Technology, 43, 517–522.
19. Kuo, C.-H., Liu, Y.-C., Chang, C.-M. J., Chen, J.-H., Chang, C., & Shieh, C.-J. (2012). Optimum conditions for lipase immobilization on chitosan-coated Fe3O4 nanoparticles. Carbohydrate Polymers, 87, 2538–2545.
20. Li, J., Cai, J., Zhong, L., & Du, Y. (2012). Immobilization of a protease on modified chitosan beads for the depolymerization of chitosan. Carbohydrate Polymers, 87, 2697–2705.
21. Liu, C. G., Desai, K. G. H., Chen, X. G., & Park, H. J. (2005). Preparation and characterization of nanoparticles containing trypsin based on hydrophobically modified chitosan. Journal of Agricultural and Food Chemistry, 53, 1728–1733.
22. Mateo, C., Palomo, J. M., Fernandez-Lorente, G.,

Guisan, J. M., & FernandezLafuente, R. (2007). Improvement of enzyme activity, stability and selectivity via immobilization techniques. Enzyme and Microbial Technology, 40, 1451–1463.

23. Mazi, H., Emregul, E., Rzaev, Z. M. O., & Kibarer, G. (2006). Preparation and properties of invertase immobilized on a poly(maleic anhydride-hexen-1) membrane. Journal of Biomaterials Science Polymer Edition, 17, 821–835.

24. Miller, G. L. (1959). Use of dinitrosalicylic acid reagent for determination of reducing sugar. Analytical Chemistry, 31, 426–428.

25. Monteiro, O. A. C., & Airoldi, C. (1999). Some studies of crosslinking chitosanglutaraldehyde interaction in a homogeneous system. International Journal of Biological Macromolecules, 26, 119–128.

26. Muzzarelli, R. A. A. (1977). Chitin. Oxford, UK: Pergamon Press.

27. Muzzarelli, R. A. A. (1980). Immobilization of enzymes on chitin and chitosan. Enzyme and Microbial Technology, 2, 177–184.

28. Muzzarelli, R. A. A. (2012). Nanochitins and nanochitosans paving the way to eco-friendly and energy-saving exploitation of marine resources. In K. Matyjaszewski, & M. Möller (Eds.), Polymer science: A comprehensive reference (pp. 153–164). Amsterdam: Elsevier BV.

29. Muzzarelli, R. A. A., Boudrant, J., Meyer, D., Manno, N., Demarchis, M., & Paoletti, M. G. (2012). Current views on fungal chitin/chitosan, human chitinases, food preservation, glucans,pectins andinulin:Atribute toHenriBraconnot,precursor of the carbohydrate polymers science on the chitin bicentennial. Carbohydrate Polymers, 87, 995–1012.

30. Orrego, C. E., Salgado, N., Valencia, J. S., Giraldo, G. I., Giraldo, O. H., & Cardona, C. A. (2010). Novel chitosan membranes as support for lipases immobilization: Characterization aspects. Carbohydrate Polymers, 79, 9–16.

31. Oztop, H. N., Hepokur, C., & Saraydin, D. (2010). Poly(acrylamide/maleic acid)–sepiolite composite hydrogels for immobilization of invertase. Polymer Bulletin (Berlin), 64, 27–40.

32. Polizzi, K. M., Bommarius, A. S., Broering, J. M., & Chaparro-Riggers, J. F. (2007). Stability of biocatalysts. Current Opinion in Chemical Biology, 11, 220–225.

33. Raj, L., Chauhan, G. S., Azmi, W., Ahn, J. H., & Manuel, J. (2011). Kinetics study of invertase covalently linked to a new functional nanogel. Bioresource Technology, 102, 2177–2184.

34. Rodrigues, R. C., Godoy, C. A., Filice, M., Bolivar, J. M., Palau-Ors, A., Garcia-Vargas, J. M., et al. (2009). Reactivation of covalently immobilized lipase from Thermomyces lanuginosus. Process Biochemistry, 44, 641–646.

35. Sanjay, G., & Sugunan, S. (2008). Acid activated montmorillonite: An efficient immobilization support for improving reusability storage stability and operational stability of enzymes. Journal of Porous Materials, 15, 359–367.

36. Shu, X. Z., & Zhu, K. J. (2002). Controlled drug release properties of ionically crosslinked chitosan beads: The influence of anion structure. International Journal of Pharmaceutics, 233, 217–225.

37. Siso, M. I. G., Lang, E., Carreno-Gómez, B., Becerra, M., Otero Espinar, F., & Blanco Méndez, J. (1997). Enzyme encapsulation on chitosan microbeads. Process Biochemistry, 32, 211–216.

38. Teodor, E., Radu, G. L., Dan, C., & Stanciu, G. A. (2006). Chitosan-hyaluronic acid films for enzyme immobilization studied by confocal scanning laser microscopy. Revista de Chimie (Bucharest Romania), 57, 211–214.

39. Turková, J. (1999). Oriented immobilization of biologically active proteins as a tool for revealing protein interactions and function. Journal of Chromatography B: Biomedical Applications, 722, 11–31.

40. Wong, L. S., Khan, F., & Micklefield, J. (2009). Selective covalent protein immobilization: Strategies and applications. Chemical Reviews, 109, 4025–4053.

41. Zhang, F., Lees, E.,Amin, F.,Rivera-Gil, P.,Yang, F.,Mulvaney, P., et al.(2011). Polymercoated nanoparticles: A universal tool for biolabelling experiments. Small, 7, 3113–3127.

CHAPTER 8

ELECTROCHEMICAL DETECTION OF CARBAMATE PESTICIDES IN FRUIT AND VEGETABLES WITH A BIOSENSOR BASED ON ACETYLCHOLINESTERASE IMMOBILISED ON A COMPOSITE OF POLYANILINE–CARBON NANOTUBES

Ivana Cesarino, Fernando C. Moraes, Marcos R.V. Lanza, and Sergio A.S. Machado

Instituto de Química de São Carlos, Universidade de São Paulo, São Carlos, São Paulo CEP 13566-590, Brazil

ABSTRACT

A sensitive electrochemical acetylcholinesterase (AChE) biosensor was successfully developed on polyaniline (PANI) and multi-walled carbon nanotubes (MWCNTs) core–shell modified glassy carbon electrode (GC), and used to detect carbamate pesticides in fruit and vegetables (apple, broccoli and cabbage). The pesticide biosensors were applied in the detection of carbaryl and methomyl pesticides in food samples using chronoamperometry (CA). The GC/MWCNT/PANI/AChE biosensor exhibited detection limits of 1.4 and 0.95 μmol L^{-1}, respectively, for carbaryl and methomyl. These detection limits were below the allowable concentrations set by Brazilian regulation standards for the samples in which these pesticides were analysed. Reproducibility and repeatability values of 2.6% and 3.2%, respectively, were obtained in the conventional procedure. The proposed biosensor was successfully applied in the determination of carbamate pesticides in cabbage, broccoli and apple samples without any spiking procedure. The obtained results were in full agreement with those from the HPLC procedure.

INTRODUCTION

The continuous monitoring for low pesticide levels in food, water, and air has become a key activity in respect to human health. Such monitoring must often be carried out in situ and in real time, which demands new analytical techniques that can provide responses very quickly and at very low concentration levels. One of the most promising research fields in this area is related to enzymatic biosensors. These devices can be easily miniaturised for in situ applications which can provide fast analytical responses and are insensitive to interference from the other components of the complex matrices.

Enzymatic biosensors used to monitor the pesticides of the carbamate class are based on the inhibition of the enzymatic reaction of acetylcholinesterase (AChE) and its substrate, acetylcholine. The carbamates bind to the active estearic site of the enzyme, inhibiting biocatalytic activity by blocking the serine residue in the catalytic triad of AChE by phosphorylation

Citation: Ivana Cesarino, Fernando C. Moraes, Marcos R.V. Lanza, Sergio A.S. Machado, Electrochemical detection of carbamate pesticides in fruit and vegetables with a biosensor based on acetylcholinesterase immobilised on a composite of polyaniline–carbon nanotubes, Food Chemistry, Volume 135, Issue 3, 1 December 2012, Pages 873-879, ISSN 0308-8146, http://dx.doi.org/10.1016/j.foodchem.2012.04.147.

or carbamylation (Singh & Agarwal, 1983). A great number of research studies report the development of biosensors for detecting carbamates based on AChE enzymatic inhibition. Recently, Cai and Du related a biosensor based on the immobilisation of acetylcholinesterase to multi-walled carbon nanotubes for the detection of carbaryl with the detection limit reaching 4.0 μg L^{-1} (Cai & Du, 2008). Pedrosa, Caetano, Machado, and Bertotti (2008) developed an amperometric biosensor based on acetylcholinesterase (AChE) immobilised onto a self-assembled modified gold electrode. The biosensor was used in the electrochemical detection of carbamate in water with the detection limit calculated at 9.0 μg L^{-1}. In another study,Upadhyay et al. (2009) immobilised the enzyme AChE on a glassy carbon surface modified with gold–platinum bimetallic nanoparticles. This biosensor was used in the electrochemical detection of carbamates and organophosphate pesticides. Other examples of biosensors for pesticides based on the inhibition of AChE have been reported (Du et al., 2010 and Viswanathan et al., 2009) confirming that this methodology constitutes one of the most important analytical branches in environmental and foodstuff monitoring.

However, despite the extensive number of publications available on the subject, there are few studies regarding the electrochemistry of enzymes immobilised at bare electrodes surfaces. Enzymes usually have short life spans and lose activity at bare electrodes (Wen, Ye, & Zhou, 1997). To overcome such limitations, studies have focused on various strategies for enzyme immobilisation, such as adsorption, covalent linkage, entrapment, and cross-linking.

To this end, the use of nanostructured materials has attracted much attention due to unique properties such as high electrical conductivity, large surface area and the presence of functional groups that can interact with the enzyme offering excellent support for immobilisation. In this context, hybrids formed with carbon nanotubes (CNTs) and conducting polymers have exhibited a synergic effect leading to significant enhancements in the electronic and mechanical characteristics of each single component (Agüí, Peña-Farfal, Yáñez-Sedeño, & Pingarrón, 2007).

Carbon nanotubes have been the subject of numerous investigations in chemical, physical and material science research owing to their extraordinary mechanical, chemical and electronic properties (Merkoçi et al., 2005). CNTs can be used as electrode materials with useful properties for electrochemical and bioelectrochemical applications (Moraes, Mascaro, Machado, & Brett, 2009). The ability of CNTs to promote electron-transfer reactions suggests interesting applications in the development of amperometric biosensors (Laschi, Bulukin, Palchetti, Cristea, & Mascini, 2008).

Polyaniline (PANI) is one of the most attractive conducting polymers due to its high conductivity, ease of preparation, good environmental stability, and extensive applications (Gerard et al., 2002 and Li et al., 2006). The presence of an extended π-conjugated system and the three-dimensional structure of the polymer allows the adsorption of the enzyme inside composite cavities promoting significant interactions between carboxyl groups of the enzyme and amino groups present in the polymer chain (Sung & Bae, 2003).

The core–shell structure in which the CNTs are covered by a compact layer of PANI (Zhou et al., 2010), exhibits intrinsic characteristics such as a redox activity at neutral pH values (Do Nascimento, Corio, Novickis, Temperini, & Dresselhaus, 2005). The electrochemical formation of polymer layers offers significant advantages such as complete coverage of the active surface, greater control over film thickness and enhanced reproducibility (Ahuja, Mir, Kumar, & Rajesh, 2007).

The objective of this study is to develop an amperometric pesticide biosensor based on acetylcholinesterase immobilised in a core–shell structure of CNTs and polyaniline (PANI) to be used to quantify methomyl and carbaryl in liquefied apple, cabbage and broccoli samples by chronoamperometry, and to compare the obtained results with those from the HPLC technique.

EXPERIMENTAL

Instrumentation

Cyclic voltammetry (CV), square-wave voltammetry (SWV) and chronoamperometry (CA) experiments were performed in a model PGSTAT 30 Autolab Electrochemical System equipped with GPES 4.9 software (Eco Chemie, Utrecht, The Netherlands). The morphology of the composite surface was characterised using a Field-Emission Gun Scanning Electron Microscope (FEG-SEM) and the images were recorded using a FEG-Zeiss model Supra 35VP (Zeiss, Germany).

The electrochemical cell was assembled in a conventional one-compartment three-electrode design: the working electrode was a glassy carbon disk modified by the core–shell multi-walled carbon nanotubes (MWCNTs) and PANI composite in which the acetylcholinesterase was immobilised; the Ag/AgCl/KCl (3.0 mol L^{-1}) was used as the reference electrode; and a Pt plate was used as the auxiliary electrode. A PTFE cover provided the access to the electrodes and to the degassing facility. All experiments were conducted at a controlled temperature of 25 °C.

Methodologies

The chronoamperometric (CA) experiments were performed by polarising the working electrode at 0.0 V for 5 s, followed by a potential step to +0.30 V for 60 s in a 0.2 mol L^{-1}phosphate buffer solution (PBS) at pH 7.0. The square-wave voltammetry (SWV) experiments were carried out using the following parameters, initially optimised: frequency of 10 Hz, pulse amplitude of 70 mV, step potential of 2 mV, in a 0.2 mol L^{-1}phosphate buffer solution (PBS) at pH 7.0.

Theanalyticalmethodologyemployedtoinvestigateenzyme inhibition by carbamate consisted of the addition of different amounts of the pesticide in the electrochemical cell containing 0.2 mol L^{-1} PBS and 4.0×10^{-5} mol L^{-1} acetylthiocholine chloride (AChC), the substrate for the enzymatic reaction. After each carbamate addition, the biosensor was immersed in the electrolyte and an induction time of 10 min was imposed to allow the competitive adsorption in the active enzymatic sites to reach equilibrium. After this, the amount of thiocholine generated in the enzymatic reaction was evaluated by CA or by SWV. After each determination, electrode activity was restored by immersion in PBS without carbamate for 10 min.

Chemicals and solutions

All solutions were prepared with water purified in a Millipore Milli-Q system (resistivity □18 MΩ cm). Aniline (Sigma–Aldrich, Germany) purification was carried out by distillation at 140 °C, using metallic zinc as a catalyst.

An amount of approximately 1.0 g of MWCNT (90% purity, Sigma–Aldrich, Germany), synthesised by thermal chemical vapour deposition, was mixed with 500 mL of a 1:3 mixture of HNO_3/H_2SO_4 for 12 h, in order to promote its functionalisation (Moraes, Cabral, Mascaro, & Machado, 2011). This was then filtered through a 0.45 μm Millipore nylon filter membrane. The resulting MWCNT was continuously washed using distilled water until the pH of the filtrate was neutral, and then dried overnight in a vacuum oven at 120 °C.

A 10 mg/mL solution of acetylcholinesterase enzyme from bovine erythrocyte (Type XII-S, 426 unit mg^{-1} of solid, Sigma–Aldrich, Germany) was prepared in 0.2 mol L^{-1} PBS pH 7.0. Acetylthiocholine chloride, carbaryl and methomyl were purchased from Sigma–Aldrich.

Preparation of the AChE biosensor

Prior to modification, the GC electrode surface was polished with 0.05 μm alumina slurries, rinsed thoroughly with double distilled water, sonicated for 3 min in alcohol followed by 3 min in water and then dried in air. The GC electrode was immersed in a solution containing 500 μL of aniline and 7.0 mg of MWCNT diluted in 25.0 mL of 0.5 mol L^{-1} of H_2SO_4 solution and using the cyclic voltammetry the hybrid was electropolymerised between −0.2 and +0.8 V *vs.* Ag/AgCl with a scan rate of 50 mV s^{-1}. The GC electrode containing the PANI/MWCNT film was removed from the solution and, after drying at a controlled temperature of 25 °C, 7.0 μL of AChE was dropped on the electrode surface. The biosensor was dried at a controlled temperature and then immersed in PBS prior to use.

Preparation of the fruit and vegetables samples and analysis of carbamate

For the analysis of carbaryl in the apple samples, four apples (from the local market) were cut into small pieces, 168.0 g of these pieces were weighed and mixed in a blender with 0.2 mol L^{-1} PBS, pH 7.0. The CA analysis of carbaryl was performed directly in the extract of the sample using the standard addition method. For the analysis of methomyl in cabbage, a bundle of leaves was cut and mixed in a blender with 0.2 mol L^{-1} PBS, pH 7.0, resulting in a final volume of 1.0 L. To prepare the samples of broccoli, 117.7 g of broccoli were weighed and mixed in a blender with 0.2 mol L^{-1} PBS, pH 7.0. The CA analysis of methomyl was carried out directly in the extracts of the cabbage and broccoli samples using the standard addition method.

HPLC measurements

The analytical curves were obtained by the standard addition method and measurements were performed in triplicate. For carbaryl, the mobile phase consisted of a mixture of acetonitrile and water 70:30 (v/v), with a flow rate of 1 mL min^{-1} under isocratic conditions. The chromatographic column was a C18 (250 × 4.6 mm) Shimadzu Shim–Pack CLC–ODS (Kyoto, Japan). The monitored wavelength was 220 nm. For methomyl, the flow rate and the chromatographic column were the same as those used for the carbaryl analysis; however the mobile phase consisted of a mixture of acetonitrile and water 50:50 (v/v) and the monitored wavelength was 230 nm. The analysis of carbamates in cabbage, broccoli and apple using the HPLC procedure was performed in the extracts of the samples described in Section 2.4. However, before analysis, the extracts were centrifuged at 3600 rpm for 15 min and the supernatant liquids were filtered using a PTFE syringe filter, 33 mm × 0.22 μm.

RESULTS AND DISCUSSION

Development and characterisation of the MWCNT/PANI/AChE biosensor

The consecutive cyclic voltammograms for the electropolymerisation of the PANI/MWCNT composite on the GC electrode from a 25 mL support electrolyte solution containing 0.5 mol L^{-1} H_2SO_4 + 500 μL of aniline and 7.0 mg of MWCNT with a scan rate of 50 mV s^{-1} showed a well-defined oxidation peak at +0.24 V (due to the polymerisation of aniline forming emeraldine salt) and this peak increase with the cyclisation. Aiming to achieve a composite film that: (a) completely covers the electrode surface with core–shell PANI/MWCNT, and (b) is sufficiently thin to avoid loss of catalytic activity by increasing the electron transfer resistance, a number of 25 voltammetric cycles in the electrodeposition step was found to be the optimum procedure. The efficiencies of such coverage were tested by the amperometric responses of the biosensor.

The morphological characteristics of composites were examined by FEG-SEM. The corresponding micrographs were presented in Fig. 1. For comparison, the MWCNT morphology can be observed in Fig. 1A, in accordance with those already presented in the literature (Hernández-Pérez et al., 2008 and Moraes et al., 2010). The complete coverage of the MWCNT with PANI is evident in Fig. 1B, pointing to the core–shell arrangement of the composite (Zhou et al., 2010). The disposition of such nanostructures on the GC electrode surfaces generates cavities, indicated by the arrows in Fig. 1C, in which the biomolecules, as enzymes, can be properly immobilised. Such tridimensional structures, plenty of nitrogen groups from PANI, are responsible for the biocompatibility between the composite and the biomolecules.

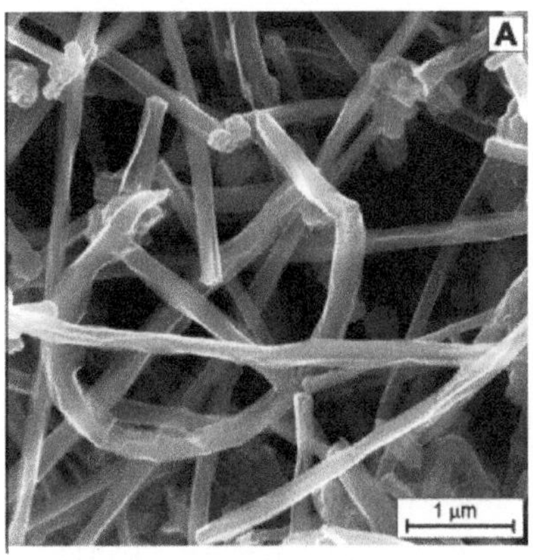

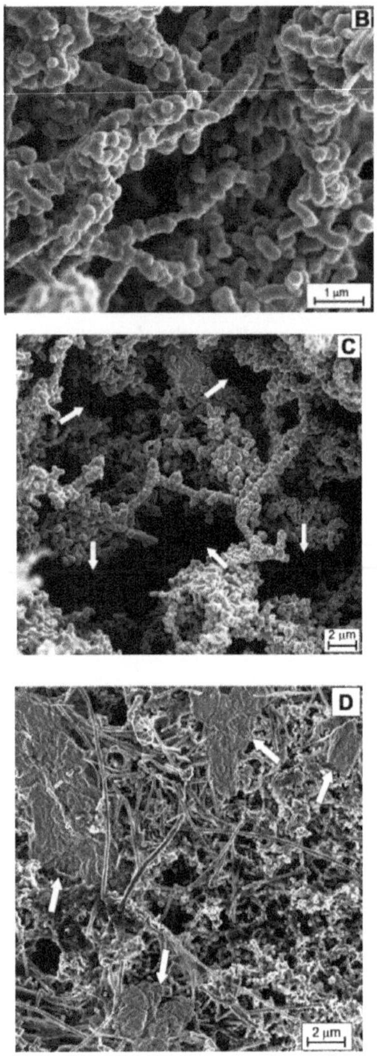

Figure 1: FEG-SEM micrographs for: (A) MWCNT (magnitude of 45,000×); (B) PANI/MWCNT core–shell (magnitude of 35,000×); (C) PANI/MWCNT cavities (magnitude of 12,000×) and (D) the anchorage of proteins in the porous of the PANI/MWCNT composite surface (magnitude of 10,000×).

As described in the Section 2, AChE immobilisation on the composite surface was carried out by dropping 7.0 µL of the buffer solution containing the enzyme on the electrode surface, which was further dried at a controlled temperature and immersed in PBS prior to use. No cross-linking agent (gluteraldehyde or diimide) was necessary. This constitutes a real contribution from the composite surface to the efficiency of the biosensor since the cross-linking agents frequently bond themselves to the active sites of enzymes, thus inhibiting their activity. The anchorage of proteins in the porous composite surface is indicated in Fig. 1D by the arrows. The amorphous material bonded to the PANI nitrogen groups is associated to the AChE.

Electrochemical response of the MWCNT/PANI/AChE biosensor

To evaluate the effect of the carbon nanotubes on the hybrid film, SWV experiments were carried out using a frequency of 10 Hz, a pulse amplitude of 70 mV, and a step potential of 2 mV in 0.2 mol L^{-1} phosphate buffer (pH 7.0) containing 20.0 µmol L^{-1} of acetylthiocholine chloride (AChC), the substrate of the enzyme. The biosensors evaluated were fabricated both in the presence of carbon nanotubes (GC/MWCNT/PANI/AChE) and in the absence of carbon nanotubes (GC/PANI/AChE).

In the SWV profiles presented in Fig. 2A the GC/MWCNT/PANI/AChE electrode (open square) showed a well-defined oxidation peak at +0.025 V. This electrochemical process has been associated to thiocholine oxidation generating the respective dimer, ditihio-bis-choline (Liu et al., 2005 and Rotariu et al., 2010), as shown in the following reactions:

$$(CH_3)_3N^+CH_2CH_2SCOCH_3 + H_2O \rightarrow (CH_3)_3N^+CH_2CH_2SH + CH_3COOH$$

$$2(CH_3)_3N^+CH_2CH_2SH \rightarrow (CH_3)_3N^+CH_2CH_2S{-}SCH_2CH_2N^+(CH_3)_3 + 2e^-$$

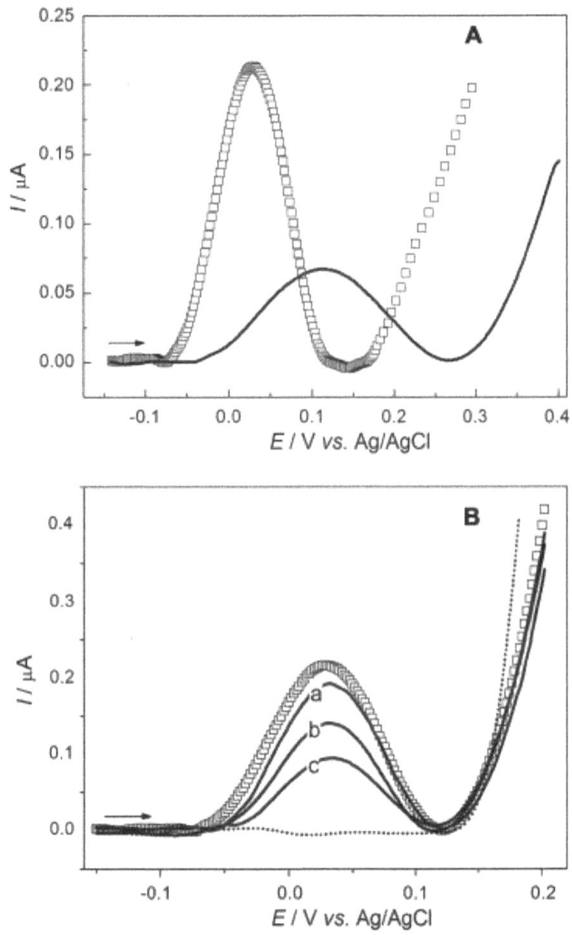

Figure 2: Square wave voltammograms in 0.2 mol L^{-1} PBS (pH 7.0) containing 20.0 µmol L^{-1} of AChC for (A) GC/MWCNT/PANI/AChE electrode (open square), GC/PANI/AChE electrode (solid line) and (B) for GC/PANI/MWCNT/AChE electrode in absence (dotted line) of AChC, presence (open square) and after addition of (a) 30, (b) 40, (c) 50 mol L^{-1} of carbaryl (solid line), with an incubation time of 10 min.

The effect of the MWCNT on the hybrid film can be observed when comparing the GC/MWCNT/PANI/AChE electrode and the biosensor fabricated in the absence of MWCNT. For the GC/PANI/AChE electrode (solid line in Fig. 2A) the dithio-bis-choline oxidation peak occurred at +0.125 V, i.e. the biosensor prepared with carbon nanotubes (GC/

MWCNT/PANI/AChE electrode) shifted in 100 mV the peak potential to more negative values, and also increased the current peak value by approximately three times. In this sense, the presence of MWCNT improved the electrochemical response of the enzymatic product and acted as an electrocatalyst.

The amount of enzyme dropped over the top of the electrode was optimised in the range of 1.0–5.0 units mg^{-1}. The current peak increased up to 3.0 unit mg^{-1} of the enzyme on the biosensor. For values greater than 3.0 unit mg^{-1} the current peak remained constant. Thus, 3.0 unit mg^{-1} was chosen as the optimal value.

The biosensor developed above was tested by immersion in a solution containing 0.2 mol L^{-1} phosphate buffer (pH 7.0) and 20.0 µmol L^{-1} of acetylthiocholine chloride (AChC). The SWV responses were collected in Fig. 2B, after the exposition of the enzyme to its substrate (open square) and only to the PBS (dotted line).

The electrochemical potential for the dimerisation reaction depends strongly on the catalytic properties of the electrode surface. To avoid the action of several species that can undergo electrochemical oxidation at positive potential values, it is convenient that thiocholine dimerisation occurs at nearly 0.0 V (*vs.* Ag/AgCl). Occasionally mediators are used to shift the potential towards less positive values. As can be observed in Fig. 2B, in this study no mediator had to be used to obtain a dimerisation potential close to 0.0 V. This observed effect is due to the extraordinary catalytic property of MWCNTs.

It is possible to observe that, after the addition of growing amounts of carbaryl (solid line inFig. 2B) and an incubation time of 10 min, the SWV response presented peak currents correspondingly lower, although at the same potential as that obtained for thiocholine oxidation. This is associated to the reversible competitive adsorption of carbaryl on the active enzyme site, inhibiting its activity (Du, Chen, Cai, & Zhang, 2007). Therefore, this phenomenon can be used to quantify the amount of carbaryl in the electrolyte.

Due to the reversible character of carbamates inhibition, the biosensor activity can be restored, after use, by simple immersion in PBS during 10 min. Further tests showed that the enzyme activity recovers its initial value, reproducing the same thiocholine oxidation measurement obtained without inhibition.

Analytical performance of the GC/MWCNT/PANI/AChE biosensor

With the voltammetric behaviour of the biosensor well settled, chronoamperometric experiments were performed, varying the carbaryl concentration, aiming to build up the calibration curves, as described in the Section 2. The variation of pesticide concentration in the electrolyte, from 9.9 to 49.6 µmol L^{-1}, promoted increasing inhibition in the diffusion currents of the chronoamperograms, similar to those demonstrated for SWV peak currents presented in Fig. 3A. The percentage of inhibition, calculated as previously reported (Caetano & Machado, 2008), exhibits a linear behaviour with the pesticide concentration, following the equation: $\%I = 3.75 + 1.10$[carbaryl], with a correlation coefficient of 0.999 ($n = 5$). A limit of detection (LOD) of 1.4 µmol L^{-1} (0.28 mg kg^{-1}) was determined using the 3 /slope ratio, where is the standard deviation of the diffusion current values of 10 chronoamperograms of the blank, as recommended by IUPAC (Blanc et al., 2000 and Currie, 1999). The same procedure was applied to the variation of methomyl concentration (Fig. 3B) and the linear response range was found to be between 4.9 and 29.2 µmol L^{-1}, with a linear relationship of $\%I = 0.91 + 2.32$[methomyl], a correlation factor of 0.998 ($n = 6$), and LOD of 0.95 µmol L^{-1} (0.15 mg kg^{-1}). Such values of the LOD are adequate to monitor contaminations with such pesticides according to the National Health Surveillance Agency (ANVISA) in Brazil, which establishes a maximum residue limit of 3.0 mg kg^{-1} for methomyl in cabbage and broccoli and 2.0 mg kg^{-1} for carbaryl in apples (Agência Nacional de Vigilância Sanitária).

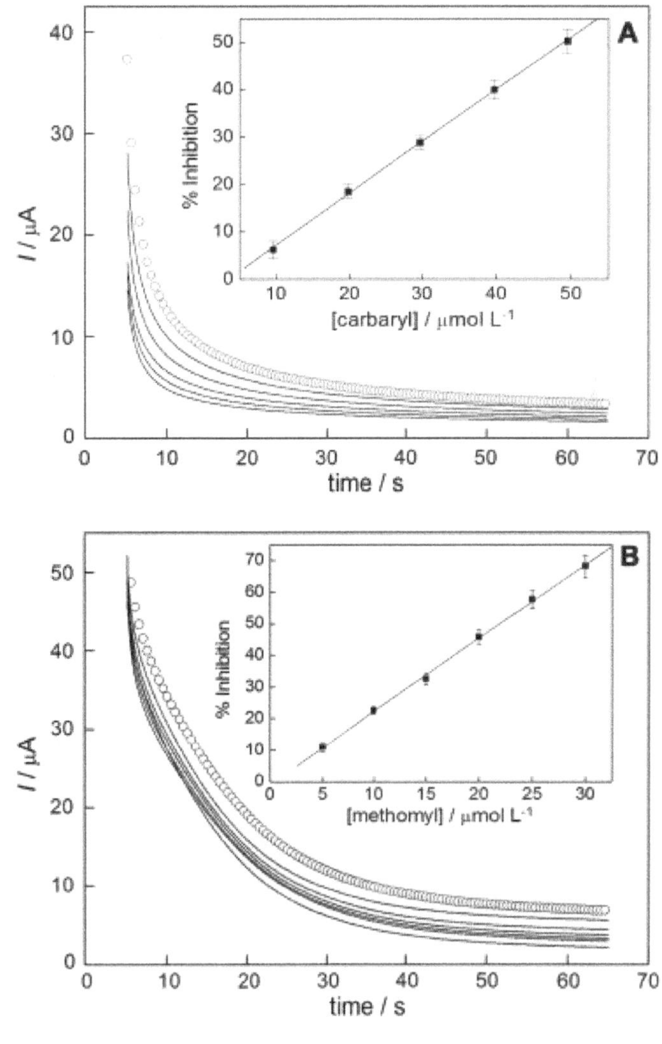

Figure 3: Chronoamperograms using GC/PANI/MWCNT/AChE electrode in 0.2 mol L^{-1} PBS (pH 7.0) containing 4.0×10^{-5} mol L^{-1} of AChC for (A) carbaryl and (B) methomyl in the absence (○) and presence (solid line) of the pesticides. Inset: linear dependence of percentage of inhibition with the concentrations of carbaryl (A) and methomyl (B).

Aiming to perform reproducibility and repeatability tests, the electrochemical measurements with thiocholine were performed 10 times each in the conventional procedure, given reproducibility and repeatability values of 2.6% and 3.2%, respectively.

In order to test the biosensor stability, it was stored in 0.20 mol L^{-1} PBS (pH 7.0) at 4 °C and used daily in experiments similar to those described above, for up to 60 days. After 2 months the electrode loses only 8.5% of its activity, which represents very reasonable stability for this kind of biosensor.

Comparison with the results obtained by high-performance liquid chromatography

It is well known that biosensors show low selectivity for a given class of carbamates or organophosphorous pesticides. In this way, during the development of the analytical procedure, the use of HPLC is convenient to identify the nature of a particular pesticide and allows the consequent monitoring in the target crop by the biosensor.

For this reason, HPLC experiments were performed to analyse carbaryl in apple samples, and methomyl in cabbage and broccoli samples, in order to compare the chromatographic results with those from the biosensor measurements. Aliquots of 20 μL of acetonitrile (analytical grade) spiked with different carbaryl or methomyl amounts were injected into the chromatograph. Fig. 4A reports the results obtained for carbaryl and Fig. 4B for methomyl. The retention times were found to be 4.0 and 3.5 min for carbaryl and methomyl, respectively. The inset in Fig. 4A shows the relationship between peak areas and carbaryl concentration. A well-defined linear relationship was observed in the range of 0.4–8.0 μmol L^{-1} with a slope of 1.46 and a correlation coefficient of 0.999. For methomyl (inset Fig. 4B), the linear response was found to be in the same concentration range as that of carbaryl, with a slope of 0.38 and a correlation coefficient of 0.999.

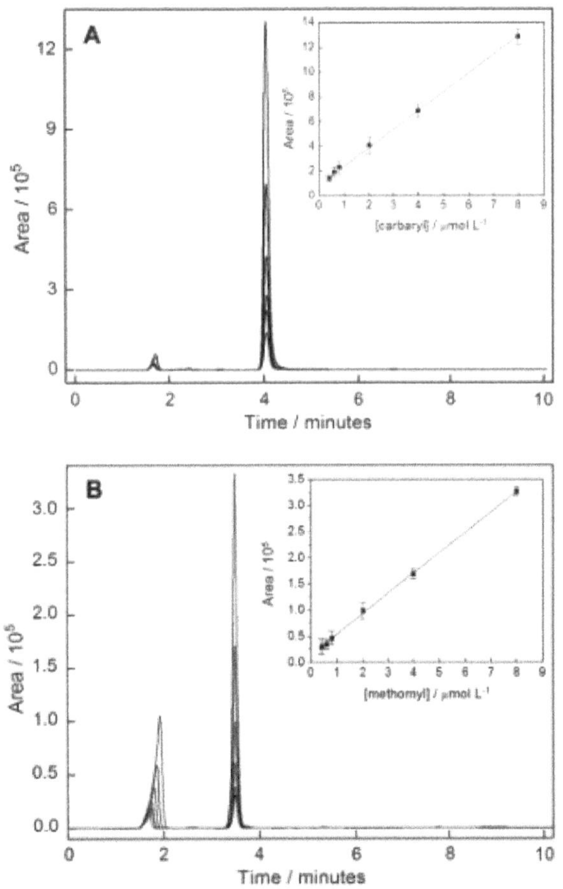

Figure 4: Chromatograms for different concentrations of (A) carbaryl and (B) methomyl. Insert: the analytical curve obtained by the areas of the chromatographic peaks.

Analysis of carbamates pesticides in fruit and vegetables

The developed biosensor was used for the quantification of carbaryl in apple samples acquired locally. Carbaryl determinations were performed in triplicate, without any further spiking procedure, using the standard addition method. Recoveries between 97.3% and 103.4% of carbaryl from apple samples were obtained for 9.9, 19.8 and 29.7 μmol L^{-1}carbaryl additions, and were extrapolated to the resulting straight line. The results obtained for the samples were compared with those for the same samples using HPLC and the calibration curve in Table 1. According to the Student's *t*-test, there were no significant differences between the HPLC and the biosensor methods at a 95% confidence level. This result indicates that the GC/MWCNT/PANI/AChE electrode can be used for the determination of carbaryl in apple samples. The average value of carbaryl detected in the apple samples by the proposed biosensor was 1.68 μmol L^{-1} (0.34 mg kg^{-1}) and this value is smaller than the maximum residue level of carbaryl allowed by the National Health Surveillance Agency in apple samples, as discussed earlier (2.0 mg kg^{-1}).

Table 1: Determination of carbaryl in the apple sample

Repetition	Carbaryl (μmol L^{-1})		
	Proposedbiosensor	HPLC	E_r (%)*
1	1.65	1.59	3.6
2	1.72		7.6
3	1.68		5.4
Mean ± SD	1.68 ± 0.10**		

* E_r = detected proposed biosensor *vs.* detected HPLC.
** $\bar{x} \pm t_{(n-1)} s/\sqrt{n}$.

The GC/MWCNT/PANI/AChE electrode was also applied, using the same methodology as used for methomyl determination in broccoli and cabbage samples. Recoveries between 94.6% and 101.2% and 95.4% and 101.9% methomyl from cabbage and broccoli, respectively, were obtained. The content of pesticides found in samples without any spiking, using the standard addition method for 4.9, 9.8, 14.7 μmol L^{-1} methomyl are presented in Table 2. The *t*-test yielded no significant differences between the HPLC and the proposed analytical method at a 95% confidence level. This result attests the accuracy of the AChE biosensor for the determination of methomyl in cabbage and broccoli samples. The average concentrations of methomyl detected in cabbage (0.38 mg kg^{-1}) and broccoli (0.30 mg kg^{-1}) samples are lower than the maximum residue level of methomyl permitted by Brazilian law, as discussed earlier (3.0 mg kg^{-1}).

Table 2: Determination of methomyl in the broccoli and cabbage samples.

Repetition	Methomyl in the cabbage sample (μmol L^{-1})			Methomyl in the broccoli sample (μmol L^{-1})		
	Proposed biosensor	HPLC	E_r*(%)	Proposed biosensor	HPLC	E_r*(%)
1	2.42	2.22	8.3	1.80		2.8
2	2.31		3.9	1.83	1.75	4.4
3	2.25		1.3	1.92		8.8
Mean ± SD	2.33 ± 0.22**			1.85 ± 0.15**		

* E_r = detected proposed biosensor *vs.* detected HPLC.
** $\bar{x} \pm t_{(n-1)} s/\sqrt{n}$.

CONCLUSIONS

The GC electrode modified with PANI/MWCNT as a core–shell structure was successfully applied for the immobilisation of the acetylcholinesterase enzyme and the determination of carbamate pesticides in fruit and vegetables. The results obtained by the proposed methodology for carbaryl in apple, and methomyl in cabbage and broccoli, were quite similar to those obtained using the HPLC procedure. Although the acetylcholinesterase biosensors are known presenting low selectivity for carbamate, which make then inappropriate for screening procedure for pesticides, they are quite sensitive. The amperometric GC/MWCNT/PANI/AChE biosensor was demonstrated to be a convenient methodology to quantify pesticides in foods, presenting long-term stability and not requiring the use of mediators (phthalocyanine) or cross-linking agents (glutaraldehide), which is promising for future applications in the area of biosensors.

ACKNOWLEDGEMENTS

Financial support from Fapesp (Grant 2010/11567-3 and 2010/11049-2) and CNPq is hereby gratefully acknowledged.

REFERENCES

1. ANVISA – Agência Nacional de Vigilância Sanitária – National Health Surveillance Agency. Available from http://www.anvisa.gov.br/divulga/consulta/62_2001.htm. Accessed June, 2010.
2. Agüí, L., Peña-Farfal, C., Yáñez-Sedeño, P., & Pingarrón, J. M. (2007). Poly-(3- methylthiophene)/carbon nanotubes hybrid composite-modified electrodes. Electrochimica Acta, 52, 7946–7952.
3. Ahuja, T., Mir, I. A., Kumar, D., & Rajesh (2007). Biomolecular immobilization on conducting polymers for biosensing applications. Biomaterials, 28, 791–805.
4. Blanc, R., González-Casado, A., Navalón, A., & Vílchez, J. L. (2000). On the estimate of blanks in differential pulse voltammetric techniques: Application to detection limits evaluation as recommended by IUPAC. Analytica Chimica Acta, 403, 117–123.
5. Caetano, J., & Machado, S. A. S. (2008). Determination of carbaryl in tomato "in natura" using an amperometric biosensor based on the inhibition of acetylcholinesterase activity. Sensors and Actuators, B: Chemical, 129, 40–46.
6. Cai, J., & Du, D. (2008). A disposable sensor based on immobilization of acetylcholinesterase to multiwall carbon nanotube modified screen-printed electrode for determination of carbaryl. Journal of Applied Electrochemistry, 38, 1217–1222.
7. Currie, L. A. (1999). Nomenclature in evaluation of analytical methods including detection and quantification capabilities: (IUPAC Recommendations, 1995). Analytica Chimica Acta, 391, 105–126.
8. Do Nascimento, G. M., Corio, P., Novickis, R. W., Temperini, M. L. A., & Dresselhaus, M. S. (2005). Synthesis and characterization of single-wall-carbon-nanotubedoped emeraldine salt and base polyaniline nanocomposites. Journal of Polymer Science: Polymer Chemistry, 43, 815–822.
9. Du, D., Chen, S., Cai, J., & Zhang, A. (2007). Immobilization of acetylcholinesterase on gold nanoparticles embedded in sol-gel film for amperometric detection of organophosphorous insecticide. Biosensors and Bioelectronics, 23, 130–134.
10. Du, D., Ye, X. X., Cai, J., Liu, J. A., & Zhang, A. D. (2010). Acetylcholinesterase biosensor design based on carbon nanotube-encapsulated polypyrrole and polyaniline copolymer for amperometric detection of organophosphates. Biosensors and Bioelectronics, 25, 2503–2508.
11. Gerard, M., Chaubey, A., & Malhotra, B. D. (2002). Application of conducting polymers to biosensors. Biosensors and Bioelectronics, 17, 345–359.
12. Hernández-Pérez, A., Avilés, F., May-Pat, A., Valadez-González, A., Herrera-Franco, P. J., & Bartolo-Pérez, P. (2008). Effective properties of multiwalled carbon nanotube/epoxy composites using two different tubes. Composites Science and Technology, 68, 1422–1431.
13. Laschi, S., Bulukin, E., Palchetti, I., Cristea, C., & Mascini, M. (2008). Disposable electrodes modified with multi-wall carbon nanotubes for biosensor applications. Ingénierie et Recherche Biomédicale, 29, 202–207.
14. Li, M., Guo, Y., Wei, Y., MacDiarmid, A. G., & Lelkes, P. I. (2006). Electrospinning polyaniline-contained gelatin nanofibers for tissue engineering applications. Biomaterials, 27, 2705–2715.
15. Liu, G., Riechers, S. L., Mellen, M. C., & Lin, Y. (2005). Sensitive electrochemical detection of enzymatically generated thiocholine at carbon nanotube modified glassy carbon electrode. Electrochemistry Communication, 7, 1163–1169.
16. Merkoçi, A., Pumera, M., Llopis, X., Pérez, B., del Valle, M., & Alegret, S. (2005). New materials for electrochemical sensing VI: Carbon nanotubes. TrAC, Trends in Analytical Chemistry, 24, 826–838.
17. Moraes, F. C., Cabral, M. F., Mascaro, L. H., & Machado, S. A. S. (2011). The electrochemical effect of acid functionalisation of carbon nanotubes to be used in sensors development. Surface Science, 605, 435–440.
18. Moraes, F. C., Golinelli, D. L. C., Mascaro, L. H., & Machado, S. A. S. (2010). Determination of epinephrine in urine using multi-walled carbon nanotube modified with cobalt phthalocyanine in a paraffin composite electrode. Sensors and Actuators, B: Chemical, 148, 492–497.
19. Moraes, F. C., Mascaro, L. H., Machado, S. A. S., & Brett, C. M. A. (2009). Direct electrochemical determination of carbaryl using a multi-walled carbon nanotube/cobalt phthalocyanine modified electrode. Talanta, 79, 1406–1411.
20. Pedrosa, V. A., Caetano, J., Machado, S. A. S., & Bertotti, M. (2008). Determination of parathion and carbaryl pesticides in water and food samples using a self assembled monolayer/acetylcholinesterase electrochemical biosensor. Sensors, 8, 4600–4610.

21. Rotariu, L., Zamfir, L. G., & Bala, C. (2010). Low potential thiocholine oxidation at carbon nanotube-ionic liquid gel sensor. Sensors and Actuators, B: Chemical, 150, 73–79.

22. Singh, D. K., & Agarwal, R. A. (1983). Inhibition kinetics of certain organophosphorus and carbamate pesticides on acetylcholinesterase from the snail Lymnaea acuminata. Toxicology Letters, 19, 313–319.

23. Sung, W. J., & Bae, Y. H. (2003). A glucose oxidase electrode based on polypyrrole with polyanion/PEG/ enzyme conjugate dopant. Biosensors and Bioelectronics, 18, 1231–1239.

24. Upadhyay, S., Rao, G. R., Sharma, M. K., Bhattacharya, B. K., Rao, V. K., & Vijayaraghavan, R. (2009). Immobilization of acetylcholineesterase–choline oxidase on a gold–platinum bimetallic nanoparticles modified glassy carbon electrode for the sensitive detection of organophosphate pesticides, carbamates and nerve agents. Biosensors and Bioelectronics, 25, 832–838.

25. Viswanathan, S., Radecka, H., & Radecki, J. (2009). Electrochemical biosensor for pesticides based on acetylcholinesterase immobilized on polyaniline deposited on vertically assembled carbon nanotubes wrapped with ssDNA. Biosensors and Bioelectronics, 24, 2772–2777.

26. Wen, Z. H., Ye, B. X., & Zhou, X. Y. (1997). Direct electron transfer reaction of glucose oxidase at bare silver electrodes and its application in analysis. Electroanalysis, 9, 641–644.

27. Zhou, Y., Qin, Z. Y., Li, L., Zhang, Y., Wei, Y. L., Wang, L. F., & Zhu, M. F. (2010). Polyaniline/multi-walled carbon nanotube composites with core-shell structures as supercapacitor electrode materials. Electrochimica Acta, 55, 3904–3908.

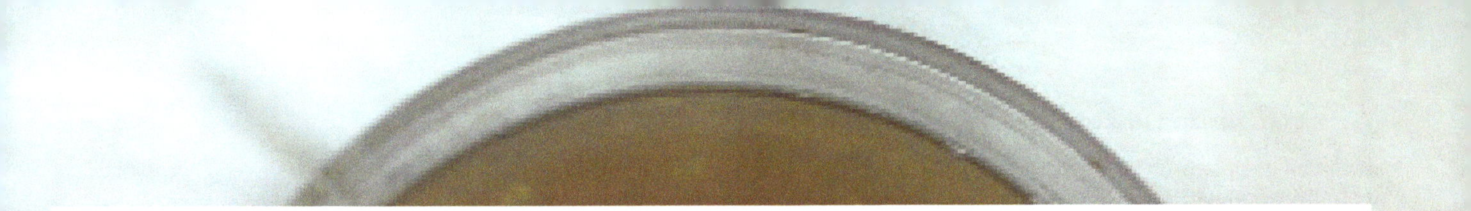

CHAPTER 9

ENZYMATIC CONVERSION OF GLYCEROL TO GLYCERIC ACID WITH IMMOBILISED LACCASE IN NA-ALGINATE MATRIX

Chi Shein Hong, Sim Yee Chin, Chin Kui Cheng, Mahmud Mohd Sabri, and Gek Kee Chua

Faculty of Chemical & Natural Resources Engineering, Universiti Malaysia Pahang, 26300 Gambang, Pahang, MALAYSIA

ABSTRACT

In this study, enzymatic oxidation of glycerol was performed for the production of glyceric acid. The effectiveness of immobilised laccase in Na-Alginate matrix was also verified. Glycerol was incubated at 25°C with 30 mM of 2,2,6,6-tetramethylpiperidine-N-oxyl (TEMPO), which acted as a mediator, in the presence of immobilized laccase from Trametes versicolor. Glyceric acid was quantified using HPLC, while the activity of enzyme was measured using ABTS assay. The yield of laccase activity was 90%. On the other hand, glyceric acid obtained was around 7 wt% from the total product at 24h. In conclusion, enzymatic oxidation is able to produce comparable amount of glyceric acid through a more environmental friendly method.

INTRODUCTION

Biodiesel has emerged as an alternative sustainable source of fuel due to the great demand and consumption of petroleum. It is estimated that the global market will produce 37 billion gallons of biodiesel with an average annual growth of 42% by 2016 [1]. Glycerol (Gly), a byproduct of biodiesel, will be generated at around 10 wt% of total biodiesel production, thus approximately 4 billion gallons of glycerol will be produced in 2016 [2].

Citation: Chi Shein Hong, Sim Yee Chin, Chin Kui Cheng, Mahmud Mohd Sabri, Gek Kee Chua, Enzymatic Conversion of Glycerol to Glyceric Acid with Immobilised Laccase in Na-Alginate Matrix, doi:10.1016/j.proche.2015.12.102

The emanation of Gly from biodiesel has flooded the market which led to a significant decrease in price for Gly. A progressive changeover of biodiesel byproduct to high value derivatives transpired to be an inevitable necessity. The development of new applications for Gly is considered important in technological progress and would be enthusiastically welcomed by the industry.

Selective oxidation of Gly produces derivatives such as glyceraldehyde (Gald), glyceric acid (G1Ac), tartronic acid and so on which serve as new chemical intermediates. Since the products are of practical value, these would address a very large chemical market [3]. In selective oxidation, primary hydroxyl groups would be oxidised to Gald, an intermediate for carbohydrate metabolism and also standard for comparison of chiral molecules (D-or L-series). G1Ac can be used for treatment of skin disorders. Additionally, GIAc in its ester form with the combination of a quaternary ammonium salt can act as a biodegradable fabric softener [4]. GIAc derivatives can be used as a multifunctional building block compound [5]. From biology perspective, the enantiometric isomer D-G1Ac has been reported to accelerate the oxidation of acetaldehyde and ethanol in vivo in rats through circumventing the rate limiting step [6]. It has liver stimulating and cholesterolytic actitivies in dog [7]. It is also acts as trypsin inhibitors [8] and building block of bio-based polyesters [5].

Selective oxidation can also be performed by using catalyst such as Pt, Pd and Au [9,10]. However, one of the drawbacks in using the supported Pt and Pd catalysts during reaction is their deactivation increase with time. Particularly, the platinum group metals based catalyst suffer oxygen poisoning which is proportional to the oxygen partial pressure [11,12]. Low partial pressure of oxygen is typically used to limit the oxygen dissolution. Gold catalysts have better resistance than platinum based catalysts towards oxygen poisoning, allowing the use of higher oxygen partial pressure [13]. Bimetallic catalysts (featuring active sites with two or more metallic elements) are introduced to enhance activity or selectivity. Comparing Au—Pd with Au and Pd catalysts in the liquid phase oxidation of glycerol, the activity is highly enhanced by the presence of alloyed phase [14]. Preparation and impregnation of these catalysts often appear to be time consuming and higher temperature and pressure are needed as well.

A greener approach- TEMPO/laccase to oxidise glycerol is being introduced [15]. Laccase (EC 1.10.3.2: benzendiol : oxygen oxidoreductases), distributed widely in bacteria, plants, fungi and insects, is a type of multi-copper containing enzymes [16]. It can catalyse oxidation of a variety of inorganic and phenolic compounds, concomitantly reduce oxygen to water without the formation of hydrogen peroxide [17,18]. Laccases have potential in chemical synthesis [19], which are limited by its redox potential. By imitating nature, mediators are added as electron shuttles for the substrate to laccase [20]. 2,2,6,6-tetramethylpiperidine- 1 -oxyl (TEMPO) and its derivatives have been used in many areas of synthetic organic chemistry with the possible achievement of chemoselectivity in the oxidation process. TEMPO acts as mediator, in which laccases oxidise it to corresponding oxoammonium cation through one electron oxidation, thus regenerate TEMPO. The oxoammonium cation, in turn, oxidise substrate such as alcohol via heterolytic pathway [21].

There are many constraints in using enzymes and the common perception is that enzymes are sensitive, unstable, not ideal as a catalyst and undesirable in most synthesis and reaction. The decrease in stability and the catalytic activity of free enzymes due to its catalytic environment seriously limit their industrial application. Immobilisation is the most adopted stabilisation method in overcoming these limitations which allows enzyme reutilisation. Laccases have been immobilised on different supports for the removal of several pollutants including dyes [22]. Entrapment is the easiest method in which it induces no structural alteration of enzymes, less tedious procedure and reduced number of chemicals used It is just the physical retention of the enzymes in porous solid matrix [23]. As reported by Niladevi and Prema [24], immobilisation of laccase in Cu alginate has immobilisation yield of 61% while immobilization for the entrapment in PVA-alginate yields 97.4% [25].

In this paper, the oxidation of Gly with an enzyme-mediator system to obtain G1Ac is shown. The study also being extended by comparing the enzymatic reaction with chemical reaction [9] and effect of different immobilisation methods in converting glycerol to G1Ac. The use of immobilised laccase allows the catalyst to be separated from reaction mixture easily for recycling purposes.

MATERIALS AND METHODS

Laccase from Trametes versicolor was purchased from Sigma Aldrich, Germany without further purification. Glycerol was obtained from Fisher Scientific. 2,2,6,6-tetramethylpiperidine-N-oxyl (TEMPO), polyethylene glycol (PEG) (MW = 4000), calcium chloride ($CaCl_2$), sodium acetate, glutaraldehyde were sought from Merck. Sodium Alginate was bought with R&M Chemicals and 2,2'-azino-bis(3-ethylbenzothiazoline-6-sulphonic acid) (ABTS) was procured from Roche. All standards for HPLC were purchased from Sigma-Aldrich.

Laccase activity assay

Laccase activity was determined by monitoring the absorbance of ABTS oxidation at 415nm wavelength using UV-VIS spectrophotometer at room temperature (light path = 1 cm). The assay mixture for the blank consists of 1.9 ml of 0.1 M acetate buffer (pH 4.5), 0.1 ml of pure water and 1.0 ml of 0.5 mM ABTS solution. Pure water was replaced with samples for determination of the laccase activity. One unit (U) of the enzyme was defined as 1 umol ABTS oxidised per minute under the stated assay condition.

Immobilisation of laccase

2.0 g of gelatine and 0.5 g of PEG was added to 100 ml of 2.0% (w/v) sodium alginate solution. 0.1 g of laccase (10U/ml) was then added and mixed thoroughly for 10 minutes at 25°C. The mixture was withdrawn by using a sterile 5 ml syringe and extruded through a 21 gauge (0.51 mm inner diameter) needle into 2.0% (w/v) $CaCl_2$ solution, beads formed immediately when touched the surface of the solution. Beads were left unstirred to harden in the $CaCl_2$ solution at 4°C for 2 hours, after which the $CaCl_2$ solution was removed and the beads were washed twice with distilled water. Then, the beads were subsequently incubated in 100 ml of 0.6% (w/v) glutaraldehyde solution and stirred at 4°C for 2 hours. The beads were washed several times with phosphate buffer (pH 7) to remove the laccase which was not entrapped into the beads. The immobilised Laccase enzymes are stored at 4°C in acetate buffer (pH 4.5) [22].

Oxidation of Glycerol

50 mL mixture of 100 mM glycerol, 30 mM TEMPO, and 1 g of immobilised laccase in 100 mM sodium acetate buffer (pH4.5) was incubated at 25°C under stirred condition. The oxidation process was run in triplicate for 24 hours and sampling was done hourly.

Analytical Method

Quantitative analysis of the oxidation products were conducted by using a high performance liquid chromatography (HPLC) equipped with DAD and RID. A 300 mm Phenomenex Rezek ROA column was used. 3 mM H2SO4 was used as mobile phase with 0.5 mL flowrate, 210 nm wavelength, injection volume of 10 'IL and the column temperature at 75 °C. The calibration plot was obtained by linear regression of peak-area ratio against concentration. Products were identified by comparison of their retention times with standards of the respective components.

RESULTS AND DISCUSSIONS

Methods of immobilisation affected the amount of laccase immobilised and influenced the final activity remained. This is due to the different in procedures of immobilisation and chemicals used, which in turn determined the types of interaction between laccase and support. The lengthiness of the procedures also contributes to the possible loss of enzyme during the process. Table 1 shows the activity of laccase after immobilisation in Na-Alginate matrix in this study was 90%. Activity of laccases, which was being immobilised using other immobilisation methods, was also depicted in Table 1. It could be seen that some of the immobilization yield were high but some were moderate.

Table 1: Immobilisation methods and yield

Author	Immobilisation method	Material	Immobilisation yield (%)
[26]	covalent bond	Chitosan	52.2
[27]	entrapment	Alginate solution cross-linked by $CuSO_4$	74.6 - 146.8
[28]	encapsulation	LbL technique (assembly of differently charged polyelectrolyte layers)	75
		on manganese carbonate ($MnCO_3$)	
[29]	adsorption	Chitosan-g-PEI	72
[30]	covalent bond	Green coconut fiber	50,70
Current study	entrapment	Na-Alginate	90

Time series profiles of the reaction are depicted in Figure 1. It can be seen from Figure 1 that in the early stage of reaction, Gald was the only product. This implied that selectivity of TEMPO is highly towards primary alcohol. Nearly 30% of Gly was converted. The concentration of Gald increased until it reached its maximum at 24h. After 3h of incubation, GIAc was detected in the reaction mixture. The concentration of GIAc increased steadily throughout the period of incubation. The final amount of GIAc obtained is about 7 wt %. In this oxidation process, Gald became substrate for TEMPO besides the initial reactant, which is Gly. G1Ac was obtained as a result of further oxidation from Gald. The oxidation of Gly does not stop at G1Ac, it can lead to other organic acids such as tartronic acid, mesoxalic acid, formic acid etc. However, the concern of this paper is only G1Ac, therefore the products appeared in extended time after 24 h were not included.

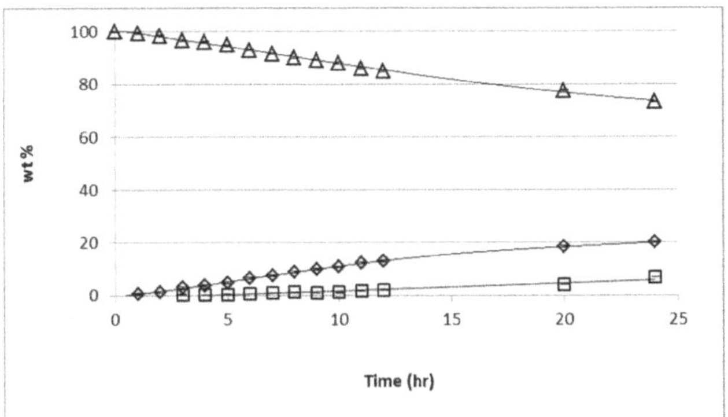

Figure 1: Oxidation of 100mM glycerol by 30mM TEMPO with ommobilized Laccase and information of product (Δ:glycerol;◊:glyceraldehyde;□:glyceric acid)

In general, nitroxyl radical plays an important role in selective oxidation of alcohol. Due to its high reaction rate and high selectivity, nitroxyl radical is employed for the modification of high molecular mass polysaccharides too [31]. Conventional methods such as manganese dioxide, dimethyl sulfoxide activated by oxalyl chloride and silver oxide are based on chromium reagents which are toxic and only applicable to lab scale usage [15]. As a result, nitroxyl radicals such as TEMPO and its derivatives that are proven to be effective in selective oxidation of alcohols [15,32] has drawn attention as a substitution. In addition, TEMPO can be recovered from reaction mixture by extracting it with organic solvents [33], thus giving an extra advantage. The role of TEMPO as mediator in enzymatic oxidation with the presence of immobilised laccase is outlined in Figure 2. In this case, TEMPO acted as electron shuttle, being oxidised by laccase, diffused away from enzymatic pocket and oxidized glycerol (substrate) which could not directly entered and reacted with laccase. Therefore, with the presence of mediator, the range of substrate is being extended, oxidation which is not available to enzyme can be performed [34].

The current study was an example of oxidation by TEMPO/laccase in which Na-Alginate as a support for laccase seemed to be a promising immobilisation method. The oxidation products by TEMPO/immobilised laccase in Na-Alginate within 24 h of reaction were shown in Figure 3. Comparing with Figure 4 which was reported by Liebminger et al. [15] (activity recovery on carrier was 68%) glyceric acid obtained by immobilised laccase on Na-Alginate was just about half of it. One of the reasons was due to difference in immobilisation method. Immobilisation of laccase in Na-Alginate was an entrapment method while immobilisation of laccase on Alumina pellet was a covalent bond technique. TEMPO had more contact with laccase in the covalent bond technique than entrapment since laccases were bonded on the surface covalently which made the oxidation reacted faster to desired product. However, laccases were trapped in the gel matrix in entrapment technique which caused diffusion limitation. TEMPO will have to take longer period of time to diffuse and react with laccase and substrate as described earlier. Thus, longer reaction period is required to obtain desired product. Another possible reason might be the difference in redox potential for laccase species. Liebminger et al. [15] used the Trametes hirsute while Trametes versicolor was being used in this study. Each and every laccase has its own intrinsic redox potential, the more positive the potential, the greater the species' affinity for electrons and tendency to be reduced. Immobilisation by using Na-Alginate not only cheaper than the compared method (immobilisation on Alumina pellet), it also took only half a day for a complete immobilisation rather than a week for the compared method.

A study involved catalytic reaction which was reported by Porta and Prati [9] was used to compare with current enzymatic study. The catalytic oxidation in which 1% Au/C was used with graphite (Lonza HSAG3OO) as the support in the following reaction condition, T = 60 'C, 10 wt% of glycerol, NaOH/ glycerol = 1 mol/mol, glycerol/Au = 500 mol/mol, the yield for GlAc was about 2.5 mole %, which is almost the same with the oxidation of Gly by using 3OmM TEMPO in the presence of immobilised laccase on Na-Alginate at 24h. The reaction was slow for TEMPO/laccase system, which might be due to the reaction that was not optimised yet (the reaction was mu under ambient temperature and pressure).

CONCLUSION

Laccase from *Trametes versicoZor* was successfully immobilised in Na-Alginate. The TEMPO/ laccase system foroxidation of glycerol to glyceric acid was reliable, achieved nearly 7 wt% of the total product at the end of reactiontime. This study proved that laccase was an effective and environmental friendly regeneration oxidant compared toother stoichiometrically used chemicals. Optimisation is recommended for future study.

REFERENCES

1. Wang, L., Du, W., Liu, D.H., Li, L.L., Dai, N.M., 2006. Lipase-catalyzed biodiesel production from soybean oil deodorizer distillate with absorbent present in tert-butanol system. Journal of Molecular Catalysis B Enzymatic, 43:29-32.
2. Anand, P. and Saxena, R.K., 2012. A comparative study of solvent-assisted pretreatment of biodiesel derived crude glycerol on growth and 1,3-propanediol production from Citrobacter freundii. New Biotechnology, 29:199-205.
3. Werpy,T. and Peterson.G., 2004. Top Value Added Chemicals from Biomass Volume I — Results of Screening for Potential Candidates from Sugars and Synthesis Gas Top Value Added Chemicals From Biomass Volume I: Results of Screening for Potential Candidates. US Department of Energy.
4. Behr, A., Jens, E., Ken, I., Julia, L. and Falk, L., 2008. Improved Utilisation of Renewable Resources: New Important Derivatives of Glycerol. Green Chemistry, 10:13-.
5. Fukuoka, T., Habe, H., Kitamoto, D. and Sakaki, K., 2011. Bioprocessing of glycerol into glyceric Acid for use in bioplastic monomer. Journal of Oleo Science, 60:69-373.
6. Eriksson, C.J.P., Saarenmaa, T.P.S., Bykov, I.L. and Heino, P.U., 2007. Acceleration of ethanol and acetaldehyde oxidation by d-glycerate in rats. Metabolism: Clinical and Experimental, 56: 895-98.
7. Handa, S., Handa, A.K., Hasegawa, P.M., Bressan, R.A., 1986. Proline accumulation and the adaptation of cultured plant cells to water stress. Plant Physiology, 80: 938-945.
8. Habe, H., Fukuoka, T., Sato, Shun., Kitamoto, D. and Sakaki, K., 2011. Synthesis and evaluation of dioleoyl glyceric acids showing antitrypsin activity. Journal of Oleo Science, 60:327-331.
9. Porta, F. and Prati, L., 2004. Selective oxidation of glycerol to sodium glycerate with gold-on-carbon catalyst: An insight into reaction selectivity. Journal of Catalysis, 224:397-403.
10. Wang, F.F., Shao., S., Liu, C.L., Xu, C.L., Yang, R.Z. and Dong, W.S., 2015. Selective oxidation of glycerol over Pt supported on mesoporous carbon nitride in base-free aqueous solution. Chemical Engineering Journal, 264:336-343.
11. Gallezot, P., 1997. Selective oxidation with air on metal catalysts. Catalysis Today, 37:405-418.
12. Fordham, P., Besson, M. and Gallezot, P., 1995. Selective catalytic oxidation of glyceric acid to tartronic and hydroxypyruvic acids. Applied Catalysis A, General, 133:179-184.
13. Demirel, S., Lehnert, K., Lucas,M. and Claus, P., 2007. Use of renewables for the production of chemicals: Glycerol oxidation over carbon supported gold catalysts. Applied Catalysis B: Environmental, 70:637-643.
14. Villa, A., Campione, C. and Prati, L., 2007. Bimetallic gold/palladium catalysts for the selective liquid phase oxidation of glycerol. Catalysis Letters, 115:133-136.
15. Liebminger, S., Siebenhofer, M. and Guebitz, G., 2009. Oxidation of glycerol by 2,2,6,6-tetramethylpiperidine-N-oxyl (TEMPO) in the presence of laccase. Bioresource Technology, 100:4541-4545.
16. Lettera, V., Piscitelli, A., Leo, G., Birolo, L., Pezzella, C. and Sannia, G., 2010. Identification of a new member of Pleurotus ostreatus laccase family from mature fruiting body. Fungal Biology, 114:724-730.
17. Bourbonnais, R. and Paice M.G., 1992. Demethylation and delignification of kraft pulp by Trametes versicolor laccase in the presence of 2,2_-azinobis-(3-ethylbenzthiazoline-6-sulphonate). Applied Microbiology and Biotechnology, 36:823-7.
18. Yaropolov, A.I., Skorobogat'ko, O.V., Vartanov, S.S. and Varfolomeyev, S.D., 1994. Laccase: properties, catalytic

mechanism, and applicability. Applied Biochemistry and Biotechnology, 49:257-280.

19. Schroeder, M. Pereira, L., Couto, S.R., Erlacher, A., Schoening, K.U., Cavaco-Paulo, A. and Guebitz, G.M., 2007. Enzymatic synthesis of Tinuvin. Enzyme and Microbial Technology, 40:1748-1752.

20. Banci, L., Ciofi-Baffoni, S. and Tien, M., 1999. Lignin and Mn peroxidase-catalyzed oxidation of phenolic lignin oligomers. Biochemistry, 38:3205-3210.

21. Arends, I.W.C.E., Li, Y.X., Ausan, R. and Sheldon, R.A., 2006. Comparison of TEMPO and its derivatives as mediators in laccase catalysed oxidation of alcohols. Tetrahedron, 62:6659-6665.

22. Wang P., Fan X., Cui L., Wang Q., Zhou A., 2008. Decolorization of reactive dyes by laccase immobilized in alginate/gelatine blent with PEG. Journal of Environmental Sciences, 20: 1519-1522.

23. Lu. L., Zhao. M. and Wang. Y., 2007. Immobilization of laccase by alginate–chitosan microcapsules and its use in dye decolorization. World Journal of Microbiology and Biotechnology. 23:159-166.

24. Niladevi K.N. and Prema P., 2008. Immobilization of laccase from Streptomyces psammoticus and its application in phenol removal using packed bed reactor. World Journal of Microbiology and Biotechnology, 24:1215-1222.

25. Mohd Zain, N.A., Mohd Suardi, S. and Idris, A., 2010. Hydrolysis of liquid pineapple waste by invertase immobilized in PVA-alginate matrix. Biochemical Engineering Journal, 50:83-89.

26. Chen, H., Zhang, J., Wang, W. and Yang, Y., 2006. Preparation and characteristics of immobilized laccase from coriolus versicolor on chitosan. Beijing Daxue Xuebao (Ziran Kexue Ban)/Acta Scientiarum Naturalium Universitatis Pekinensis, 42:254-258.

27. Teerapatsakul. C., Bucke. C., Parra. R., Keshavarz. T. and Chitradon. L., 2008. Dye decolorisation by laccase entrapped in copper alginate. World Journal of Microbiology and Biotechnology, 24:1367-1374.

28. Cretini, C., Perazzini, R. and Saladino, R., 2010. Oxidative Functionalisation of lignin by layer-by-layer immobilized laccases and laccase microcapsules. Applied Catalysis A: General, 372: 115-123.

29. Metin, A.U., 2013. Immobilization of Laccase onto Polyethylenimine Grafted Chitosan Films: Effects of System Parameters. Macromolecular Research. 21:1145-1152.

30. Cristóvão, R.O., Silvério, S.C., Tavares, A.P. M., Brígida, A.I. S., Loureiro, J.M., Boaventura, R.A.R., Macedo, E.A. and Coelho, M.A. Z., 2012. Green coconut fiber: A novel carrier for the immobilization of commercial laccase by covalent attachment for textile dyes decolourization. World Journal of Microbiology and Biotechnology, 28:2827-2838.

31. Bragd, P.L., Bekkum, V.H. and Besemer, A.C., 2004. TEMPO-mediated oxidation of polysaccharides: survey of methods and applications. Topics in Catalysis, 27:49-66.

32. Gamez, P., Arends, I.W.C.E. and Sheldon, R.A., 2004. Room temperature aerobic copper-catalysed selective oxidation of primary alcohols to aldehydes. Advanced Synthesis and Catalysis 346:805-811.

33. Nooy, A.E.J. De, Besemer, A.C. and Bekkum, H. Van, 1996. On the Use of Stable Organic Nitroxyl Radicals for the Oxidation of Primary and Secondary Alcohols. Synthesis, 10:1153-1176.

34. Hildén, L., Johansson, G., Pettersson, G., Li, J.B., Ljungquist, P., Henriksson, G., 2000. Do the extracellular enzymes cellobiose dehydrogenase and manganese peroxidase form a pathway in lignin biodegradation? Federation of European Biochemical Societies Letters, 477:79-83.

CHAPTER 10

LACCASE IMMOBILIZATION BY CHELATED METAL ION COORDINATION CHEMISTRY

Qingqing Wang, Jing Cui, Guohui Li, Jinning Zhang, Fenglin Huang, and Qufu Wei

Key Laboratory of Eco-Textiles, Ministry of Education, Jiangnan University, Wuxi 214122, China

ABSTRACT

In this work, amidoxime polyacrylonitrile (AOPAN) nanofibrous membrane was prepared by a reaction between PAN nanofibers and hydroxylamine hydrochloride. The AOPAN nanofibrous membranes were used for four metal ions (Fe^{3+}, Cu^{2+}, Ni^{2+}, Cd^{2+}) chelation under different conditions. Further, the competition of different metal ions coordinating with AOPAN nanofibrous membrane was also studied. The AOPAN chelated with individual metal ion (Fe^{3+}, Cu^{2+}, Ni^{2+}, Cd^{2+}) and also the four mixed metal ions were further used for laccase (Lac) immobilization. Compared with free laccase, the immobilized laccase showed better resistance to pH and temperature changes as well as improved storage stability. Among the four individual metal ion chelated membranes, the stability of the immobilized enzymes generally followed the order as Fe–AOPAN–Lac > Cu–AOPAN–Lac > Ni–AOPAN–Lac > Cd–AOPAN–Lac. In addition, the immobilized enzyme on the carrier of AOPAN chelated with four mixed metal ions showed the best properties.

INTRODUCTION

Laccase is a copper-containing enzyme that can catalyze the oxidation of a wide variety of substrates with the concomitant four electron reduction of oxygen to water [1,2]. Due to its high efficiency, low cost, and good availability, it has been widely used in various applications including paper manufacturing [3], wood processing [4], environmental bioremediation [5,6], the food industry [7], as well as textile engineering [8]. However, it has been realized that free enzymes have the weakness of non-recyclability, and reusability in industrial applications. These problems can be well solved by enzyme immobilization.

Various methods concerning laccase immobilization have been reported, including entrapment, encapsulation, adsorption, covalent binding, as well as self-immobilization [9]. Among all these methods, adsorption is relatively simple and less expensive, so it may have a higher commercial potential than the other methodologies. Some studies have demonstrated that adsorption is preferable to other techniques for the immobilization of some particular laccases [10]. Recently, metal-based supports have been receiving emerging

Citation: Qingqing Wang, Jing Cui, Guohui Li, Jinning Zhang, Fenglin Huang, and Qufu Wei, Laccase Immobilization by Chelated Metal Ion Coordination Chemistry, doi:10.3390/polym6092357.

attention for metal-chelated enzyme adsorption [9]. Enzyme immobilization on metal chelated support is based on multipoint interactions between chelated metal ions on the support and the imidazole group of histidine, the indole group of tryptophan and the thiol group of cysteine [11]. The transition metal ions are considered strong Lewis acids and interact with strong Lewis bases such as nitrogen and oxygen [12]. In addition, Cu-, Ca-, Al-, Zr-based supports have been studied for the laccase immobilization [12,13,14,15]. A general improvement in laccase activity and stability was observed with all these metal-based supports.

Electrospun nanofibers have been recognized as excellent supports for enzyme immobilization since they have a large surface area, high recovery and reusability of the enzymes [16]. Polyacrylonitrile (PAN) nanofibers after surface modification have been studied and proved to be ideal support for laccase immobilization [17]. Amidoxime polyacrylonitrile (AOPAN) prepared by the chemical modification of the nitrile group was known as good adsorbent for metal ions [18]. Feng [19] has proved that after immobilized onto the surface of Fe(III) chelated AOPAN nanofibrous membrane, catalase showed better stabilities. However, as far as we are concerned, metal ion chelated AOPAN nanofibers have not been used for laccase immobilization yet.

In this work, AOPAN nanofibrous membranes were obtained by a facile electrospinning process and followed by treatment in the aqueous solution of hydroxylamine hydrochloride. Four different metal ions (Fe^{3+}, Cu^{2+}, Ni^{2+}, Cd^{2+}) were then chelated by AOPAN nanofibrous membranes. In addition, the metal ion chelated AOPAN nanofibrous membranes were then used to immobilize laccase, the mechanism of which was shown in Figure 1. The properties of the immobilized enzyme including optimum pH and temperature, storage stability as well as operational stability were investigated. Moreover, the AOPAN nanofibrous membrane chelated with these four mixed metal ions mentioned above was also used as substrate to evaluate its ability in enzyme immobilization.

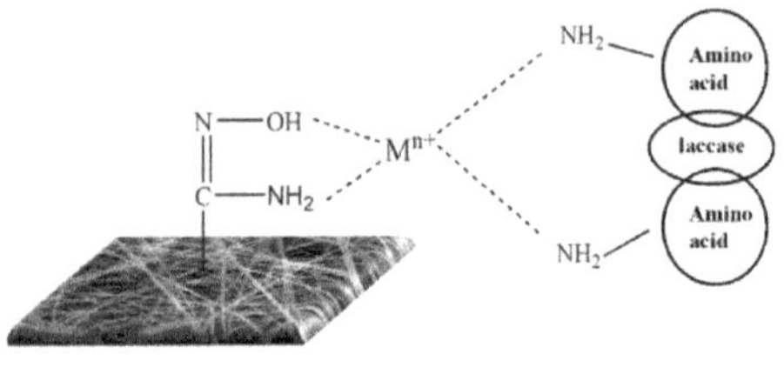

Figure 1: Schematic illustration of laccase immobilized by amidoxime polyacrylonitrile (AOPAN)-metal chelated nanofibrous membrane.

EXPERIMENTAL SECTION

Materials

The laccase produced by *Ganoderma lucidum* using submerged fermentation was donated by Ding, who has reported the process of the enzyme production and characterization in their work [20]. 2,2›-azino-bis-(3-ethylbenzothiazoline-6-sulfonic acid) (ABTS) was obtained from Richu Biosciences Co., Ltd. (Shanghai, China). PAN (M_w = 50,000 g·mol^{-1}) was purchased from Shangyu Wu and Yue Economic and Trade Co. Ltd. (Zhejiang, China). All the other reagents used in this work were of analytical grade and were purchased from Sinopharm Chemical Reagent Co., Ltd. (Shanghai, China).

Preparation of Metal Chelated Amidoxime PAN Nanofibrous Membranes

Three grams PAN powders were added into 20 mL *N,N*-dimethylformamide (DMF) and magnetically stirred for about 24 h until a homogeneous solution was obtained. Then the solutions were electrospun at a positive voltage of 15 kV with a working distance of 15 cm, and a flow rate of 0.5 mL·h^{-1}.

The as-prepared electrospun nanofibers (about 0.2 g) were then immersed in 50 mL of 20 g·L^{-1} hydroxylamine hydrochloride aqueous solution. The pH of the reaction solution was adjusted to 7 by adding sodium carbonate solution. The reaction was carried out at 50–60 °C for 2 h. After being thoroughly washed, the AOPAN nanofibrous membranes were dried in vacuum oven at 50 °C. The conversion of nitrile group in the PAN was calculated as follows [21]:

$$C = \frac{(W_1 - W_0)\times 53}{W_0 \times 33}\times 100 \qquad (1)$$

where C (%) is the conversion of the nitrile group into amidoxime group in the PAN, W_0 is the weight of the initial PAN nanofiber membranes, W_1 is the weight of the modified PAN nanofiber membranes, 53 and 33 are the molecular weights of acrylonitrile monomer and hydroxylamine, respectively. The conversion rate could be affected by temperature, reaction time and also the drying process, so it would be variable in different batches. Certain amounts of ferric chloride, copper chloride dihydrate, nickel chloride, cadmium chloride were dissolved in deionized water with the final concentration of 1 g·L^{-1}. The pH of the solution was adjusted to 4 by HCl (1 mM). Then the dried AOPAN nanofibrous membranes were separately put into a beaker with 50 mL metal salt solutions. After reaction for 72 h at 30 °C, the membranes were taken out. After being thoroughly washed with deionized water, the membranes were dried in vacuum oven at 50 °C.

Immobilization of Laccase on Metal-Chelated AOPAN Nanofibrous Membranes

The laccase was dissolved in acetic acid/sodium acetate buffer (pH = 4.5) solution at the concentration of 3 g·L^{-1} by magnetic stirring for 20 min in ice bath. The supernatant was collected by centrifuging for 5 min and used for the following immobilization process.

The nanofibrous membrane (100 mg accurately weighed) was placed into the centrifuge tubes and then enzyme solution (8 mL per tube) was distributed into it. The immobilized laccase on AOPAN nanofibrous membrane before metal-chelation was described in our previous work [22]. The immobilization process was conducted in the refrigerator at 4 °C for 12 h. After that, the membranes were taken out and washed thoroughly with buffer solution until no enzyme can be detected in the washing solution. The activity of the washing solution was determined by adding ABTS into the specimen taken from the washing water, the absorbance of which at the wavelength of 420 nm after reaction for 5 min was recorded. If the absorbance value was less than 0.1, it could be considered as no enzyme in the washing water.

Structural Characterization

The morphology of the electrospun PAN nanofibers before and after amidoxime modification as well as metal ion chelation was characterized by scanning electron microscope (SEM) (Quanta 200, Holland FEI Company, Eindhoven, The Netherlands. The samples were sputter coated with a thin layer of Au nanoparticles to reduce the charging effects.

Fourier Transform Infrared Spectroscopy (FTIR) (Nicolet Nexus, Thermo Electron Corporation, Waltham, MA, USA) was employed here to analyze the chemical structure of PAN and AOPAN nanofibrous membrane in the range of 4000–400 cm^{-1}, using an Attenuated total reflectance (ATR) FTIR method. The spectra were recorded with 16 scans with a resolution of 4 cm^{-1}.

Energy dispersive X-ray analysis (EDX) (Hitachi H-7500 transmission electron microscope, Tokyo, Japan) was used here to further confirm the existence of the metal ions chelated by the AOPAN nanofibrous membranes. Besides, the concentration of the metal ions was determined by atomic absorption spectroscopy (AAS) (AA-220, Varian, Palo Alto, CA, USA).

Activity Assays of Free and Immobilized Laccase

The immobilized enzyme activity was assayed at 30 °C using ABTS as the substrate. The detailed process was reported in our previous paper [23].

Kinetic tests were carried out at 30 °C in 100 mM sodium acetate (pH = 4.5) buffer using ABTS as the substrate, with the substrate concentration varied from 0.1 to 1 mM. The kinetic parameters of K_m and V_{max} were calculated according to the Lineweaver–Burk double reciprocal models [24].

To determine the optimum pH, the immobilized enzymes were incubated in 100 mM sodium acetate buffers with pH ranging from 2 to 7 at 4 °C for 2 h and then assayed for activity, while the optimum temperature was determined by the activities of the immobilized enzymes incubated in buffers (pH 4.5) for 5 min at specific temperatures varying from 30 to 65 °C before adding ABTS.

The storage stability of the immobilized enzyme was determined by the activity retention ratio during storage at 4 °C in 100 mM sodium acetate buffer solution (pH 4.5), at a regular intervals up to 20 days. Besides, the operational stability was studied by repeated usage for about 10 times, and the relative enzyme activity was recorded. The experiments were carried out at 30 °C, pH 4.5.

RESULTS AND DISCUSSION

Characterization of the Amidoximized PAN

The SEM images of the PAN and AOPAN nanofibers are shown in Figure 2a,b, respectively. Compared with PAN nanofibers, the AOPAN nanofibers (51.7% conversion) showed an almost similar morphology and no cracks or physical degradation was seen. However, the overall fibrous structure became more compact and the fibers were curved and entangled with each other, which could explain the reason why after amidoxime modification, the nanofibrous membrane became stiff and brittle. The FTIR spectra of PAN and AOPAN nanofibrous membranes are also presented in Figure 2c. Compared with the pristine PAN nanofibrous membrane, the FTIR spectrum of the AOPAN nanofibrous membrane showed some new peaks. The characteristic bands located at 3500–3100 cm^{-1}, 1577 cm^{-1}, 1640 cm^{-1}, 1161 cm^{-1}, and 987 cm^{-1} could be respectively attributed to the stretching vibration of O–H, C=N, N–H, C–N and N–O, indicating the introduction of amidoxime group [21,25].

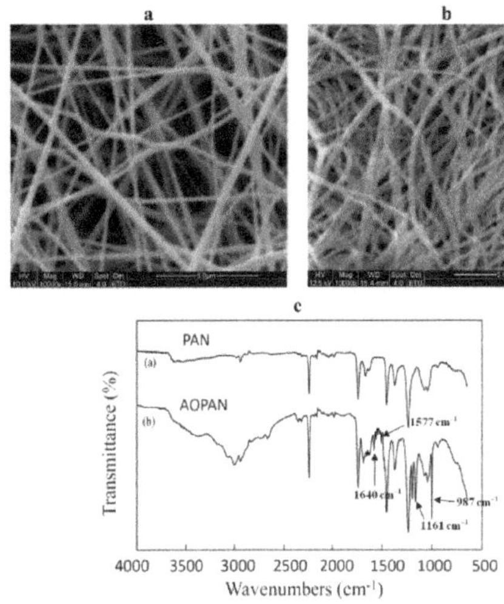

Figure 2: SEM Images of (**a**) pristine PAN; and (**b**) AOPAN nanofibrous membrane; and (**c**) FTIR spectra of PAN and AOPAN nanofibrous membrane.

Characterization of the Metal Ion Chelated PAN Nanofibrous Membranes

The morphology and the respective elemental analysis of the metal ion chelated AOPAN nanofibrous membranes (33% conversion) were recorded by SEM (Figure 3) and EDX (Figure 4). From the SEM images (Figure 3a–d), it can be observed that the metal ion chelated nanofibers showed a similar structure as that of the AOPAN. The fibers were distorted and entangled together, and no sign of physical cracks was seen. Besides, the results of the EDX (Figure 4a–e) confirmed the existence of the metal ions.

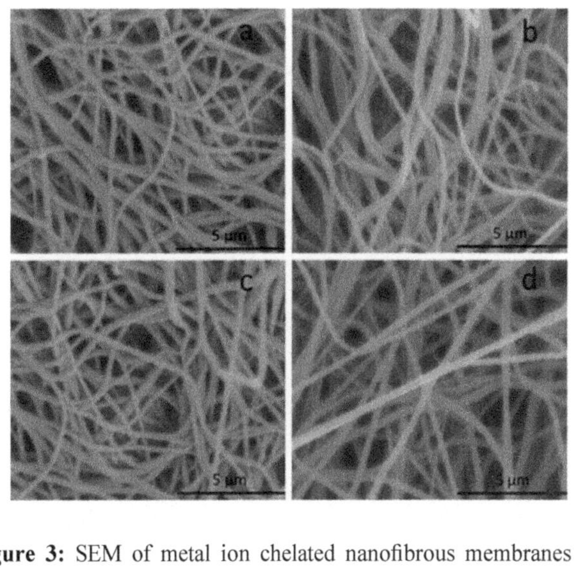

Figure 3: SEM of metal ion chelated nanofibrous membranes (**a**) Fe(III)-AOPAN; (**b**) Cu(II)–AOPAN; (**c**) Ni(II)–AOPAN; and (**d**) Cd(II)–AOPAN.

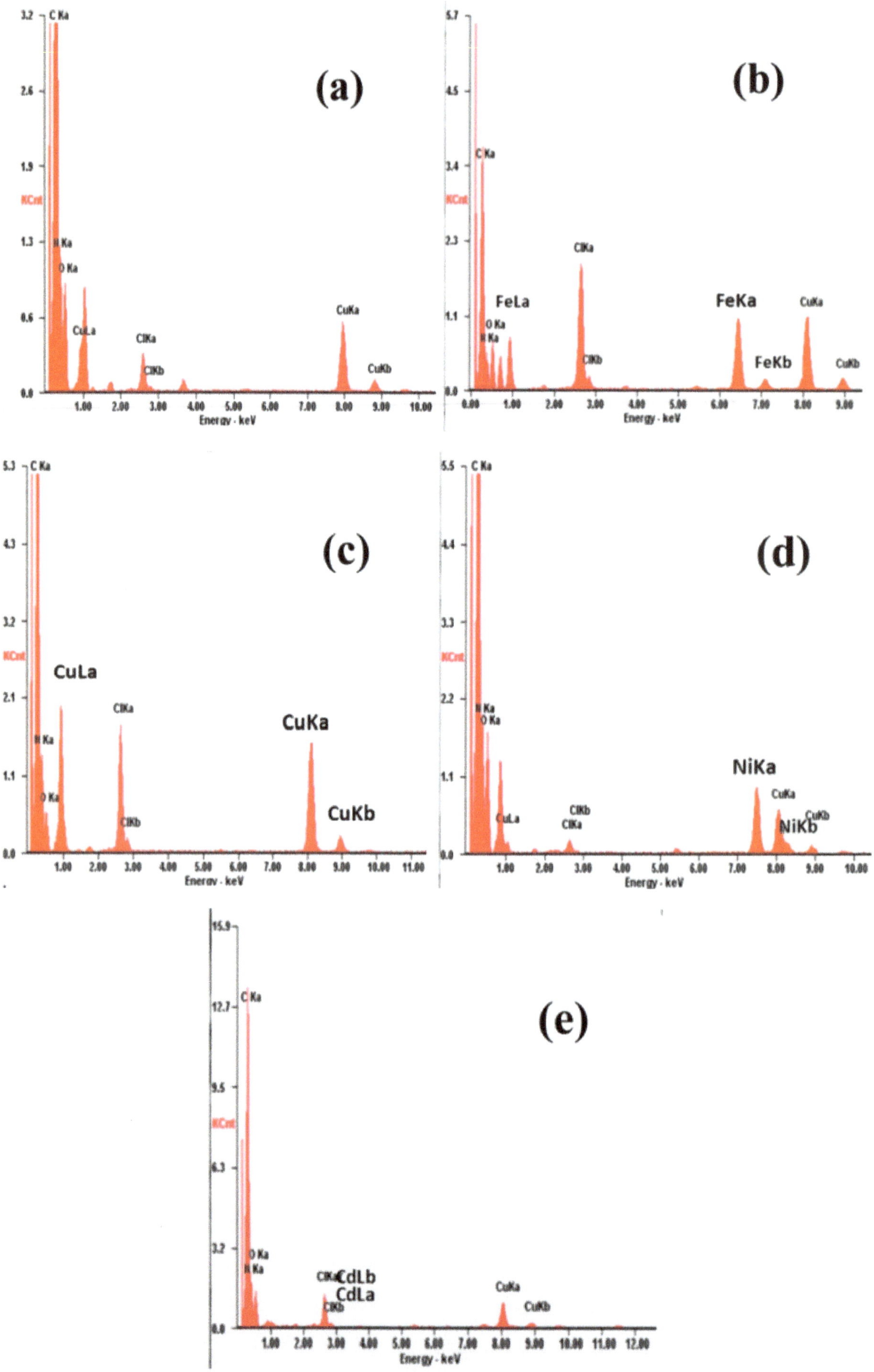

Figure 4: EDX of metal ion chelated nanofibrous membranes: (**a**) AOPAN; (**b**) Fe(III)–AOPAN; (**c**) Cu(II)–AOPAN; (**d**) Ni(II)–AOPAN; and (**e**) Cd(II)–AOPAN.

The Coordination of Metal Ions and Their Properties in Enzyme Immobilization

The amount of metal ions adsorbed onto AOPAN nanofibrous membranes (45% conversion) in three different concentrations (100 mg·L^{-1}, 500 mg·L^{-1}, 1000 mg·L^{-1}) were shown in Figure 5a. It could be clearly seen that with the increase of metal ion concentration, the amount of the chelated metal ions by AOPAN nanofibrous membrane was also increased. Besides, the four metal ions at the same concentration showed different behavior in the chelation process. The adsorption capacity followed the order as $Fe^{3+} > Cu^{2+} > Ni^{2+} > Cd^{2+}$. The adsorption behavior *versus* time was also studied and the results were presented in Figure 5b. The AOPAN nanofibrous membrane with a conversion rate of 39% was chosen for this experiment. During the first 6 h, the adsorbed metal ions were sharply increased. After that, it achieved equilibrium value.

The amount of the adsorbed enzymes on four different metal ion chelated membranes was calculated by the method of Bradford [23], and the results were illustrated by Figure 5c. As can be seen, the Fe–AOPAN and Cu–AOPAN adsorbed the most enzymes, Ni–AOPAN adsorbed relatively less enzymes, and the Cd–AOPAN showed the least, from which it could be inferred that the amount rather than the category of metal ion determined the final amount of adsorbed enzyme. After enzyme immobilization, the surface of the nanofibers are generally uniform while in some region small enzyme aggregates can be seen (Figure 5d).

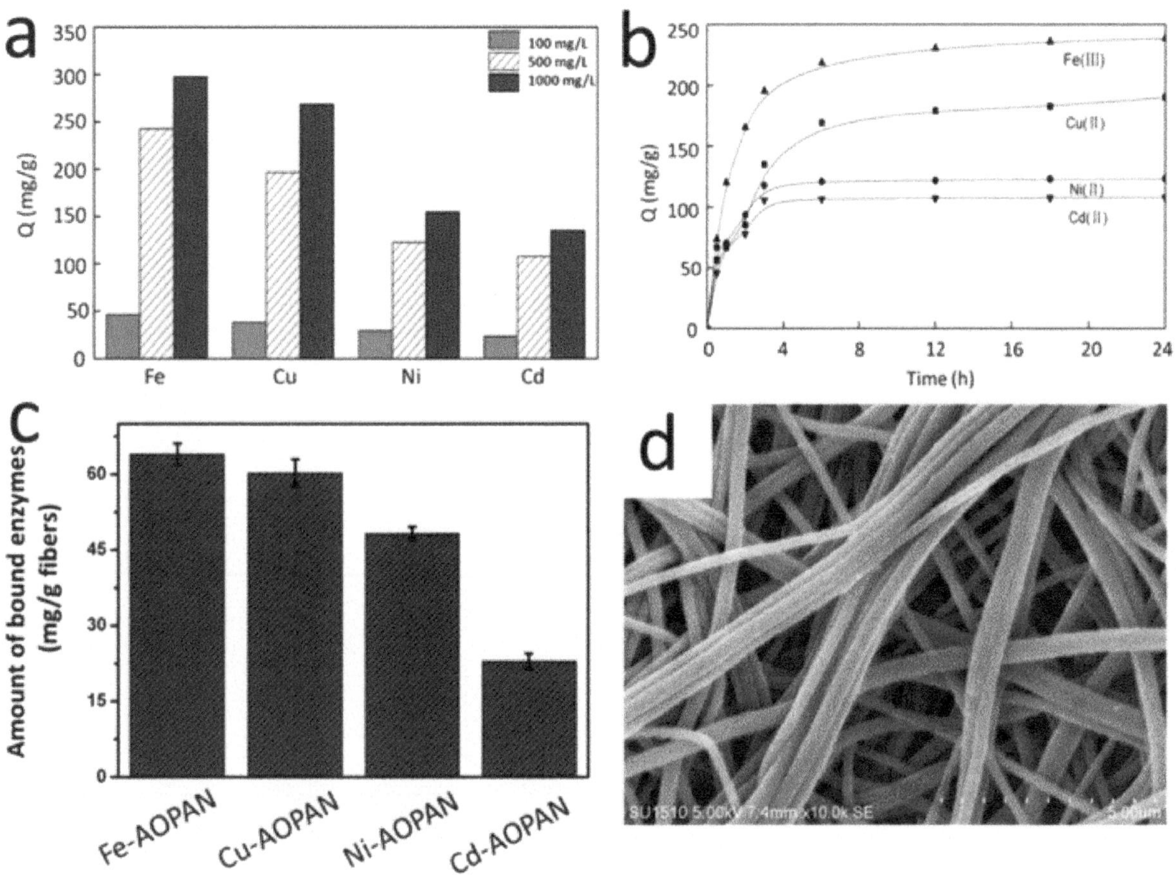

Figure 5: (**a**) Coordination quantities of metal ions chelated by AOPAN nanofibrous membranes (pH = 4, temperature at 30 °C, reaction time = 72 h); (**b**) Relationship between reaction time and coordination quantities (the concentration of the solution was 500 mg·L^{-1}, pH = 4, temperature at 30 °C); (**c**) Amount of bound enzymes on different metal ion chelated AOPAN nanofibers; and (**d**) SEM of Fe–AOPAN nanofibrous membrane after enzyme immobilization.

Properties of the Immobilized Enzyme

Kinetic Properties

The K_m values for free and immobilized laccase (Fe–AOPAN–Lac, Cu–AOPAN–Lac, Ni–AOPAN–Lac, Cd–AOPAN–Lac) were 0.114, 2.713, 0.84, 1.53, 1.21 mM·L^{-1}, respectively. After enzyme immobilization, the K_m values were generally increased; indicating a lower affinity for the substrate caused by diffusion limitations and decreased protein flexibility after immobilization [15].

Optimum pH

Effect of pH on the activity of the free and immobilized laccase on AOPAN chelated with different metal ions (Fe^{3+}, Cu^{2+}, Ni^{2+}, Cd^{2+}) was investigated at different pH values ranging from 2 to 7 (Figure 6a). Relative activity was defined as the percentage of the activity relative to the maximal activity which was set as 100%. The maximal activity of the free and immobilized laccase was exhibited at pH 3.5 and pH 4.0, respectively. This peak-shift phenomenon can be explained by the electrostatic interaction influenced by the carrier microenvironment [26]. Similar behavior was also observed in other studies [26,27]. As seen in Figure 5a, the immobilized laccase generally showed a broader pH-activity profile than that of the free laccase, indicating that the immobilization methods preserved the enzyme activity in a wider pH range [28]. Among the four immobilized laccase, the Fe–AOPAN–Lac showed the best resistance to pH changes, while the Cd–AOPAN–Lac showed the least. Besides, the activity retention during pH changes of Cu–AOPAN–Lac and Ni–AOPAN–Lac was relatively lower as compared to that of Fe–AOPAN–Lac, following the same trend.

Optimum Temperature

The influence of temperature on the activity of the free and immobilized laccase is shown in Figure 6b. The immobilized enzymes were incubated in buffers (pH 4.5) for 5 min at different temperatures varying from 30 to 65 °C before adding ABTS. It could be observed that activities of both free and immobilized laccase were strongly dependent on temperature, and the maximal activity was exhibited at 50 °C. Compared with free laccase, the immobilized enzyme showed a broader temperature profile. The improvement in resistance against temperature was probably due to a reduction in molecular mobility and conformational changes by coordination into the metal chelated support. The immobilized enzymes generally showed a similar profile trend of temperature stability in the range of 30–65 °C. Among all the immobilized enzymes, the Fe–AOPAN–Lac showed the best temperature stability over a broad range, Cd–AOPAN–Lac had a very close trend to the free enzyme, while Cu–AOPAN–Lac and Ni–AOPAN–Lac showed relatively better temperature stability than that of free laccase with a quite similar trend.

Storage Stability

The storage stability of the immobilized enzymes was also studied and the results are presented in Figure 6c. The immobilized enzymes showed better storage stability than that of the free enzyme. The laccase immobilized on Fe–AOPAN, Cu–AOPAN and Ni–AOPAN composite nanofibers showed relatively higher storage stability than that of Cd–AOPAN. After being stored for 20 days, the Cu–AOPAN–Lac retained more than 65% of its initial activity, which was higher than the results we previously reported [19].

Operational Stability

The operational stability of the immobilized laccase is presented in Figure 6d. Decrease in the enzyme activity upon repeated usage was expected due to the fact that enzymes may denature or desorbed during repeated usage. After 10 repeated usage, the activity retention ratio of Fe–AOPAN–Lac, Cu–AOPAN–Lac, Ni–AOPAN–Lac, Cd–AOPAN–Lac were 41.25%, 49.17%, 26.40%, 0.23%, respectively. Among the four immobilized enzymes, Fe–AOPAN–Lac and Cu–AOPAN–Lac well retained their activity during repeated usage, showing their potential value in future application. However, Cd–AOPAN–Lac had lost all the activity since the 6th operation process.

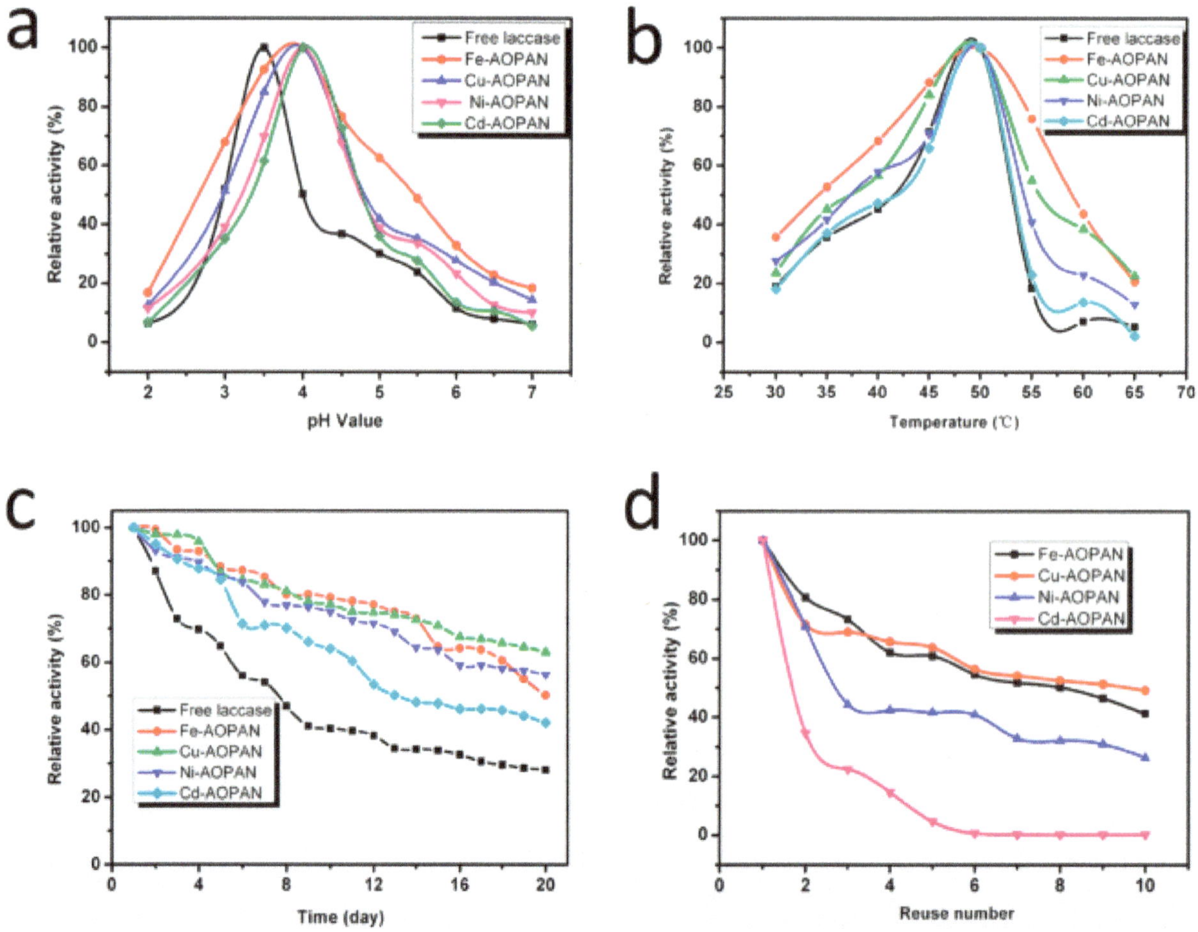

Figure 6: The properties of both free and immobilized enzymes: (**a**) optimum pH; (**b**) optimum temperature; (**c**) storage stability; and (**d**) operational stability.

Those results of the immobilized enzyme stability presented hereinbefore can be interpreted in terms of the size of the metal ions and electric density which could further determine the coordination capability between the chelate fiber ligand and the metal ions. According to Lewis theory [29] and Pearson rules [30], metal ions, which have small size, high positive electric, low polarization ability, and the property of being difficult to be deformed, are called hard acid. Hard acid such as Fe^{3+} easily accept the electron pair provided by the ligand. In contrast, those metal ions with large size, low positive electric, high polarization ability are called soft acid, which cannot easily accept the electric pair from ligand. For example, Cd^{2+} is a kind of soft acid. Besides, the properties of some other metal ions fall in between of hard acid and soft acid, such as Cu^{2+} and Ni^{2+}. The amino group existed in both AOPAN and enzyme is a kind of hard alkali, which has strong electronegativity, and the polarization capability of the electron pair provider is low. According to Pearson, hard acid prefers to combine with hard alkali, while soft alkali tends to react with soft alkali. In addition, the coordination compound formed between hard acid and hard alkali, soft acid and soft alkali are relatively stable, so the coordination between the AOPAN and Fe^{3+}, enzyme and Fe^{3+} is relatively more stable than that of AOPAN and Cd^{2+}, enzyme and Cd^{2+}. Same explanation applies to that of Cu^{2+} and Ni^{2+}. Therefore, the general enzyme stability immobilized onto different supports follows the order as Fe–AOPAN–Lac > Cu–AOPAN–Lac > Ni–AOPAN–Lac > Cd–AOPAN–Lac.

Enzyme Immobilization on the AOPAN Chelated with Four Metal Ions

The metal ion chelated AOPAN nanofibrous membranes used as enzyme carrier can improve the properties of the immobilized enzyme to different extent. Therefore, in this work, we further explored the properties of the immobilized enzyme on the AOPAN nanofibrous membrane chelated with the four metal ions mentioned above.

Coordination with Mixed Metal Ions by AO-PAN Nanofibers

Table 1 showed the results of the concentration of the metal ions before and after coordination process in a mixed solution with four metal ions. In a mixed solution, the Fe^{3+} and Cu^{2+} still showed better coordination properties than Ni^{2+} and Cd^{2+}, which was in correspondence with the results of the individual metal ions adsorption.

Table 1: Coordination characteristics of chelating nanofibers with mixed metal ions

Metal Ions	Before Coordination Reaction ($mg \cdot L^{-1}$)	After Coordination Reaction ($mg \cdot L^{-1}$)
Fe^{3+}	2.640	0.920
Cu^{2+}	2.899	0.613
Ni^{2+}	7.290	7.080
Cd^{2+}	5.560	5.360

Properties of the Immobilized Enzyme on the Mixed Metal Ion Chelated AOPAN Nanofibrous Membrane

The properties of the immobilized enzymes on mixed metal ion chelated AOPAN nanofibrous membrane were studied and presented in Figure 7. The optimum pH and temperature of the immobilized enzyme were at 4.0 and 50 °C (Figure 7a,b), respectively, which was in consistent with that of the immobilized enzymes on individual metal ion chelated membranes. Moreover, the resistance to pH and temperature changes was generally increased. Besides, the storage stability and the reusability (Figure 7c,d) were similar to that of the Fe–AOPAN–Lac.

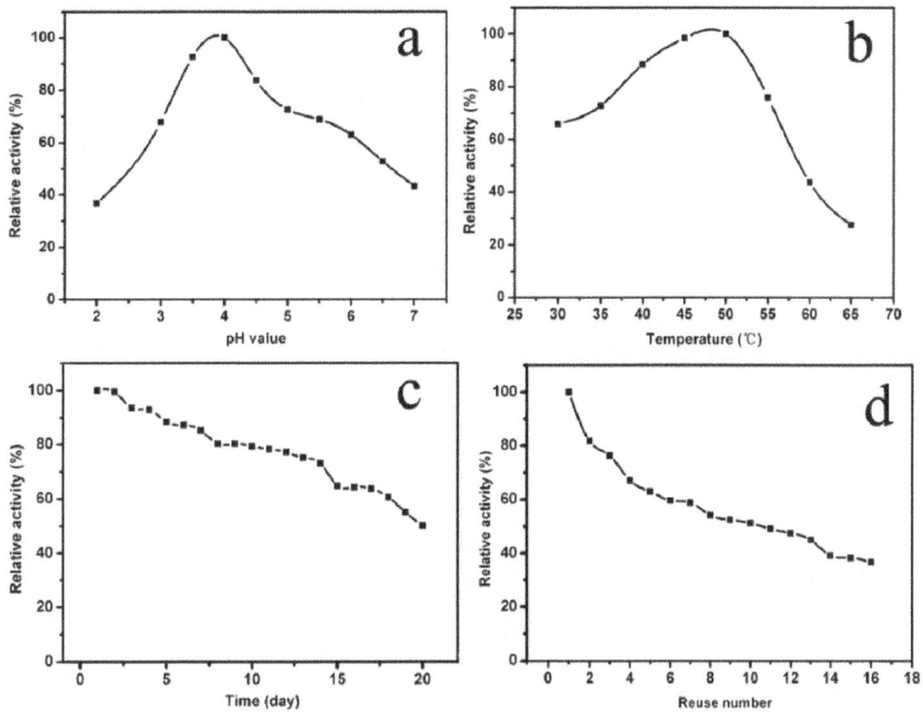

Figure 7: The properties of the immobilized enzyme on mixed metal ion chelated AOPAN nanofibrous membrane: (**a**) optimum pH; (**b**) optimum temperature; (**c**) storage stability; and (**d**) operational stability.

CONCLUSIONS

AOPAN nanofibrous membranes chelated with different metal ions including Fe^{3+}, Cu^{2+}, Ni^{2+}, Cd^{2+} as well as the combination of these four metal ions were successfully prepared and used as the carrier for laccase. The immobilization process enhanced the enzyme stability against pH and temperature changes as well as the storage stability. As for the reusability, the Fe–AOPAN–Lac and Cu–AOPAN–Lac as well as the mixed metal ion–AOPAN–Lac showed ideal results. In conclusion, all of the four metal ions contributed to improving the stability of the immobilized enzyme, and the properties of the immobilized enzymes were more dependent on the quantity of the chelated metal ions than on the metal ion itself. Therefore AOPAN nanofibrous membranes chelated with various metal ions can be an ideal support for enzyme immobilization.

ACKNOWLEDGMENTS

This work was financially supported by the National High-tech R&D Program of China (No.2012AA030313), Changjiang Scholars and Innovative Research Team in University (No. IRT1135), National Natural Science Foundation of China (No.51006046 and No.51163014) and the Priority Academic Program Development of Jiangsu Higher Education Institutions.

AUTHOR CONTRIBUTIONS

All listed authors contributed to this work. Qingqing Wang was responsible for the preparation of the membrane, the enzyme immobilization, analyzing the data as well as writing this manuscript; Jing Cui, Jinning Zhang, Guohui Li did the optimum pH, temperature, storage stability and operational stability. Qufu Wei and Fenglin Huang designed this work and obtained financial support.

REFERENCES

1. Galhaup, C.; Goller, S.; Peterbauer, C.K.; Strauss, J.; Haltrich, D. Characterization of the major laccase isoenzyme from trametes pubescens and regulation of its synthesis by metal ions. *Microbiology* 2002, *148*, 2159–2169.
2. Madhavi, V.; Lele, S.S. Laccase: Properties and applications. *Bioresources* 2009, *4*, 1694–1717.
3. Flory, A.R.; Requesens, D.V.; Devaiah, S.P.; Teoh, K.T.; Mansfield, S.D.; Hood, E.E. Development of a green binder system for paper products. *BMC Biotechnol.* 2013, *13*.
4. Fackler, K.; Kuncinger, T.; Ters, T.; Srebotnik, E. Laccase-catalyzed functionalization with 4-hydroxy-3-methoxybenzylurea significantly improves internal bond of particle boards. *Holzforschung* 2008, *62*, 223–229.
5. Ashrafi, S.D.; Rezaei, S.; Forootanfar, H.; Mahvi, A.H.; Faramarzi, M.A. The enzymatic decolorization and detoxification of synthetic dyes by the laccase from a soil-isolated ascomycete, paraconiothyrium variabile. *Int. Biodeterior. Biodegrad.* 2013, *85*, 173–181.
6. Zhang, Y.; Zeng, Z.T.; Zeng, G.M.; Liu, X.M.; Liu, Z.F.; Chen, M.; Liu, L.F.; Li, J.B.; Xie, G.X. Effect of triton x-100 on the removal of aqueous phenol by laccase analyzed with a combined approach of experiments and molecular docking. *Colloid Surf. B* 2012, *97*, 7–12.
7. Dhillon, G.S.; Kaur, S.; Brar, S.K.; Verma, M. Flocculation and haze removal from crude beer using in-house produced laccase from trametes versicolor cultured on brewer's spent grain. *J. Agric. Food Chem.* 2012, *60*, 7895–7904.
8. Basto, C.; Tzanov, T.; Cavaco-Paulo, A. Combined ultrasound-laccase assisted bleaching of cotton. *Ultrason. Sonochem.* 2007, *14*, 350–354.
9. Fernandez-Fernandez, M.; Sanroman, M.A.; Moldes, D. Recent developments and applications of immobilized laccase. *Biotechnol. Adv.* 2013, *31*, 1808–1825.
10. Qiu, H.J.; Xu, C.X.; Huang, X.R.; Ding, Y.; Qu, Y.B.; Gao, P.J. Immobilization of laccase on nanoporous gold: Comparative studies on the immobilization strategies and the particle size effects. *J. Phys. Chem. C* 2009, *113*, 2521–2525.
11. Feng, Q.A.; Xia, X.; Wei, A.F.; Wang, X.Q.; Wei, Q.F.; Huo, D.Y.; Wei, A.J. Preparation of Cu(II)-chelated poly(vinyl alcohol) nanofibrous membranes for catalase immobilization. *J. Appl. Polym. Sci.* 2011, *120*, 3291–3296.
12. Bayramoglu, G.; Yilmaz, M.; Arica, M.Y. Reversible immobilization of laccase to poly(4-vinylpyridine) grafted and Cu(II) chelated magnetic beads: Biodegradation of reactive dyes. *Bioresour. Technol.* 2010, *101*, 6615–6621.
13. Mazur, M.; Krysinski, P.; Michota-Kaminska, A.; Bukowska, J.; Rogalski, J.; Blanchard, G.J. Immobilization of laccase on gold, silver and indium tin oxide by zirconium-phosphonate-carboxylate (ZPC) coordination chemistry. *Bioelectrochemistry* 2007, *71*, 15–22.
14. Teerapatsakul, C.; Bucke, C.; Parra, R.; Keshavarz, T.; Chitradon, L. Dye decolorisation by laccase entrapped in copper alginate. *World J. Microbiol. Biotechnol.* 2008, *24*, 1367–1374.
15. Lu, L.; Zhao, M.; Wang, Y. Immobilization of laccase by alginate-chitosan microcapsules and its use in dye decolorization. *World J. Microbiol. Biotechnol.* 2007, *23*, 159–166.
16. Wang, Z.G.; Wan, L.S.; Liu, Z.M.; Huang, X.J.; Xu, Z.K. Enzyme immobilization on electrospun polymer nanofibers: An overview. *J. Mol. Catal. B Enzym.* 2009, *56*, 189–195.
17. Xu, R.; Chi, C.L.; Li, F.T.; Zhang, B.R. Laccase-polyacrylonitrile nanofibrous membrane: Highly immobilized, stable, reusable, and efficacious for 2,4,6-trichlorophenol removal. *ACS Appl. Mater. Int.* 2013, *5*, 12554–12560.
18. Saeed, K.; Haider, S.; Oh, T.J.; Park, S.Y. Preparation of amidoxime-modified polyacrylonitrile (pan-oxime) nanofibers and their applications to metal ions adsorption. *J. Membr. Sci.* 2008, *322*, 400–405.
19. Feng, Q.; Wang, Q.Q.; Tang, B.; Wei, A.F.; Wang, X.Q.; Wei, Q.F.; Huang, F.L.; Cai, Y.B.; Hou, D.Y.; Bi, S.M. Immobilization of catalases on amidoxime

polyacrylonitrile nanofibrous membranes. *Polym. Int.* 2013, *62*, 251–256.

20. Ding, Z.Y.; Peng, L.; Chen, Y.Z.; Zhang, L.; Gu, Z.H.; Shi, G.Y.; Zhang, K.C. Production and characterization of thermostable laccase from the mushroom, ganoderma lucidum, using submerged fermentation. *Afr. J. Microbiol. Res.* 2012, *6*, 1147–1157.
21. Huang, F.L.; Xu, Y.F.; Liao, S.Q.; Yang, D.W.; Hsieh, Y.L.; Wei, Q.F. Preparation of amidoxime polyacrylonitrile chelating nanofibers and their application for adsorption of metal ions. *Materials* 2013, *6*, 969–980.
22. Zhang, P.; Wang, Q.Q.; Zhang, J.N.; Li, G.H.; Wei, Q.F. Preparation of amidoxime-modified polyacrylonitrile nanofibers immobilized with laccase for dye degradation. *Fiber. Polym.* 2014, *15*, 30–34.
23. Wang, Q.Q.; Peng, L.; Du, Y.Z.; Xu, J.; Cai, Y.B.; Feng, Q.; Huang, F.L.; Wei, Q.F. Fabrication of hydrophilic nanoporous PMMA/*o*-MMT composite microfibrous membrane and its use in enzyme immobilization. *J. Porous Mater.* 2013, *20*, 457–464.
24. Wang, Q.Q.; Peng, L.; Li, G.H.; Zhang, P.; Li, D.W.; Huang, F.L.; Wei, Q.F. Activity of laccase immobilized on TiO_2-montmorillonite complexes. *Int. J. Mol. Sci.* 2013, *14*, 12520–12532.
25. Bilba, N.; Bilba, D.; Moroi, G. Synthesis of a polyacrylamidoxime chelating fiber and its efficiency in the retention of palladium ions. *J. Appl. Polym. Sci.* 2004, *92*, 3730–3735.
26. Arica, M.Y. Immobilization of polyphenol oxidase on carboxymethylcellulose hydrogel beads: Preparation and characterization. *Polym. Int.* 2000, *49*, 775–781.
27. Liu, Y.Y.; Zeng, Z.T.; Zeng, G.M.; Tang, L.; Pang, Y.; Li, Z.; Liu, C.; Lei, X.X.; Wu, M.S.; Ren, P.Y.; *et al.* Immobilization of laccase on magnetic bimodal mesoporous carbon and the application in the removal of phenolic compounds. *Bioresour. Technol.* 2012, *115*, 21–26.
28. Bayramoglu, G.; Yilmaz, M.; Senel, A.U.; Arica, M.Y. Preparation of nanofibrous polymer grafted magnetic poly(GMA-MMA)-*g*-MAA beads for immobilization of trypsin via adsorption. *Biochem. Eng. J.* 2008, *40*, 262–274.
29. Zhang, L.H.; Zhang, X.S.; Li, P.P.; Zhang, W.Q. Effective Cd^{2+} chelating fiber based on polyacrylonitrile. *React. Funct. Polym.* 2009, *69*, 48–54.
30. Horzum, N.; Boyaci, E.; Eroglu, A.E.; Shahwan, T.; Demir, M.M. Sorption efficiency of chitosan nanofibers toward metal ions at low concentrations. *Biomacromolecules* 2010, *11*, 3301–3308.

CHAPTER 11

ENZYMES ENCAPSULATION WITHIN CALCIUM ALGINATE-CLAY BEADS: CHARACTERIZATION AND APPLICATION FOR CASSAVA SLURRY SACCHARIFICATION

Siti Noraida Abd Rahim[1], Alawi Sulaiman[1], Fazlena Hamzah[2], Ku Halim Ku Hamid[2], Miradatul Najwa Muhd Rodhi[2] , Mohibbah Musa[2], and Nurul Aini Edama[1]

[1]Tropical Agro-Biomass Research Group, Faculty of Plantation and Agrotechnology, Universiti Teknologi MARA, 40450 Shah Alam, Malaysia

[2]Faculty of Chemical Engineering, Universiti Teknologi MARA, 40450 Shah Alam, Malaysia

ABSTRACT

Multi-enzymes (alpha-amylase, glucoamylase and cellulase) were successfully encapsulated in calcium alginate-clay beads to hydrolyze cassava roots for glucose production. Under the optimal conditions (2% clay solution and 0.2 M $CaCl_2$), the loading efficiency and the immobilization yield of enzymes were calculated to be 97.07% and 52.14%, respectively. The calciumalginate-clay beads have been analyzed using Fourier Transform Infrared Spectroscopy (FTIR), Field Emission Scanning Electron Microscopy (FESEM) and Energy Dispersive X-Ray (EDX). FTIR results showed that asymmetric and symmetric COO^- and OH^- group shifted to 1621, 1419 and 3324 cm^{-1} and proved strong interaction between alginate and clay of the beads. Observation under FESEM showed that the beads have a rough surface and the enzymes were successfully encapsulated in the beads. The presence of the clay elements in the beads was confirmed by EDX analysis where the beads consist of 13.97% C, 43.69% O, 28.96% Ca, 6.46% Al and 6.91% Si. The beads were later used for the saccharification of cassava slurry into glucose. Results showed that the process was successful and the encapsulated enzymes in calcium alginate-clay beads have enhanced the enzymes reusability in the process compared to pure alginate beads where it retained 51.77% of its activity after seven hydrolysis cycles.

INTRODUCTION

Cassava root is a very popular in the tropical countries such as Brazil, Thailand, Malaysia, Indonesia and some parts of Africa [1]. It is an inexpensive material that often used as a feedstock for the production of syrups containing maltose, glucose or fructose. For the glucose syrup, it can be further fermented to produce bioethanol which is widely used in food and pharmaceutical industries [2]. There are a number of enzymes that are commonly used in the production of glucose.

Alpha-amylase is usually used to hydrolyze starch by cleaving α-(1→4) glycosidic bonds to produce maltodextrin while glucoamylase is able to hydrolyze α-(1→4) and α-(1→6) glycosidic bonds in starch to produce glucose [3]. Besides that, cellulase can be used to promote the hydrolysis of cassava

Citation: Siti Noraida Abd Rahim, Alawi Sulaiman, Fazlena Hamzah, Ku Halim Ku Hamid, Miradatul Najwa Muhd Rodhi, Mohibbah Musa, and Nurul Aini Edama, Enzymes encapsulation within calcium alginate-clay beads: Characterization and application for cassava slurry saccharification, doi: 10.1016/j.proeng.2013.12.200

roots due to it has ability to attack β-(1→4) linkages in fibre to produce glucose [4]. Due to these features, these three enzymes are normally used to produce glucose from materials that contain both starch and fibre such as cassava root as reported in the literature [5].

Nowadays, the immobilization of enzymes is widely studied to replace the free system that often hampered by a lack of availability, high price and limited stability under operational conditions. The immobilization makes possible for easier separation between the enzyme and product and therefore the enzymes can be reused [6, 7]. Though enzyme immobilization has several advantages, the selection of support material and method of immobilization are important in order to obtain higher performance of enzymatic reaction. Encapsulation is one of the immobilization techniques which can be defined as the process by which the enzyme is enclosed physically or chemically within semi-permeable membrane [8]. The alginate bead is commonly used in the encapsulation process because it is easy for formulation, mild gelation conditions, non-toxic, biocompatibility, low cost and resistance to microbial attack. However, alginate beads have also several disadvantages such as low mechanical strength, large pore size and high leakage of enzyme from beads [9, 10]. To reduce this problem, alginate can be blended with clay to improve stability of the beads [11].

Kaolinite clay is negatively charged aluminosilicate crystalline structure consisting of a three-dimensional arrangement of SiO_2 and AlO_4^- tetrahedrally linked to each other by a shared oxygen atom. This clay has a good potential as supporting material for encapsulation purposes because it is stable, relatively good pore aperture and posses good mechanical strength and thermostability as reported in the previous studies [12, 13]. Besides that, kaolinite clay has a strong interest in many modern industrial applications due to it is natural source, low cost and also readily available [14]. Therefore, in this study, the kaolinite clay material obtained from Sg. Sayong, Perak, Malaysia is used and blended together with sodium alginate to prepare calcium aginate-clay beads. In the study, the enzymes (alpha-amylase, glucoamylase and cellulase) were encapsulated in calcium alginateclay beads. The effects of immobilization conditions such as concentration of clay and calcium chloride were evaluated. The characterizations of the beads were performed using Fourier transform infrared spectroscopy (FTIR), Field Emission Scanning Electron Microscopy (FESEM) and Energy Dispersive X-Ray (EDX). The reusability of the immobilized enzymes was also studied.

MATERIALS AND METHODS

Materials

Alpha-amylase from Bacillus subtilis, glucoamylase from Rhizopus niveus Lyophilized and cellulase from Aspergillus niger were purchased from MP Biomedicals, United States. All other chemicals reagent used were analytical grade.

Preparation of tapioca slurry

Cassava root was purchased from local market as a substrate. It was cleaned, hand peeled and sliced to small pieces. The chips were dried in oven at 65 o C for 24 hr to remove moisture and then ground into powder. 1 % (w/v) of tapioca slurry was prepared using 0.1 M citrate phosphate buffer (pH 5) solution and then boiled in the water bath with continuously agitated for 1 hr to gelatinize the starch before further use.

Preparation of clay powder

The local kaolinite clay was obtained from Sg. Sayong, Perak, Malaysia. Before use, it was dried at 90°C for 24 hr, ground into fine powder and sieved through 150 μm mesh.

Immobilization of enzymes within calcium alginate-clay beads

The encapsulation of enzymes in alginate-clay beads was carried out by following method reported by Adzmi et al., [11]. 2.5 g (2.5 % w/v) of previously prepared kaolinite clay was dissolved in 100 mL of 0.1 M sodium acetate buffer (pH 5) solution and stirred for 1 hr at room temperature. Then, 2.5 g of alginate powder was added to the clay solution and stirred for 4 hr. Next, 2.5 mL of glycerol was added to the alginate-clay solution. 1 mL of each enzyme solution (1 mg solid/mL of alpha-amylase, glucoamylase and cellulase) was mixed with 12 mL of alginate-clay solution which produced the final clay concentration 2 % w/v. The mixture solution was stirred thoroughly to ensure complete mixing and dropped into 0.2 M CaCl2 solution by using syringe. After 3 hr of hardening, the beads were collected by filtration and then washed with buffer solution several times to remove any unbound enzymes. Finally, the beads were stored in buffer solution at 4 o C until use. The enzyme loading efficiency and immobilization yield was determined by the following Equation 1and 2:

$$\text{Loading efficiency (\%)} = \frac{C_iV_i - C_fV_f}{C_iV_i} \times 100\% \quad (1)$$

where C_i is the initial protein concentration (mg/mL), V_i is the initial volume of enzyme solution (mL), C_f is the protein concentration in the total filtrate (mg/mL) and V_f is the total volume of filtrate (mL).

$$\text{Immobilization yield (\%)} = \left(\frac{A_{imm}}{A_{free}}\right) \times 100\% \quad (2)$$

where A_{imm} is specific activity of immobilized enzyme and A_{free} is specific activity of free enzyme.

Characterization techniques

The infrared spectra of the clay, alginate beads and alginate-clay beads were recorded with a FTIR spectrophotometer (Perkin-Elmer Spectrum One, USA) and scanned from 4000 to 450 cm^{-1} at ambient temperature. FESEM analysis was used to study the surface and cross-section image of the beads. The beads were mounted on stubs using double sided adhesive tape and sputter coated with gold using a sputter coater. And then, the coated beads were examined under FESEM (ZEISS Supra 40VP, USA). Additionally, the elemental composition of the beads was investigated by EDX that incorporated with FESEM.

Estimation of protein

The amount of protein before and after immobilization was

determined by Lowry's procedure modified by Hartree [15] using bovine serum albumin (BSA) as a standard.

Estimation of enzyme activities

The activity of enzymes was measured by using spectrophotometer at 540 nm with 3,5-dinitrosalicylic acid (DNS) as an indicator [16]. Theoretically one unit of alpha-amylase activity is defined as the quantity of enzyme that releases 1 mg of reducing sugar as maltose per minute at pH 6.6 and 30 °C. One unit of glucoamylase is expressed as the amount of enzyme releasing 10 mg of reducing sugar (glucose) per minute at pH 4.5 and 40 °C. One unit of cellulase activity is liberated as the amount of enzyme that produces 1 □mole of reducing sugar (glucose) at pH 5 and 37 °C in 1 min.

Reusability of encapsulated enzymes

To evaluate the reusability of the immobilized enzymes, the alginate and alginate-clay beads were used several times for the hydrolysis reaction. After 1 hr of hydrolysis, the beads were separated and washed with 0.1 M sodium acetate buffer (pH 5). Next, the fresh tapioca slurry solution was added to start the next round. The activity of each round was compared with the first run which was defined as 100 %.

RESULTS AND DISCUSSION

Characterization

The average diameter of alginate and alginate-clay beads after encapsulation was 2.6 mm. These beads were characterized by FTIR, FESEM and EDX analysis. The FTIR spectra of clay, alginate beads and alginate-clay beads are shown in Fig. 1. The clay spectrum showed the characteristic peaks at 3836 cm^{-1} (Si-O-H stretching), 3234 cm^{-1} (OH- stretching), 1030 and 1007 cm^{-1} (Si-O-Si stretching), 912 cm^{-1} (Al-O-H stretching) and 750 cm-1 (Si-O stretching). While in the FTIR spectrum of alginate beads, the peak appearing at 1615 and 1417 cm^{-1} could be assigned to COO^- asymmetric and symmetric stretching, 3270 cm^{-1} (OH^- stretching) and 1084 and 1027 cm^{-1} (C-OC stretching). When the clay was added to alginate, it can be observed that the asymmetric and symmetric of COOwas shifted to 1621 and 1419 cm^{-1}. This is due to the negative charge of the carboxyl group in the alginate have electrostatic interaction with the positive charge of the clay [11, 17]. Moreover, the silanol group (Si-O-H stretching) of clay was disappeared in the alginate-clay beads and OH- stretching was shifted to a higher wavenumber. With these results, the intermolecular hydrogen bonding and electrostatic forces between alginate and clay was existence and the alginate-clay beads could be used in the encapsulation process.

The surface and cross-section image of alginate and alginate-clay beads was investigated using FESEM (Fig. 2). It was found that alginate-clay beads have a different structure compared to alginate beads where the bead has a very rough surface. The rough surface of the beads does not mean the leakage of enzymes was occurring but it showed that the presence of physical barrier like clay in the alginate beads may prevent the leakage and keep the enzyme activity and it was supported by Gülay and ùanlÕ-Mohamed [18]. For cross-sectional images (Fig. 2b and d), the enzymes have been successfully encapsulated and the enzymes was very clearly seen in the beads. As reported in the previous study [12], SiO_2 and Al_2O_3 are the major component of the kaolinite clay. Therefore, the elemental composition in the alginate-clay beads was studied using EDX analysis to confirm the existence of clay elements (Fig. 3). The results revealed that the alginate-clay beads composed of 13.97 % C, 43.69 % O, 28.96 % Ca, 6.46 % Al and 6.91 % Si which is demonstrated that these beads contains SiO_2 and Al_2O_3 that derived from clay material.

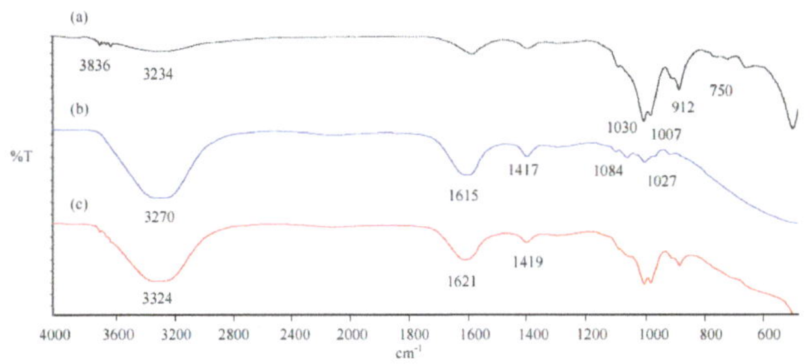

Figure 1: FTIR spectra of (a) clay, (b) alginate beads and (c) alginate-clay beads

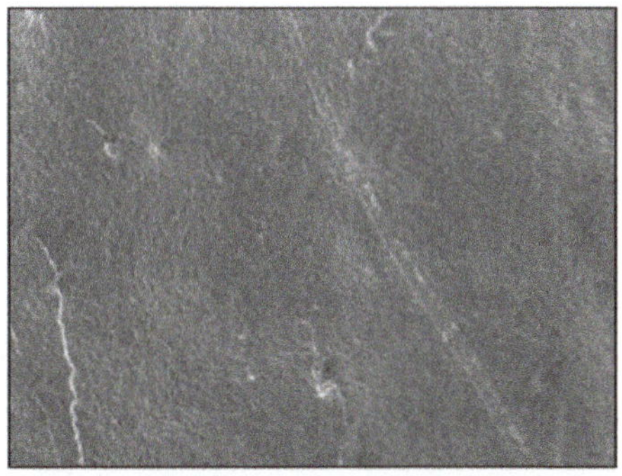
(a)

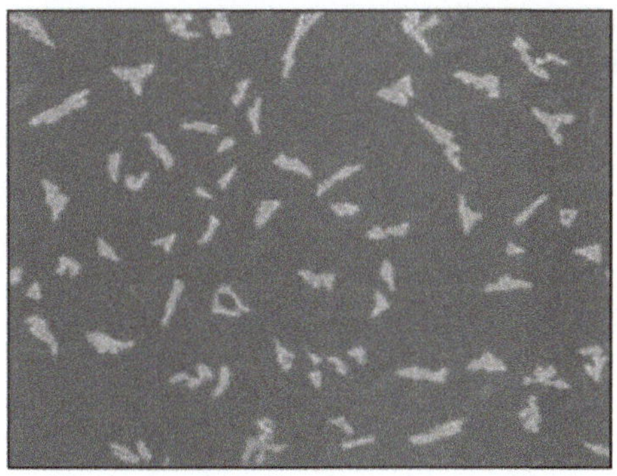
(b)

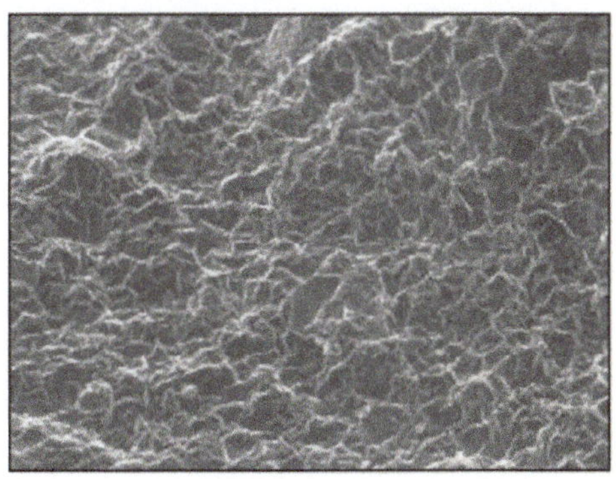
(c)

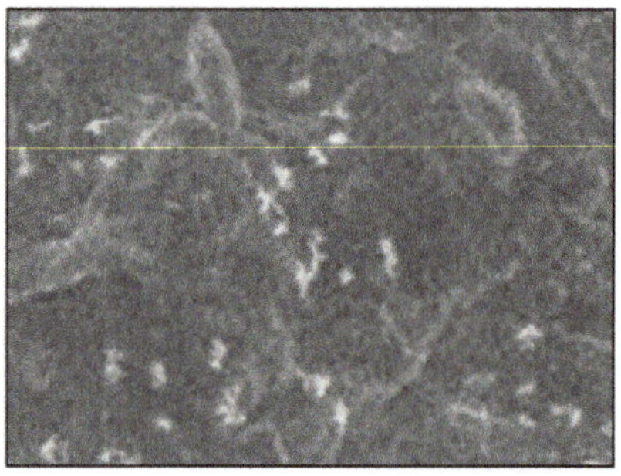
(d)

Figure 2: FESEM micrographs (1000x magnification) (a) surface image of alginate beads. (c) surface image of alginate-clay beads. (b) cross-section image of alginate beads. (d) cross-section image of alginate-clay beads.

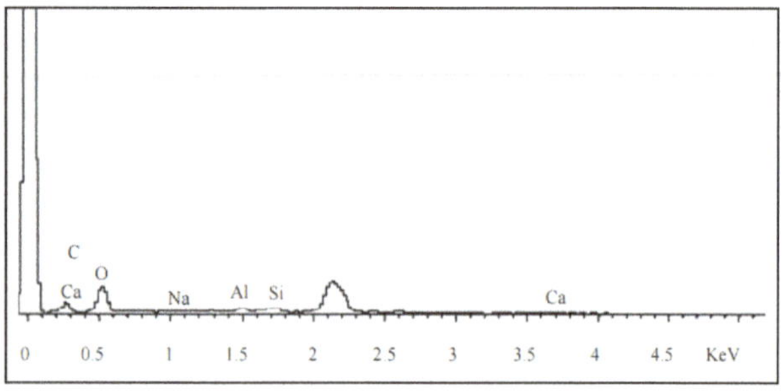

Figure 3: EDX spectrum of alginate-clay beads

Immobilization process

Sodium alginate and $CaCl_2$ concentration are the critical parameters for enzyme encapsulation in gel beads due to gelation formation by cross-linking between alginate and Ca^{2+} ions. The various sodium alginate concentrations were examined in order to obtain a maximum immobilization yield. In several studies, the maximum yield was reported at 2 % (w/v) sodium alginate [18-20]. Therefore, encapsulation enzyme within 2 % alginate concentration was selected and the loading efficiency and immobilization yield obtained was 95.57 % and 64.02 %, respectively. However, the alginate beads showed poor mechanical stability with high leakage of enzyme through the polymeric membrane layer. This phenomenon was attributed to the porous structure of alginate [9, 10, 18]. In order to solve this problem, Sg Sayong clay was mixed with sodium alginate to improve the stability of the encapsulated enzymes. As can be seen in Table 1, the addition of clay into the alginate beads has improved the enzyme loading efficiency whereas the best yield of immobilization was obtained at 2 % (w/v) clay solution. The increase in loading efficiency of enzymes with high clay concentration may be due to the stronger and tightly beads formation. When the concentration of clay solution was increased, the percent immobilization yield of enzymes reduced. It could be explained due to diffusion limitation of substrate and the encapsulated enzymes [21].

In the second step, the effect of $CaCl_2$ solution on loading efficiency and immobilization yield was studied from 0.1 to 0.5 M at 2 % clay concentration. The highest enzyme loading efficiency and immobilization yield was determined to be 97.07 % and 52.14 %, respectively at 0.2 M $CaCl_2$ solution. The immobilization yield was small at below 0.2 M $CaCl_2$ concentration due to unstable cross-linked gel formation and dispersion of beads in short times [18]. When the concentration of $CaCl_2$ increased beyond 0.2 M, the immobilization yield of enzymes was decreased. The relative enzyme activity has decreased due to the pH of $CaCl_2$ solution changes with its concentration [7, 19].

Table 1: Effect of clay and CaCl2 concentration on the encapsulation enzymes

	Loading efficiency (%)	Immobilization yield (%)
Clay concentration (%) (0.2 M $CaCl_2$)		
2	97.07	52.14
4	97.21	39.52
6	97.48	35.97
8	97.68	31.84
10	98.16	25.16
$CaCl_2$ concentration (M) (2 % clay)		
0.1	94.48	41.84
0.2	97.07	52.14
0.3	96.46	31.50
0.4	96.73	19.65
0.5	96.93	14.05

Reusability

The operational stability of encapsulated enzymes in pure alginate and alginate-clay beads was observed by reuse these beads for the possible number of cycles (Fig. 4). After seven cycles, more than 51.77 % of the enzyme activity in alginate-clay beads was retained that indicating a good operational stability of the immobilized enzyme. Meanwhile, the encapsulated enzymes in alginate beads retained only 20.37 %. Similar results were observed with other immobilized enzymes that reported the soft surface of alginate beads may be the reason for the high leakage of enzymes from the beads [18]. Thus, the results proved that the alginate-clay beads will enhance the operational stability by reducing the loss of enzyme activity.

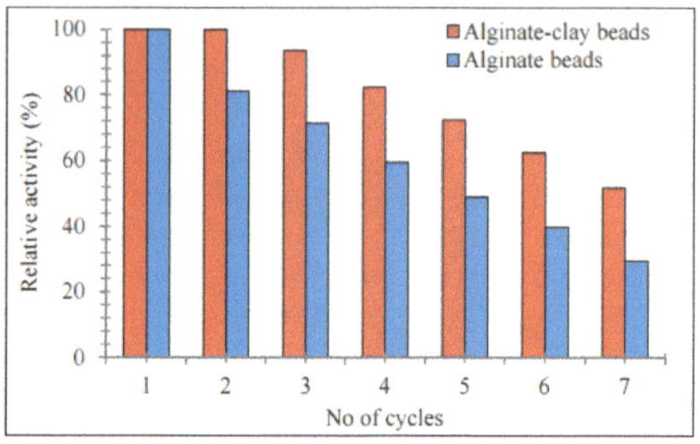

Figure 4: Reusability of encapsulated enzymes in both alginate and alginate-clay beads

CONCLUSIONS

In this work, the characterization of calcium alginate-clay beads encapsulated with alpha-amylase, glucoamylase and cellulase and its performance on the saccharafication of cassava slurry was successfully studied. The beads were prepared by combining sodium alginate and Sg. Sayong kaolinite clay solution and then cross-linked with $CaCl_2$ solution. The best enzymes loading efficiency and immobilization yield were found at 2 % (w/v) of clay and 0.2 M $CaCl_2$ solutions. The alginate-clay beads were characterized using FTIR, FESEM and EDX. FTIR analysis showed that the functional group of alginate and clay has a strong interaction with each other. Cross-section image from FESEM analysis, showed the enzymes was successfully encapsulated in the beads. Besides that, EDX analysis incorporated with FESEM analysis proved that the clay elements were detected within the beads. Reusability of encapsulated enzymes in the alginate-clay beads have also been evaluated and compared with pure alginate beads. The alginate-clay beads were able to maintain 51.77% of the residual enzymes activity after seven hydrolysis cycles. This high operational stability indicates that the encapsulated enzymes within alginate-clay beads is a promising technology for the future enzymatic reaction application for converting cassava slurry into glucose in a sustainable, cheaper and productive manner.

ACKNOWLEDGEMENTS

The authors are grateful for the financial support provided by Universiti Teknologi MARA (UiTM) Excellent Funds (600-RMI/ST/DANA 5/3/Dst 495/2011 & 475/2011) and Ministry of Higher Education, Fundamental Research Grant Scheme (600-RMI/ST/FRGS 5/3 Fst 2/4/2010).

REFERENCES

1. García, N. L., Famá, L., Dufresne, A., Aranguren, M. and Goyanes, S., 2009. A comparison between the physicochemical properties of tuber and cereal starches, Food Research International 42, pp. 976-982.
2. Roy, I. and Gupta, M. N., 2004. Hydrolysis of starch by a mixture of glucoamylase and pullulanase entrapped individually in calcium alginate beads, Enzyme and Microbial Technology 34, pp. 26-32.
3. van der Maarel, M. J. E. C., van der Veen, B., Uitdehaag, J. C. M., Leemhuis, H. and Dijkhuizen, L., 2002. Properties and applications of starch-converting enzymes of the ɭ-amylase family, Journal of Biotechnology 94, pp. 137-155.
4. Sukumaran, R. K., Singhania, R. R. and Pandey, A., 2005. Microbial cellulases - Production, applications and challenges, Journal of Scientific & Industrial Research 64, pp. 832-844.
5. Kearsley, M. W. and Dziedzic, S. Z., 1995. Starch Hydrolysis Products and Their Derivatives, Blackie Academic & Professional, Springer, U. K., pp 9-10.
6. Won, K., Kim, S., Kim, K.-J., Park, H. W. and Moon, S.-J., 2005. Optimization of lipase entrapment in Ca-alginate gel beads, Process Biochemistry 40, pp. 2149-2154.
7. Talekar, S. and Chavare, S., 2012. Optimization of immobilization of ɭ-amylase in alginate gel and its comparative biochemical studies with free ɭ-amylase, Recent Research in Science and Technology 4, pp. 01-05.
8. Cao, L. and Schmid, R. D., 2006. Carrier-bound Immobilized Enzymes: Principles, Application and Design: Wiley, Germany, pp. 398.
9. Konsoula, Z. and Liakopoulou-Kyriakides, M., 2006. Starch hydrolysis by the action of an entrapped in alginate capsules ɭ-amylase from Bacillus subtilis, Process Biochemistry 41, pp. 343-349.
10. DeGroot, A. R. and Neufeld, R. J., 2001. Encapsulation of urease in alginate beads and protection from ɭ-chymotrypsin with chitosan membranes, Enzyme and Microbial Technology 29, pp. 321-327.
11. Adzmi, F., Meon, S., Musa, M. H. and Yusuf, N. A., 2012. Preparation, characterisation and viability of encapsulated Trichoderma harzianum UPM40 in alginate-montmorillonite clay, Journal of Microencapsulation 29, pp. 205-210.
12. Edama, N. A., Sulaiman, A., Ku Hamid, K. H., Muhd Rodhi, M. N., Musa, M. and Abd Rahim, S. N., 2013. Preparation and Characterization of Sg. Sayong Clay Material for Biocatalyst Immobilization, Materials Science Forum 737, pp. 145-152.
13. Ajayi, O. A., Nok, A. J. and Adefila, S. S., 2012. Immobilization of Cassava Linamarase on Kankara Kaolinite Clay, Journal of Natural Sciences Research 2, pp. 55-62.
14. Abdul Rahman, M. B., Tajudin, S. M., Hussein, M. Z., Abdul Rahman, R. N. Z. R., Salleh, A. B. and Basri, M., 2005. Application of natural kaolin as support for the immobilization of lipase from Candida rugosa as biocatalsyt for effective esterification, Applied Clay Science 29, pp. 111-116.
15. Hartree, E. F., 1972. Determination of protein: A modification of the lowry method that gives a linear photometric response, Analytical Biochemistry 48, pp. 422-427.
16. Bernfeld, P., 1955. Amylases alpha and beta. Methods in Enzymology, pp. 149-158.
17. Pongjanyakul, T. and Puttipipatkhachorn, S., 2007. Sodium alginate-magnesium aluminum silicate composite gels: characterization of flow behavior, microviscosity, and drug diffusivity, AAPS PharmSciTech 8, pp. E1-E8.
18. Gülay, S. and ùanlÕ-Mohamed, G., 2012. Immobilization of thermoalkalophilic recombinant esterase enzyme by entrapment in silicate coated Ca-alginate beads and its hydrolytic properties, International Journal of Biological Macromolecules 50, pp. 545-551.
19. Anwar, A., Ali, S., Qader, U., Raiz, A., Iqbal, S. and Azhar, A., 2009. Calcium Alginate: A Support Material for Immobilization of Proteases from Newly Isolated Strain of Bacillus subtilis KIBGE-HAS, World Applied Sciences Journal 7, pp. 1281-1286.
20. Quiroga, E., Illanes, C. O., Ochoa, N. A., and Barberis, S., 2011. Performance improvement of araujian, a cystein phytoprotease, by immobilization within calcium alginate beads, Process Biochemistry 46, pp. 1029-1034.
21. Kumar, S., Dwevedi, A. and Kayastha, A. M., 2009. Immobilization of soybean (Glycine max) urease on alginate and chitosan beads showing improved stability: Analytical applications, Journal of Molecular B: Catalysis Enzymtic 55, pp. 138-145.

CHAPTER 12

CONTINUOUS DEGRADATION OF MALTOSE BY ENZYME ENTRAPMENT TECHNOLOGY USING CALCIUM ALGINATE BEADS AS A MATRIX

Muhammad Asif Nawaz, Haneef Ur Rehman, Zainab Bibi, Afsheen Aman, and Shah Ali Ul Qader

The Karachi Institute of Biotechnology and Genetic Engineering (KIBGE), University of Karachi, Karachi 75270, Pakistan

ABSTRACT

Maltase from *Bacillus licheniformis* KIBGE-IB4 was immobilized within calcium alginate beads using entrapment technique. Immobilized maltase showed maximum immobilization yield with 4% sodium alginate and 0.2 M calcium chloride within 90.0 min of curing time. Entrapment increases the enzyme–substrate reaction time and temperature from 5.0 to 10.0 min and 45 °C to 50 °C, respectively as compared to its free counterpart. However, pH optima remained same for maltose hydrolysis. Diffusional limitation of substrate (maltose) caused a declined in *Vmax* of immobilized enzyme from 8411.0 to 4919.0 U ml^{-1} min^{-1} whereas, *Km* apparently increased from 1.71 to 3.17 mM ml^{-1}. Immobilization also increased the stability of free maltase against a broad temperature range and enzyme retained 45% and 32% activity at 55 °C and 60 °C, respectively after 90.0 min. Immobilized enzyme also exhibited recycling efficiency more than six cycles and retained 17% of its initial activity even after 6th cycles. Immobilized enzyme showed relatively better storage stability at 4 °C and 30 °C after 60.0 days as compared to free enzyme.

INTRODUCTION

Maltase [E.C 3.2.1.20] is an enzyme that catalyzes the hydrolytic process of α-1→4 glycosidic linkages and yields α-D-glucose as an end product. Maltase is used in different industries like food, brewing, distilling and pharmaceutical industry [1]. Catalytic property of an enzyme is one of the critical aspects for its commercialization and in some cases free enzymes are unstable to fulfill the requirement of industries due to low operational stability and recovery [2]. Various techniques such as protein engineering, chemical modification and immobilization have been investigated to overcome the limitation of enzymes for industrial bioreactors [3], [4] and [5]. Immobilization is an encouraging approach that not only enhances the stability of enzymes but also ensures the reusability of enzymes for various industrial bioprocesses [6], [7], [8] and [9]. Different immobilization techniques including covalent binding, adsorption, crosslinking and entrapment are used for continuous use of different enzymes on industrial

Citation: Muhammad Asif Nawaz, Haneef Ur Rehman, Zainab Bibi, Afsheen Aman, Shah Ali Ul Qader, Continuous degradation of maltose by enzyme entrapment technology using calcium alginate beads as a matrix, Biochemistry and Biophysics Reports, Volume 4, December 2015, Pages 250-256, ISSN 2405-5808, http://dx.doi.org/10.1016/j.bbrep.2015.09.025.

scale. Among them, entrapment confines the enzyme within the structured matrix space and create negligible impact on its catalytic properties [10] and [11]. There are various matrices such as alginate, polyacrylamide and agar–agar are employed for enzyme immobilization. Alginate is an anionic polysaccharide distributed widely in the cell walls of brown algae and commonly used for the formation of beads capable of entrapping different macromolecules in the form of calcium-alginate beads [12], [13] and [14]. Calcium alginate is a cost effective and biocompatible matrix for the entrapment of different enzymes including α-amylase, protease and pectinase [15] and [16].

In the current study, maltase from *B. licheniformis* KIBGE-IB4 was immobilized using entrapment technique within calcium alginate beads.

MATERIALS AND METHODS

Induction of bacterial isolate for maltase production

Wheat starch was used for the induction of previously isolated *B. licheniformis* KIBGE-IB4 for extracellular maltase production. Culture was incubated in optimized fermentation medium: wheat starch (2.5%), peptone (1.0%), yeast extract (0.2%), meat extract (0.4%), K_2HPO_4 (0.3%) and KH_2PO_4 (0.1%) at 37 °C and pH-7.0 for 48 h [17]. The cells were harvested by centrifugation at 40,248×*g* for 15.0 min at 4 °C. Cell free filtrate (CFF) was precipitated using 40% saturation of ammonium sulfate on ice bath and kept for 24 h at 4 °C. The obtained precipitates were dissolved in potassium phosphate buffer (100.0 mM, pH-6.5) and dialyzed against same buffer for 24 h at 4 °C and used for immobilization within calcium alginate beads.

Maltase immobilized calcium alginate beads formation

Partially purified maltase was mixed with equal volume (1:1) of sodium alginate (4.0%) and added drop wise into calcium chloride solution (0.2 M) with constant shaking on ice bath. Immobilized enzyme and control beads were hardened by storing them into fresh 0.2 M calcium chloride solution for 90.0 min at 4 °C. The beads were washed with deionized water and potassium phosphate buffer to remove any unbound enzyme molecules.

Enzyme assay for free and immobilized maltase

The enzyme activity of both free and immobilized maltase was determined by GOD–PAP method [18] and [19] using glucose and maltose as a standard and substrate respectively.

One unit of maltase was defined as the "*amount of enzyme required to release 1.0 μmol of glucose per minute under the standard assay conditions* ".

Effect of sodium alginate and calcium chloride concentration on beads formation

Different concentrations of sodium alginate (1.0–6.0%) and calcium chloride (0.1–0.5 M) were used to attain the maximum immobilization yield of maltase with stable beads structure.

Effect of curing time on percent immobilization yield of maltase

The influence of curing time on percent immobilization of enzyme was examined by keeping beads in 0.2 M calcium chloride solution for different time intervals ranging from 30.0 to 180.0 min. Beads were washed with double deionized water and potassium phosphate buffer (pH-6.5).

Effect of bead size on the activity of immobilized maltase

The effect of beads size on the catalytic activity of immobilized maltase was determined by preparing beads having different diameters ranging from 1 to 5 mm.

Effect of reaction time on the catalytic activity of free and immobilized enzyme

The effect of reaction time on the catalytic activity of free and immobilized maltase was studied by performing the enzyme assay at different time intervals keeping the reaction temperature, pH and substrate constant.

Effect of temperature on the catalytic activity of free and immobilized enzyme

The effect of temperature on the catalytic activity of free and immobilized maltase was studied by varying the enzyme–substrate reaction temperature ranging from 30 °C to 60 °C, keeping other parameters constant.

Effect of pH on the catalytic activity of free and immobilized enzyme

The influence of pH on the catalytic activity of free and immobilized maltase was investigated by performing the enzyme assay at different pH ranging from 5.0 to 8.0. Different buffers such as acetate buffer (pH: 5.0–6.0), citrate buffer (pH: 5.0–6.0) and potassium phosphate buffer (pH: 7.0–8.0) with same ionic strength (100.0 mM) were used.

Substrate kinetics of free and immobilized enzyme

The kinetic parameters such as *Km* and *Vmax* of free and immobilized maltase were determined using Lineweaver–Burk plot by performing the enzyme assay at different substrate concentrations (2.5–50 mM).

Thermal stability of free and immobilized enzyme

Thermal stability of free and immobilized maltase were examined by keeping them at different temperatures such as 40 °C, 45 °C, 50 °C, 55 °C and 60 °C for different time intervals (30.0–180.0 min). Sample aliquots were retrieved after every 30.0 min and enzyme assay was performed at 50 °C for 10 min.

Storage stability of free and immobilized enzyme

Storage stability of free and immobilized maltase was determined by storing them at different temperatures (4 °C and 30 °C) for 60 days. Control and immobilized enzyme beads were placed in 0.1 M potassium phosphate buffer (pH-6.5) and stored at aforementioned temperatures. Percent residual activity was calculated by considering activity of control as 100%.

Recycling efficiency of immobilized enzyme

Reusability of immobilized maltase was investigated by using 0.5 g of beads for several reaction cycles. Beads were washed with phosphate buffer (pH-7.0) after each reaction cycle and fresh substrate was mixed for next reaction. Percent residual activity was calculated by considering the activity of first run as 100%.

Scanning electron microscopy of beads with and without immobilized enzyme

Scanning electron microscopy was used to study the morphological changes of beads with and without immobilized maltase. For this purpose, beads were allowed to dry at 37 °C for 24 h and coated with gold using auto coater (Jeol Japan, Model JFC-1500). These gold coated samples were subjected to scanning electron microscope (SEM) (Jeol Japan, Model JSM 6380 A) for morphological analysis at different magnification power range (2500× and 5000×).

RESULTS AND DISCUSSION

Bacterial isolates are preferred to use at commercial level due to short fermentation time and selective product formation but free enzyme always faces problem due to difficult recovery from reaction mixture which also decreases the product purity. Immobilization is a supportive technique which not only permits easy recovery of product and enzyme from reaction mixture but also improves the thermal and operational stability of an enzyme. Effect of sodium alginate concentration on entrapment of maltase was studied different concentrations of sodium alginate. It was observed that the percent immobilization yield of maltase was found maximum at 4.0% sodium alginate (Fig. 1A) and as the concentration of sodium alginate increased from maximum level, immobilization efficiency decreased which might be due to the formation of beads having small pore size that causes hindrance to enter the substrate in calcium alginate beads and to react with immobilized enzyme. However, low concentration of sodium alginate resulted in the formation of soft and fragile beads with large pore size which resulted in the increased enzyme leakage during washing step [20] and [21]. Therefore proper concentration of sodium alginate is necessary for maximum immobilization of enzyme.

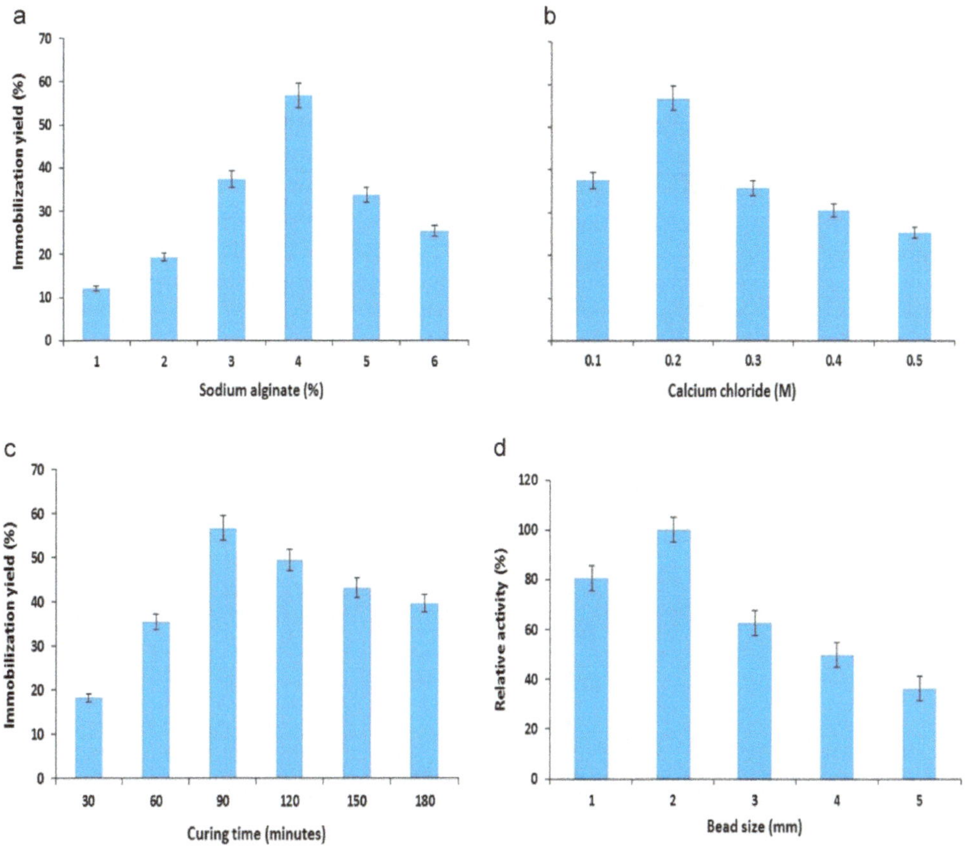

Figure 1: Effect of sodium alginate (A), calcium chloride (B), concentrations and (C) curing time (D) beads size on the immobilization of maltase from *B. licheniformis* KIBGE-IB4 (means±S.E., *n*=6).

It was observed that calcium chloride (0.2 M) provided more stabilized beads with maximum percent immobilization yield (Fig. 1B). The percent immobilization of maltase was decreased when concentration of calcium chloride increased or decreased beyond optimum level. High concentration of calcium chloride causes change in pH which might be responsible to decrease the catalytic activity of enzyme in microenvironment of beads. Immobilized maltase showed relatively high immobilization yield as compared to immobilized pectinase, while lower than immobilized α-amylase within same matrix[3] and [15].

The curing time of calcium alginate beads was also investigated for synthesis of stable beads structure. It was found that the percent immobilization yield increased with time and maximum immobilization yield was achieved after 90 min (Fig. 1C). It was previously reported that the probability of cross linkages formation increased between the matrix and enzyme when beads exposed with calcium chloride solution for relatively long time period [22]. However, it was also found that immobilization yield reduced due to prolonged curing time of beads which might be due to the enzyme leakage. Fragile beads are formed when beads are not exposed to calcium chloride solution for optimum time [14]. An appropriate bead size is also significant factor that affects the catalytic efficiency of immobilized enzyme and determines the suitability of reactor configuration [23]. It was observed that beads having 2.0 mm diameter size showed maximum catalytic activity as compared to other beads sizes (Fig. 1D). Catalytic activity was declined as diameter size of beads increased or decreased beyond optimum size. It was assumed that the beads size of 2.0 mm might provide more favorable conditions in terms of substrate and product diffusion from the matrix as compared to other sizes.

The impact of enzyme–substrate reaction time was also investigated on the catalytic activity of free and immobilized maltase. It was found that reaction time of immobilized maltase increased and maximum activity was achieved after 10.0 min as compared to free enzyme having maximum activity at 5.0 min at 50 °C (Fig. 2). This shift in reaction time might be due to the diffusion limitation of substrate into microenvironment of immobilized beads in order to reach the active site of enzyme. Similar findings have been reported in case of dextransucrase when immobilized within same support [24].

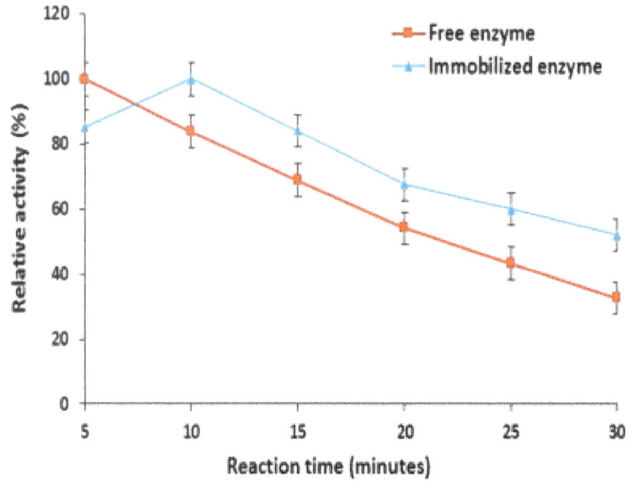

Figure 2: Effect of reaction time on the catalytic activity of calcium alginate immobilized maltase (means±S.E.,*n*=6).

Effect of reaction temperature on the catalytic activity of free and immobilized maltase was carried out at 30–60 °C for 10 min and it was revealed that the immobilized maltase showed maximum relative activity at 50 °C as compared to free maltase having optimum activity at 45 °C (Fig. 3). This shift in temperature optima might occur due to the impairment of hydrophobic and secondary interactions of the enzyme molecules within the matrix which improved at higher temperature to attain the maximum catalytic activity. It has been reported earlier that the immobilized endo-β-1,4-xylanase showed maximum relative activity at 60 °C which was 10 °C increase in temperature as compared to free enzyme [25]. Same finding was reported when endo (1→4) β-D-glucanase immobilized within agar–agar matrix using same technique where temperature optima was shifted from 60 °C to 70 °C after immobilization [26].

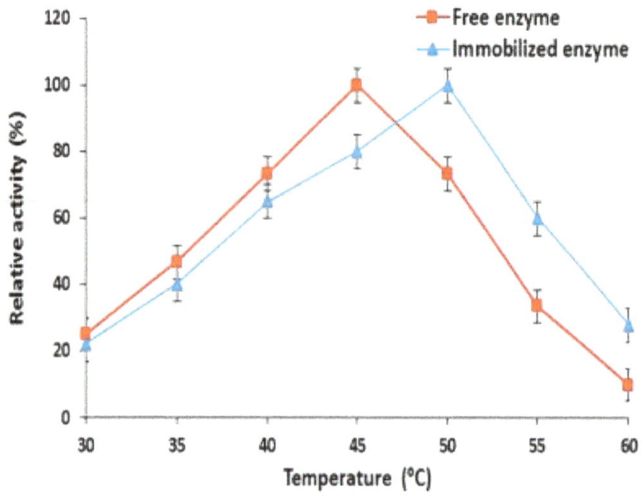

Figure 3: Effect of incubation temperature on the catalytic activity of calcium alginate immobilized maltase (means±S.E., *n*=6).

Effect of pH on the catalytic activity of free and immobilized maltase was performed at different pH levels (5.0–8.0) and it was observed that entrapment of maltase did not change the optimum pH of maltase activity and maximum relative activity was obtained at pH-6.5 (Fig. 4). It was noted that immobilized maltase extended its activity at broad pH range as compared to free enzyme.

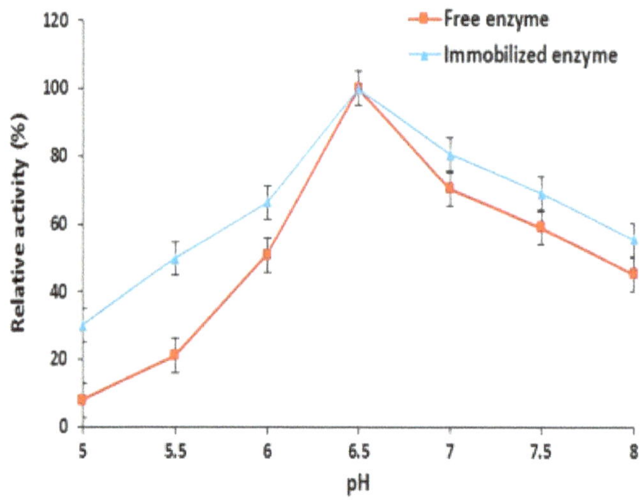

Figure 4: Effect of reaction pH on the catalytic activity of calcium alginate immobilized maltase (means±S.E.,*n*=6).

Km and V_{max} values of both free and immobilized maltase were calculated using Lineweaver–Burk plot by performing the enzymatic assay using different substrate concentrations. It was found that the K_m value of maltase was apparently increased from 1.71 to 3.17 mM ml^{-1} and V_{max} decreased from

8411 U ml^{-1} min^{-1} to 4919 U ml^{-1} min^{-1}after immobilization (Fig. 5). Increase in K_m and decrease in V_{max} values might be due to the hindrance of substrate to penetrate in matrix for enzyme–substrate reaction. Similar finding has been reported when maltase was immobilized within agar–agar support using same technique [27].

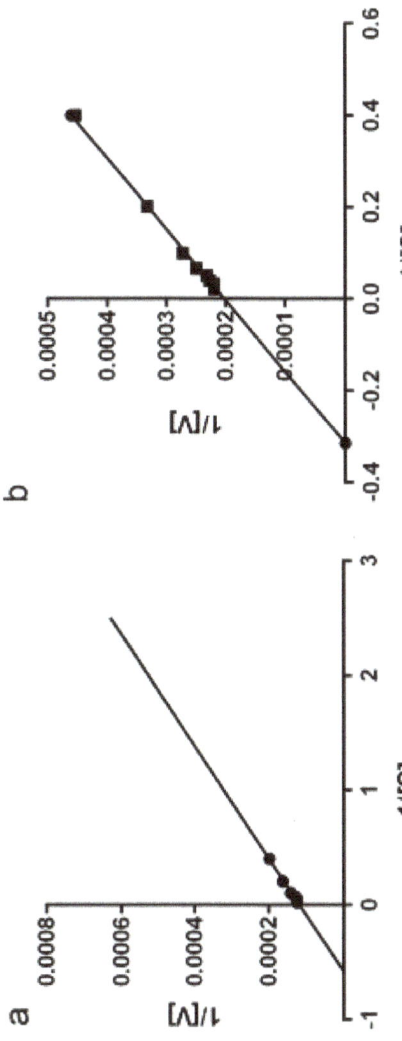

Figure 5: Lineweaver–Burk plots of (a) soluble and (b) immobilized maltase.

Thermal stability of free and immobilized maltase was also analyzed by keeping them at various temperatures ranging from 40 °C to 60 °C for different time intervals. Immobilized maltase showed broad stability range at different temperatures as compared to free enzyme (Fig. 6). Immobilized maltase retained 75% and 55% activity at 40 °C and 45 °C, respectively after 3.0 h whereas, free enzyme showed only 45% and 25% residual activity at the same temperatures. Free maltase lost its complete activity at 55 °C after 90.0 min. However, immobilized maltase retained its activity even after 180.0 min under same conditions. It was also observed that immobilized enzyme also retained 67% activity at 60 °C even after 30.0 min. Enzyme rigidity can be enhanced by immobilizing it in various supports where interaction between enzyme and matrix plays significant role to prevent the major conformational changes induced by the heat [28]. It was reported that various factors including number of bonds formed between enzyme and support, the nature of the bonds, the confinement degree of an enzyme within support and the environmental conditions of an enzyme may affects the stability of immobilized enzyme [29].

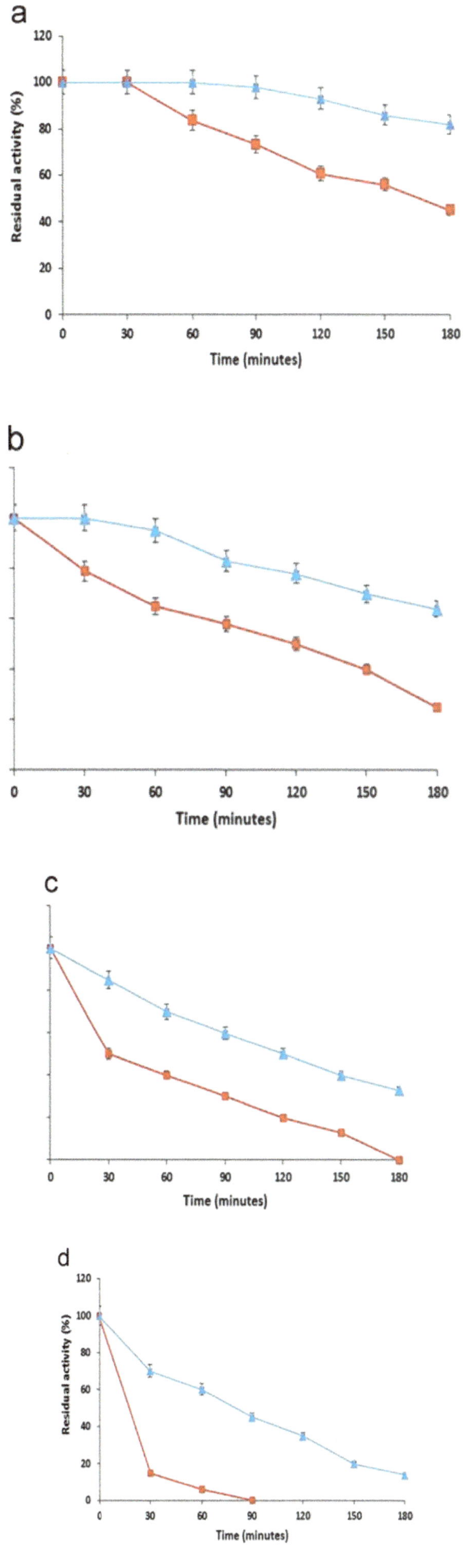

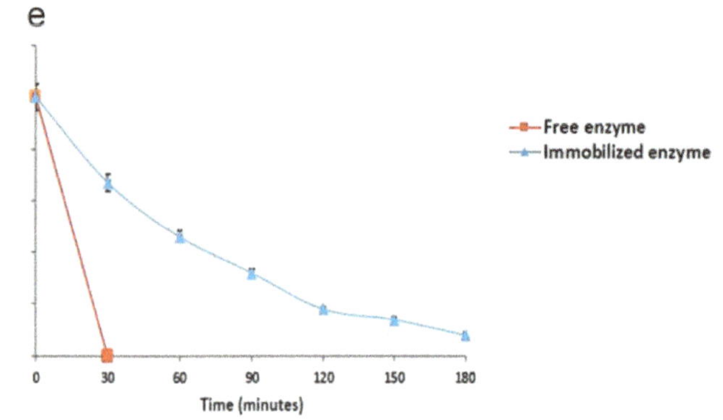

Figure 6: Thermal stability of free and calcium alginate immobilized maltase. (a) 40 °C, (b) 45 °C, (c) 50 °C, (d) 55 °C, and (e) 60 °C (means±S.E., *n*=6).

Free and immobilized maltase were stored up to 60 days at different temperatures (4 °C and 30 °C) and residual activity was calculated. It was found that immobilized maltase showed broad storage stability range at 4 °C and 30 °C as compared to free enzyme (Fig. 7). Free maltase retained 52% and 37% residual activity at 4 °C and 30 °C, respectively. However, immobilized maltase showed 58% and 49% residual activity at same aforementioned storage conditions. Several research groups have reported that enzyme immobilization within hydrogel carriers like alginate, poly (hydroxyethylmethyacrylate-coglycidylmethacrylate), gelatin and polyacrylamide exhibited high stability due to the protective microenvironment provide by the gel matrix [3], [30] and [31]. The current finding suggested that calcium alginate matrix generates positive impact by forming different interactions with an enzyme which also provides conformational stability to immobilized maltase.

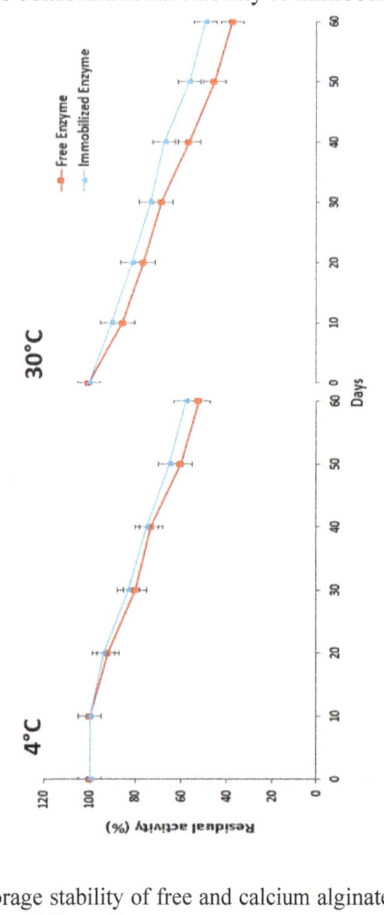

Figure 7: Storage stability of free and calcium alginate immobilized maltase (means±S.E., *n*=6).

Immobilized maltase retained more than 80% and 67% recycling efficiency after 2nd and 3rd cycle. It was also observed that the immobilized maltase showed 17% residual activity even after 6th cycles (Fig. 8). The residual activity of immobilized maltase was decreased progressively as the number of cycles increased and complete loss of activity was found after six reaction cycles which might be due to the leakage of enzyme from support as a result of excessive washing of beads after each cycle. Similar results have been reported when enzyme was immobilized using calcium alginate as a support [15].

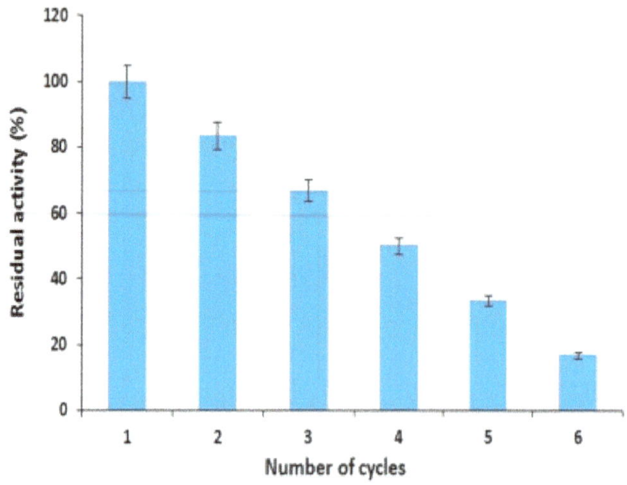

Figure 8: Recycling efficiency of calcium alginate immobilized maltase (means±S.E., *n*=6).

The surface topology of calcium alginate beads before and after immobilization of maltase was examined by using scanning electron microscope (SEM). Compact pores was observed on the surface of beads having no enzyme (Fig. 9A and C) whereas, for maltase immobilized beads small particles with irregular shape were observed on the surface (Fig. 9B and D). Morphological changes on the surface of immobilized laccase were also observed where TiO_2–montmorillonite complex was used as matrix [32].

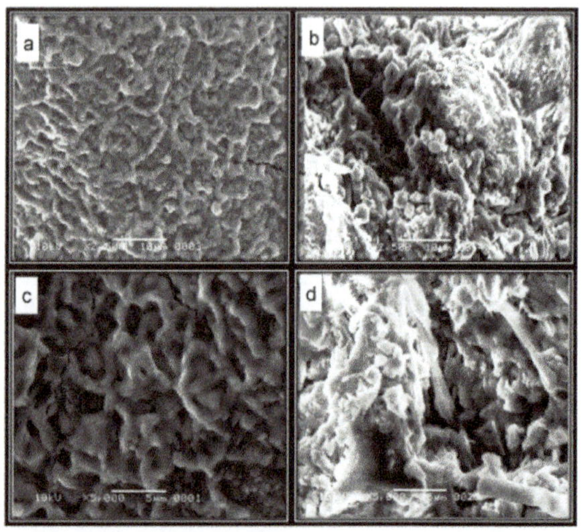

Figure 9: Scanning electron micrographs of calcium alginate beads before and after immobilization of maltase. (A and C) SEM micrographs of calcium alginate beads at 2500× and 500× of magnification before immobilization respectively. (B and D) SEM micrographs calcium alginate beads after immobilization of maltase 2500× and 5000× of magnification respectively.

CONCLUSIONS

This study justifies the use of calcium alginate as an appropriate support for the immobilization of maltase through entrapment technique. Immobilized enzyme showed remarkable thermal stability and recycling efficiency. All parameters of maltase were improved after immobilization which showed that enzyme stability and biocompatibility within calcium alginate matrix. Such improved catalytic efficiency of immobilized maltase is significant for various commercial applications.

ACKNOWLEDGMENTS

This research was supported by the Karachi Institute of Biotechnology and Genetic Engineering (KIBGE), University of Karachi, Pakistan.

REFERENCES

1. R.A. Mohamed, A.B. Salleh, R.N.Z.R.A. Rahman, M. Basri, T.C. Leow, Afr. J. Microbiol. Res. 6 (2012) 2909–2917.
2. B. Sarrouh, T.M. Santos, A. Miyoshi, R. Dias, V. Azevedo, J. Bioprocess. Biotech S4 (2012) 002, http://dx.doi.org/10.4172/2155-9821.S4-002.
3. G. Dey, B. Singh, R. Banerjee, Braz. Arch. Biol. Technol. 46 (2003) 167–176.
4. A.R.M. Doaa, W.A. Helmy, J. Appl. Sci. Res. 5 (2009) 2466–2476.
5. A.A. Homaei, R. Sariri, F. Vianello, R. Stevanatom, J. Chem. Biol. 6 (2013) 185–205.
6. S.B. Jadhav, R.S. Singhal, Carbohydr. Polym. 105 (2014) 49–56.
7. S. Talekar, V. Ghodake, A. Kate, N. Samant, C. Kumar, S. Gadagkar, Aust. J. Basic Appl. Sci. 4 (2010) 4760–4765.
8. F.M. Bautista, M.C. Bravo, J.M. Campelo, A. Garcia, D. Luna, J.M. Marinas, A.A. Romero, J. Chem. Technol. Biotechnol. 72 (1998) 249–254.
9. M.J.J. Litjens, K.Q. Le, A.J.J. Straathof, J.A. Jongejan, J.J. Heijnen, Biocatal. Biotransform. 19 (2001) 1–19.
10. M.H. Kafshgari, M. Khorram, M. Mansouri, A. Samimi, S. Osfouri, Iran. Polym. J. 21 (2012) 99–107.
11. S. Kumari, P.S. Panesar, M.B. Bera, H.K. Chopra, Iran. Polym. J. 23 (2014) 111–119.
12. C. Le-Tien, M. Millette, M. Lacriox, M.C. Mateescu, Biotechnol. Appl. Biochem. 39 (2004) 189–198.
13. M. Matto, Q. Hussain, J. Mol. Catal. B: Enzym. 57 (2009) 164–170.
14. R. Satar, M. Matto, Q. Husain, J. Sci. Ind. Res. 67 (2008) 609–615.
15. H.U. Rehman, A. Aman, A. Silipo, S.A.U. Qader, A. Molinaro, A. Ansari, Food Chem. 139 (2013) 1081–1086.
16. A. Riaz, S.A.U. Qader, A. Anwar, S. Iqbal, Aust. J. Basic Appl. Sci. 3 (2009) 2883–2887.
17. M.A. Nawaz, Z. Bibi, A. Aman, R.R. Zohra, S.A.U. Qader, Pak. J. Pharm. Sci. 27 (2014) 1437–1442.
18. P. Trinder, J. Clin. Pathol. 22 (1969) 246.
19. P. Trinder, Ann. Clin. Biochem. 6 (1969) 24–28.
20. M.A. Longo, I.S. Novella., L.A. Garcia, M. Daz, Enzyme Microb. Tech. 14 (1992) 585–590.
21. S.A. Ahmed, Ramadan A. Al Domany, Nefisa M.A. El-Shayeb, H.H. Radwan, S.A. Saleh, S.A. Res. J. Agric. Biol. Sci. 4 (2008) 434–466.
22. A.S. Waldman, L. Schechinger, G. Govindarajoo, J.S. Nowick, L.H. Pignolet, J. Chem. Educ. 75 (1998) 1430–1431.
23. F. Ertan, H. Yagar, Prep. Biochem. Biotechnol. 37 (2007) 195–204.
24. S.A.U. Qader, A. Aman, N. Syed, S. Bano, A. Azhar, Ital. J. Biochem. 56 (2007) 158–162.
25. Z. Bibi, F. Shahid, S.A.U. Qader, A. Aman, Int. J. Biol. Macromol. 75 (2015) 121–127.
26. A. Karim, S.A.U. Qader, A. Nawaz, A. Aman, Chem. Eng. Trans. 38 (2014) 409–414.
27. M.A. Nawaz, A. Karim, A. Aman, R. Marchetti, S.A.U. Qader, A. Molinaro, Bioprocess Biosyst. Eng. (2014), http://dx.doi.org/10.1007/s00449-014-1302-6.
28. A.M. Klibanov, Adv. Appl. Microbiol. 29 (1982) 1–28.
29. M.C. Chia, R.C. Lyub, L.L. Linb, H.B. Huangc, Biochem. Eng. J. 39 (2008) 376–382.
30. F. Sahin, G. Demirel, H. Tumturk, Int. J. Biol. Macromol. 37 (2005) 148–153.
31. A. Yahsi, F. Sahin, G. Demirel, H. Tumturk, Int. J. Biol. Macromol. 36 (2005) 253–258.
32. Q. Wang, L. Peng, G. Li, P. Zhang, D. Li, F. Huang, Q. Wei, Int. J. Mol. Sci. 14 (2013) 12520–12532.

CHAPTER 13

CHARACTERIZATION AND MULTI-STEP TRANSKETOLASE-Ω-TRANSAMINASE BIOCONVERSIONS IN AN IMMOBILIZED ENZYME MICROREACTOR (IEMR) WITH PACKED TUBE

Amanatuzzakiah Abdul Halim[1,2], Nicolas Szita[1], and Frank Baganz [1]

[1]The Advanced Centre for Biochemical Engineering, University College London, Torrington Place, London WC1E 7JE, United Kingdom

[2]Biotechnology Engineering Department, International Islamic University Malaysia (IIUM), 50728 Kuala Lumpur, Malaysia

ABSTRACT

The concept of *de novo* metabolic engineering through novel synthetic pathways offers new directions for multi-step enzymatic synthesis of complex molecules. This has been complemented by recent progress in performing enzymatic reactions using immobilized enzyme microreactors (IEMR). This work is concerned with the construction of *de novo*designed enzyme pathways in a microreactor synthesizing chiral molecules. An interesting compound, commonly used as the building block in several pharmaceutical syntheses, is a single diastereoisomer of 2-amino-1,3,4-butanetriol (ABT). This chiral amino alcohol can be synthesized from simple achiral substrates using two enzymes, transketolase (TK) and transaminase (TAm). Here we describe the development of an IEMR using His_6-tagged TK and TAm immobilized onto Ni-NTA agarose beads and packed into tubes to enable multi-step enzyme reactions. The kinetic parameters of both enzymes were first determined using single IEMRs evaluated by a kinetic model developed for packed bed reactors. The $Km_{(app)}$ for both enzymes appeared to be flow rate dependent, while the turnover number k_{cat} was reduced 3 fold compared to solution-phase TK and TAm reactions. For the multi-step enzyme reaction, single IEMRs were cascaded in series, whereby the first enzyme, TK, catalyzed a model reaction of lithium-hydroxypyruvate (HPA) and glycolaldehyde (GA) to l-erythrulose (ERY), and the second unit of the IEMR with immobilized TAm converted ERY into ABT using (*S*)-α-methylbenzylamine (MBA) as amine donor. With initial 60 mM (HPA and GA each) and 6 mM (MBA) substrate concentration mixture, the coupled reaction reached approximately 83% conversion in 20 min at the lowest flow rate. The ability to synthesize a chiral pharmaceutical intermediate, ABT in relatively short time proves this IEMR system as a powerful tool for construction and evaluation of *de novo* pathways as well as for determination of enzyme kinetics.

Citation: Amanatuzzakiah Abdul Halim, Nicolas Szita, Frank Baganz, Characterization and multi-step transketolase-ω-transaminase bioconversions in an immobilized enzyme microreactor (IEMR) with packed tube, Journal of Biotechnology, Volume 168, Issue 4, December 2013, Pages 567-575, ISSN 0168-1656, http://dx.doi.org/10.1016/j.jbiotec.2013.09.001.

INTRODUCTION

Microreactor technology has immense potential to notably improve chemical and biological analyses as well as abundant applications in medical diagnostics and therapeutic devices. Owing to their miniature size, they require only a small volume of reagent, shorter processing time and can be parallelized using lab-on-a-chip technology (Hong et al., 2009 and Mark et al., 2010).

The integration of immobilized enzymes in microreactor systems has great impact in many emerging technologies employing biocatalytic and biorecognition events. Several reports show that a microreactor can be an ideal platform for enzyme reactions (Hickey et al., 2007, Hickey et al., 2009, Ngamsom et al., 2010, Alam et al., 2011 and Matsuura et al., 2011). Rapid screening can be done with the capacity of multiple usage of the immobilized enzyme (Forsberg et al., 2004 and Matosevic et al., 2011a). In addition to that, enzymes are found to gain higher stability facilitating their application under harsh environmental conditions of temperature, pH and organic solvents (Sakai-Kato et al., 2002, Nomura et al., 2004 and Berne et al., 2006). Enzyme immobilization may also result in changes in enzymatic activity, optimum pH or affinity for the substrate (Krenková and Foret, 2004 and Ladero et al., 2006) and must be considered when designing enzyme microreactors.

Several enzyme immobilization methods are available. Nidetzky (2013) have summarized typical techniques for preparation of enzymes immobilized in microreactors. A generically applicable and well established approach for the immobilization of oriented protein molecules is the use of recombinant tags, particularly poly-histidine tags (His-tag). This technique stemmed from immobilized metal ion affinity chromatography (IMAC) and has been applied extensively to protein immobilization (Block et al., 2009 and Cheung et al., 2012). It offers fast and specific binding and allows protein immobilization without pre-purification steps. Moreover, the His-tag does not interfere with the structure or function of proteins and does not affect the secretion, compartmentalization, or folding of fusion proteins within cells (Cha et al., 2004). Application of the immobilization of protein molecules *via* His-tag has been successfully demonstrated in numerous studies. Site-specific immobilization which orientates all proteins uniformly was performed by Abad et al. (2005), whereby His-tagged horseradish peroxidase and ferredoxin-NADP+ reductase confirmed full functional attachment of the protein to the NTA-Co(II)-matrix. Immobilization of His-tagged *Rhodotorula gracilis*d-amino acid oxidase (*Rg*DAO) on Ni-NTA magnetic beads was demonstrated by Kuan et al. (2008). The immobilization of*Rg*DAO enhanced its stability against thermal inactivation where the relative activity of the immobilized enzyme was 56% while the free enzyme was completely inactivated after incubation at 50 °C for 1 h. The residual activity and the storage stability were also improved in the immobilized form of the enzyme. The work by Cha et al. (2005) showed that sulfotransferases enzymes immobilized through poly-His tag exhibited similar activities as in the free solution as a result of highly specific orientation offered by the His-tag linkage.

We recently developed an IEMR based on the reversible immobilization of His-tagged enzyme *via* Ni-NTA linkage to surface derivatized silica (Matosevic et al., 2009). This IEMR was first characterized and used to quantify the immobilized transketolase kinetics in a stop-flow mode. In the current work, we focus on a single diastereoisomer of 2-amino-1,3,4-butanetriol (ABT). This chiral amino alcohol is one of the building blocks of several antibiotics, antivirals and sphingolipids (Contestabile and John, 1996). ABT can be produced using two enzymes, transketolase (TK) and transaminase (TAm), that catalyze asymmetric carbon-carbon bond formation and amine group addition, respectively (Fig. 1a). Studies have been carried out by our research group to screen for different enzyme combinations to synthesize the optically pure amino alcohol. Ingram et al. (2007), demonstrated the synthesis of ABT, where *E. coli* TK converts achiral lithium-hydroxypyruvate (HPA) and glycolaldehyde (GA) into l-erythrulose (ERY), and subsequently β-alanine:pyruvate TAm from *Pseudomonas aeruginosa* catalyzes the reaction between (*S*)-α-methylbenzylamine (MBA) and Ery, producing ABT in a one-pot synthesis. In addition, Rios-Solis et al. (2011) used wild-type *E. coli* TK paired with an ω-TAm from *Chromobacterium violaceum* 2025 (Kaulmann et al., 2007) to synthesize ABT in a whole-cell system. And more recently Matosevic et al. (2011b), further investigated this reaction in an IEMR. To demonstrate the feasibility of performing the continuous flow TK–TAm enzymatic reaction using immobilized enzymes Matosevic et al. (2011b)derivatized the inner surface of a fused silica capillary ("open-tubular" reactor).

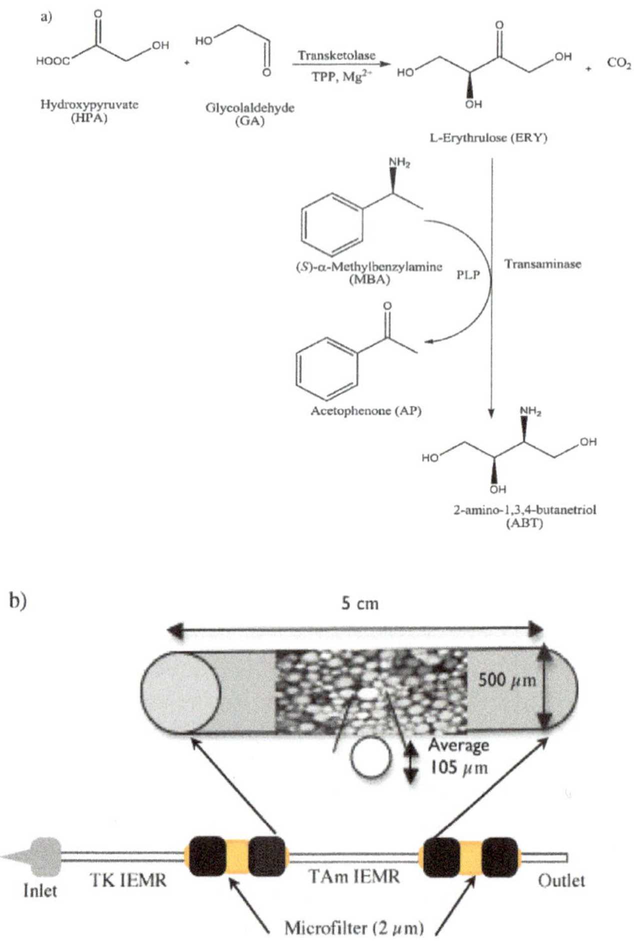

Figure 1: (a) Reaction scheme for the synthesis of 2-amino-1,3,4-butanetriol (ABT) by the TK–ω-TAm enzyme pathway. Dual enzyme

pathway showing the TK-catalyzed production of l-erythrulose (ERY) followed by the TAm-catalyzed production of ABT. (b) Schematic illustration of the IEMR used for analysis of TK and TAm reactions. An IEMR with inner diameter (ID) of 500 μm and length of 5 cm gives a total reaction volume of ~10 μL. The prototype consisted of single IEMRs that were cascaded in series to allow multi-step enzymatic reaction. TK catalyzed the reaction of hydroxypyruvate (HPA) and glycolaldehyde (GA) to l-erythrulose (ERY), and the second unit of the IEMR with immobilized TAm converted ERY into ABT using (*S*)-α-methylbenzylamine (MBA) as amine donor.

In this work, we characterize a packed tube IEMR prototype to further facilitate the *in vitro*study of this dual enzyme pathway in a continuous flow system. This microreactor system, based on the reversible immobilization of His_6-tagged enzymes was implemented for the quantitative evaluation of TK–TAm bioconversion kinetics. Higher loading of enzyme was achieved using microbeads compared to the previous system (Matosevic et al., 2011b) which were packed into a tube that works as a packed bed reactor. The strong coupling method produces a highly stable and reusable IEMR. First, the stability of both enzymes on the immobilized form was evaluated and reusability of the microreactor was demonstrated over several cycles of immobilization and elution. The IEMR prototype was then used for determining the kinetic parameters for both TK and TAm enzymes before demonstrating the synthesis of the chiral amino alcohol ABT by coupling the TK and TAm IEMRs.

MATERIALS AND METHODS

Reagents and chemicals

All reagents were of analytical grade and were obtained from Sigma–Aldrich (Gillingham, UK) unless stated otherwise. Reverse osmosis (RO) water was used in all of the experiments.

Biocatalyst production

Inoculum preparation

Stocks of *E. coli* BL21gold (DE3) containing plasmid pQR791 (His_6-TK) (Martinez-Torres et al., 2007) or pQR801 (His_6-TAm) (Kaulmann et al., 2007) (previously cloned from expression constructs provided by Prof. John Ward, Department of Biochemical Engineering, UCL) were used. Cells were grown for 16 h at 37 °C on agar plate containing LB-glycerol nutrient broth (yeast extract 5 g/L, tryptone 10 g/L, NaCl 10 g/L, glycerol 10 g/L), 15 g/L agar and 0.15 g/L selection antibiotic ampicillin (TK cultures) or kanamycin (TAm cultures). A single colony was inoculated into a sterile 250 mL shake flask containing 20 mL of the LB-glycerol medium and grown on an orbital shaker (Stuart Scientific, Redhill, UK) at 250 rpm for 12–16 h at 37 °C.

Enzyme production

20 mL of inoculum prepared earlier was transferred into 1 L shake flask containing 180 mL of LB-glycerol with 0.15 g/L antibiotic. The culture was incubated for 7–8 h at 37 °C and 250 rpm on an orbital shaker. For TAm production isopropyl β-d-1-thiogalactopyranoside (IPTG) of 0.2 mM was added after 3 h of fermentation at optical density (OD): 0.6. The $OD_{600\ nm}$ of the solution was measured every hour to check bacterial growth until it reached stationary phase. Following the fermentation, culture broth was centrifuged at 4000 rpm for 10 min at 4 °C. The supernatant was discarded and the pellets were stored at −20 °C.

Quantification of protein concentration

Sodium dodecyl sulfate polyacrylamide gel electrophoresis (SDS-PAGE) and densitometry analysis

Pre-cast gels (1.0 mm x 12 wells) of SDS 12% Tris-Glycine (Novex, Paisley, UK) were used for SDS-PAGE protein analysis as described by King and Laemmli (King and Laemmli, 1971). Gels were run in Tris-Glycine buffer (Novex, Paisley, UK) using a Mini-155 Protean II system (Bio-Rad Laboratories Inc., Hemel Hempstead, UK) at a power of 80–150 V for about 90 min. The gel was then rinsed twice for 5 min with 100 mL of RO water. Next, the gel was stained with sufficient SimplyBlue SafeStain (~20 mL) (Novex, Paisley, UK) to cover the gel for 1 h at room temperature with gentle shaking. Then the gel was de-stained with de-staining buffer (40%, v/v methanol, 10%, v/v acetic acid and 50%, v/v RO water) for 2–3 h. Finally, the mini gel was washed with 100 mL of RO water for 1–3 h.

Protein concentration determination

Quantification of protein was carried out by UV absorbance of the purified enzyme at 280 nm using ATI Unicam UV/VIS spectrophotometer (Spectronic, Leeds, UK). The extinction coefficient used were 93,905 M cm^{-1} for transketolase and 78,000 M cm^{-1} for transaminase. The Bradford protein assay was carried out based on the method of Bradford (Bradford, 1976). Bradford Reagent was used with bovine serum albumin (BSA) as the standard protein. Absorbance measurements at 595 nm were carried out on an ATI Unicam UV/VIS spectrophotometer (Spectronic, Leeds, UK).

Purification of His_6-tagged enzymes

Frozen cell pellets from the fermentation were thawed at room temperature. The pellet was resuspended in a small quantity of 50 mM Tris-HCl (1:25, v/v) and vortexed. The suspension was sonicated on ice using a Soniprep 150 sonicator (MSE, Sanyo, Japan, 10 cycles of 10 s ON, 10 s OFF). The solution was centrifuged using a Hettich Universal 320R mid-bench centrifuge (DJB Labcare, Newport Pagnell, UK) for 10 min at 4000 rpm and 4 °C. The supernatant was decanted and the pellets were discarded. The supernatant was filtered using a 0.2 μm pore filter (GE Healthcare, Buckinghamshire, UK). Purification of His_6-tagged enzymes was performed using Ni-NTA superflow cartridges (Qiagen, Sussex, UK). The cartridge was equilibrated with 10 mL of binding buffer (50 mM NaH_2PO_4, 300 mM NaCl and 10 mM imidazole). The clarified lysate was loaded into the cartridge and subsequently washed with 20 mL of wash buffer (50 mM NaH_2PO_4, 300 mM NaCl and 20 mM imidazole). The enzyme was eluted with 2.5 mL of elution buffer (50 mM NaH_2PO_4, 300 mM NaCl and 250 mM imidazole).

Microreactor set-up and bioconversions

Packing of a FEP tube with microbeads

HIS-Select Nickel Affinity agarose beads (Sigma–Aldrich, UK) with diameter of 45–165 μm were used as the packing material. They were first washed with 50 mM of Tris-HCl and left in suspension in the same buffer. A Nanobaume (Presearch Ltd., UK) pressurized container was used to pack the beads in the tube. The bead slurry containing 5% (v/v) beads was placed in a glass vial with total volume of 1.5 mL. A stir bar was put in the vial and the vial was placed into the pressurized container. The container was then set on top of a magnetic stirrer. The fluorinated ethylene propylene (FEP) tube (500 μm ID, Presearch Ltd., UK) was attached to the container while a microfilter assembly (2 μm, Presearch Ltd., UK) was connected at the end of the tube. A two-way valve was connected to the pressurized container while the other end was connected to a nitrogen gas cylinder. A pressure of 5 bar was applied to the system and the pressure drop (0.6–1 bar) was monitored by the pressure gauge installed. The packing took 5–10 min.

Enzyme immobilization and elution

The packed tube IEMR was first equilibrated with binding buffer (500 mM NaCl, 20 mM Tris-HCl, 5 mM imidazole; pH 7.0 or 7.5) at 5 μL/min for 15 min. Enzyme concentrations used were 0.05–2.0 mg/mL. The purified enzyme (100 μL) was loaded into the tube at 5 μL/min and was left to incubate for an additional 1 h following immobilization. Non-specific binding of the enzyme on the beads was removed by treating the tube with washing buffer (50 mM Tris-HCl; pH 7.0 or 7.5) at 25 μL/min for 20 min. Bound enzyme was removed by treating the tube with elution buffer (1 M imidazole, 500 mM NaCl and 20 mM Tris-HCl; pH 7.0 or 7.5) at 5 μL/min for 10 min.

Transketolase reaction in IEMR

The packed tube IEMR (5 cm length) was connected to a KDS100 syringe pump (KD Scientific, Holliston, MA, USA) on one end and connected to a microfilter assembly (2 μm, Presearch Ltd., UK) at the outlet. The TK reaction was carried out at room temperature in 50 mM Tris-HCl buffer (pH 7.0). One unit of TK enzyme was defined as the amount that catalyses the formation of 1 μmol of erythrulose (ERY) per minute using 50 mM of glycolaldehyde (GA) and lithium-hydroxypyruvate (HPA) at pH 7.0. TK reactions were conducted using model substrates, 50 mM of HPA and 50 mM of GA unless stated otherwise. The enzyme was incubated with cofactors, thiamine pyrophosphate (TPP) of 2.4 mM and $MgCl_2$ of 9 mM at 5 μL/min for 30 min to reconstitute the holoenzyme. Substrate solutions of HPA and GA were loaded continuously at various flow rates and 20 μL samples were taken from the outlet of the tube at regular time intervals. The samples were quenched with 0.1% trifluoroacetic acid (TFA) and analyzed by HPLC. Reactions were monitored for EYR production by HPLC assay.

Transaminase reaction in IEMR

The reaction was carried out at room temperature in 50 mM Tris-HCl buffer (pH 7.5). One unit of TAm enzyme was defined as the amount that catalyses the production of 1 μmol of acetophenone (AP) per minute using 100 mM l-erythrulose (ERY) and 10 mM (*S*)-α-methylbenzylamine (MBA) at pH 7.5. TAm reactions were conducted using model substrates, 10 mM of MBA and 100 mM of ERY unless stated otherwise. Substrate solutions of MBA and ERY and 0.2 mM cofactor pyridoxal-5′-phosphate (PLP) were loaded continuously at various flow rates and 20 μL samples were taken from the outlet of the tube at regular time intervals. The samples were quenched with 0.1% TFA and analyzed by HPLC. Reactions were monitored for acetophenone (AP) production by HPLC assay.

Coupled TK–TAm reaction in IEMR

The set-up of the IEMR cascade is shown in Fig. 1b. The enzyme immobilization of individual IEMRs was conducted separately before assembly. Two separate (2 cm × 5 cm) packed tubes of IEMR were connected in series where the beads were retained using a microfilter assembly (2 μm, Presearch Ltd., UK). The second packed tube IEMR was connected to the first using a microferrule (Presearch Ltd.). The two-tube system was connected to a KDS100 syringe pump (KD Scientific, Holliston, MA, USA) so that the inlet of the first tube was directly connected to the pump while its outlet was used to collect the samples for further off-line HPLC analysis. Coupled reactions were carried out using the TK–TAm model reaction. Solution substrates of GA, HPA and MBA at desired concentrations and cofactors for both enzymes at pH 7.5 were loaded together into the first tube containing packed beads immobilized with TK. The erythrulose required by the immobilized TAm was provided by the TK reaction step. The reaction was carried out continuously at flow rates between 1 and 30 μL/min. 20 μL samples were taken at different time intervals and quenched with 0.1% TFA. Reactions were monitored for ABT and AP production by HPLC assay.

HPLC analysis

HPLC assay for TK

HPLC analysis was conducted using a microbore HPLC system (Dionex, Surrey, UK) controlled by Chromeleon 6.8 software with a multiwell plate autosampler (Spark, Emmen, Holland). The HPLC system contained an Aminex 87H column (Bio-Rad, Hertfordshire, UK) and analysis of ERY was conducted at 60 °C using a mobile phase of 0.1% (v/v) TFA in water at 0.6 mL/min. Detection was at 210 nm and the retention time was around 11.5 min for ERY.

HPLC assay for TAm

AP and ABT were analyzed using an ACE5 C18 reverse phase column (150 mm × 4.6 mm, Advanced Chromatography Technologies, Aberdeen, UK) on the Dionex HPLC system. The mobile phase for AP detection comprised of 0.1% TFA at 1.0 mL/min with a gradient of acetonitrile from 15% to 72% over 9 min. The gradient was followed by a 2 min equilibration. UV detection was carried out at 250 nm with retention time for AP was around 7.3 min. For ABT analysis, the samples were derivatized by addition of an excess of 6-aminoquinolyl-N-hydroxysuccinimidyl carbamate. The derivatizing reagent was made in house based on Cohen and Michaud (1993) protocol and the HPLC method reported by Ingram et al. (2007) was used.

RESULTS AND DISCUSSION

Microreactor set-up and enzyme immobilization

A schematic illustration of the IEMR used in this study is shown in Fig. 1b. Packing of the tube was carried out using a Nanobaume pressurized container. This method produced consistent and reproducibly packed tubes according to microscope images. The dense packing in the tube was formed as a result of the compressibility of the agarose beads used.

Purified enzymes were used in all immobilization processes. Samples collected from the feed and elution steps were ran on SDS-PAGE and quantified by the Bradford assay. The eluted enzyme purity was >95%, observed as a single band on the gel for TK and TAm (Fig. 2). Specific activity of purified TK and TAm calculated were 20.3 ± 4.5 U/mg and 0.13 ± 0.025 U/mg respectively. Following immobilization of 0.2 mg/mL of the enzyme feed, 0.1 ± 0.01 mg/mL (10 μg) of the enzyme was recovered during the elution process. This shows approximately 50% of the enzyme feed was retained inside the microreactor while the remaining proportion was removed during the washing process. In order to test if the maximum enzyme loading capacity was reached, the concentration of TAm in the feed was increased to 1.9 mg/mL, which resulted in a 10-fold increase of immobilized enzyme in the microreactor (1.1 ± 0.15 mg/mL). This suggests that enzyme loading increases gradually with higher enzyme concentration in the feed until it reaches the maximum capacity of the beads. The IEMR with higher TAm loading was later used in the dual enzyme reaction. In addition, no adsorption of protein on the surface of the FEP tube was detected when a similar immobilization-washing-elution process was conducted on an empty tube. This eliminates any errors that might mask the actual enzyme activity in the IEMR.

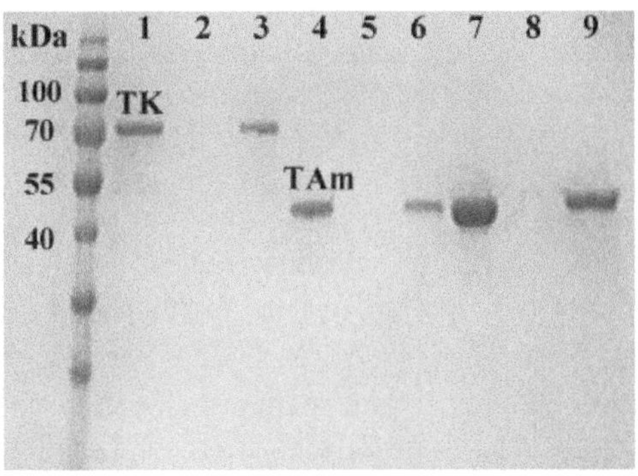

Figure 2: SDS-PAGE gel of enzyme immobilization-elution process. Purification performed as described in Section 2.3. (1) Purified TK (0.2 mg/mL); (2) flow through from the IEMR; (3) eluted TK (0.1 mg/mL); (4) purified TAm (0.2 mg/mL); (5) flow through from the IEMR; (6) eluted TAm (0.1 mg/mL); (7) purified TAm (1.9 mg/mL); (8) flow through from the IEMR; (9) eluted TAm (1.1 mg/mL).

Operational and storage stability of IEMR

The operational stability of the immobilized TK was evaluated continuously for 48 h at a flow rate of 1 μL/min. Samples were taken every 12 h to measure the conversion of HPA and GA to ERY product. Immobilized TK was found to remain stable during the first 12 h of reaction with 100% productivity (Fig. 3a). However the productivity dropped to 76% after 48 h of reaction. A similar test was performed with immobilized TAm where the productivity was determined continuously every hour at 1 μL/min. In contrast to TK, after only 8 h of continuous reaction, the productivity of the immobilized TAm was reduced to 60% (Fig. 3b).

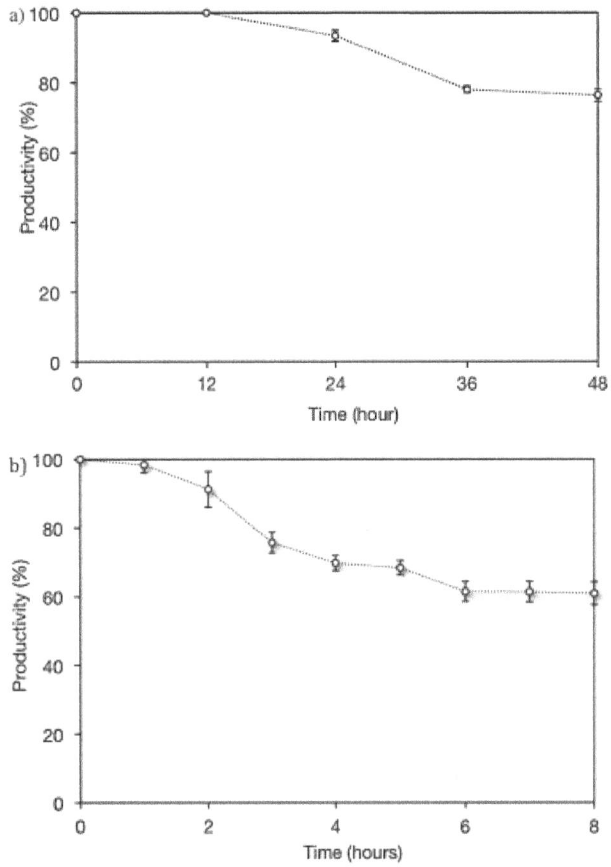

Figure 3: Operational stability of TK and TAm in IEMR. The productivity is defined as the amount of product generated over 12 h (TK) or 1 h (TAm) (per amount of enzyme) divided by the product formed of the starting enzyme activity determined in the same way. A productivity of 100% indicates no deactivation and degradation of the immobilized enzyme in the reactor. (a) Continuous reaction of TK in packed tube IEMR at 1 μL/min for 48 h. (b) Continuous reaction of TAm in packed tube IEMR at 1 μL/min for 8 h. Concentration of immobilized TK and TAm was ~0.2 mg/mL and 0.3 mg/mL respectively. Error bars represent one standard deviation about the mean ($n = 3$).

The storage stability of immobilized TK in the packed tube stored at 4 °C in 50 mM of Tris-HCl was measured over a period of 16 days. Immobilized TK retained almost 100% productivity after 16 days, when compared to the first day of reaction (Fig. 4a). This consistency in activity of the immobilized TK enzyme, proved the high stability of TK in the immobilized form whilst ~8% of activity drop was observed with free enzyme (data not shown). By contrast, immobilized TAm only retained ~7% of its initial productivity after 5 days (Fig. 4b) compared to ~85% residual activity in free enzyme (data not shown). The sharp decrease in productivity showed that immobilization significantly affect enzyme stability (Krenková and Foret, 2004 and Ladero et al., 2006). This low catalytic stability of TAm was most likely due to the alteration in the three-dimensional conformation of the protein upon immobilization, which led to rapid inactivation of TAm after repeated use. Besides that, the stability of an enzyme was found to be highly dependent on the type of immobilization support used. ω-TAm from *Vibrio fluvialis* immobilized on chitosan beads exhibited higher stability compared to Eupergit® C, whereby 70% of residual activity was retained after 3.5 weeks and only 10% for the latter one (Yi et al., 2007).

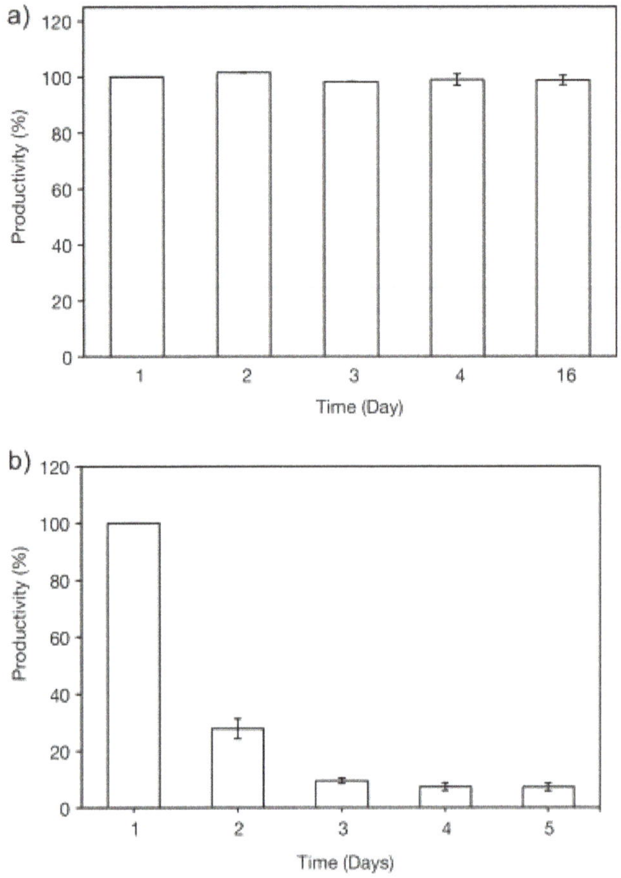

Figure 4: Storage stability at 4 °C in 50 mM Tris-HCl of (a) TK IEMR and (b) TAm IEMR. The productivity of the IEMR was determined according to the standard TK and TAm assays (Sections 2.4.3 and 2.4.4). The productivity is defined as the amount of product generated (per amount of enzyme) divided by the product formed of the starting enzyme activity determined in the same way. Slight fluctuations may be due to small differences in the concentration of the immobilized enzyme. Concentration of immobilized TK and TAm was ~0.2 mg/mL and 0.3 mg/mL respectively. Error bars represent one standard deviation about the mean ($n = 3$).

Reusability of IEMR

The reusability of the microreactor was investigated by using the same packed tube reactor in seven TK immobilization-elution processes. The activity of the immobilized enzyme was then tested to verify that the repeated immobilization-elution process in the microreactor does not affect the activity of the enzyme. Based on the results shown in Fig. 5, the productivity of the IEMR was retained over seven cycles of the immobilization-elution process within a 4-week period. This is similar to the findings from work ofMatosevic et al. (2009), however the reusability of the packed tube was made simpler compared to the surface derivatized capillary as it does not require a regeneration step before the subsequent immobilization process. The same concentration of enzyme was used throughout the experiments to allow a direct comparison between each cycle of the immobilization.

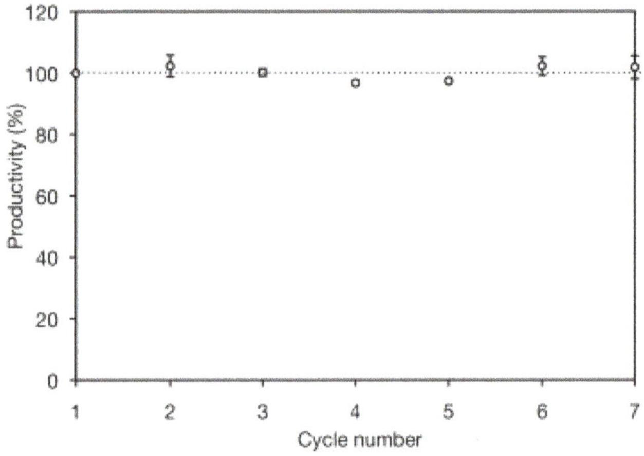

Figure 5: Reusability test of the IEMR. Reusability of packed tube was measured by productivity of the immobilized TK over seven cycles of immobilization-enzyme reaction-elution processes using a single IEMR. Productivity of the IEMR was measured according to the standard TK assay (Section2.4.3) over a 4-week period. Slight fluctuations may be due to small differences in the concentration of the immobilized enzyme. Concentration of immobilized TK and TAm was ~0.2 mg/mL and 0.3 mg/mL respectively. Error bars represent one standard deviation about the mean ($n = 3$).

TK and TAm kinetics in IEMR

The kinetic parameters of the immobilized enzyme may result in different values compared to the solution-phase reaction due to mass transfer effects and enzyme deactivation (Kerby et al., 2006 and Ozdural et al., 2001). Under these conditions, apparent kinetic constants are used to evaluate the kinetics of the immobilized enzyme. The Lilly et al. (1966) model is commonly employed to determine kinetic parameters for continuous packed-bed reactors. Furthermore, the model is able to account for any mass transfer effects that may be present and can alter the overall enzyme kinetics. Under steady state conditions the apparent kinetic parameters in a packed-bed enzyme reactor can be determined by the following equation:

$$f[A_0] = K_{m(app)}\ln(1-f) + \frac{C}{Q}$$

where f is the fraction of substrate reacted, $[A_0]$ is the initial substrate concentration,$Km_{(}app_{)}$ is the apparent Michaelis constant, C is the reactor capacity and Q is the flow rate.

Determination of kinetic parameters of the continuous flow mode packed tube IEMR was conducted for both enzymes at flow rates ranging from 2 to 30 μL/min. For TK, the concentration of GA was varied from 10 to 50 mM while HPA was kept at 100 mM and for TAm the concentration of ERY was varied from 50 to 300 mM with MBA kept at 10 mM. The progression curve for each substrate concentration was shown to follow the standard TK and TAm reaction profile in solution-phase, with conversion yield increasing proportionately to substrate concentration (Fig. 6). Data from these graphs were then used to plot the Lilly et al. (1966) kinetic model and determine the kinetic parameters for the continuous flow reaction (Fig. 7). The slopes of the fitted lines were evaluated to give the $Km_{(}app_{)}$ for each flow rate. From Fig.

8, a pronounced effect of flow rate was observed on $Km_{(app)}$, where the values increased with lower flow rates. This can be explained by the presence of a diffusion layer surrounding the immobilized enzyme (Lilly et al., 1966). At high flow rates, the mass transfer effect is less pronounced as the diffusion layer around the bead gets thinner, while the effect is more significant at lower flow rates, which could be observed as an elevation of the $Km_{(app)}$ values. This diffusion layer theory fits the trend displayed in Fig. 8, where an approximately linear decrease was seen for $Km_{(app)}$ between 2 and 10 µL/min, and then the values started to level-off between 10 and 30 µL/min. Toda (1975) addressed the problem of inter-particle mass transfer in connection with the overall kinetics of gel-entrapped enzymes packed in a fixed bed reactor. He suggested that the increase in the $Km_{(app)}$ value at a low flow rate is due to film diffusion or liquid channelling factor. In addition, several experimental studies have shown similar flow dependency of $Km_{(app)}$ (Lilly et al., 1966, Ozdural et al., 2001 and Shiraishi et al., 1996). They explained the importance of mass transfer outside a solid catalyst with the fact that the apparent Michaelis constant determined in a packed column reactor varied with the flow rate of the liquid. On the contrary, Matosevic et al. (2011b) found an increase in $Km_{(app)}$ with increasing flow rate of the substrate through a surface derivatised capillary containing immobilized TK enzyme. With lower operating flow rates of 0.33–3 µL/min and reduced enzyme concentration compared to the packed tube reactor, the mass transfer effect observed was possibly not as pronounced as in our system.

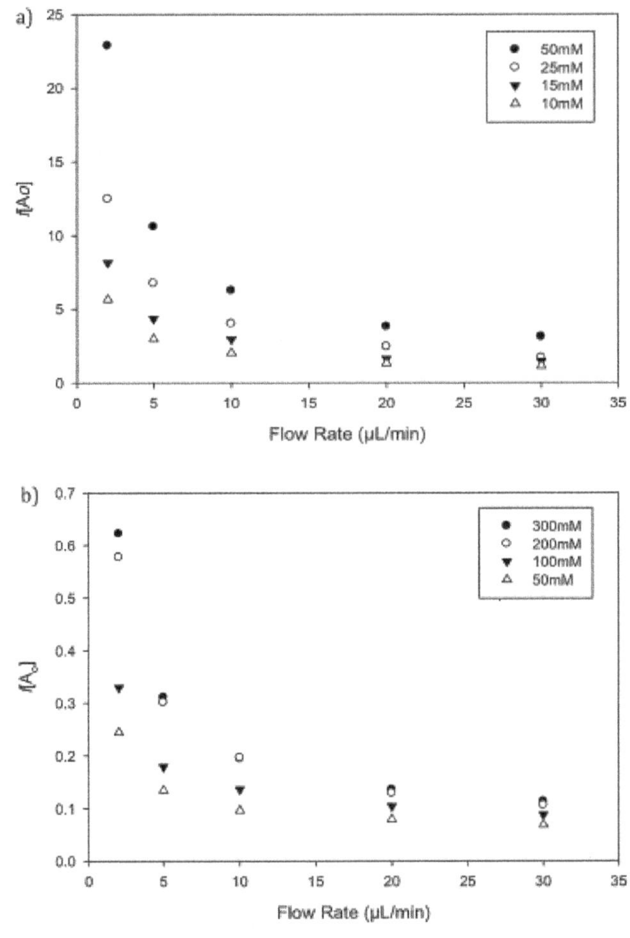

Figure 6: Bioconversions using different substrate concentration at flow rates ranging from 2 µL/min to 30 µL/min. (a) Continuous-flow TK-catalyzed reaction at different GA concentration (10–50 mM) with HPA concentration was kept at 100 mM. (b) Continuous-flow TAm-catalyzed reaction at different ERY concentration (50–300 mM) with MBA concentration was kept at 10 mM. Concentration of immobilized TK and TAm was ~0.05 mg/mL and 0.4 mg/mL respectively.

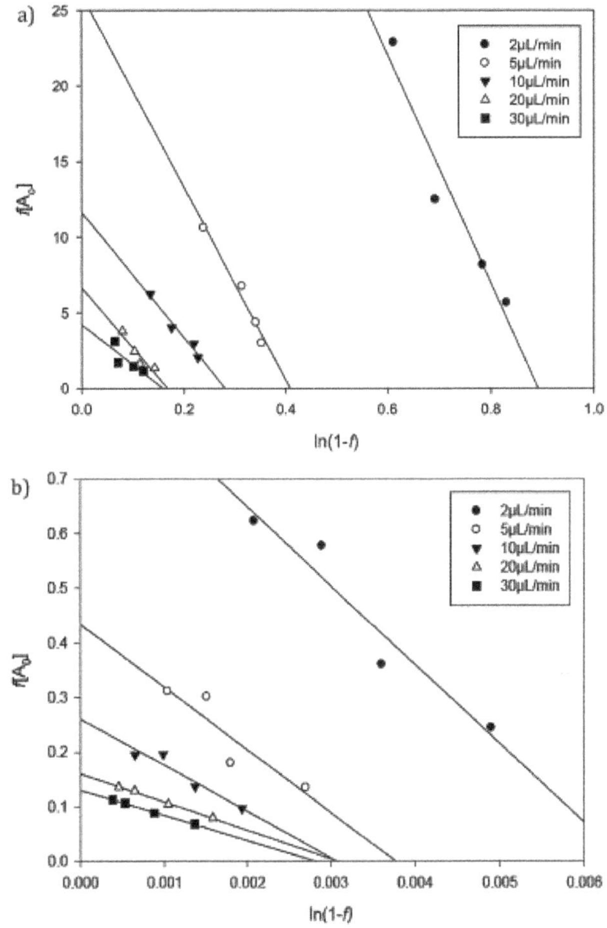

Figure 7: Kinetics determination using Lilly–Hornby model for continuous-flow mode reaction. Data to obtain these plots was taken from Fig. 6. (a) TK-catalyzed reaction and (b) TAm-catalyzed reaction. Solid lines fitted by linear regression. Values of $Km_{(app)}$ and C for the continuous flow reactions could be obtained from these plots.

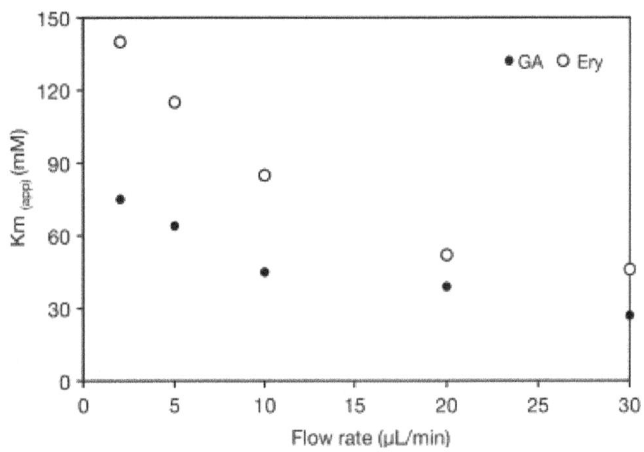

Figure 8: $Km_{(app)}$ values for GA (TK-catalyzed reaction) and ERY (TAm-catalyzed reaction) ranging from 2 µL/min-30 µL/min. Data to obtain this plot was taken from Fig. 7.

The reactor capacity (*C*) values obtained were nearly constant for all experiments, with an average of 12.88 ± 0.27 mM/min and 0.28 ± 0.027 mM/min for GA and ERY, respectively. These values were then used to determine the turnover number, k_{cat} which was 32 s^{-1} for TK and 0.058 s^{-1} for TAm. These values were approximately 3 times less than k_{cat} values in solution-phase (data not shown). This apparent loss in activity may be due to the inaccessibility of some enzyme molecules that were immobilized in the packed tube, to the substrate molecules. The intrinsic compressibility of porous agarose beads resulted in highly compact packing, which may have created a barrier to the accessibility of the substrate to those enzyme molecules located within the beads packing. In addition, non-uniform enzyme distributions within the support would also affect the apparent kinetic parameters where the catalyst immobilized on the surface can render far greater enzyme utilization efficiency compared to the catalyst in the interior core configuration (Dalvie and Baltus, 1992 and Juang and Weng, 1984). Furthermore, altered conformation of enzyme and steric hindrance could also contribute to the loss of activity of the immobilized enzyme (Delouise and Miller, 2005, Kerby et al., 2006 and Krenková and Foret, 2004). Overall, further investigations are required to evaluate the effect of the external and internal mass transfer on the enzyme behaviour in this packed tube IEMR.

Demonstration of dual enzyme reaction: TK–TAm bioconversion

The multi-step enzyme reaction was executed by connecting single TK and TAm IEMR in series (see Fig. 1b). The substrates used for each enzyme were a mixture of 60 mM (GA and HPA, each) and 6 mM (MBA), were fed together in the presence of cofactors. The coupled reaction was performed at pH 7.5, which was previously shown to be suitable for both enzymes (Kaulmann et al., 2007 and Mitra et al., 1998). In the first reaction, TK catalyzed the conversion of the achiral substrates HPA and GA to form ERY as a product, which was then fed to the second IEMR. The dual reaction was completed with the bioconversion of ERY and MBA by TAm into AP and ABT.

With initial substrate concentration of 6 mM MBA, the coupled reaction reached approximately 83% conversion in 20 min at the lowest flow rate (Fig. 9a). This corresponds to a specific activity of 0.03 $\mu mol/min/mg_{TAm}$. However, this analysis only gives information on the final product yield of the coupled reaction. In order to estimate the conversion yield of GA and HPA to ERY in the first reaction a separate experiment was conducted. A single IEMR was prepared with immobilized TK. Similar enzyme and substrate concentrations were used to match the condition in the coupled reaction. Based on Fig. 9b, the reaction reached a conversion of 90% in 10 min at the lowest flow rate with a specific activity of 8.20 $\mu mol/min/mg_{TK}$. This suggested that a total of 54 mM ERY was fed to the second IEMR in the coupled reaction. Therefore, about 9% of the relative amount of ERY supplied by the TK step was aminated by TAm in the second reaction. Based on these results, it is apparent that transaminase has a much slower conversion rate compared to the transketolase enzyme. A similar trend was also observed in the rates of catalysis of the two enzymes in batch solution as well as studies with whole cells (Ingram et al., 2007; Matosevic et al., 2011; Rios-Solis et al., 2011). Nevertheless, low specific activity of TAm was not only due to the slow amination rate of the enzyme but also due to the presence of side reactions with residual HPA and GA. Kaulmann et al. (2007)reported that TAm has the affinity to catalyze the amination of both HPA and GA leading to the reduced formation of ABT at higher flow rates when higher residual concentrations of HPA and GA from the TK-catalyzed reaction were present.

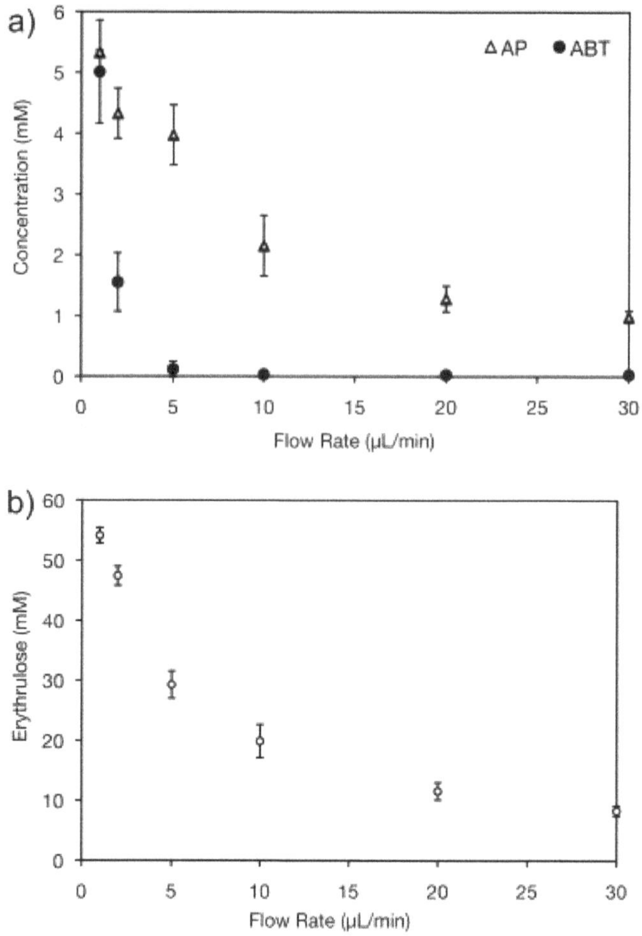

Figure 9: Multi-step TK–TAm reaction in IEMR. (a) Multi-step TK–TAm reaction with initial substrate concentration of 60 mM (HPA and GA) and 6 mM (MBA), and (b) single TK IEMR running with similar condition as in coupled reaction. The concentration of immobilized enzyme was 1.1 ± 0.15 mg/mL for TAm and 0.15 ± 0.006 mg/mL for TK. Error bars represent one standard deviation about the mean ($n = 3$).

CONCLUSION

The packed tube microreactor system described here represents a prototype IEMR that allows multi-step enzymatic synthesis of complex molecules such as ABT. Characterization of the IEMR was first demonstrated using TK and TAm enzymes. Storage stability tests of both enzymes were performed and results showed that immobilized TK retained its activity up to 16 days, while immobilized TAm lost ~93% of its activity after 5 days. Furthermore, the reusability of the IEMR was tested, whereby a single IEMR could be used for at least seven TK immobilization-bioconversion-elution processes without having any reduction in binding property and enzyme activity.

In addition, kinetic properties of immobilized TK and TAm were determined in continuous flow mode at flow rates ranging from 2 μL/min to 30 μL/min. The kinetic analysis relied on the Lilly et al. (1966) model, which has been previously used to evaluate apparent kinetic parameters in packed-bed enzyme reactors under steady-state conditions. $Km_{(app)}$ for all substrates appeared to be flow rate dependent indicating that for these heterogeneous enzyme reactions, the $Km_{(app)}$ value may be affected by mass transfer resistance that masked the actual enzyme kinetics. The increase in $Km_{(app)}$ at low flow rates was possibly due to the diffusional resistance imposed by the liquid layer surrounding each particle. Besides that, the turnover rate k_{cat} was found to be significantly lower than that in solution-phase. Imperfect immobilization, altered enzyme conformation and steric hindrance were suggested to be the causes for the observed difference.

In the final part of the work, we demonstrated the TK–TAm coupled reaction in continuous flow mode to synthesize the chiral amino alcohol, ABT. Based on the solution-phase data, specific activity of TAm was found to be significantly lower than that of TK. Based on previous results (Matosevic et al., 2011b), it was expected that the amination rate of TAm would be slower compared to the TK catalyzed reaction in the two-step enzymatic conversion. Therefore a 7-fold higher TAm concentration (1.1 ± 0.15 mg/mL) was loaded compared to 0.15 ± 0.006 mg/mL of TK to compensate for the much lower reaction rate of TAm, which improved the yield of the coupled reaction synthesizing ABT. Moreover, while the overall reaction was mainly affected by the low catalytic activity of TAm, the amount of ERY produced in the first step of the reaction was also affecting the rate of amination in the TAm step as well as minimizing the side reaction with residual substrates from the TK step. Therefore, identifying the best matching enzyme concentration for loading would be a suitable strategy to optimize this dual-step enzyme reaction.

The packed tube IEMR described here has shown the ability to synthesize the chiral pharmaceutical intermediate, ABT. The facile His tag-immobilization method provides a stable heterogeneous catalyst, which can be easily incorporated into a FEP tube and is cost-effective due the low material cost and reusability property of the IEMR. The versatility of the immobilization method makes the system amenable to a wide range of biocatalysts to be integrated in order to create a variety of synthetic pathways for rapid synthesis and screening purposes. Finally, this packed tube IEMR system could potentially be used as model for a microfluidic reactor on a chip (Boehm et al., 2013).

ACKNOWLEDGEMENTS

The authors wish to acknowledge the UK Engineering and Physical Sciences Research Council (EPSRC) for support of the multidisciplinary Biocatalysis Integrated with Chemistry and Engineering (BiCE) programme. AAH would like to thank the Ministry of Higher Education Malaysia (MOHE) for financial support.

REFERENCES

1. Abad, J.M., Mertens, S.F.L., Pita, M., Fernández, V.M., Schiffrin, D.J., 2005. Functionalization of thioctic acid-capped gold nanoparticles for specific immobilization of histidine-tagged proteins. J. Am. Chem. Soc. 127, 5689–5694.
2. Alam, M.N.H.Z., Pinelo, M., Samanta, K., Jonsson, G., Meyer, A., Gernaey, K.V., 2011. A continuous membrane microbioreactor system for development of integrated pectin modification and separation processes. Chem. Eng. J. 167, 418–426
3. Berne, C., Betancor, L., Luckarift, H.R., Spain, J.C., 2006. Application of a microfluidic reactor for screening cancer prodrug activation using silica-immobilized nitrobenzene nitroreductase. Biomacromolecules 7, 2631–2636.
4. Block, H., Maertens, B., Spriestersbach, A., Brinker, N., Kubicek, J., Fabis, R., Labahn, J., Schäfer, F., 2009. Immobilized-metal affinity chromatography (IMAC): a review. Enzymol. Methods Enzymol., 439–473.
5. Boehm, C.R., Freemont, P.S., Ces, O., 2013. Design of a prototype flow microreactor for synthetic biology in vitro. Lab. Chip 13, 3426–3432.
6. Bradford, M.M., 1976. A rapid and sensitive method for the quantitation of microgram quantities of protein using the principle of protein dye binding. Anal. Biochem. 72, 248–254.
7. Cha, T., Guo, A., Jun, Y., Pei, D., Zhu, X.-Y., 2004. Immobilization of oriented protein molecules on poly(ethylene glycol)-coated Si(1 1 1). Proteomics 4, 1965–1976
8. Cha, T., Guo, A., Zhu, X.-Y., 2005. Enzymatic activity on a chip: the critical role of protein orientation. Proteomics 5, 416–419
9. Cheung, R., Wong, J., Ng, T., 2012. Immobilized metal ion affinity chromatography: a review on its applications. Appl. Microbiol. Biotechnol. 96, 1411–1420.
10. Cohen, S.A., Michaud, D.P., 1993. Synthesis of a fluorescent derivatizing reagent, 6-aminoquinolyl-N-hydroxysuccinimidyl carbamate, and its application for the analysis of hydrolysate amino acids via high-performance liquid chromatography. Anal. Biochem. 211, 279–287.
11. Contestabile, R., John, R.A., 1996. The mechanism of high-yielding chiral syntheses catalysed by wild-type and mutant forms of aspartate aminotransferase. FEBS J. 240, 150–155.
12. Dalvie, S.K., Baltus, R.E., 1992. Distribution of immobilized enzymes on porous membranes. Biotechnol. Bioeng. 40, 1173–1180.
13. Delouise, L.a, Miller, B.L., 2005. Enzyme immobilization in porous silicon: quantitative analysis of the kinetic parameters for glutathione-S-transferases. Anal. Chem. 77, 1950–1956.
14. Forsberg, E.M., Green, J.R.A., Brennan, J.D., 2004. Continuous flow immobilized enzyme reactor-tandem mass spectrometry for screening of AChE inhibitors in complex mixtures. Anal. Chem. 83, 5230–5236.
15. Hickey, A.M., Marle, L., McCreedy, T., Watts, P., Greenway, G.M., Littlechild, J.A., 2007. Immobilization of thermophilic enzymes in miniaturized flow reactors. Biochem. Soc. Trans. 35, 1621–1623.
16. Hickey, A.M., Ngamsom, B., Wiles, C., Greenway, G.M., Watts, P., Littlechild, J.A., 2009. A microreactor for the study of biotransformations by a cross-linked -lactamase enzyme. Biotechnol. J. 4, 510–516
17. Hong, J., Edel, J.B., DeMello, A.J., 2009. Micro- and

nanofluidic systems for highthroughput biological screening. Drug Discov. Today 14, 134–146.

18. Ingram, C.U., Bommer, M., Smith, M.E.B., Dalby, P.A., Ward, J.M., Hailes, H.C., Lye, G.J., 2007. One-pot synthesis of amino-alcohols using a de-novo transketolase and -alanine: pyruvate transaminase pathway in Escherichia coli. Biotechnol. Bioeng. 96, 559–569.

19. Juang, H.D., Weng, H.S., 1984. Performance of biocatalysts with nonuniformly distributed immobilized enzyme. Biotechnol. Bioeng. 26, 623–626.

20. Kaulmann, U., Smithies,K., Smith, M., Hailes, H.,Ward, J., 2007. Substrate spectrum of -transaminase from Chromobacterium violaceum DSM30191 and its potential for biocatalysis. Enzyme Microb. Technol. 41, 628–637.

21. Kerby, M.B., Legge, R.S., Tripathi, A., 2006. Measurements of kinetic parameters in a microfluidic reactor. Anal. Chem. 78, 8273–8280.

22. King, J., Laemmli, U.K., 1971. Polypeptides of the tail fibres of bacteriophage T4. J. Mol. Biol. 62, 465–477.

23. Krenková, J., Foret, F., 2004. Immobilized microfluidic enzymatic reactors. Electrophoresis 25, 3550–3563.

24. Kuan, I., Liao, R., Hsieh, H., Chen, K., Yu, C., 2008. Properties of Rhodotorula gracilis d-amino acid oxidase immobilized on magnetic beads through His-tag. J. Biosci. Bioeng. 105, 110–115.

25. Ladero, M., Ruiz, G., Pessela, B.,Vian,A., Santos,A., Garciaochoa, F., 2006. Thermal and pH inactivation of an immobilized thermostable -galactosidase from Thermus sp. strain T2: comparison to the free enzyme. Biochem. Eng. J. 31, 14–24.

26. Lilly, M.D., Hornby, W.E., Crook, E.M., 1966. The kinetics of carboxymethylcellulose- ficin in packed beds Na-benzoylarginine. Biochem. J. 100, 718–723.

27. Mark, D., Haeberle, S., Roth, G., Von Stetten, F., Zengerle, R., 2010. Microfluidic labon-a-chip platforms: requirements, characteristics and applications. Chem. Soc. Rev. 39, 1153–1182.

28. Martinez-Torres, R., Aucamp, J., George, R., Dalby, P., 2007. Structural stability of E. colitransketolase to urea denaturation. Enzyme Microb. Technol. 41, 653–662.

29. Matosevic, S., Lye, G.J., Baganz, F., 2009. Design and characterization of a prototype enzyme microreactor: quantification of immobilized transketolase kinetics. Biotechnol. Prog. 26, 118–126.

30. Matosevic, S., Szita, N., Baganz, F., 2011a. Fundamentals and applications of immobilized microfluidic enzymatic reactors. J. Chem. Technol. Biotechnol. 86, 325–334.

31. Matosevic, S., Lye, G.J., Baganz, F., 2011b. Immobilised enzyme microreactor for screening of multi-step bioconversions: characterisation of a de novo transketolase--transaminase pathway to synthesise chiral amino alcohols. J. Biotechnol. 155, 320–329.

32. Matsuura, S., Ishii, R., Itoh, T., Hamakawa, S., Tsunoda, T., Hanaoka, T., Mizukami, F., 2011. Immobilization of enzyme-encapsulated nanoporous material in a microreactor and reaction analysis. Chem. Eng. J. 167, 744–749.

33. Mitra, R., Woodley, J., Lilly, M., 1998. Escherichia coli transketolase-catalysed carbon-carbon bond formation: biotransformation characterisation for reactor evaluation and selection. Enzyme Microb. Technol. 22, 64–70.

34. Ngamsom, B., Hickey, A.M., Greenway, G.M., Littlechild, J.A., Watts, P., Wiles, C., 2010. Development of a high throughput screening tool for biotransformations utilising a thermophilic l-aminoacylase enzyme. J. Mol. Catal. B: Enzymes 63, 81–86.

35. Nidetzky, B., 2013. Smart enzyme immobilization in microstructured reactors. Chem. Today 31 (3), 50–55.

36. Nomura, A., Shin, S., Mehdi, O.O., Kauffmann, J.-M., 2004. Preparation, characterization, and application of an enzyme-immobilized magnetic microreactor for flow injection analysis. Anal. Chem. 76, 5498–5502.

37. Ozdural, A.R., Tanyolac¸ , D., Demircan, Z., Boyaci, ˙ I.H., Mutlu, M., Webb, C., 2001. A new method for determination of apparent kinetics parameters in recirculating packed-bed immobilized enzyme reactors. Chem. Eng. Sci. 56, 3483–3490

38. Rios-Solis, L., Halim, M., Cázares, A., Morris, P., Ward, J.M., Hailes, H.C., Dalby, P.A., Baganz, F., Lye, G.J., 2011. A toolbox approach for the rapid evaluation of multistep enzymatic syntheses comprising a mix and match E. coli expression system with microscale experimentation. Biocatal. Biotransform. 29, 192–203.

39. Sakai-Kato, K., Kato, M., Toyo'oka, T., 2002. On-line trypsin-encapsulated enzyme reactor by the sol–gel method integrated into capillary electrophoresis. Anal. Chem. 74, 2943–2949.

40. Shiraishi, F., Hasegawa, T., Kasai, S., Makishita, N., Miyakawa, H., 1996. Bed immobilized enzyme reactor. Chem. Eng. Sci. 51, 2847–2852.

41. Toda,I., 1975.Interparticlemass transfer study withapackedcolumnofimmobilized microbes. Biotechnol. Bioeng. XVII, 1729–1747.

42. Yi, S.-S., Lee, C., Kim, J., Kyung, D., Kim, B.-G., Lee, Y.-S., 2007. Covalentimmobilization of -transaminase from Vibrio fluvialis JS17 on chitosan beads. Process. Biochem. 42, 895–898.

CHAPTER 14

EVALUATION OF IMMOBILIZED LIPASES ON POLY-HYDROXY-BUTYRATE BEADS TO CATALYZE BIODIESEL SYNTHESIS

Adriano A. Mendes [1,2], Pedro C. Oliveira[3], Ana M. Vélez [2], Roberto C. Giordano [2], Raquel de L.C. Giordano [2], and Heizir F. de Castro[3]

[1]Laboratory of Biocatalysis, Federal University of São João del Rei, 35701-970, Sete Lagoas, MG, Brazil

[2]Department of Chemical Engineering, Federal University of São Carlos, 13565-905, São Carlos, SP, Brazil

[3]Department of Chemical Engineering, Engineering School of Lorena, University of São Paulo, 12602-810, Lorena, SP, Brazil

ABSTRACT

Five microbial lipase preparations from several sources were immobilized by hydrophobic adsorption on small or large poly-hydroxybutyrate (PHB) beads and the effect of the support particle size on the biocatalyst activity was assessed in the hydrolysis of olive oil, esterification of butyric acid with butanol and transesterification of babassu oil (*Orbignya*sp.) with ethanol. The catalytic activity of the immobilized lipases in both olive oil hydrolysis and biodiesel synthesis was influenced by the particle size of PHB and lipase source. In the esterification reaction such influence was not observed. *Geobacillus thermocatenulatus* lipase (BTL2) was considered to be inadequate to catalyze biodiesel synthesis, but displayed high esterification activity. Butyl butyrate synthesis catalyzed by BTL2 immobilized on small PHB beads gave the highest yield (≈90 mmol L^{-1}). In biodiesel synthesis, the catalytic activity of the immobilized lipases was significantly increased in comparison to the free lipases. Full conversion of babassu oil into ethyl esters was achieved at 72 h in the presence of *Pseudozyma antarctica* type B (CALB),*Thermomyces lanuginosus* lipase (Lipex® 100L) immobilized on either small or large PHB beads and *Pseudomonas fluorescens* (PFL) immobilized on large PHB beads. The latter preparation presented the highest productivity (40.9 mg of ethyl esters mg^{-1}immobilized protein h^{-1}).

INTRODUCTION

Lipases (triacylglycerol ester hydrolases, EC 3.1.1.3) are enzymes that in nature catalyze the cleavage of ester bonds in tri-, di-, and monoacylglycerols to glycerol and free fatty acids [1]. In organic medium, lipases also catalyze esterifications, transesterifications and interesterifications [1], [2], [3] and [4]. These enzymes are applied in several industrial processes including synthesis of biopolymers, production of pharmaceuticals, agrochemicals, cosmetics, flavors, treatment of waste rich in lipids and biodiesel synthesis by transesterification of triglycerides with short-chain alcohols [1], [2], [3], [4],[5], [6], [7], [8], [9], [10], [11], [12], [13], [14], [15], [16] and [17].

With respect to the biodiesel synthesis, and despite the fact that an enzymatic process is still not commercially developed, a number of articles have shown that enzyme holds promise as catalyst [6], [7] and [18]. These studies consist

Citation: Adriano A. Mendes, Pedro C. Oliveira, Ana M. Vélez, Roberto C. Giordano, Raquel de L.C. Giordano, Heizir F. de Castro, Evaluation of immobilized lipases on poly-hydroxybutyrate beads to catalyze biodiesel synthesis, International Journal of Biological Macromolecules, Volume 50, Issue 3, 1 April 2012, Pages 503-511, ISSN 0141-8130, http://dx.doi.org/10.1016/j.ijbiomac.2012.01.020.

mainly in optimizing the reaction conditions (vegetable oil, alcohol, molar ratio, solvent, temperature, lipase source and immobilizing support) in order to establish the characteristics for industrial applications [6], [7], [8], [9], [10], [11], [12], [13] and [14].

Lipases have been immobilized in different organic/ inorganic supports by physical adsorption, encapsulation, covalent attachment or cross-linking [8], [9], [10], [11], [12],[13], [14], [15], [16], [17], [18], [19], [20], [21], [22], [23], [24], [25], [26], [27], [28], [29], [30],[31], [32], [33] and [34]. For application in an organic medium, a strong enzyme-support interaction is not required. Under these conditions, the enzyme is insoluble in the apolar medium and physical adsorption may be a suitable method of immobilization. This method is easy, cheap and allows a simple recycling of the support at the end of the enzyme's life. However, the chance of desorption of the enzyme is the main disadvantage of this immobilization procedure [19].

The immobilization by physical adsorption on hydrophobic supports is explained through an important property of the lipases, namely interfacial activation in the presence of hydrophobic interfaces. Lipases have an oligopeptide chain "lid" that covers their active site making them inaccessible to substrates. In the absence of an interface, the active site is secluded from the reaction medium showing "closed conformation". However, in the presence of a hydrophobic interface, important conformational rearrangements take place resulting in the "open conformation" [35], [36] and [37]. In this case, lipases are strongly adsorbed to hydrophobic interfaces through the hydrophobic face of the "lid" that cover their active site, an hydrophobic surrounding of the active center of the lipase. Hence, lipases recognize these surfaces as similar to those of their natural substrates (drops of oil), yielding "immobilized open structures" [30], [31], [32], [33] and [34].

Commercially available immobilized lipases are, in general, immobilized by physical adsorption on organic matrices such as *Pseudozyma antarctica* type B immobilized on Lewatit VPOC 1600 that consists of poly(methyl methacrylate-*co*-divinylbenzene) (Novozym® 435) and *Mucor miehei* immobilized on macroporous anion exchange resin (Lipozyme® RM IM) from Novozymes [38], [39] and [40]. These biocatalysts have high costs, mainly associated with the support price [20]. Therefore, the use of cheaper supports in the lipase immobilization has been proposed as an alternative to the expensive supports for further applications in large scale processes. Several inexpensive hydrophobic supports have been already used in the lipase immobilization by physical adsorption such as egg shell [21], phyllosilicates [10] and [22], bentonite [23], polypropylene [12], [13], [24], [25], [26] and [27], and porous styrene-divinylbenzene[28] and [29].A promising hydrophobic support that can be also used is the poly-hydroxybutyrate (PHB) beads, a polymer derivative of poly-hydroxyalkanoates (PHAs) that is produced as an energy storage material by many bacteria. Physical and mechanical properties of this polymer are similar to those of synthetic thermoplastics such as polypropylene [41]. The glass transition temperature is in the range from 0 to 30 °C while the melting temperature is near 180 °C. The densities of crystalline and amorphous PHB are 1.26 and 1.18 g cm^{-3}, respectively, and present Young's modulus of 3.5 GPa and the tensile strength of 40 MPa. PHB is highly soluble in halogenated solvents such as chloroform and dichloromethane [41] and [42]. As PHAs have been demonstrated to be a family of biopolymers with good biodegradability and biocompatibility [43], PHB has been investigated for use in surgical suture [44], nerve repairer [45] or soft tissue repairer [46] and matrices as cell growth supporting materials[47]. Recently, it has become of industrial interest to evaluate these polyesters for a wide range of applications [47]. However, their use as a support for lipase immobilization still has been scarcely reported in the literature.

The aim of this work was to immobilize five microbial lipase preparations from several sources such as two enzymatic preparations from *Thermomyces lanuginosus*commercially available as Lipolase® (TLL1) and Lipex® 100L (TLL2), *P. antarctica* type B (CALB, also known as lipase B from *Candida antarctica*), *Geobacillus thermocatenulatus*(named BTL2) and *Pseudomonas fluorescens* (PFL) by physical adsorption on small or large PHB beads to catalyze the transesterification of babassu oil with ethanol in a solvent-free medium. Immobilization parameters were estimated in the hydrolysis of olive oil and butyl butyrate synthesis (pineapple flavor). The ability of the immobilized BTL2 on SPHB to catalyze aliphatic esters from carboxylic acids (butyric, lauric and oleic) with short-chain alcohols (ethanol and butanol) was also investigated.

The main desirable characteristic of lipases for biodiesel production from triacylglycerols is the tolerance towards short-chain alcohols such as methanol and ethanol, high catalytic activity and broad specificity [6], [8], [9], [10], [11], [12], [48] and [49]. *P. antarctica* type B (CALB) has a very small lid and does not suffer from an increase in activity by interfacial activation [50]. It is constituted of 317 amino acids with a molecular weight of 33 kDa[39] and [51], dimension of 30 Å × 40 Å × 50 Å and isoelectric point of 6.0 [39] and [52]. CALB has been the most studied lipase for biodiesel production in various reaction systems. It also catalyzes acyl transfer reactions of various oils and acyl acceptors (alcohols or esters) showing high resistance to organic solvents, high thermal stability and broad substrate specificity [48] and [49]. Lipolase® (TLL1) and Lipex® 100L (TLL2) are alkalophilic lipase preparations from *T. lanuginosus* [53] and generally regarded as region-specific for *sn*-1,3 positions [54]. However, their specificity can be influenced by many issues such as media, substrates, and even acyl migration from the *sn*-2 position to the *sn*-1 or *sn*-3 positions in partial glyceride [55] and [56]. Both of them are single chain protein consisting of 269 amino acids with a molecular weight of 31.7 kDa, isoelectric point of 4.4 and a size of 35 Å × 45 Å × 50 Å [54] and [57]. *P. fluorescens* lipase (PFL) is a non-specific enzyme with a molecular weight of 33 kDa and isoelectric point at pH 4.0[58] and [59]. This lipase also displays high catalytic activity at alkaline region [60]. Both PFL and TLL present a true lid which isolates the active center of the lipases from the medium (closed form). It has been found that these lipase preparations are able to aggregate into bimolecular structures even at moderate enzyme concentrations[50] and [59]. These two lipase preparations also have been reported to be efficient biocatalysts for the transesterification reaction with high alcohol:oil molar ratio in solvent-free systems [8], [9], [10], [11] and [12]. These properties make these lipase preparations good biocatalysts for biodiesel synthesis by transesterification of babassu oil in solvent-free systems. *G. thermocatenulatus* (BTL2) is a thermoalkalophilic and non-specific lipase consisting of 389 amino acids with molecular weight of 43 kDa and isoelectric point of 7.4 [61], [62] and [63]. BTL2 is the first crystallized lipase with double lid, which implies in a complex catalytic mechanism [63]. It has a dimension based on its crystal structure of 73.07 Å × 129.08 Å × 127.49 Å [63] and [64]. Although BTL2 is highly active in aqueous systems [61], [62] and [65], its use in organic systems to produce alkyl esters by transesterification of triglycerides with short-chain alcohols (biodiesel) and direct esterification of carboxylic acids with short-chain alcohols has not been reported in the literature yet.

MATERIALS AND METHODS

Materials

Lipase preparations from *P. antarctica* type B (CALB), *T. lanuginosus* named Lipolase®(TLL1), Lipex® 100L (TLL2) from Novozymes (Araucária, Brazil) and *P. fluorescens*(PFL) from Amano Enzyme (Nagoya, Japan) were used without further treatment. Lipase from *G. thermocatenulatus* cloned in *E. coli* (BTL2) was produced as previously described [62]. The catalytic properties of the lipase preparations are presented in Table 1. SPHB (average particle diameters of 75–90 μm) and LPHB (average particle diameters of 750–1180 μm) were acquired from PHB Industrial (São Paulo, Brazil). Anhydrous ethanol was purchased from Cromoline (São Paulo, Brazil). Butanol and carboxylic acids (butyric, lauric and oleic) were purchased from Merck (Darmstadt, Germany). Babassu oil was courtesy of Cognis (São Paulo, Brazil) having the following composition in fatty acids (w/v): 3.50% caprylic, 4.50% capric, 44.7% lauric, 17.5% myristic, 9.70% palmitic, 3.1% stearic, 15.2 oleic, and 1.80% linoleic, with 709.4 g mol^{-1}average molecular weight. All the other reagents were of analytical grade.

Table 1: Catalytic properties of lipase preparations used in the present work

Lipase	Source organism	Designation	Supplier	Protein ($mg\,g^{-1}$)	Activity[c] ($IU\,g^{-1}$)	Specific activity ($IU\,mg^{-1}$ of protein)
CALB[a]	*P. antarctica*	Lipozyme® CALB	Novozymes	3.5	278.6	79.6
TLL1[a]	*T. lanuginosus*	Lipolase®	Novozymes	18.7	6705.8	358.6
TLL2[a]	*T. lanuginosus*	Lipex® 100L	Novozymes	25.8	10,784.4	418.0
PFL[b]	*P. fluorescens*	AK	Amano Enzyme	20.0	6092.0	304.6
BTL2[a]	*G. thermocatenulatus*	-	-	0.67	109.1	162.8

[a] Liquid lipase preparations.
[b] Powder lipase preparation.
[c] Activity measured on the olive oil emulsion hydrolysis.

Determination of hydrolytic activity

Hydrolytic activities (HA) of soluble and immobilized lipases were assayed by the hydrolysis of olive oil emulsion, according to the methodology described by Soares et al.[66], with slight modifications. The substrate was prepared by mixing 50 g of olive oil with 150 g of gum arabic solution (3 wt.%). The reaction mixture containing 5 mL of the emulsion, 5 mL of 100 mmol L^{-1} phosphate buffer (pH 8.0) and soluble (0.2 mL, 0.5 mg mL^{-1}) or immobilized (0.1 g) lipase was incubated for 5 min at 37 °C. The reaction was stopped by the addition of 10 mL of commercial ethanol. The liberated fatty acids were titrated with 20 mmol L^{-1} sodium hydroxide solution in the presence of phenolphthalein as indicator. One international unit (IU) of activity was defined as the amount of enzyme required to liberate 1 μmol of free fatty acid per minute under the conditions described above.

Determination of protein

Protein was determined according to methodology described by Bradford using bovine serum albumin (BSA) as standard [67]. All solutions were prepared using Milli-Q water.

Immobilization of lipases on small or large PHB beads

Initially, 10 g of SPHB or LPHB was soaked into 50 mL of anhydrous ethanol for 2 h at room temperature, following previous published procedure [68]. After swelling step, the excess of ethanol was removed and washed with 50 mL of 5 mmol L^{-1} sodium phosphate buffer at pH 7.0 under vacuum filtration. The immobilization technique was performed using previously described protocol, based on the interfacial activation of lipases on hydrophobic supports at low ionic strength [30], [31], [32], [33] and [34]. For all commercially lipases preparations (CALB, TLL1, TLL2 and PFL) a protein loading of 5 mg g^{-1} of support were offered. For the BTL2, the protein loading was increased to 10 mg g^{-1} of support. Both lipase preparations were diluted in 180 mL of 5 mmol L^{-1}sodium phosphate buffer at pH 7.0 and then, PHB beads were added in each enzymatic solution and kept under mild stirring in shaker during 12 h (SPHB) and 18 h (LPHB) at 25 °C. The resulting derivatives were filtered (Whatman filter paper 41) and washed with Milli-Q water. Analyses of the hydrolytic activities and protein concentration were carried out on initial and spent lipase solutions.

Immobilization parameters

Immobilization yield (IY) was calculated by measuring the units of hydrolytic activity in the supernatant before and after immobilization procedure. Specific activity (SA) was then calculated as the hydrolytic activity of derivatives per milligram of immobilized protein (IU $mgIP^{-1}$). Recovered activity percentage (RA) was calculated after determining the activity of the immobilized enzyme (apparent hydrolytic activity) and comparing with the number of enzyme units that disappeared from the supernatant (theoretically immobilized).

Esterification reactions

A mixture consisting of butanol (100 mmol L^{-1}) and butyric acid (100 mmol L^{-1}) diluted in heptane was used as substrate. The reaction medium (10 mL) was incubated with 0.5 g of the immobilized derivatives at 37 °C during 24 h with continuous agitation at 150 rpm, as described by Soares et al. [66]. The yield into butyl butyrate was quantified by measurements of the concentration of residual butyric acid in the reaction mixture. Samples were withdrawn (150–200 mg), diluted in 10 mL of an ethanol/acetone 50:50 (v/v) mixture and titrated with ethanol KOH solution (20 mmol L^{-1}) using phenolphthalein as the end-point indicator. BTL2 immobilized on SPHB was used as biocatalyst in the esterification reaction of carboxilic acids (butyric, lauric and oleic) with ethanol or butanol in the same conditions as above described, varying molar ratio of alcohol to carboxylic acid (1:1 and 1:9).

Transesterification reactions

Reactions were performed in 25 mL closed flasks containing 20 g of substrate consisting of babassu oil (12.6 g) and anhydrous ethanol (7.4 g) without the addition of solvents, at fixed molar ratio oil to alcohol (1:9) [11]. The mixtures were incubated with free (12.6 mg protein) and immobilized lipases on PHB beads at proportion of 10 wt.% in relation to the total weight of reactants involved in the reaction media. Reactions were performed for a maximum period of 96 h under constant magnetic agitation of 180 rpm at 45 °C. For the time course studies, an aliquot of reaction medium was taken at various time intervals and diluted in hexane for gas chromatograph analysis.

Gas chromatograph analysis

Ethyl esters derived from ethanolysis of babassu oil were analyzed by gas chromatograph using a Varian CG 3800 model (Varian, Inc. Corporate Headquarters, Palo Alto, CA, USA) equipped with flame–ionization detector and 5% DEGS on Chromosorb WHP 80/100 mesh (6 ft 2.0 mm ID) in stainless steel packed column (Restek, Frankel Commerce of Analytic Instruments Ltd., SP, Brazil). Nitrogen was used as the carrier gas with a flow rate of 25 mL min^{-1}. Temperature programming was performed. The column temperature was kept at 90 °C for 3 min, heated to 120 °C at 25 °C min^{-1} and kept constant for 10 min. Then, the temperature was programmed at 25 °C min^{-1} to 170 °C and kept constant for 15 min. The temperatures of the injector and detector were set at 250 °C. Data collection and analyses were performed using the software Galaxie Chromatography Data System version 1.9. Calibration curves were built from standard ethyl esters (ethyl caproate, caprate, laurate, myristate, palmitate, stearate, oleate, linoleate) using hexanol as internal standard. The ethyl ester yield was calculated by taking into account the mass of ester content obtained by GC analysis and the total theoretical ester mass based on the reaction molar ratio [8], [9], [11] and [14].

Ethyl esters viscosity determination

The kinematic viscosity of ethyl esters (biodiesel) was determined with LVDV-II cone and plate spindle Brookfield viscosimeter (Brookfield Viscometers Ltd., U.K.) using a CP 42 cone. A circulating water bath was used to maintain the temperature at 40 °C during the assays. The shear stress measurements were taken as a function of shear rate, and the dynamic viscosity was determined as a slope constant. Biodiesel samples of 0.5 mL were used and the measurements were replicated three times.

RESULTS AND DISCUSSION

Immobilization of lipases on SPHB and LPHB beads

Results displayed in Table 2 showed that CALB, TLL1 and TLL2 were completely immobilized on either small or large PHB beads. For these three lipase preparations, the particle size of the support did not exert any influence on the immobilization yield (IY), ratio between activity immobilized and activity initially offered. Generally, during the immobilization by physical adsorption on highly hydrophobic matrices, the lipase undergoes changes in its conformation in the presence of the support surface and is adsorbed in the open conformation (monomolecular structure) [12]. According to the results reported here, the incubation of lipase preparations from *T. lanuginosus* (TLL1 and TLL2) in the presence of small or large PHB beads shifted the equilibrium between close (bimolecular aggregates) and open conformation towards monomolecular structures (open conformation), allowing complete enzyme immobilization. CALB also was completely immobilized on either small or large PHB beads, however, no bimolecular aggregates was identified even at high enzyme concentrations [50]. However, for PFL and BTL2 a strong influence of the particle size on the immobilization yield was verified. For PFL and using SPHB, around 78.7 ± 4.6% of lipase activity initially offered was immobilized, and using LPHB the immobilization yield decreased to 59.4 ± 3.8%. BTL2 was also completely immobilized on SPHB. On the other hand, replacing SPHB by LPHB, only 67.2 ± 2.7% of the enzymatic activity initially offered was immobilized. This strong influence of the particle size on the immobilization yield of PFL and BTL2 may be attributed to diffusion problems of the bimolecular aggregates of lipases that may reduce the immobilization rate when the particle size of the support increases. Generally, the specific surface area of small particles is higher than large particles [23] and [26]. Hence, more surface area on small PHB beads is free for the immobilization of the lipases.

From the data presented in Table 2, the hydrolytic activity, specific activity and recovered activity of the biocatalysts were also strongly influenced by the particle size of the support. SPHB derivatives presented higher hydrolytic activities than derivatives from LPHB, as well as higher specific activity and recovered activity. The derivatives with highest hydrolytic activity were obtained by using lipase preparations from *T. lanuginosus* TLL2 (1717.5 ± 23.0 IU g^{-1}) and TLL1 (1558.1 ± 17.6 IU g^{-1}), followed by PFL (1099.6 ± 100 IU g^{-1}), BTL2 (256.6 ± 19.3 IU g^{-1}) and CALB (135.5 ± 6.0 IU g^{-1}) immobilized on SPHB. Replacing SPHB by LPHB, a strong decrease in the hydrolytic activity of the biocatalysts was observed. These results can be explained by steric effects of large PHB beads (LPHB) which had significant influence on the lipase conformation, leading to the changes of the enzyme activity or diffusional limitations that restricted the contact between the active site of the lipase and substrate (olive oil).

It is well known that diffusional limitations become more a restraining factor as the diameter of the support beads increases [13], [26], [27] and [69]. Results displayed in Table 2 show that the specific activity of the biocatalysts, ratio between hydrolytic activity and immobilized protein amount, also decreased after immobilization on LPHB. The biocatalysts prepared by immobilizing on SPHB beads followed the specific activity of the free lipases: TLL2 > TLL1 > PFL > BTL2 > CALB (see Table 1). However, the immobilization of lipases on LPHB beads led to conformational changes of the immobilized lipases. Among the biocatalysts prepared, PFL displayed the highest specific activity (75.7 ± 15.0 IU $mgIP^{-1}$), almost twice as high as the lipase preparations TLL1 and TLL2 (around of 40.0 IU $mgIP^{-1}$). The immobilization on PHB beads reduced drastically the specific activity of CALB and BTL2, in comparison to the specific activity values for free enzymes (see Table 1). For CALB, specific activity varied from 27.1 ± 1.20 to 8.68 ± 0.88 IU $mgIP^{-1}$, while for BTL2 the specific activity 34.4 ± 2.40 to 13.9 ± 1.80% by immobilizing on SPHB and LPHB, respectively. The conformational change on the three-dimensional structure of lipase (that influences its catalytic activity) depends on the structure of the enzyme and nature of the surface and pores of the support [70]. Among the biocatalysts prepared, the immobilization of TLL2 on small PHB beads displayed highest specific activity, while the immobilization of PFL on large PHB beads led to better conformation of the enzyme, rendering the highest specific activity among the biocatalysts prepared on LPHB.

Table 2: Immobilization parameters of lipase preparations from several sources on small and large PHB beads. The parameters were measured on the hydrolysis of olive oil emulsion at pH 8.0 and 37 °C

Lipase	PHB	Immobilization parameters			
		IY[a] (%)	HA[b] (IU g^{-1} of support)	SA[c] (IU $mgIP^{-1}$)	RA[d] (%)
CALB	SPHB	100	135.5 ± 6.0	27.1 ± 1.20	34.0 ± 1.5
	LPHB	100	43.0 ± 4.8	8.68 ± 0.88	10.8 ± 1.2
TLL1	SPHB	100	1558.1 ± 17.6	313.2 ± 2.00	86.9 ± 1.0
	LPHB	100	200.0 ± 29.5	40.0 ± 5.90	11.2 ± 1.6
TLL2	SPHB	100	1717.5 ± 23.0	343.5 ± 4.60	82.2 ± 1.1
	LPHB	100	190.2 ± 13.8	38.0 ± 2.76	9.10 ± 0.7
PFL	SPHB	78.7 ± 4.57	1099.6 ± 100	284.9 ± 25.9	91.7 ± 2.9
	LPHB	59.4 ± 3.78	212.8 ± 42.2	75.7 ± 15.0	23.3 ± 3.2
BTL2	SPHB	100	256.6 ± 19.3	34.4 ± 2.40	15.8 ± 1.1
	LPHB	67.2 ± 2.70	69.6 ± 10.8	13.9 ± 1.80	6.30 ± 0.7

[a] IY - immobilization yield (%).
[b] HA - hydrolytic activity (IU g^{-1} of support).
[c] SA - specific activity (IU $mgIP^{-1}$).
[d] RA - recovered activity (%).

The immobilization of PFL on small PHB beads (SPHB) allowed obtaining a biocatalyst with the highest recovered activity (91.7 ± 2.9%), followed by TLL1 (86.9 ± 1.0%), TLL2 (82.2 ± 1.1%), BTL2 (34.4 ± 2.40%) and CALB (27.1 ± 1.20%). Interestingly, the immobilization of lipase molecules on highly hydrophobic supports shows a dramatic hyperactivation [31], [34], [65] and [71]. However, a slight reduction of the recovered activity of lipases immobilized on PHB beads was verified due to low accessibility of the large substrate (olive oil) to the active center of the lipase (diffusional limitations). Palomo et al. evaluated the influence of different substrates such as ethyl butyrate and *p*-nitrophenyl propionate (*p*NPP) on the relative activity of microbial lipases such as *P. antarctica* (CALB), *Candida rugosa* (CRL) and *M. miehei* (MML) immobilized by physical adsorption on octadecyl-Sepabeads beads [31]. The authors observed the highest relative activity for the biocatalysts when performed with smaller substrate (ethyl butyrate) which can easily accessed the active site of the immobilized enzyme. Another possible explanation is the distortion of a certain fraction of lipase molecules caused by the bad orientation between the enzyme and support during the immobilization step (randomic immobilization) by offering low enzyme loading [24] and [70]. Drastic reduction on the recovered activity was found for lipase preparations from *T. lanuginosus* (TLL1 and TLL2). The recovered activity of TLL1 and TLL2 immobilized on LPHB was reduced by a factor eight and nine, respectively, in relation to the SPHB derivatives. This strong interaction between *T. lanuginosus* lipases (TLL1 and TLL2) and PHB beads had a positive effect on the immobilization yield, but also caused strong diffusional limitation of the substrate to active site of the immobilized lipases and a possible distortion of its native structure. However, for CALB and BTL2 this reduction was around threefold. Recovered activity of PFL immobilized on LPHB was reduced in the order of fourfold, however the biocatalyst displayed the highest recovered activity among the biocatalysts (23.3 ± 3.2%), in agreement to the results of specific activity. Based on these results, lipases from several sources showed very different mechanisms of adsorption and orientation when immobilized on the two conformations of hydrophobic PHB beads (small or large PHB beads), which allowed the modulation of the catalytic properties of lipases.

Butyl butyrate synthesis

The catalytic activity of the lipase preparation derivatives was also investigated in an organic medium in the butyl butyrate synthesis (pineapple flavor) via direct esterification (Fig. 1). The highest butyl butyrate concentration (>90 mmol L^{-1}) was found in the reactions mediated by commercial lipase preparations. Commercial lipase preparations immobilized in both PHB beads showed good activity and the particle size of the support had no influence on the esterification activity of the derivatives. Lipase preparations from *T. lanuginosus* (TLL1 and TLL2) and *P. fluorescens* (PFL) were more active than *P. antarctica* type B (CALB) in aqueous medium (olive oil emulsion hydrolysis). However, the catalytic activity of the four commercial lipase preparations displayed similar performance on butyl butyrate synthesis. According to data reported in the literature, lipases from *T. lanuginosus* and *P. fluorescens* are very efficient in emulsified triglycerides hydrolysis, but the enzymes could be inactivated in organic medium [8], [11], [16] and [25]. These lipase preparations are "typical lipases" with a large lid domain that covers their active sites [25]. Synthetic activity is highly dependent on the degree of hydration and a minimum amount of water is necessary to maintain the optimal conformation of the enzyme [15]. Conversion between two conformations (open and closed) of *T. lanuginosus*(TLL1 and TLL2) and *P. fluorescens* (PFL) involves large internal movements and catalysis probably requires more protein mobility than in CALB, which does not have a typical lid region.

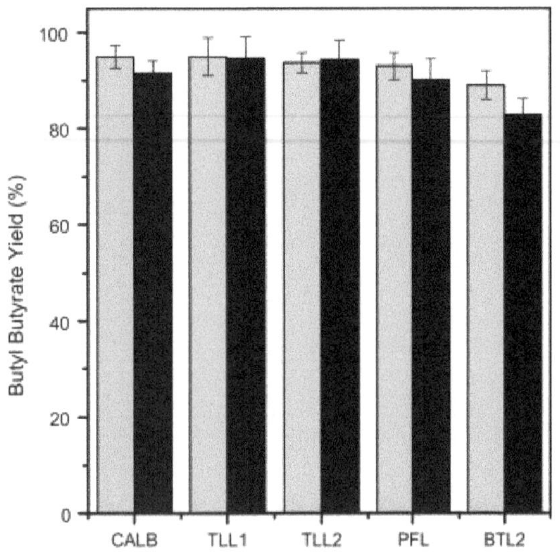

Figure 1: Butyl butyrate yield catalyzed by lipase preparations immobilized on small (gray) and large (black) PHB beads. The reactions were performed at 37 °C and 24 h, using a molar ratio butanol and butyric acid (1:1) in heptane system and 10 wt.% immobilized lipases.

Butyl butyrate synthesis mediated by immobilized BTL2 attained high yield, ranging from 80 to 90 mmol L^{-1} at 24 h, a slightly lower performance than that of the immobilized commercial lipase preparations. These results contrast with the low activity in the hydrolysis reaction achieved with this preparation, which may be attributed to diffusional limitations due to the strong interaction between the two lids of the lipase with the surface of the support. Thus, a reduction on the access of large substrate to its active site (olive oil) was verified, but this allowed a good diffusion of short-chain substrates (butanol and butyric acid). BTL2 has been broadly applied in aqueous media in the enantio- and regiospecific hydrolysis of esters and triacylglycerols [33], [61], [62] and [65]. However, its application in biotransformation reactions in organic medium has still few reports in published literature. Based on these results, the lipase showed promising results in esterification reactions which play an important role in the production of flavor esters.

Biodiesel synthesis

Evaluation of lipases was performed to find the immobilized biocatalyst with the best catalytic activity in the transesterification of babassu oil with ethanol in solvent-free systems. Fig. 2(a)–(d) shows the ethyl esters yield and the viscosity reduction versus time reaction. In all reaction systems, a good dispersion of the biocatalyst in the reaction medium was verified as well as an easiness of the product separation (biodiesel) at the end of the reaction.

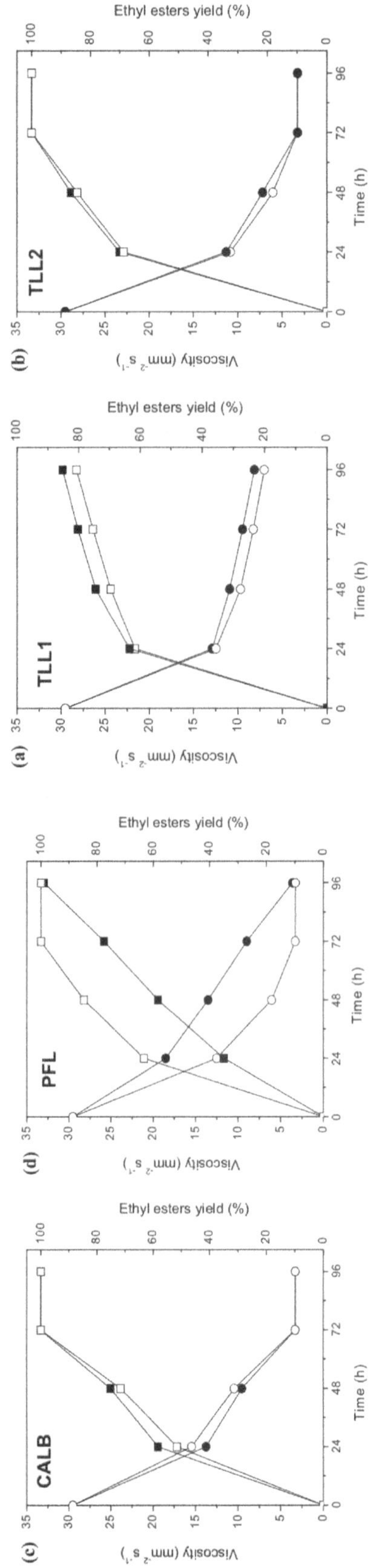

Figure 2: Relation between viscosity (circle) and ethyl esters yield (square) for transesterification catalyzed by lipase preparations TLL1 (a), TLL2 (b), CALB (c) and PFL (d) immobilized on small (full) and large (open) PHB beads. The reactions were performed in solvent-free systems at 45 °C, using a molar ratio of babassu oil:ethanol (1:9) and 10 wt.% immobilized lipases.

The observation of Fig. 2(a)–(d) indicates that all tested lipases were able to form ethyl esters from babassu oil, except BTL2 that did not display any transesterification activity. Full conversion into ethyl esters (biodiesel) was achieved at 72 h for transesterification reaction catalyzed by CALB and TLL2 immobilized on either SPHB or LPHB and PFL immobilized on LPHB. The maximum ethyl esters yield catalyzed by TLL1 ranged from 80 to 85% after 96 h reaction. Although both TLL1 and TLL2 are commercial lipase preparations from *T. lanuginosus* commercially available from Novozymes, TLL2 (Lipex®100L) presented higher hydrolytic and transesterification activities than TLL1 (Lipolase®). These lipase preparations were also tested in the modification of poly(ethylene terephtalate) (PET) fabrics in aqueous systems, followed by a grafting process with acrylic acid (AA) to improve its absorption properties [72]. The results showed that TLL2 also displayed highest catalytic activity on the modification of PET fabrics, in agreement with our results. Among them, the catalytic properties of TLL1 have been well documented in the literature [9], [11] and [54]. However, technical information regarding the lipase preparation Lipex® 100L (TLL2) is still seldom reported and thus speculations about its structure/activity can exist presently. The different performance of these two lipase preparations could be attributed to different additives and preserving agents present in the formulations which play an important role on its catalytic properties [73] or that TLL2 is a variation of TLL1 obtained by genetic modification.

For lipase preparations from *T. lanuginosus* (TLL1 and TLL2) and *P. antarctica* type B (CALB), the particle size of the support did not have influence on the transesterification reaction. On the other hand, the particle size of the support exerted strong influence on the transesterification activity of PFL. In aqueous medium (olive oil emulsion hydrolysis), PFL immobilized on SPHB presented higher catalytic activity than immobilized on LPHB. In this medium, the catalytic activity of PFL immobilized on SPHB was better adapted than the LPHB. Therefore, PFL derivatives displayed a different performance in organic medium. For butyl butyrate synthesis, the particle size of the support exhibits no influence on catalytic activity of the lipase preparation. Conversely, biodiesel synthesis catalyzed by PFL immobilized on LPHB presented higher transesterification activity than SPHB derivative. It can be seen that, full conversion into ethyl esters (biodiesel) catalyzed by PFL immobilized on LPHB and SPHB was reached at 72 and 96 h of reaction, respectively. According to the results, the increase of the particle size of PHB increased the transesterification yield of *P. fluorescens* lipase (PFL). Among the biocatalysts produced by immobilizing on large PHB beads, PFL displayed the highest specific activity and recovered activity values (see Table 2). Thus, the biocatalyst was also more active in transesterification reactions due to the formation of a conformation more active of the enzyme in comparison to other four lipase preparations. The effect of the glycerol in the core of the biocatalyst also could explain these results. Glycerol could be more easily adsorbed on the surface of small PHB beads (highest surface area) inducing a drastic decrease of the catalytic activity of PFL due to formation of a hydrophilic layer around the enzyme resulting in diffusional limitations during triglyceride transfer to the active site of the lipase [13]. Therefore, this effect was not verified for the other lipase preparations due to the different features of each lipase and the interaction with the support. Microbial lipases are generally inactivated by short-chains such as methanol and ethanol [6],[7] and [10]. Therefore, the inactivation of PFL by incubating at substrate containing high ethanol concentration may be discharged. This lipase preparation showed to have high ethanol tolerance as already described in the literature [8] and [11]. It seems that small and large PHB beads protected all lipase preparations, including PFL, by the inhibiting action of ethanol. Physical adsorption of lipase from *P. antarctica* type B on polypropylene-based hydrophobic matrix Accurel MP1001 with similar particle size (≤1000 μm) also displayed the highest catalytic activity on butyl oleate synthesis [13]. In another study, Foresti and Ferreira verified that the highest reaction rate in ethyl oleate synthesis was attained by *C. rugosa* lipase immobilized by physical adsorption on highly-hydrophobic polypropylene particles having large particles size, which is in agreement with our findings [27].

Transesterification of babassu oil with ethanol in a solvent-free system catalyzed by lipase preparations immobilized on PHB beads with different particle sizes aimed at obtaining ethyl esters with kinematic viscosity values in accordance with specifications recommended by the American Society for Testing and Materials (ASTM) D6751 for biofuel, except ethyl ester synthesis catalyzed by TLL1 immobilized on LPHB. As can be seen in Fig. 2, the viscosity of ethyl esters produced by transesterification of babassu oil catalyzed by PFL immobilized on LPHB, CALB and TLL2 derivatives have acceptable values, in the range limit from 3.0 to 6.0 $mm^2\ s^{-1}$, after 72 h of reaction (≈3.5 $mm^2\ s^{-1}$). At 96 h reaction, the viscosity of the biodiesel produced by TLL1 immobilized on SPHB was found to be 5.9 $mm^2\ s^{-1}$, except the transesterification catalyzed by LPHB derivative (6.8 $mm^2\ s^{-1}$), which is outside the specifications for biofuel. For PFL immobilized on SPHB, an acceptable level of viscosity was also found after 96 h of reaction (in the order of 3.6 $mm^2\ s^{-1}$). According to Fig. 2(a)–(d), the transesterification reaction decreased the viscosity of babassu oil, thus enhancing its fluidity in diesel engines.

The concentration of free and immobilized protein offered, yield and productivity into ethyl esters at 24 h are displayed in Table 3. Immobilized protein (IP) concentrations offered, based on the amount of immobilized lipase to catalyze biodiesel synthesis by CALB, TLL1, TLL2, immobilized on either SPHB or LPHB, and BTL2 immobilized on LPHB, was found to be 0.79 mg g^{-1} of oil. BTL2 immobilized on SPHB was the system with the highest immobilized lipase concentration offered (1.17 mg g^{-1} of oil). The immobilization of lipases from commercial preparations or crude extracts on hydrophobic supports has also been broadly used in the purification of these enzymes by selective adsorption [30],[32], [33], [34], [65] and [74]. However, other hydrophobic proteins present in the crude extract could be also adsorbed [65]. Thus, the biocatalysts prepared from BTL2 crude extract may contain lipase and other hydrophobic proteins. The concentration of immobilized protein offered in the transesterification of babassu oil catalyzed by PFL were lower than other lipase preparations tested; 0.61 and 0.45 mg g^{-1} of oil for lipase immobilized on SPHB and LPHB, respectively. In this set of experiments, free lipases were also used for comparative purposes by offering 1 mg protein g^{-1} of oil. It is expected that the immobilization process may bring about conformational changes to the tridimensional structure of the enzymes and the adsorption of water or substrate through support could change its properties and consequently the ethyl ester yield.

Table 3: Yield and productivity in ethyl esters catalyzed by lipase preparations immobilized on small and large PHB beads. All reactions were performed in solvent-free systems at 45 °C, using a molar ratio of babassu oil:ethanol (1:9)

Lipase	PHB	IP amount g^{-1} of oil (mg g^{-1})[a]	Yield (%)	Time of reaction (h)	Productivity (mg biodiesel mgIP^{-1} h^{-1})$^{24\,h}$
CALB	Free	1.00	0	96	0
	SPHB	0.79	100	72	21.4
	LPHB	0.79	100	72	18.9
TLL1	Free	1.00	54.0	96	16.8
	SPHB	0.79	85.2	96	22.8
	LPHB	0.79	80.7	96	23.4
TLL2	Free	1.00	5.80	96	0
	SPHB	0.79	100	72	25.7
	LPHB	0.79	100	72	25.1
PFL	Free	1.00	64.0	96	14.9
	SPHB	0.61	100	96	15.8
	LPHB	0.45	100	72	40.9
BTL2	Free	1.00	0	96	0
	SPHB	1.17	0	96	0
	LPHB	0.79	0	96	0

[a]IP – immobilized protein (mg g^{-1}).

Immobilized lipases were consistently more active than the free form in the transesterification of babassu oil, in agreement to the results reported by Iso et al. [10]. Under these conditions, the aggregates of lipases formed by strong interaction of hydrophobic pockets surrounding the active site of lipase molecules in open conformation were displaced to monomeric form. Strong inactivation of CALB and TLL2 was found, while free TLL1 and PFL yielded 54 and 61% of ethyl esters after 96 h of reaction, respectively. The productivity into ethyl esters was found to be 16.8 and 14.9 mg biodiesel mgIP^{-1} h^{-1}, for TLL1 and PFL, slightly less than immobilized lipases. PFL immobilized on LPHB presented the highest productivity into ethyl esters at 24 h (40.9 mg biodiesel mgIP^{-1} h^{-1}), almost three times higher than the immobilized derivative on SPHB (15.8 mg biodiesel mgIP^{-1} h^{-1}). Derivatives from *T. lanuginosus* (TLL1 and TLL2) and CALB yielded productivity into ethyl esters at 24 h varying from 18.7 to 25.7 mg biodiesel mgIP^{-1} h^{-1}.

Based on these results, PFL immobilized by physical adsorption on LPHB was the most suitable biocatalyst for biodiesel synthesis by transesterification of babassu oil with ethanol in solvent-free systems, followed by TLL2 and CALB immobilized on either SPHB or LPHB. This implied that for PFL a less immobilized protein was required to achieve optimum activity. It is clear that this would have important economic implication on the selection of the lipase preparation for industrial biodiesel synthesis. In agreement with these results, previous screening of commercial lipases reported in the literature show that PFL was the immobilized biocatalyst with the highest transesterification activity [8],[10] and [12]. Moreira et al. [8] investigated the transesterification of palm oil with ethanol in a solvent-free system catalyzed by lipases from several sources immobilized by covalent attachment on hybrid support polysiloxane-polyvinyl alcohol. The best performance on the biodiesel synthesis was attained with *P. fluorescens* lipase that reached almost full conversion (≈98%) in less than 24 h of reaction. Salis et al. [12]screened eight microbial lipase preparations immobilized by hydrophobic adsorption on macroporous polypropylene in the biodiesel synthesis from different feedstocks and the highest yield and productivity into methyl esters also was attained by *P. fluorescens*lipase.

Esterification reactions catalyzed by immobilized BTL2 on SPHB

In spite of their transesterification activities, BTL2 immobilized derivatives displayed high esterification activities as determined in the butyl butyrate synthesis (see Fig. 2). Therefore, the ability of the BTL2 immobilized derivative on small PHB beads (SPHB) to catalyze other aliphatic esters was investigated. In this set of experiments, BTL2 was tested in the esterification reactions of butyric (BA), lauric (LA) and oleic acid (OA) with ethanol (Et) and butanol (But) at various molar ratios. Fig. 3 displays the molar yields into aliphatic esters.

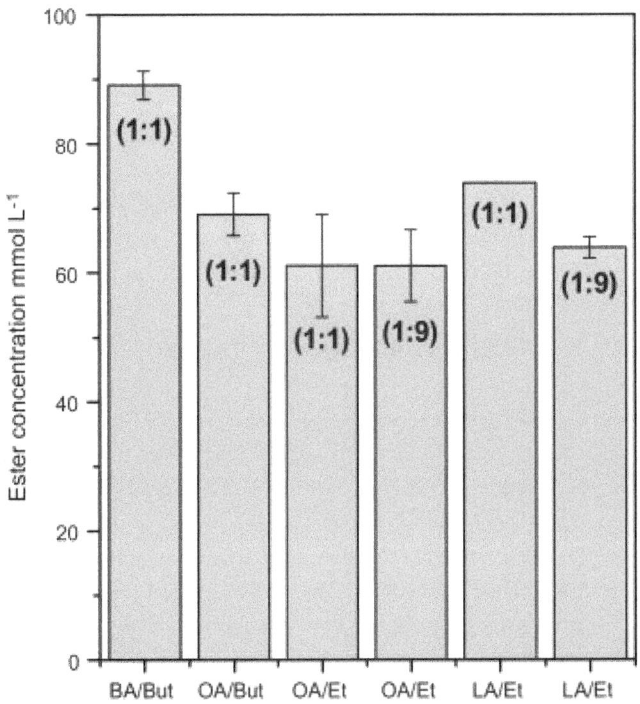

Figure 3: Esterification of butyric (BA), lauric (LA) and oleic acids (OA) with ethanol (Et) and butanol (But) catalyzed by BTL2 immobilized on small PHB beads. The reactions were performed at 37 °C and 24 h, using a molar alcohol and carboxylic acid (1:1) in heptane system and 10 wt.% derivative BTL2-SPHB.

The maximum yield was reached for the esterification of butyric acid with butanol (BA/But), rendering butyl butyrate concentrations around 90 mmol L^{-1}. Esterification of oleic acid with butanol (OA/But) yielded 70 mmol L^{-1} of butyl oleate. Replacing butanol by ethanol, ethyl oleate concentration was found to be 60 mmol L^{-1} and using ethanol in excess (1:9) the ester concentration was similar to that one attaining in the reaction carried out under equimolar concentration (1:1), showing no influence of the ethanol concentration on the ethyl oleate synthesis. In the case of esterification of lauric acid with ethanol (LA/Et), when conducted with equimolar concentration (1:1), a slight increase on the yield in ester was verified in relation to the assay performed in excess of ethanol (1:9). According to the results, the BTL2 showed high catalytic activity in organic medium in the esterification reactions of carboxylic acids with short-chain alcohols such as ethanol and butanol.

In an organic medium, BTL2 did not display catalytic activity in the transesterification reaction. On the other hand, the lipase preparation showed to be active in the esterification of short-chain alcohols with carboxylic acids. According to the results, BTL2 displayed a promising performance in aliphatic esters synthesis that have vast applications in cosmetic, pharmaceutical and food industries as flavors in fruit juices, cheeses, beverages and ice creams.

CONCLUSIONS

PHB was found to be a suitable support for lipase immobilization, enabling the preparation of highly active immobilized lipases in organic (ester and biodiesel synthesis) or aqueous (olive oil emulsion hydrolysis) media. Hydrolytic activity, specific activity and recovered activity were drastically reduced with lipases immobilized on large PHB beads (LPHB) due to the mass transfer limitations of substrate to the active site of the lipases. In the butyl butyrate synthesis, the particle size of the support did not exert any influence on the catalytic activity of the derivatives. The catalytic activity of immobilized lipases was highly increased in comparison with free lipases in biodiesel synthesis, mainly for TLL2 and CALB in which were completely inactivated under the experimental conditions. PFL immobilized on LPHB beads presented the highest productivity in ethyl esters, almost threefold higher than the prepared immobilized derivative in small PHB beads (SPHB). Lipex® 100L (TLL2) and CALB immobilized on either SPHB or LPHB beads also presented high catalytic activity on transesterification of babassu oil with ethanol in solvent-free system. The kinematic viscosity of the ethyl esters (3.4–5.9 $mm^2\ s^{-1}$) are in accordance with specifications recommended by the ASTM D6751 to be used as a fuel, except for the biodiesel sample obtained in the synthesis catalyzed by TLL1 immobilized on LPHB beads. Although immobilized BTL2 has displayed high catalytic activity on aliphatic esters synthesis, this lipase preparation was considered to be inadequate to catalyze biodiesel synthesis. Here, we have demonstrated that PHB beads, a cheap support, is compatible with lipase preparations from several sources, rendering biocatalysts with interesting properties that can be used in both organic and aqueous media.

ACKNOWLEDGMENTS

The authors gratefully acknowledge to FAPESP (Brazil) for financial support (Project 04/14593-4) and Novozymes S.A. (Araucária-PR, Brazil) for the donation of the enzyme preparations TLL1, TLL2 and CALB used in this work.

REFERENCES

1. H.F. de Castro, A.A. Mendes, J.C. Santos, C.L. Aguiar, Quim. Nova 27 (2004) 146–156.
2. P. Villeneuve, J.M. Muderhwa, J. Graille, M.J. Haas, J. Mol. Catal. B: Enzym. 9 (2000) 113–148.
3. R. Sharma, Y. Chisti, U.C. Banerjee, Biotechnol. Adv. 19 (2001) 627–662.
4. F. Hasan, A.A. Shah, A. Hameed, Enzyme Microb. Technol. 39 (2006) 235–251.
5. R. Dalla-Vecchia, M.G. Nascimento, V. Soldi, Quim. Nova 27 (2004) 623–630.
6. M.S. Antczak, A. Kubiak, T. Antczak, S. Bielecki, Renew. Energy 34 (2009) 1185–1194.
7. T. Tan, J. Lu, K. Nie, L. Deng, F. Wang, Biotechnol. Adv. 28 (2010) 628–634.
8. A.B.R. Moreira, V.H. Perez, G.M. Zanin, H.F. de Castro, Energy Fuels 21 (2007) 3689–3694.
9. A.A. Mendes, H.F. de Castro, D.S. Rodrigues, W.S. Adriano, P.W. Tardioli, E.J. Mammarella, R.C. Giordano, R.L.C. Giordano, J. Ind. Microbiol. Biotechnol. 38 (2011) 1055–1066.
10. M. Iso, B. Chen, M. Eguchi, T. Kudo, S. Shrestha, J. Mol. Catal. B: Enzym. 16 (2001) 53–58.
11. A.A. Mendes, R.C. Giordano, R.L.C. Giordano, H.F. de Castro, J. Mol. Catal. B: Enzym. 68 (2011) 109–115.

12. A. Salis, M. Pinna, M. Monduzzi, V. Solinas, J. Mol. Catal. B: Enzym. 54 (2008) 19–26.
13. E. Séverac, O. Galy, E. Turon, C.A. Pantel, J.S. Condoret, P. Monsan, A. Marty, Enzyme Microb. Technol. 48 (2011) 61–70.
14. P.C.M. Da Rós, G.A.M. Silva, A.A. Mendes, J.C. Santos, H.F. de Castro, Bioresour. Technol. 101 (2010) 5508–5516.
15. J.M.S. Rocha, M.H. Gil, F.A.P. Garcia, J. Chem. Technol. Biotechnol. 74 (1999) 607–612.
16. A.R. Rufino, F.C. Biaggio, J.C. Santos, H.F. de Castro, Int. J. Biol. Macromol. 47 (2010) 5–9.
17. S. Gupta, K. Singh, A. Bhattacharya, Int. J. Biol. Macromol. 46 (2010) 445–450.
18. C. Garcia-Galan, A. Berenguer-Murcia, R. Fernandez-Lafuente, R.C. Rodrigues, Adv. Synth. Catal. 353 (2011) 2885–2904.
19. F. Secundo, J. Miehé-Brendlé, C. Chelaru, E.E. Ferrandi, E. Dumitriu, Microporous Mesoporous Mater. 109 (2008) 350–361.
20. M.W. Christensen, L. Andersen, T.L. Husum, O. Kirk, Eur. J. Lipid Sci. Technol. 105 (2003) 318–321.
21. G. Vemuri, R. Banerjee, B.C. Bhattacharryya, Bioprocess. Eng. 19 (1998) 111–114.
22. I.E. de Fuentes, C.A. Viseras, D. Ubiali, M. Terreni, A.R. Alcantara, J. Mol. Catal. B: Enzym. 11 (2001) 657–663.
23. Y. Yes¸ iloglu, ˘ Process Biochem. 40 (2005) 2155–2159.
24. J. Bosley, E.A.D. Peilow, J. Am. Oil Chem. Soc. 74 (1997) 107–111.
25. A. Salis, I. Svensson, M. Monduzzi, V. Solinas, P. Adlercreutz, Biochim. Biophys. Acta 1646 (2003) 145–151.
26. B. Al-Duri, Y.P. Yong, J. Mol. Catal. B: Enzym. 3 (1997) 177–188.
27. M.L. Foresti, M.L. Ferreira, Appl. Surf. Sci. 238 (2004) 86–90.
28. P.C. de Oliveira, G.M. Alves, H.F. de Castro, Biochem. Eng. J. 5 (2000) 63–71.
29. K. Hernández, C. Garcia-Galan, R. Fernández-Lafuente, Enzyme Microb. Technol. 49 (2011) 72–78.
30. D.S. Rodrigues, A.A. Mendes, M. Filice, R. Fernández-Lafuente, J.M. Guisán, J.M. Palomo, J. Mol. Catal. B: Enzym. 58 (2009) 36–40.
31. J.M. Palomo, G. Munoz, ˜ G. Fernández-Lorente, C. Mateo, R. Fernández-Lafuente, J.M. Guisán, J. Mol. Catal. B: Enzym. 19–20 (2002) 279–286.
32. A.A. Mendes, D.S. Rodrigues, M. Filice, R. Fernández-Lafuente, J.M. Guisán, J.M. Palomo, Tetrahedron 64 (2008) 10721–10727.
33. G. Fernández-Lorente, Z. Cabrera, C.A. Godoy, R. Fernández-Lafuente, J.M. Palomo, J.M. Guisán, Process Biochem. 43 (2008) 1061–1067.
34. A. Bastida, P. Sabuquillo, P. Armisén, R. Fernández-Lafuente, J. Huguet, J.M. Guisán, Biotechnol. Bioeng. 58 (1998) 486–493.
35. L. Sarda, P. Desnuelle, Biochim. Biophys. Acta 30 (1958) 513–521.
36. R. Verger, Trends Biotechnol. 15 (1997) 32–38.
37. P. Reis, K. Holmberg, H. Watzke, M.E. Leser, R. Miller, Adv. Colloid Interface Sci. 147–148 (2009) 237–250.
38. B.Chen,J. Hu, E.M.Miller,W.Xie,M.Cai,R.A. Gross,Biomacromolecules 9 (2008) 463–471.
39. Z. Cabrera, G. Fernández-Lorente, R. Fernández-Lafuente, J.M. Palomo, J.M. Guisán, J. Mol. Catal. B: Enzym. 57 (2009) 171–176.
40. Y. Dudal, R. Lortie, Biotechnol. Bioeng. 45 (1995) 129–134.
41. Z. Spitalsky, I. Lacik, E. Lathova, I. Janigova, I. Chodak, Polym. Degrad. Stab. 91 (2006) 856–861.
42. N. Jacquel, C.W. Lo, H.S. Wu, Y.H. Wei, S.S. Wang, AIChE J. 53 (2007) 2704–2714.
43. S.F. Williams, D.P. Martin, D.M. Horowitz, O.P. Peoples, Int. J. Biol. Macromol. 25 (1999) 111–121.
44. E.I. Shishatskaya, T.G. Volova, A.P. Puzyr, O.A. Mogilnaya, S.N. Efremov, J. Mater. Sci. Mater. Med. 15 (2004) 719–728.
45. A. Mosahebi, P. Fuller, M.Wiberg, G. Terenghi, Exp. Neurol. 173 (2002) 213–223.
46. T. Freier, C. Kunze, C. Nischan, S. Kramer, K. Sternberg, M. Saß, U.T. Hopt, K.P. Schmitz, Biomaterials 23 (2002) 2649–2657.
47. X.T. Li, Y. Zhang, G.Q. Chen, Biomaterials 29 (2008) 3720–3728.
48. N. Lukovic, Z. Knezevic-Jugovic, D. Bezbradica, in: M. Manzanera (Ed.), Alternative Fuel, Intech Open Acess Publisher, 2011, pp. 47–72.
49. A. Salis, M. Monduzzi, V. Solinas, in: J. Polaina, A.P. MacCabe (Eds.), Industrial Enzymes, Springer, Dordrecht, 2007, pp. 317–339.
50. J.M. Palomo, M. Fuentes, G. Fernández-Lorente, C. Mateo, J.M. Guisán, R. Fernández-Lafuente, Biomacromolecules 4 (2003) 1–6.
51. V. Gotor-Fernández, E. Busto, V. Gotor, Adv. Synth. Catal. 348 (2006) 797–812.
52. J. Uppenberg, M.T. Hansen, S. Patkar, T.A. Jones, Structure 2 (1994) 293–308.
53. A. Houde, A. Kademi, D. Leblanc, Appl. Biochem. Biotechnol. 118 (2004) 155–170.
54. R. Fernández-Lafuente, J. Mol. Catal. B: Enzym. 62 (2010) 197–212.
55. M. Kaieda, T. Samukaw, T. Matsumoto, K. Ban, A. Kondo, Y. Shimada, H. Noda, F. Nomoto, K. Ohtsuka, E. Izumoto, J. Biosci. Bioeng. 88 (1999) 627–631.
56. Y. Xu, W. Du, J. Zeng, D. Liu, Biocatal. Biotransform. 22 (2004) 45–48.
57. B.K. Jha, M. Svensson, B. Kronberg, K. Holmberg, J. Colloid Interface Sci. 213 (1999) 262–264.
58. Y.Kojima,M.Yokoe, T.Mase,Biosci.Biotechnol. Biochem. 58 (1994) 1564–1568.
59. G. Fernández-Lorente, J.M. Palomo, M. Fuentes, C. Mateo, J.M. Guisán, R. Fernández-Lafuente, Biotechnol. Bioeng. 82 (2003) 232–237.

60. J.C. Santos, A.V. Paula, G.F.M. Nunes, H.F. de Castro, J. Mol. Catal. B: Enzym. 52–53 (2008) 49–57.

61. C. Schmidt-Dannert, M.L. Rúa, H. Atomi, R.D. Schmid, Biochim. Biophys. Acta 1301 (1996) 105–114.

62. M.L. Rúa, C. Schmidt-Dannert, S. Wahl, A. Sprauer, R.D. Schmid, J. Biotechnol. 56 (1997) 89–102.

63. C. Carrasco-López, C. Godoy, B. de las Rivas, G. Fernández-Lorente, J.M. Palomo, J.M. Guisán, R. Fernández-Lafuente, M. Martínez-Ripoll, J.A. Hermoso, J. Biol. Chem. 284 (2009) 4365–4372.

64. C. Carrasco-López, C. Godoy, B. de las Rivas, G. Fernández-Lorente, J.M. Palomo, J.M. Guisán, R. Fernández-Lafuente, M. Martínez-Ripoll, J.A. Hermoso, Acta Crystallogr. F64 (2008) 1043–1045.

65. G. Fernández-Lorente, C.A. Godoy, A.A. Mendes, F. López-Gallego, V. Grazu, B. de las Rivas, J.M. Palomo, J. Hermoso, R. Fernández-Lafuente, J.M. Guisán, Biomacromolecules 9 (2008) 2553–2561.

66. C.M.F. Soares, H.F. de Castro, G.M. Zanin, F.F. de Moraes, Appl. Biochem. Biotechnol. 77/79 (1999) 745–757.

67. M.M. Bradford, Anal. Biochem. 72 (1976) 248–254.

68. P. Adlercreutz, in: J.M. Guisán (Ed.), Immobilization of Enzymes and Cells, Humana Press, Totowa, 2006, pp. 251–256.

69. A. Illanes, J.M. González, J.M. Gómez, P. Valencia, L. Wilson, Electron. J. Biotechnol. 13 (2010) 1–9.

70. S. Sabbani, E. Hedenström, O. Nordin, J. Mol. Catal. B: Enzym. 42 (2006) 1–9.

71. Y. Li, F. Gao, W. Wei, J.B. Qu, G.H. Ma, W.Q. Zhou, J. Mol. Catal. B: Enzym. 66 (2010) 182–189.

72. F. Dadashian, M. Montazer, S. Ferdowsi, J. Appl. Polym. Sci. 116 (2010) 203–209.

73. N. López,R. Pérez, F.Vásquez, F.Valero,A. Sánchez,J. Chem. Technol. Biotechnol. 77 (2002) 175–182.

74. J.M. Palomo, C. Ortiz, M. Fuentes, G. Fernández-Lorente, J.M. Guisán, R. Fernández-Lafuente, J. Chromatogr. A 1038 (2004) 267–273.

CHAPTER 15

IMPROVED ACTIVITY OF LIPASE IMMOBILIZED IN MICROEMULSION-BASED ORGANOGELS FOR (R, S)-KETOPROFEN ESTER RESOLUTION: LONG-TERM STABILITY AND REUSABILITY

Wei-Wei Zhang [1], Jun-Qi Jia[1], Na Wang[1], Cheng-Li Hu[1], Sheng-Yong Yang[2], Xiao-Qi Yu [1]

[1]Key Laboratory of Green Chemistry & Technology, Ministry of Education, College of Chemistry, Sichuan University, Chengdu 610064, PR China

[2]State Key Laboratory of Biotherapy and Cancer Center, West China Hospital, West China Medical School, Sichuan University, Chengdu 610064, PR China

ABSTRACT

Microemulsion-based organogels (MBGs) were effectively employed for the immobilization of four commonly used lipases. During the asymmetric hydrolysis of ketoprofen vinyl ester at 30 °C for 24 h, lipase from *Rhizomucor miehei* and *Mucor javanicus* immobilized in microemulsion-based organogels (RML MBGs and MJL MBGs) maintained good enantioselectivities (ee_p were 86.2% and 99.2%, respectively), and their activities increased 12.8-fold and 7.8-fold, respectively, compared with their free forms. They gave higher yields compared with other lipase MBGs and exhibited better enantioselectivity than commercial immobilized lipases. Immobilization considerably increased the tolerance to organic solvents and high temperature. Both MJL MBGs and RML MBGs showed excellent reusability during 30 cycles of repeated 24 h reactions at 30 °C (over 40 days). The system maintained yields of greater than 50%, while the ee_s values of RML MBGs and MJL MBGs remained nearly constant at 95% and 88%, respectively.

INTRODUCTION

Immobilization is often the key to optimizing the operational performance of an enzyme in an industrial process, especially for use in non-aqueous media. For lipases in particular, immobilization often provides the extra advantage of increasing the catalytic activity of the lipase compared with free enzymes [1] and [2]. Microemulsion-based organogels (MBGs) have become an attractive approach for enzyme immobilization to facilitate enzymatic catalysis in non-conventional media, because of their long-term stabilities [3], [4] and [5]. Moreover, compared with other methods, the use of MBGs is a relatively inexpensive and facile method of enzyme immobilization. Although successful biotransformations employing enzymes immobilized in MBGs have been well documented [6] and [7] and were highlighted in our previous study [8], few reports have focused on the application of this method to kinetic resolution reactions.

Ketoprofen (2-(3-benzoylphenyl) propionic acid) is an important nonsteroidal antiinflammatory drug (NSAID) and is widely used as a racemic mixture to reduce inflammation and relieve pain [9]. However, these healing properties are mainly exerted by the (*S*)-enantiomer, whereas the (*R*)-enantiomer

Citation: Wei-Wei Zhang, Jun-Qi Jia, Na Wang, Cheng-Li Hu, Sheng-Yong Yang, Xiao-Qi Yu, Improved activity of lipase immobilized in microemulsion-based organogels for (R, S)-ketoprofen ester resolution: Long-term stability and reusability, Biotechnology Reports, Volume 7, September 2015, Pages 1-8, ISSN 2215-017X, http://dx.doi.org/10.1016/j.btre.2015.04.003.

can be used as a toothpaste additive to prevent periodontal disease. Thus, efforts to obtain single enantiomer of ketoprofen are important and are expected to receive increased attention because of the pharmacological benefits of both enantiomers.

In recent years, enzymatic kinetic resolution has gained increasing importance as a versatile method for accessing optically active fine chemicals, particularly pharmaceuticals [10]. Indeed, hydrolase-catalyzed kinetic resolution remains an attractive option, because of the simplicity of the process and the high enantioselectivity of biocatalysis under mild conditions [11].

To obtain optically pure ketoprofen, commercial lipases such as Novozyme 435 (from*Candida antarctica*) and Lipozyme immobilized from *Mucor miehei* [12] and [13] have been used to resolve enantiomers of ketoprofen via biocatalytic resolution. However, in many cases, it has been difficult to find a commercial enzyme that exhibits both satisfactory enantioselectivity and activity for the kinetic resolution of unnatural substrates, including ketoprofen. Moreover, the high cost of nonrenewable catalysis further restricts its industrial application. Therefore, different strategies have been proposed to enhance the activity and enantioselectivity of existing commercial enzymes, including optimization of reaction conditions [14], modification of the substrate and/or enzyme [15], and improvement of mass transfer in the reaction [16] and [17]. Enzyme immobilization may facilitate the sustainability of this process and present numerous advantages for industrial applications, such as allowing repeated usage of an enzyme, simplification of product separation and improvement of enzyme stability, among other benefits [18] and [19].

Although aqueous solutions are frequently used in the resolution of ketoprofen, they can cause difficulties in substrate diffusion, product separation and enzyme recovery. In an attempt to ameliorate these issues, approaches involving surfactants and non-aqueous systems have been explored [17]. The practical application of surfactants has been hindered by the need to separate the surfactant from the reaction products. The use of non-aqueous solutions has been less studied because many enzymes are readily denatured in organic solvents, and enzyme-catalyzed hydrolysis generally requires the use of aqueous solutions. In this study, several lipases were immobilized in MBGs, and their activity and enantioselectivity in the hydrolysis of ketoprofen vinyl ester in a non-aqueous solvent were examined. Lipase from *Rhizomucor miehei* immobilized in microemulsion-based organogels (RML MBGs) and lipase from *Mucor javanicus* immobilized in microemulsion-based organogels (MJL MBGs) were selected to further study the activity, selectivity and stability under various experiment conditions. The tolerance of the immobilized lipases to various organic solvents, temperatures and repeated long use was evaluated. Furthermore, the gram-scale resolutions of ketoprofen vinyl esters were tested to determine the possibility of applying the lipase MBGs in industrial chemical synthesis and other bulk applications.

MATERIALS AND METHODS

Materials

Sodium bis-(2-ethylhexyl) sulfosuccinate (AOT) was obtained from Acros Organics (New Jersey, USA). Lipase B acrylic resin from *C. antarctica* (CAL-B, ≥5000 U/g, Shanghai), lipase from *Candida rugosa* (CRL, Type VII, white powder, 739 U/mg, Japan), lipase from*M. javanicus* (MJL, Amano, white powder, ≥10,000 U/g, Japan), Lipozyme immobilized from *M. miehei* (MML, 102.5 U/g, Switzerland), lipase from *R. miehei* (RML, slightly yellow transparent liquid, ≥20,000 U/g, Denmark), lipase from *Pseudomonas cepacia* (PCL, Amano, white powder, ≥50,000 U/g, Japan) and gelatin (from porcine skin, Type A, Shanghai) were purchased from Sigma–Aldrich. Ketoprofen was obtained from Wuhan Gang Zheng Biology Technology (China). Profen vinyl esters were synthesized and purified as described by Wang et al. [20]. All of the organic solvents and other chemicals used were of analytical reagent grade from local manufacturers in China and used without further dehydration. Double distilled water was employed throughout the experiments.

Preparation of lipase MBGs

The method used for the preparation of lipase MBGs was similar to that described in a previous paper by our group [8]. Lipase (15 mg/mL) was incubated overnight at 4 °C in phosphate buffer (pH 7.0, 0.1 M) in the presence of 1 mM AOT. After 24 h, the lipase solution was added to reverse micellar solutions of AOT/buffer/isooctane (Wo = 60). A thermodynamically stable reverse micellar solution was prepared by mixing each component in a suitable ratio. Lipase-containing microemulsions were obtained by adding the previously prepared AOT-coated lipase solution to the AOT reverse micellar solution with the appropriate water content in isooctane. This solution was shaken briefly, then immediately added to a second solution of 14% (w/v) gelatin in phosphate buffer (pH 7.0, 0.1 M) at 55 °C, stirred vigorously until reaching homogeneity, and cooled to 25 °C. The gelatin solution was prepared by dissolving gelatin obtained from porcine skin in phosphate buffer, followed by autoclaving and cooling to 55 °C before use. The obtained gel was then poured into plastic plates and left overnight to air-dry. The following day, the dried gel was cut into small pieces (approximately 1–2 mm^2 in size) and refrigerated at −18 °C for later use.

Enzymatic reaction conditions

In addition to the solvent experiments, additional experiments were performed in isopropyl ether. Lipase MBGs were weighed and added to 2 mL of isopropyl ether containing 10 mg ketoprofen vinyl ester, unless otherwise noted. Each reaction mixture was incubated at 200 rpm in a temperature–controlled shaker at the desired temperature. The reaction was initiated by the addition of 10 mg of lipase or lipase MBGs containing 10 mg lipase. Samples were withdrawn from the reaction medium at regular intervals and analyzed via high-performance liquid chromatography (HPLC). The hydrolysis reaction was conducted at a larger scale using 5 g of ketoprofen vinyl ester, as described in detail in Section 3.5. All of the experiments were repeated at least three times.

Tolerance of lipase MBGs to solvents and temperatures

RML MBGs and MJL MBGs were used for the hydrolysis of ketoprofen vinyl ester in various organic solvents. Free lipase served as the control. Hexane (water content ≤ 0.05%), isopropyl ether (water content 0.1%), *tert*-butyl methyl ether (water content ≤ 0.05%), toluene (water content 0.03%), isopropanol (water content ≤ 0.2%), tetrahydrofuran (THF, water content ≤ 0.01%), 2-methyl tetrahydrofuran (2-MeTHF, water content ≤ 0.05%), 1, 4-dioxane (water content ≤ 0.1%), acetone (water content ≤ 0.3%) and acetonitrile (water content ≤ 0.01%) were selected as solvents.

Tolerance of temperature changes was examined by performing the above reaction at three different temperatures: 20, 30 and 50 °C. A control setup was included at each temperature, with free lipase as the catalyst.

Reusability of RML MBGs and MJL MBGs

To analyze the reusability of lipase MBGs, the hydrolysis reaction was performed as described above for ketoprofen vinyl ester. Upon the completion of one cycle, the immobilized enzyme was recovered via filtration. The recovered lipase MBGs were washed three times with isopropyl ether (3 mL × 3) to ensure complete removal of the product and substrates. The residual solvent was subsequently removed using N_2, and fresh solvent was reintroduced into the system. This procedure was repeated for several cycles.

Analytical procedures

Quantitative analysis of the samples was performed via HPLC through a CHIRALPAK column (Chiral AD-H, 5 μm, 4.6 mm × 250 mm; Daicel Chemical) using a Shimadzu LC-2010A HT apparatus equipped with a 254 nm UV detector. Hexane with 10% (v/v) isopropanol was employed as the mobile phase with a split flow rate of 0.5 mL/min. Retention times were as follows: (*R*)-ketoprofen vinyl ester, 12.24 min; (*S*)-ketoprofen vinyl ester, 13.00 min; (*R*)-ketoprofen, 25.18 min; and (*S*)-ketoprofen, 29.14 min.

Computational methods

All molecular simulations were performed using Discovery Studio 3.1 (Accelrys, San Diego, CA, USA). Accelrys Discovery Studio 3.1 is available from Accelrys Inc.-San Diego, CA 92,121, USA. Protein structures were obtained from the Protein Data Bank (PDB). All protein structures were prepared before molecular docking. Active sites were defined using "From Current Selection" tools based on active residues including catalytic triad residues and oxyanion hole residues. All chemical compounds were constructed manually using Discovery Studio Visualizer and were subjected to the "Minimization" module for full structural refinement with 5000 steps of the steepest descent algorithm, followed by 2000 steps of the conjugate gradient algorithm energy minimization, utilizing the generalized born implicit solvent model and the CHARMM forcefield. Molecular docking was then performed using the "Flexible Docking" module [21] implemented in Discovery Studio 3.1 [22]. Finally, conformations with highest -CDocker interaction energy in each docking process were analyzed and visualized in Discovery Studio (detailed in Supporting information). For the substrate binding step, -CDocker Interaction Energy (-CDIE) and -CDocker Energy (-CDE) were used to evaluate the interaction energy and enzyme-substrate complex stability, respectively.

RESULTS AND DISCUSSION

Different lipases immobilized in MBGs

In a preliminary study, we demonstrated that entrapment within microemulsion-based organogels was an efficient and facile method for immobilizing lipase from *C. rugosa*, as immobilized *C. rugosa* lipase showed remarkable stability in organic solvents and at high temperatures [8]. To extend this work, experiments were performed to investigate the differences among different lipases immobilized in MBGs. Changes in the activity and selectivity of the enzymes following immobilization were detected through the hydrolysis of ketoprofen vinyl ester. Initially, to investigate the role of the immobilization system in the hydrolysis reaction, the control experiments were performed using single gelatin, surfactant, reverse micelle solution and MBGs without lipase as catalyst. The reaction without lipase was investigated as well. There was no product observed in all these reactions even after 48 h, suggesting that the catalytic effect of the polymeric support was excluded. In the present work, four lipases that are differ in species, taxa and structural characteristics were selected. Lipases from *M. javanicus* (MJL) and *R. miehei* (RML) were selected as representatives of the fungal taxon. Lipases from *C. rugosa* (CRL) and*P. cepacia* (PCL) were chosen to represent the yeast and prokaryotic bacterial taxa, respectively. The results are shown in Table 1. After 72 h, most of the free lipases showed very low activity in isopropyl ether, with a maximum yield of 11.8%. Additionally, only MJL and RML exhibited good enantioselectivity. Following immobilization, MJL MBGs and RML MBGs demonstrated maximum increases in activities and gave higher yields compared with other lipases. After 24 h, the activities of MJL MBGs and RML MBGs increased 7.8-fold and 12.8-fold, respectively, compared with their free forms. In addition, MJL MBGs exhibited a high *E*-value greater than 200 after 24 h, whereas RML MBGs maintained a moderate *E*-value. In the case of CRL, both activity and enantioselectivity improved 1.5-fold after immobilization in MBGs. Moreover, PCL MBGs showed almost constant activity and decreased enantioselectivity.

Table 1: Kinetic resolutions of ketoprofen vinyl ester catalyzed by free or immobilized lipases from different sources[a].

Source	Name	Abbreviation	Time	Yield[b] (%)	ee_p[b] (%)	ee_s[b] (%)	E[c]
Fungus	*Mucor javanicus* lipase	MJL	24 h	3.7 ± 0.5	99 ± 0.2	6 ± 0.5	>200
			48 h	10.9 ± 1.3	99 ± 0.2	14 ± 0.7	>200
			72 h	11.8 ± 0.7	83 ± 1.2	20 ± 1.3	13.4
		MJL MBGs	24 h	29.0 ± 1.1	99 ± 0.5	54 ± 0.9	>200
			48 h	36.7 ± 0.8	95 ± 1.1	87 ± 1.3	104.7
			72 h	43.1 ± 0.6	88 ± 2.1	98 ± 1.2	71.6
	Rhizomucor miehei lipase	RML	24 h	3.2 ± 0.3	99 ± 1.7	6 ± 1.4	>200
			48 h	7.3 ± 0.6	99 ± 1.3	9 ± 0.5	>200
			72 h	7.3 ± 0.7	98 ± 1.1	11 ± 1.6	116.5
		RML MBGs	24 h	41.1 ± 2.4	86 ± 0.6	88 ± 2.1	35.2
			48 h	44.9 ± 1.3	77 ± 1.8	95 ± 1.7	27.9
			72 h	48.9 ± 0.8	67 ± 1.3	97 ± 2.6	17.8
	Mucor miehei lipase	MML	24 h	25.0 ± 1.5	68 ± 2.3	86 ± 1.6	14.5
			48 h	52.0 ± 2.3	45 ± 1.9	91 ± 3.1	7.7
			72 h	66.1 ± 1.4	28 ± 1.2	89 ± 1.9	4.7
Yeast	*Candida rugosa* lipase	CRL	24 h	2.0 ± 0.4	36 ± 2.4	1 ± 0.3	2.1
			48 h	3.8 ± 0.7	35 ± 2.2	1 ± 0.5	2.1
			72 h	4.0 ± 1.1	26 ± 1.9	3 ± 1.4	1.8
		CRL MBGs	24 h	3.0 ± 1.1	57 ± 0.7	2 ± 0.9	3.8
			48 h	4.9 ± 1.4	64 ± 1.6	5 ± 1.3	4.8
			72 h	6.0 ± 1.7	70 ± 1.5	6 ± 1.1	6.1
	Candida antarctica lipase B	CAL-B	24 h	75.7 ± 1.7	0 ± 0.0	9 ± 0.5	1.1
			48 h	80.6 ± 2.6	8 ± 0.4	5 ± 0.7	1.2
			72 h	85.5 ± 3.5	5 ± 0.3	2 ± 0.6	1.1
Bacteria	*Pseudomonas cepacia* lipase	PCL	24 h	1.1 ± 1.5	49 ± 1.6	0 ± 0.0	2.9
			48 h	1.4 ± 1.7	31 ± 1.4	0 ± 0.0	1.8
			72 h	2.8 ± 1.1	15 ± 0.7	0 ± 0.1	1.4
		PCL MBGs	24 h	1.2 ± 0.5	12 ± 0.6	0 ± 0.0	1.3
			48 h	1.8 ± 0.9	18 ± 1.2	0 ± 0.1	1.4
			72 h	2.4 ± 0.7	35 ± 1.5	1 ± 0.4	2.1

[a] Reactions were performed with 10 mg ketoprofen vinyl ester and 10 mg lipase or lipase immobilized in microemulsion-based organogels (MBGs) containing 10 mg lipase in 2 mL isopropyl ether at 30 °C for 24 h, 48 h and 72 h.
[b] Yield and *ee* were determined via high-performance liquid chromatography (HPLC) using a Chiral AD-H column.
[c] $E = \ln[1 - c(1 + ee_p)]/\ln[1 - c(1 - ee_p)]$, $c = ee_s/(ee_s + ee_p)$ [23].

To compare the performance of the lipase MBGs with that of commercial immobilized lipases addressed in the literature, we also measured the activities and enantioselectivities of Novozyme 435 (*C. antarctica*) and Lipozyme MML. MML showed high activity and low *E*-values, whereas CAL-B exhibited almost no selectivity. Compared with other processes using commercial or laboratorial immobilized lipases (Table 2), RML MBGs and MJL MBGs showed obvious superiority in the asymmetric hydrolysis of ketoprofen vinyl ester at 30 °C. The biotransformation catalyzed by MBGs resulted in improved enzymatic conversion efficiency and good enantioselectivity. Further advantages of MBGs reaction systems without the addition of water are improved solubility of aromatic substrates and products, excellent recyclability and low enzyme loading.

Table 2: Comparison of the efficiency of ketoprofen kinetic resolution using different immobilized lipases.

Enzyme	Reaction conditions	Product	Reaction efficiency			Ref.
			Yield (%)	*ee* (%)	*E*	
Lipozyme IM (commercial immobilized RML)	Esterification isopropyl ether, 37 °C	*R*-Ketoprofen 1-butanol ester	20 (72 h)	9.5	2.4	LoÅpez -Belmonte et al. [13]
Immobilized MJL	Hydrolysis pH 7.0 buffer, 30 °C (containing 30% v/v acetone) (2 mg substrate versus 30 mg MJL)	*R*-ketoprofen (after 5 cycles remaining 50% activity)	53	86	55	Kato et al. [24]
Immobilized CRL	Hydrolysis pH 3.5 buffer with Tween 80 as additive, 30 °C	*S*-Ketoprofen	22.3 (72 h)	94	50	Liu et al. [25]
Immobilized CAL-B in an enzymatic membrane reactor	Esterification dichloropropane: hexane = 20: 80, 40 °C	*R*-ketoprofen 1-butanol ester (after 5 cycles remaining 50% activity)	73 (24 h)	ee_p 87 ee_s 57	27	Ong et al. [15]
RML MBGs	Hydrolysis acetate vinyl ester, 30°C	*R*-Ketoprofen	41 (24 h)	ee_p 86 ee_s 88	35	This study
			53 (after 30 cycles)	ee_p 89 ee_s 91	67	
MJL MBGs			29 (24 h)	ee_p 99 ee_s 54	>200	
			55 (after 30 cycles)	ee_p 90 ee_s 81	48	

These results may be attributed to the specific active sites and spatial conformations of the different lipases. The conformational behaviors of the tested lipases indicate remarkable differences among them, not only in terms of the accessibility of the active site but also regarding the modification of the geometry of the catalytic machinery.

A unique property that distinguishes lipases from esterases is the enhanced activity of lipases at or near the lipid/water interface [26], which is related to the presence of a flexible protein domain called the lid [27]. The lid opens upon contact of the lipase with an interface. This leads to the restructuring of the lipase by creating an electrophilic region (the oxyanion hole) around the serine residue by exposing hydrophobic residues and burying hydrophilic ones. This process stabilizes the transition state intermediate during catalysis [28] and [29].

It has been reported that PCL exhibits a broad lid domain and a stable open conformation, even in aqueous environments [30]. Therefore, the MBG system likely has little effect on the conformation of this lipase. In contrast to PCL, the lids of RML and MJL are much smaller and, consequently, can be easily altered in the immobilization environment. In addition, CRL has an intermediate-sized lid domain, between the sizes of the wide lid of the *Pseudomonas* lipase and the small lid observed in the fungal RML and MJL lipases, and it displayed intermediate increases in activity and selectivity. Finally, CAL-B represents an atypical case in these experiments because this enzyme has a very small lid, which is not able to cover the active site. Accordingly, CAL-B showed the highest activity but the lowest enantioselectivity.

Computational simulations

In light of these clear differences in activity and selectivity among lipases from different species, we then performed computational docking simulations to consider the variations occurring at the level of the catalytic machinery of each enzyme in more detail. The first priority was to identify the residues forming the catalytic machinery of each considered enzyme, namely the catalytic triad and the oxyanion hole (Table 3).

Table 3: Residues forming the catalytic machinery of the tested lipases.

Enzyme	PDB code	Catalytic triad	Oxyanion hole
RML	4TGL	Ser144, His257, Asp203	Ser82, Leu145
CAL-B	1TCA	Ser105, His224, Asp187	Thr40, Gln106
CRL	1CRL	Ser209, His449, Glu341	Gly123, Gly124, Ala210
PCL	3LIP	Ser87, His286, Asp264	Leu17, Glu88

It is well established that there are two steps required for enzyme catalysis: connection between substrate and enzyme and catalytic transformation from substrate to product through catalytic residues. The ability to bind and the efficiency of catalysis are of equal importance and are therefore considered in molecular docking analyses. For the substrate binding step, we used -CDocker Interaction Energy (-CDIE) and -CDocker Energy (-CDE) to evaluate the interaction energy and enzyme-substrate complex stability, respectively. The ability of an enzyme to bind substrates can be estimated using -CDIE. Higher -CDIE values indicate better binding abilities, whereas higher -CDE values indicate a greater ability of stable substrates to remain in the correct position. For the enzyme catalysis step, the distances between the substrate atoms involved in catalysis and the catalytic residues were considered, including the catalytic triad and oxyanion hole residues. In particular, the distance between the carbon atom of the carbonyl in the vinyl ester and the hydroxyl oxygen of the nucleophilic serine side chain (D_1) and the distance between the oxygen atom of the carbonyl and the closest hydrogen donor of the oxyanion hole (D_2) were analyzed (Fig. 1).

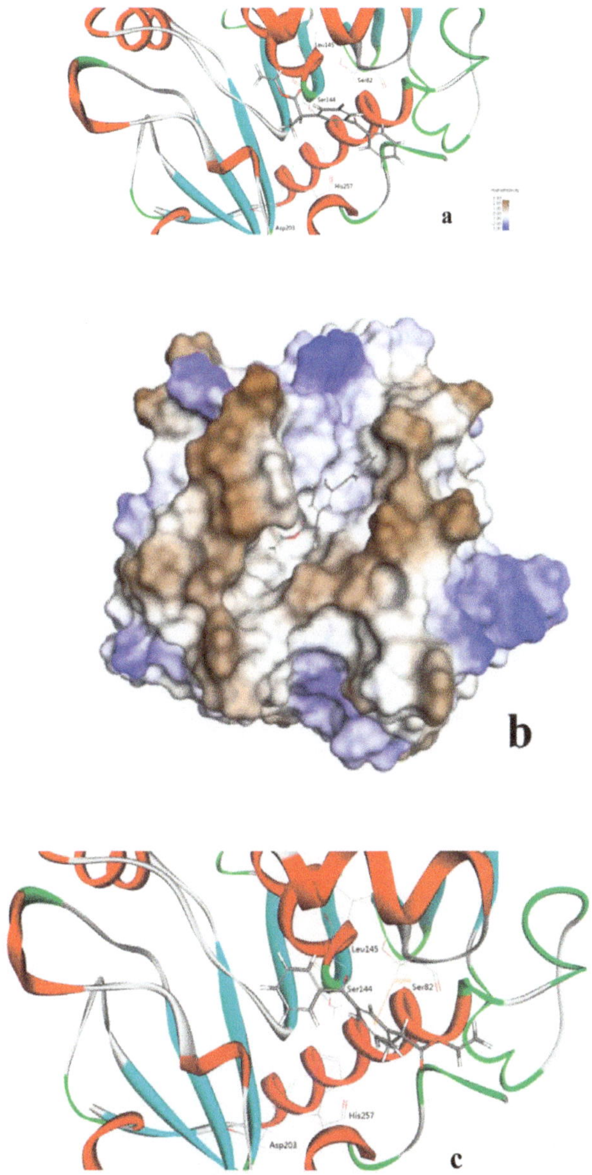

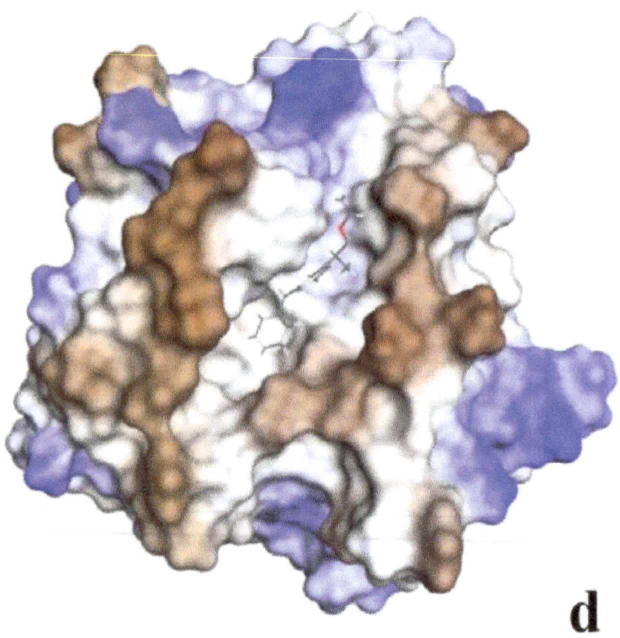

Figure 1: Comparison between (*R*)- and (*S*)-ketoprofen vinyl ester binding with lipase from *Rhizomucor miehei*(RML). (a) Three-dimensional structure of the binding between (*R*)-enantiomer and RML. (b) Binding surfaces of the complex of (*R*)-enantiomer and RML. The hydrophilic areas are displayed in blue and the hydrophobic areas are visualized in brown. (c) Three-dimensional structure of the binding between (*S*)-enantiomer and RML. (d) Binding surfaces of the complex of (*S*)-enantiomer and RML. The hydrophilic areas are displayed in blue and the hydrophobic areas are visualized in brown.

As shown in Table 4, the -CDIE and -CDE values between RML and the (*R*)-enantiomer were 34.9154 kcal/mol and 26.4078 kcal/mol, respectively, indicating that RML is apt to interact with the (*R*)-enantiomer and that the interaction is stable. Although both the -CDIE and -CDE values of RML and the (*S*)-enantiomer were greater than those of RML and the (*R*)-enantiomer, the carbonyl oxygen of the (*S*)-enantiomer is far away from the catalytic triad and oxyanion hole residues (Fig. 1). Moreover, as can be seen in Fig. 1(c), the phenyl ring in (*S*)-enantiomer can form a Pi-sigma interaction with Ser82 residue, which enhanced the stability of RML-substrate complex and hindered catalysis. This might account for the low activity and high enantioselectivity of the native RML. After activated in the immobilization environment, lipase was stabilized in an open form. Therefore, RML MBGs showed efficiently improved yield while maintained good enantioselectivities. CAL-B showed high -CDIE and -CDE values to both enantiomers, which explained the ability of CAL-B to interact efficiently. Additionally, the oxygen atom of the carbonyl in (*R*)-ketoprofen formed three hydrogen bonds with the residues in the oxyanion hole in CAL-B, which enhanced the interaction between the (*R*)-enantiomer and CAL-B and aligned the carbonyl group in the catalytic sites to facilitate subsequent catalysis. However, because of the similar D_1 and D_2 values of the (*R*)- and (*S*)-enantiomers, CAL-B showed almost no selectivity. The enantioselectivity of CRL was poor because of the similar connection ability and activity to both enantiomers. Higher -CDIE and -CDE values of the (*S*)-enantiomer indicated the binding of the (*S*)-enantiomer with PCL was more efficient. However, the incorrect orientation of the (*S*)-enantiomer towards the oxyanion hole inhibited catalysis. Therefore, PCL showed intermediate enantioselectivity and afforded low yields in the reaction. Furthermore, PCL showed little interfacial activation, thus performed limited improvement in activity after immobilization.

Table 4: The docking results of different lipases[a].

Lipase	Chirality	Distance 1 (Å)	Distance 2 (Å)	-CDIE (kcal/mol)	-CDE (kcal/mol)
RML	R	3.20	3.05	34.92	26.41
	S	6.51	8.71	39.64	28.91
CAL-B	R	4.05	2.31 (3*HB[b])	43.17	32.56
	S	4.04	1.97(1*HB[c])	40.96	28.81
CRL	R	3.05	2.01 (1*HB[d])	29.88	20.20
	S	3.63	3.26	30.56	21.48
PCL	R	3.52	2.98	35.89	24.78
	S	4.61	6.83	37.10	28.86

[a] All molecular simulations were performed using Discovery Studio 3.1 (Accelrys, San Diego, CA, USA). Protein structures were obtained from Protein Data Bank. PDB codes for lipase from *Rhizomucor miehei* (RML), lipase B from *Candida antarctica* (CAL-B), lipase from *Candida rugosa* (CRL) and lipase from *Pseudomonas cepacia* (PCL) are 4TGL, 1TCA, 1CRL and 3LIP, respectively. All protein structures were prepared before molecular docking.
[b] The oxygen atom of the carbonyl in (*R*)-ketoprofen formed three hydrogen bonds with the residues in the oxyanion hole in CAL-B.
[c] The oxygen atom of the carbonyl in (*S*)-ketoprofen formed a hydrogen bond with the residues in the oxyanion hole in CAL-B.
[d] The oxygen atom of the carbonyl in (*R*)-ketoprofen formed a hydrogen bond with the residues in the oxyanion hole in CRL.

Although it is difficult to design a reasonable experimental scheme to verify the influence of organic solvents in the molecular simulations using the "Flexible Docking" module, because of the complexity of the lipase/MBGs/solvent system, the docking results can help explain the low catalytic efficiency of these lipases in their native forms during the hydrolysis of ketoprofen vinyl ester. As a consequence, by combining the conformational behaviors of these lipases and our experimental data, this experiment could demonstrate that MBGs can efficiently improve the activity of RML after immobilization.

Tolerance of organic solvents and temperatures

To meet the demands of a technical process, the possibility of adapting an enzyme to various solvents is crucial to exploring novel applications and optimization for industrial needs. Therefore, various organic solvents were examined to determine their effects on the activity and stability of MJL MBGs, RML MBGs and their original forms, and these results are presented in Fig. 2. Ketoprofen vinyl ester is soluble under all experimental conditions. In all of the solvents except for hexane, the activity and selectivity of the lipases immobilized in MBGs were improved, most notably for RML MBGs. After 24 h, RML MBGs produced yields greater than 20% in all of the solvents except for toxic toluene and viscous isopropanol, whereas little product was observed when free RML was used. The relative lower yields of immobilized lipase in more polar solvents might also because of their deleterious effects to gelatin MBGs. These results suggested that the stability of these enzymes in the presence of organic solvents depends on both the enzyme and the nature of the organic solvent being used. Notably, the RML MBGs produced a 21.7% yield after 24 h in the bio-based green solvent 2-methyl-tetrahydrofuran (2-MeTHF), which is easily biodegradable. Concerning the environmental characteristics of these processes, the use of green organic solvents that combine an awareness of the ecological footprint with enzymatic efficiency may allow more environmentally friendly measures to be applied in industrial syntheses. When hexane were used as the solvent, it can be observed from the HPLC chromatograms that the peak of (*R*)-ketoprofen vinyl ester reduced obviously, whereas (*R*)-ketoprofen had a slender peak (Figs. A. 1 and A. 2 of Supplementary information). As shown in Fig. 2, the MBGs in hexane exhibited high ee_s values and low yields. Therefore, we tested mixtures of different ratios of hexane and isopropanol as solvents to investigate the resultant reactions (Fig. A. 1 of Supplementary information). In hexane with 20% isopropanol, the RML MBGs afforded a 33.6% yield with a 70.7% ee_s, whereas a 50% isopropanol mixture gave a 17.9% yield with a 47.7% ee_s. These results could indicate that the diffusion of ketoprofen in the MBGs was hindered in non-polar hexane, because of the strong polarity of this compound.

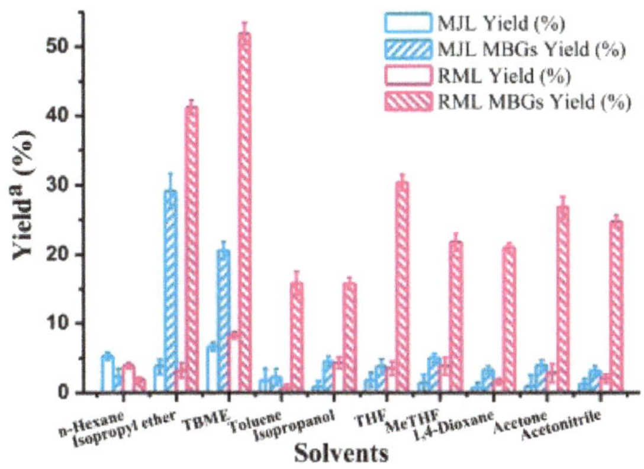

(a). Yields of hydrolysis of ketoprofen vinyl ester.

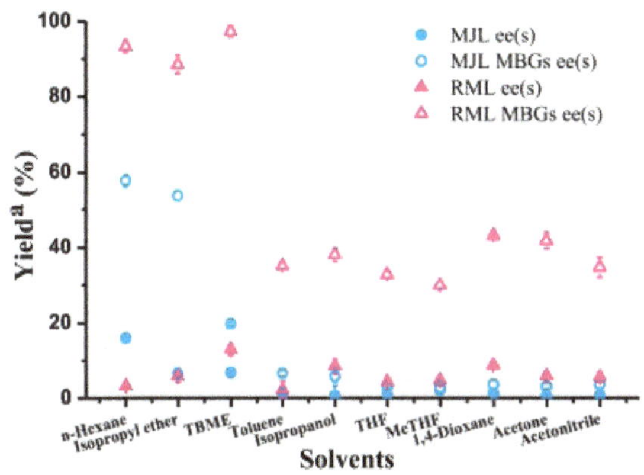

(b). ee_s of hydrolysis of ketoprofen vinyl ester.

Figure 2: Tolerance to solvents of lipase from *Rhizomucor miehei* immobilized in microemulsion-based organogels (RML MBGs) and lipase from *Mucor javanicus* immobilized in microemulsion-based organogels (MJL MBGs). (a) Yields of hydrolysis of ketoprofen vinyl ester. (b) ee_s of hydrolysis of ketoprofen vinyl ester. a. Reactions performed with 10 mg ketoprofen vinyl ester and 10 mg lipase or lipase MBGs containing 10 mg lipase in 2 mL solvent at 30 °C and 200 rpm for 24 h.

Because of their high tolerance to organic solvents, the RML MBGs were then tested to determine their thermostability. The influence of temperature on the activities of native and immobilized RML was studied using three different temperatures, as shown in Fig. 3. The activity and selectivity of free RML were low at all temperatures tested and showed limited increases over time. However, it is noteworthy that after being immobilized in MBGs, RML showed a significant enhancement of activity and selectivity at all temperatures, most notably at 30 °C. The activity of the RML MBGs increased significantly as the temperature increased, and the reaction yield was greater than 50% following the incubation at 50 °C for 24 h. Thus, ee_p decreased over time when the reaction was conducted at 50 °C. These results demonstrate that MBGs offer a more conducive microenvironment at high temperatures and might therefore protect the correct lipase conformation from destruction at high temperatures.

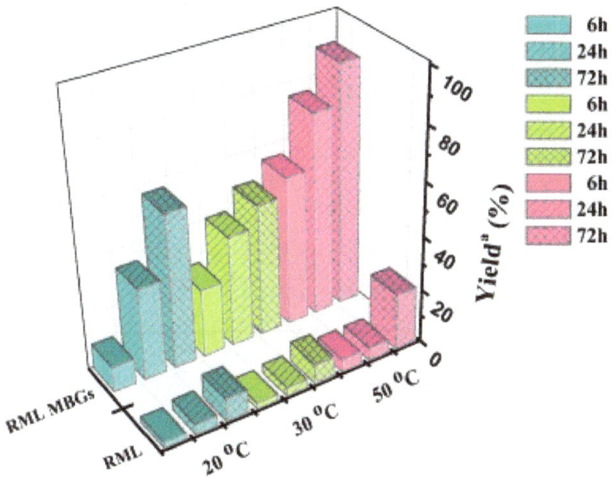

(a). Yields of hydrolysis of ketoprofen vinyl ester.

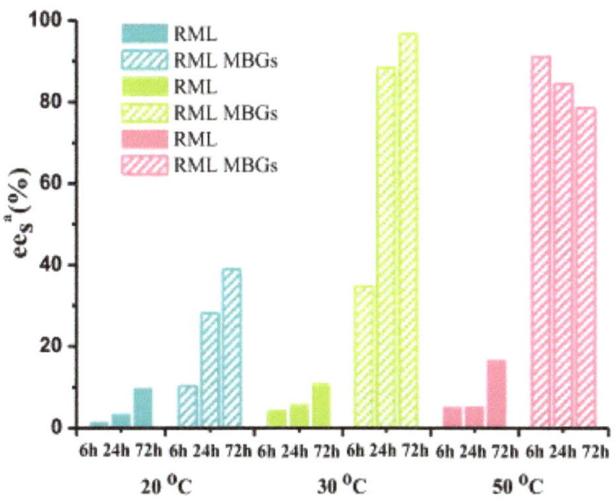

(b). ee_s of hydrolysis of ketoprofen vinyl ester.

Figure 3: Tolerance to temperatures of lipase from *Rhizomucor miehei* immobilized in microemulsion-based organogels (RML MBGs). (a)

Yields of hydrolysis of ketoprofen vinyl ester. (b) ee_s of hydrolysis of ketoprofen vinyl ester. a. Reactions performed with 10 mg ketoprofen vinyl ester and 10 mg RML or RML MBGs containing 10 mg RML in 2 mL isopropyl ether at various temperatures and 200 rpm for 24 h.

Reusability

One of the greatest advantages of immobilization is to facilitate an enzyme's reusability, which is essential to render the process technologically and economically viable. The stability of immobilized RML and MJL was assessed in repeated 24 h long runs of ketoprofen vinyl ester hydrolysis with excessive washings between the runs. Fig. 4 shows the impressive stability of the catalyst during repeated use in our system. In the first three runs, there is an obvious increase in yield and ee_s of RML MBGs and MJL MBGs, because of the imprinting effect of the residual substrate molecules in immobilized lipase[31]. One of the most successful strategies to enhance enzyme activity in organic solvents involves tuning the enzyme active site by molecular imprinting with substrates or their analogues [32]. By combining the imprinting of the lipase surface with surfactant coating and reverse micellar activation, known as dual bioimprinting [33], the RML MBGs and MJL MBGs showed high activity and no significant reduction in activity or enantioselectivity over 30 cycles. The system maintained yields of greater than 50%, even after 30 cycles of repeated use. For the RML MBGs, the ee_s values remained nearly constant at 95%, whereas the ee_s values for the MJL MBGs fluctuated by approximately 88%. Together, these results suggest that immobilized lipase demonstrates good durability and reusability. The proven high stability of the enzyme during 30 cycles (over 40 days) shows promise for effective catalyst recycling.

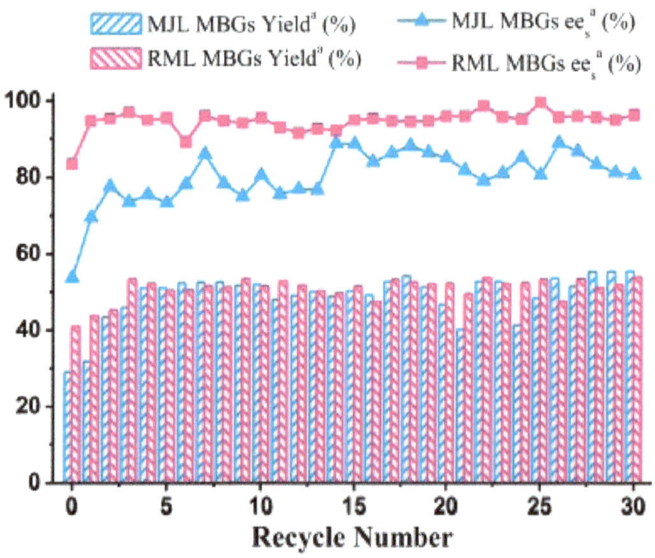

Figure 4: Reusability of lipase from *Rhizomucor miehei* immobilized in microemulsion-based organogels (RML MBGs) and lipase from *Mucor javanicus* immobilized in microemulsion-based organogels (MJL MBGs). a. Reactions performed with 10 mg ketoprofen vinyl ester and 10 mg lipase or lipase MBGs containing 10 mg lipase in 2 mL isopropyl ether at 30 °C and 200 rpm for 24 h.

Gram-scale experiment

To investigate the effect of substrate structure on this process, we synthesized several ketoprofen esters for hydrolysis, including the methyl ester, ethyl ester, 1-butyl ester, trifluoroethyl ester, 2-chloroethyl ester, benzyloxy ester and phenylethyl ester. Free RML showed trace activity for many different substrates with different leaving groups, whereas the RML MBGs were significantly more active. However, the activity of the RML MBGs was significantly affected by the properties of the leaving groups in the substrates. After 72 h, trifluoroethyl ester and 2-chloroethyl ester showed 47.8% and 35.1% yields, while maintaining 69% and 87% ee_p, respectively, because of their good leaving groups.

To verify the feasibility of the proposed process on a gram-scale, RML MBGs with 100 mg of RML were added to a mixture of 5 g of ketoprofen vinyl ester in 100 mL isopropyl ether and shaken at 30 °C. The isolated yield from the hydrolysis, as measured by column chromatography, reached 46.8% after 72 h with a 91% ee_p. The significant increase in the enantioselectivity factor (*E*) from 35.2 in analytical scale trials to 54.1 in the gram-scale experiment also underscores the possibility of using RML MBGs in industrial chemical syntheses and other bulk applications.

CONCLUSIONS

In summary, these results demonstrate that MBGs protect the correct lipase conformation from destruction by polar solvents and high temperatures. The findings of this study clearly indicate that MBGs is an ideal candidate for RML and MJL immobilization. Moreover, this work expands the application of RML MBGs and MJL MBGs to the asymmetric synthesis of (*S*)-ketoprofen. The ease of preparation and operation, the high activity of several lipases and the ability to conduct straightforward product separation make this process promising for industrial applications.

ACKNOWLEDGMENTS

This work was financially supported by the National Program on Key Basic Research Project of China (973 Program, 2013CB328900), the National Natural Science Foundation of China (Nos. 21232005 and 21321061), and the Program for Changjiang Scholars in China. We also thank the Sichuan University Analytical & Testing Center for SEM analysis.

REFERENCES

1. P. Adlercreutz, Chem. Soc. Rev. 42 (2013) 6406–6436.
2. R.C. Rodrigues, C. Ortiz, A. Berenguer-Murcia, R. Torres, R. Ferna'ndez-Lafuente, Chem. Soc. Rev. 42 (2013) 6290–6307.
3. M. Zoumpanioti, H. Stamatis, A. Xenakis, Biotechnol. Adv. 28 (2010) 395–406.
4. K. Nagayama, K. Karaiwa, T. Doi, M. Imai, Biochem. Eng. J. 2 (1998) 121–126.
5. T. Garcia, N. Sanchez, M. Martinez, J. Aracil, Enzyme Microb. Technol. 25 (1999) 591–597.
6. I.V. Pavlidis, K. Tzafestas, H. Stamatis, Biotechnol. J. 5 (2010) 805–812.
7. T. Raghavendra, D. Sayania, D. Madamwar, J. Mol. Catal. B: Enzym. 63 (2010) 31–38.
8. W.W. Zhang, N. Wang, Y.J. Zhou, T. He, X.Q. Yu, J. Mol. Catal. B: Enzym. 78 (2012) 65–71.

9. R. Kourist, P.D. de Mar'ia, K. Miyamoto, Green Chem. 13 (2011) 2607–2618.
10. S. Yoon, S. Kim, S. Park, E. Hong, J. Kim, S. Kim, T.H. Yoo, Y. Ryu, J Mol. Catal. B: Enzyme 100 (2014) 25–31.
11. M. Schober, K. Faber, Trends Biotechnol. 31 (2013) 468–478.
12. X.Q. Cai, N. Wang, X.F. Lin, Polymer 47 (2006) 6491–6495.
13. M.T. López-Belmonte, A.R. Alcántara, J.V. Sinisterra, J. Org. Chem. 62 (1997) 1831–1840.
14. X. Jin, B.K. Liu, Z. Ni, Q. Wu, X.F. Lin, Enzyme Microb. Technol. 48 (2011) 454–457.
15. A.L. Ong, A.H. Kamaruddin, S. Bhatia, H.Y. Aboul-Enein, J. Sep. Sci. 31 (2008) 2476–2485.
16. X. Li, T. Liu, L. Xu, X.H. Gui, F. Su, Y.J. Yan, Biotechnol. Bioprocess Eng. 17 (2012) 1147–1155.
17. Y.Y. Zhang, J.H. Liu, Biochem. Eng. J. 54 (2011) 40–46.
18. A. Illanes, A. Cauerhff, L. Wilson, G.R. Castro, Bioresour. Technol. 115 (2012) 48–57.
19. D.Y. Xu, Z. Yang, Chemosphere 92 (2013) 391–398.
20. Y.F. Wang, J.J. Lalonde, M. Momongan, D.E. Bergbreiter, C.H. Wong, J. Am. Chem. Soc. 110 (1988) 7200–7205.
21. J. Koska, V.Z. Spassov, A.J. Maynard, L. Yan, N. Austin, P.K. Flook, C.M. Venkatachalam, J. Chem. Inf. Model. 48 (10) (2008) 1965–1973.
22. V.Z. Spassov, L. Yan, P.K. Flook, Protein Sci. 16 (3) (2007) 494–506.
23. S. Joly, M.S. Nair, J. Mol. Catal. B: Enzyme 22 (2003) 151–160.
24. K. Kato, Y. Gong, T. Satto, H. Kimoto, J. Biosci. Bioeng. 90 (2000) 332–334.
25. Y.Y. Liu, J.H. Xu, H.Y. Wu, D. Shen, J. Biotechnol. 110 (2004) 209–217.
26. M. Moniruzzaman, Y. Hayashi, M.M.R. Talukder, E. Saito, T. Kawanishi, Biochem. Eng. J. 30 (2006) 237–244.
27. K.E. Jaeger, M.T. Reetz, Trends Biotechnol. 16 (1998) 396–403.
28. Z.S. Derewenda, A.M. Sharp, Trends Biochem. Sci. 205 (1993) 20–25.
29. H. Stamatis, A. Xenakis, F.N. Kolisis, Biotechnol. Adv. 17 (1999) 293–318.
30. V. Ferrario, C. Ebert, L. Knapic, D. Fattor, A. Basso, P. Spizzo, L. Gardossi, Adv. Synth. Catal. 353 (2011) 2466–2480.
31. G. Hellner, Z. Boros, A. Tomin, L. Poppe, Adv. Synth. Catal. 353 (2011) 2481–2491.
32. J.O. Rich, V. Mozhaev, J.S. Dordick, D.S. Clark, Y.L. Khmelnitsky, J. Am. Chem. Soc. 124 (2002) 5254–5255.
33. J.Y. Yan, Y.J. Yan, J.K. Yang, L. Xu, Y. Liu, Process Biochem. 44 (2009) 1128–1132.

CHAPTER 16

IMPROVED ACTIVITY OF LIPASE IMMOBILIZED IN MICROEMUL IMPACT OF PHYSICOCHEMICAL CHARACTER OF HYDROPHOBIC POROUS CARRIER ON REACTIVITY OF IMMOBILIZED LIPASE PROGRESSING TOWARD HIGHER REACTION RATE AND HIGH YIELD IN REPEATED USE

M. Naya, M. Imai

Course in Bioresource Utilization Sciences, Graduate School of Bioresource Sciences, Nihon University, 1866 Kameino, Fujisawa, Kanagawa 252-0880, Japan

ABSTRACT

Impact of physicochemical character of hydrophobic porous carrier on reactivity of immobilized lipase from Candida rugosa, Rhizopus arrhizus, and Wheat Germ was investigated by Accurel MP100 carrier employed. Accurel MP100 was made of a polypropylene-based hydrophobic porous support. Lipase was adsorbed and immobilized on the Accurel pores with the aid of cross-linking agent glutaraldehyde. Immobilized yield attended satisfactorily high level (98%). Effectiveness factor of immobilized lipase was improved to use reduced size Accurel particles. In the case of Candida rugosa used (particle size 500-840 µm), the effectiveness factor exceeded 0.5. Suitable carrier size to realize sufficient high reaction rate was practically depended on lipase species used. Repeated use of immobilized lipase was attractively performed and stable. The productivity was remained almost 50%. Accurel MP100 was expected as hydrophobic friendly lipase carrier progressing toward higher reaction rate and high yield in repeated use.

Citation: M. Naya, M. Imai, Impact of physicochemical character of hydrophobic porous carrier on reactivity of immobilized lipase progressing toward higher reaction rate and high yield in repeated use, doi: 10.1016/j.proeng.2012.07.492

INTRODUCTION

Lipases have been paid attention to lipid modifications [1-3]. They catalyzed hydrolysis, alcoholysis, acidolysis, amidolysis, and esterification in food and pharmaceutical industry [4-9]. Immobilized lipases have been reported to apply in various hydrophobic reactions [3, 10-12]. Reactivity of immobilized lipase was affected by physicochemical factors in reaction media [13, 14]. Especially for quick initiation of hydrophobic enzymatic reaction, a hydrophobic material is most favorable to easy diffusion of substrate in inner pore of carrier. Previously, gels and solid porous carriers having hydrophilic character were often employed even for hydorophobic substrate reaction. Detail technical data focused on carrier to realize a quick initiation of hydrophobic enzymatic reaction and a high yield repeated use of immobilized enzyme are strategically indispensable to industrial applicable practical design of hydrophobic enzyme reaction [15-17]. Authors examine impact of physicochemical character of hydrophobic porous carrier on reactivity of immobilized lipase progressing toward higher reaction rate and high yield in repeated use [18]. Effect of the resources of lipase and the geometrical size was mainly demonstrated.

Nomenclature

W: Molar ratio of the water to amphiphile [(mol-H_2O)·(mol-amphiphile)$^{-1}$]

η_{th}: Theoretical effectiveness factor calculated from Thiele modulus defined [-]

φ: Thiele modulus [-]

MATERIAL AND METHODS

Materials

Accurel MP100, provided by Membrana GmbH (Obernburg, Germany), is made of a polypropylene based hydrophobic porous support. The original shape Accurel looks like as a half cylinder. The volume and shape of Accurel particles are well preserved in various organic solvents at experimental temperatures and pressures. Lipase from Candida rugosa, Rhizopus arrhizus, and Wheat Germ was purchased from Sigma-Aldrich (St. Louis, Mo, USA). The organic phase was composed of 2,2,4-tri methylpentane (isooctane) as the main solvent and 1-butanol as a co-solvent. In this study, a volumetric ratio of 3.5(v/v)% of 1-butanol was employed as an organic phase. Triolein was employed as a typical lipid substrate. Therefore, oleic acid was assumed to be the major product. The reactivity of immobilized lipase was determined from the initial production rate of oleic acid.

Sugar ester was used as an amphiphilic component. It was necessary to form a W/O microemulsion phase in which the reaction was quickly initiated. DK-ester-F-110 (average MW = 703, Lot. No. 349519) was provided by Dai-ichi Kogyo Seiyaku, Ltd. (Kyoto, Japan). The hydrophilic-lipophilic balance (HLB) of DK-ester-F-110 was reported as 11, and was thus hydrophilic [13, 14]. The guaranteed main component was monostearic acid ester, with a minimum assay of 33.4%.

Physicochemical Character of Hydrophobic Porous Carrier

Molecular structure of immobilized enzyme was directly influenced by physicochemical factors in reaction media, opening of pore, geometrical structure of pore channel, surface hydrophobicity of carrier and hydrophobicity of organic solvent. Surface hydrophobicity of Accurel was conveniently examined by adsorption of fluorescence probe Nile Red. Accurel was immersed in Nile Red ethanol solution (100μM) for 1 hour. Fluorescence image was measured by the fluorescent microscope (KEYENCE BZ-9000) using excitation wavelength 560 nm.

Scanning electron microscopy (SEM) was employed to examine Accurel particle morphology. Prior to analysis, all samples were coated with a thin layer of Pt using a sputter coater (HITACHI E-1010 ION SPUTTER (Tokyo, Japan)). They were then analyzed by SEM (HITACHI Miniscope TM-1000 (Tokyo, Japan)).

Lipase Immobilization Method

First, Accurel MP100 was immersed and shaken in ethanol (99.5%) for 2 hours as pretreatment. This process led to hydrophilic surface condition in hydrophobic porous supports. Lipase was adsorbed onto the pore surface of Accurel particles before cross-linked immobilization. The desired amount of lipase powder was added to the acetate buffer solution (pH 6.5). The solution was centrifuged at 3000min^{-1} for 5min to remove insoluble substances. The Accurel particles were placed into the lipase buffer solution and then shaken to adsorb lipase for at least 24h at 298K. The change of aqueous phase concentration leveled off after 24h of shaking, and then adsorption equilibrium was attained. The unwashed wetted Accurel was dried moderately in desiccators for two days at room temperature. The adsorbed amount of lipase was evaluated from the decreasing lipase concentration of the buffer solution.

Next, the dried Accurel carrier with the adsorbed lipase was shaken in 3% glutaraldehyde (GA) aqueous solution for 60min at 298K to immobilize the lipase on the Accurel pore surfaces. The desorbed amount of lipase in the GA solution was then measured. The effective amount of immobilized lipase was estimated from the difference between the adsorbed amount of lipase and the desorbed amount of lipase. The immobilized lipase carrier was briefly washed in pure water and dried moderately in the desiccators for two days at room temperature.

HYDROLYSIS OF LIPID USING IMMOBILIZED LIPASE IN ACCUREL PARTICLES

Sugar ester at a concentration of 20g-(L-org)$^{-1}$ was quickly dissolved into the organic phase composed of 1-butanal/isooctane at 313K. 50mL of the organic phase was placed in a glass vessel, after which the desired amount of acetic acid buffer aqueous solution was added. The water content (W) was defined as the molar ratio of water to the amphiphilic component. Previously, the reactivity of lipase in the W/O microemulsion phase strongly depended on the value of W. The authors adopted an optimal the value of W of 3.0 based on our previous data [14]. The amphiphilic concentration for W was based on the monostearic acid ester content (33.4%) in DK-ester-F-110. Triolein with a concentration of 16mM was also added to the organic phase. The aqueous buffer solution and the organic phase containing triolein were stirred magnetically for 10min to produce an optically transparent and mature W/O microemulsion phase. The reaction was initiated when the immobilized lipase on Accurel particles was added to the W/O microemulsion phase. The concentration of produced fatty acid (oleic acid) was determined by the Lowry-Tinsley colorimetric method [19]. The calibration line was based on oleic acid.

RESULT AND DISCUSSION

Physicochemical Character of Hydrophobic Porous Carrier

Fluorescence image by Nile Red was presented in Fig.1. Fluorescence excited by 560 nm was strongly appeared from entire surface of Accurel. Hydrophobic character of polymer frame of Accurel was discernible.

Figure 1: Fluorescence microscopic view of original Accurel (×40). The fluorescence microscopic view depicts that the polymer frame of Accurel was entirely hydrophobic.

Fig.2 depicts SEM image of Accurel. Circular like pores were opened on entire surface of Accurel. The inlet diameter of pores was from 1μm-3μm. Mean void fraction of Accurel was 0.78.

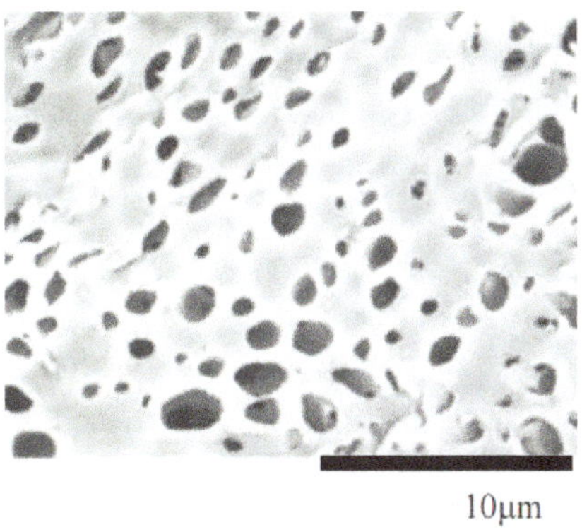

Figure 2: SEM image of original Accurel surface. The indicator bar means 10μm.

Amount of Immobilized Lipase

Pretreatment of carrier by immersion in ethanol was an important process to modify of porous surface of Accurel. As presented in Fig.3, effect of pretreatment evaluated from the amount of adsorbed Candida rugosa lipase per unit mass of Accurel.

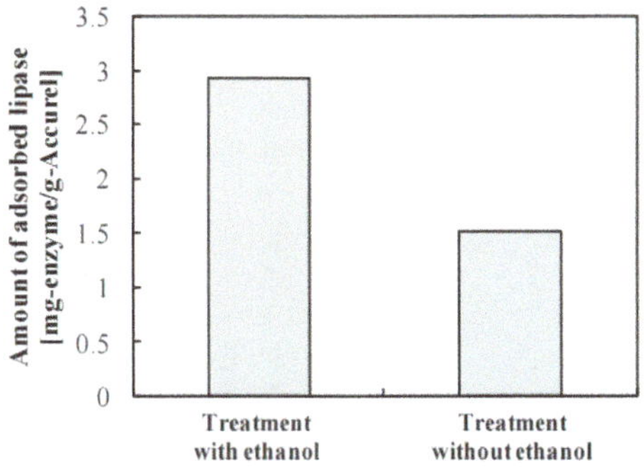

Figure 3: Amount of adsorbed Candida rugosa lipase per unit mass of original Accurel. With pretreatment by ethanol, lipase was successfully adsorbed in hydrophile phase in spite of hydrophobic material using.

The amount of adsorbed lipase onto pretreated by ethanol was much 2-fold than that of without pretreatment. With pretreatment by ethanol, lipase was successfully adsorbed on hydrophilic modified surface in spite of hydrophobic material used.

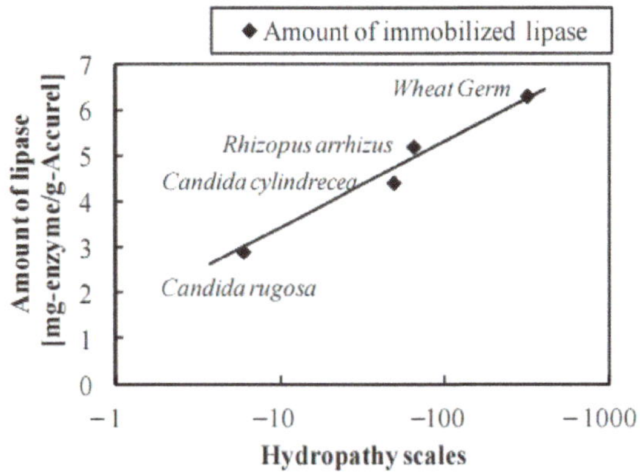

Figure 4: Comparison of amount of lipase from different species, indicate the hydropathy scales course. Hydropathy scales are enzyme hydrophobic scale by Kyte and Doolittle's evaluation [20], the amount of immobilized lipase was depended on lipase species. In 4 species, Wheat Germ lipase immobilized more the amount of lipase immobilization. High hydrophile lipase was also successfully immobilized in hydrophobic material

Cross-linking of lipase by glutaraldehyde (GA) was very beneficial for attaining a high yield of immobilization. The immobilized lipase yield based on the adsorbed amount was exceeding 98% in every class of particle size. The stabilization by GA treatment depended on the materials of the support and enzyme species. Fig.4 depicts the immobilized amount of enzyme was well correlated to hydrophilic character of enzyme.

Hydropathy scale of enzyme was accorded to Kyte and Doolittle's evaluation [20]. The amount of immobilized lipase was linearly depended on lipase species. Among tested lipase, the Wheat Germ lipase was favorably immobilized on ethanol pretreated Accurel pores. Even higher hydrophile lipase was immobilized successfully in hydrophobic material.

Reactivity in W/O microemulsion

As basically demonstration, the comparison of free lipase reactivity from various species in W/O microemulsion system was indicated in Fig.5. The water content (W) was major factor to determine reactivity of lipase in W/O microemulsion system. In this paper, it was set at 3.0 referenced from our previous study [14]. The reaction activity of the Rhizopus arrhizus lipase was more remarkable even if it has relatively hydrophilic character. No lag time was appeared in initial reaction range. According to this result, Rhizopus arrhizus and Candida rugosa lipases were mainly investigated practically.

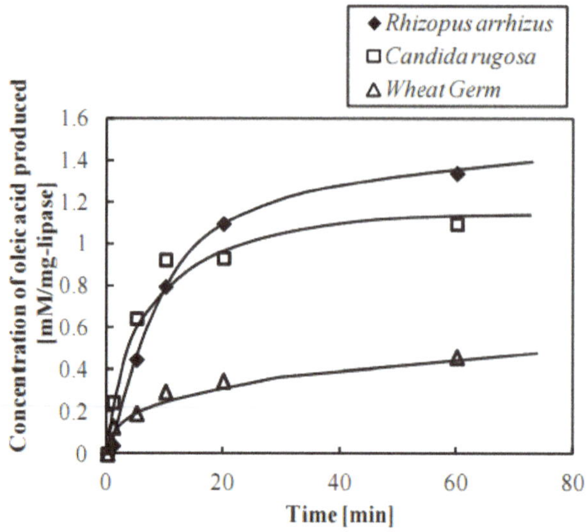

Figure 5: Comparison of free lipase reactivity from various species in W/O microemulsion system, indicate the time course of oleic acid with different specificity. The reaction activity of the Rhizopus arrhizus lipase was remarkable.

Hydrolysis of Triolein by Immobilized Lipase

Fig.6 (a) presents reactivity of free lipase (Rhizopus arrhizus) and immobilized one. Fig.6 (b) also presents reactivity of Candida rugosa lipase.

Hydrolysis reaction was quickly initiated by using immobilized lipase while no effect of diffusion limiting condition. Reaction productivity of Immobilized lipase successfully kept over 50% that of W/O microemulsion used. This suggested that diffusion of substrate and product in pore was remarkably improved by using hydrophobic carrier. The reactivity of the immobilized Rhizopus arrhizus lipase was superior to Candida rugosa lipase system.

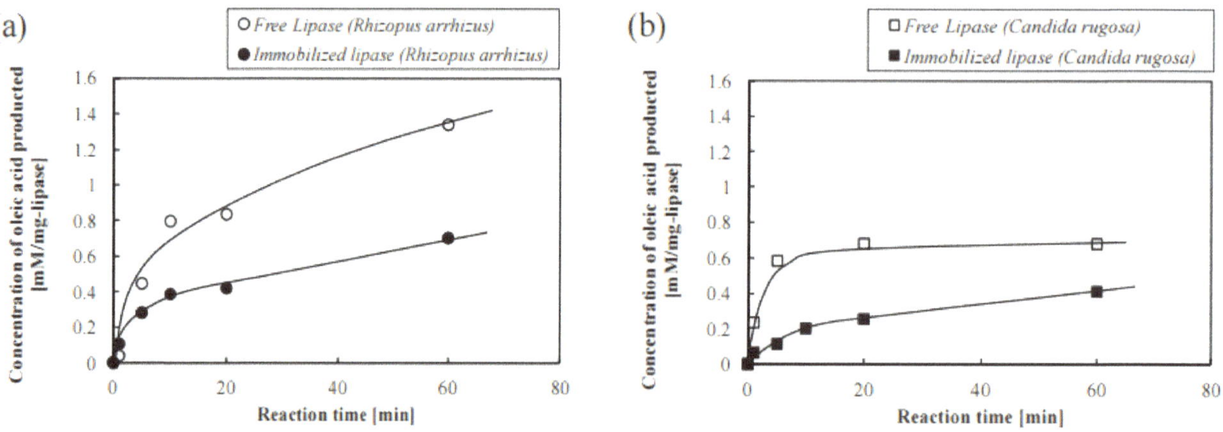

Figure 6: (a) Reactivity of free lipase (Rhizopus arrhizus) and immobilized one. Immobilized lipase productivity successfully kept over 50% of W/O microemulsion. Reaction was quickly initiated by using immobilized lipase; (b) Reactivity of free lipase (Candida rugosa) and immobilized one.

The theoretical effectiveness factor ηth is calculated from the Thiele modulus presented by Fig.7. The effectiveness factor of Rhizopus arrhizus using original particle size of Accurel attractively approached to rate-limiting condition, not limited by diffusion processes. It was superior to that of Candida rugosa Lipase (about 1.3 times).

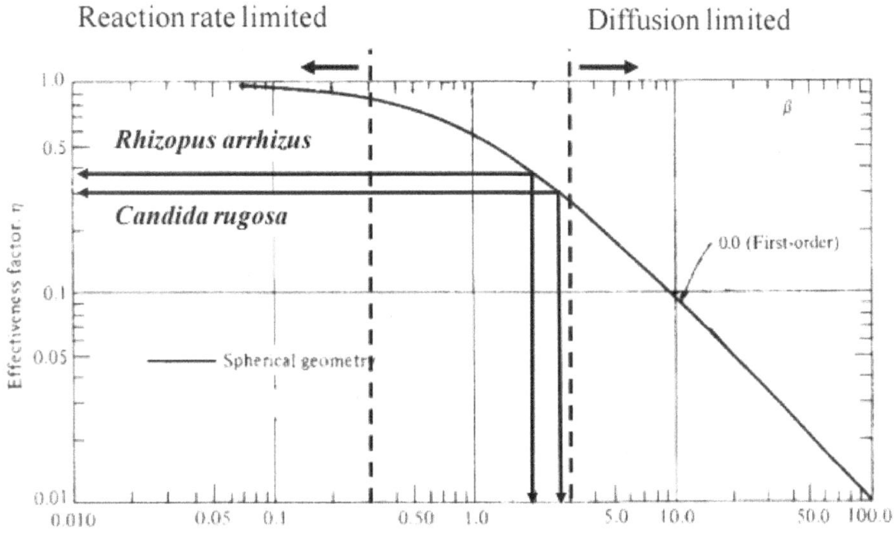

Figure 7: Relationship between effectiveness factor and Thiele modulus. The effectiveness factor of Rhizopus arrhizus lipase was superior to that of Candida rugosa lipase.

Figure 8 illustrated the effect of particle size of Accurel on the effectiveness factor is the case of Candida rugosa lipase. The effectiveness factor of smaller Accurel particles (500-840 μm) exceeded 0.5, which expected to approach for rate-limiting condition. In the case of Candida rugosa used system, smaller carrier size was favorable to obtain higher reaction rate and reaction yield. Suitable carrier size to realize sufficient high reaction rate was practically depended on lipase species used.

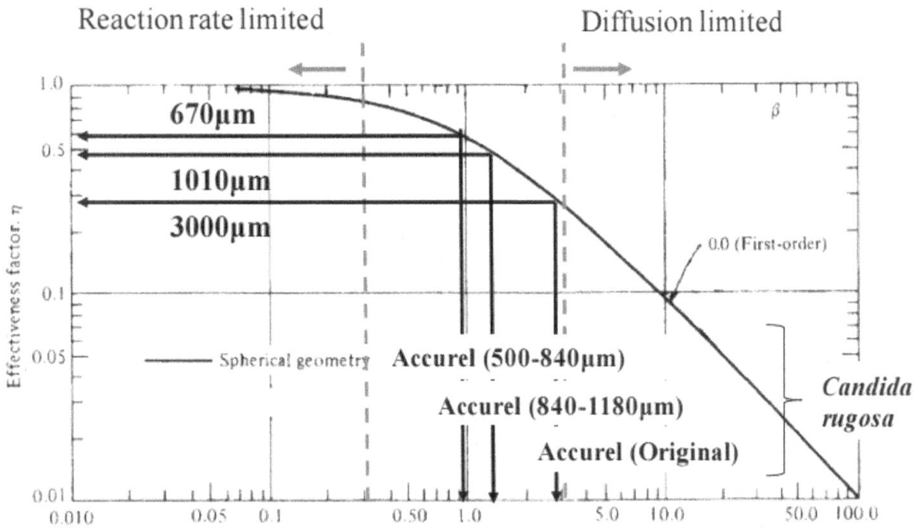

Figure 8: Relationship between effectiveness factor and Thiele modulus (Candida rugosa). The effectiveness factor of smaller Accurel particles (500-840 μm) approached to rate-limiting condition, not to diffusion limitation. Smaller size Accurel particles are favorable to obtain higher reaction rate and reaction yield.

Figure 9 (a) indicates the effect of Accurel particle size on the apparent reaction rate in the case of Candida rugosa used. The ordinate represents the concentration of produced oleic acid per unit mass of immobilized lipase. The 500 to 840μm particles evidently performed outstanding reactivity compared to that of 840 to 1180μm particles and original Accurel. Figure 9 (b) compared the production of oleic acid. The reaction yield was higher when smaller particles were used. The production of oleic acid strongly depended on the particle size, suggesting that the reaction occurred under diffusion-limiting conditions.

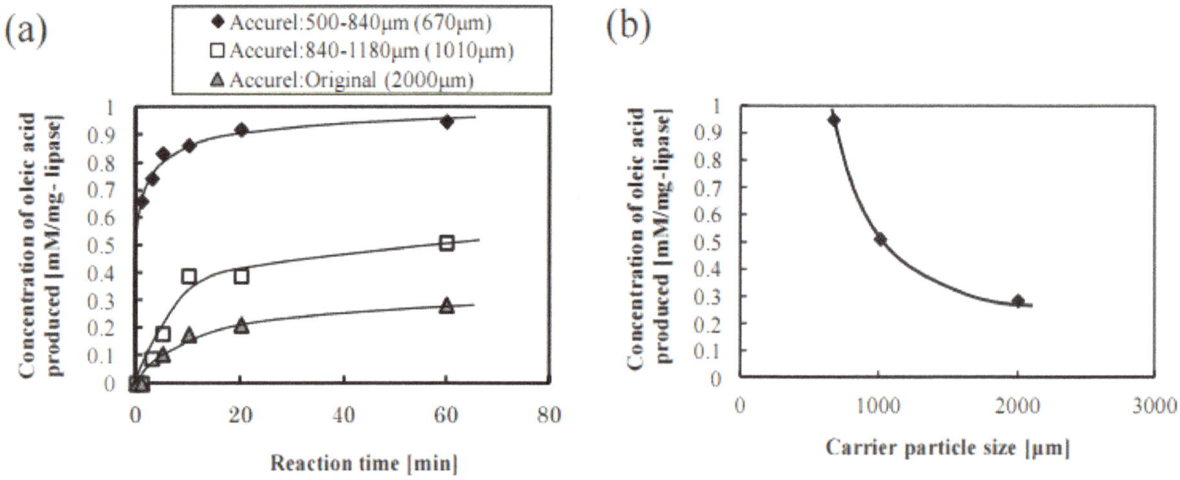

Figure 9: (a) Comparison of reactivity of immobilized Candida rugosa lipase for various particle sizes of Accurel [18]. The amount of immobilized lipase was normalized by the unit mass of immobilized lipase. The productivity depended remarkably on particle size. The 500-840μm particles evidently performed outstanding reactivity; (b) Comparison of reactivity of immobilized lipase for various particle sizes of Accurel, indicates the immobilized carrier particle size course of oleic acid produced. The amount of immobilized lipase was normalized by the unit mass of immobilized lipase.

Reusability

Figure 10 (a) and (b) presented the comparison of the reaction yields of hydrolysis by reusing immobilized lipase (a; Rhizopus arrhizus, b; Candida rugosa). To examine the stability of repeated use of immobilized lipase, the immobilized lipase was separated from the bulk reaction phase at 60 min after the reaction initiation, and then it was washed by organic phase. Immobilized lipase was added again to the next batch reaction with fresh substrates.

Productivity of oleic acid by immobilized Rhizopus arrhizus lipase at the first run was remained almost 50%. Even if the case of Candida rugosa lipase, it was still over 30% remained. The stability of production of oleic acid was attractively stable in repeated use (3 times). The accumulated amount of oleic acid by three times repeated use was superior to the once use of free lipase in W/O microemulsion phase. Economical merit of lipase immobilization was anticipated by employing hydrophobic enzyme carrier.

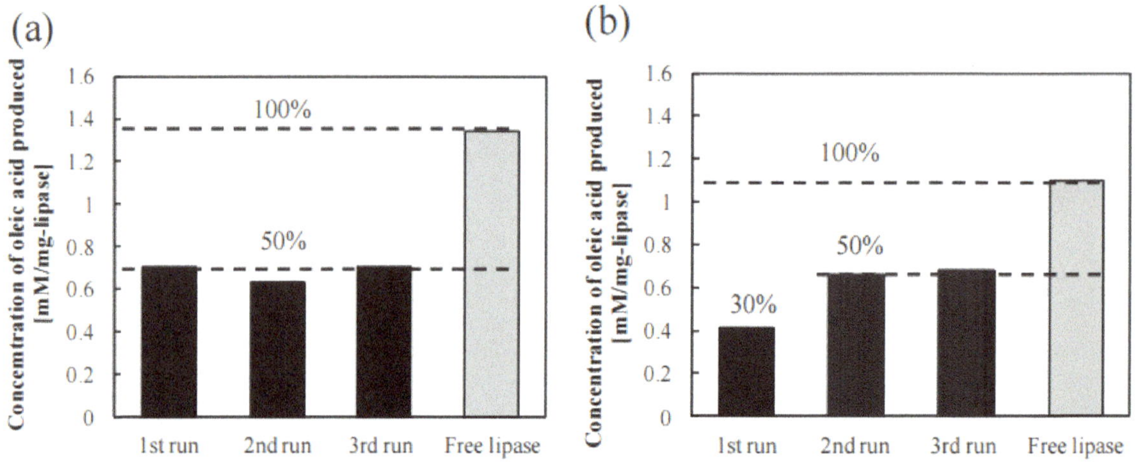

Figure 10: (a) The comparison of the reaction yields of hydrolysis by using immobilized lipase (Rhizopus arrhizus lipase). Reactivity of immobilized lipase is keeping constant level; (b) The comparison of the reaction yields of hydrolysis by using immobilized lipase (Candida rugosa lipase). Reactivity of immobilized lipase is stable for repeated use (3 times).

CONCLUSION

Impact of physicochemical character of hydrophobic porous carrier on reactivity of immobilized lipase from Candida rugosa, Rhizopus arrhizus, and Wheat Germ was investigated by Accurel MP100 carrier employed.

Accurel MP100 was made of a polypropylene-based hydrophobic porous support and provided circular diameter of inlet pores 1-3μm. Lipase was adsorbed on Accurel pores and immobilized by cross-linking agent glutaraldehyde. Immobilized yield attended satisfactorily high level (98%). Productivity of hydrolysis reaction using immobilized lipase was 50% compared with that of free lipase in W/O microemulsion system. Effectiveness factor of immobilized lipase was improved to use reduced size

Accurel particles. In the case of Candida rugosa used system, Accurel particle size 500-840 μm was expected to approach for rate-limiting condition. The effectiveness factor exceeded 0.5. Suitable carrier size to realize sufficient high reaction rate was practically depended on the lipase species used. Repeated use of immobilized lipase was attractively performed and stable. The productivity was remained almost 50%. Accurel MP100 was a potential immobilized carrier for lipase progressing toward higher reaction rate and high yield in repeated use.

ACKNOWLEDGEMENTS

DK-ester-F-110 was donated by Daiichi Kogyo Seiyaku Ltd. (Tokyo, Japan). Accurel was provided by Membrana GmbH Accurel Systems (Obernburg, Germany). Dr. Kei Tao of the Nihon University provided technical assistance on SEM photograph. The authors sincerely appreciate their kind cooperation.

REFERENCES

1. Cao L, Bornscheuer UT, Schmid RD. Lipase-catalyzed solid-phase synthesis of sugar esters. Influence of immobilization on productivity and stability of the enzyme. J Mol Catal B: Enzymatic 1999;6:279-285.
2. Kiatsimkul PP, Sutterlin WR, Suppes GJ. Selective hydrolysis of epoxidized soybean oil by commercially available lipases: Effects of epoxy group on the enzymatic hydrolysis. J Mol Catal B: Enzymatic 2006;41:55-60.
3. Virto DM, Agud I., Montero S, Blanco A, Solozabal R, Lascaray JM et al. Hydrolysis of animal fats by immobilized Candida rugosa lipase. Enzyme Microb Tech 1994;16:61-65.
4. Hita E, Robles A, Camacho B, Gonzalez PA, Esteban L, Jimenez MJ et al. Production of structured triacylglycerols by acidolysis catalyzed by lipases immobilized in a packed bed reactor. Biochem Eng J 2009;46:257-264.
5. Jimenez MJ, Esteban L, Robles A, Hita E, Gonzalez PA, Munio MM et al. Production of triacylglycerols rich in palmitic acid at sn-2 position by lipase-catalyzed acidolysis. Biochem Eng J 2010;51:172-179.
6. Pilarek M, Szewczyk KW. Kinetic model of 1,3-specific triacylglycerols alcoholysis catalyzed by lipases. J Biotechnol 2007;127:736-744.
7. Salis A, Sanjust E, Solinas V, Monduzzi M. Characterisation of Accurel MP1004 polypropylene powder and its use as a support for lipase immobilisation. J Mol Catal B: Enzymatic 2003;24-25:75-82.
8. Watanabe Y, Shimada Y, Sugihara A, Tominaga Y. Conversion of degummed soybean oil to biodiesel fuel with immobilized Candida antarctica lipase. J Mol Catal B: Enzymatic 2002;17:151-155.
9. Xie W, Ma N. Enzymatic transesterification of soybean oil by using immobilized lipase on magnetic nano-particles. Biomass Bioenergy 2010;34:890-896.
10. Salis A, Svensson I, Monduzzi M, Solinas V, Adlercreutz P. The atypical lipase B from Candida antarctica is better adapted for organic media than the typical lipase from Thermomyces lanuginose. Biochim Biophys Acta 2003;1646:145-151.
11. Palomo JM, Fernandez-Lorente G, Mateo C, Ortiz C, Fernandez-Lafuente R, Guisan JM. Modulation of the enantioselectivity of lipases via controlled immobilization and medium engineering: hydrolytic resolution of mandelic acid esters. Enzyme Microb Tech 2002;31:775-783.
12. Persson M, Mladenoska I, Wehtje E, Adlercreutz P. Preparation of lipases for use in organic solvents. Enzyme Microb Tech 2002;31:833-841.
13. Naoe K, Ohsa T, Kawagoe M, Imai M. Esterification by Rhizopus delemar lipase in organic solvent using sugar ester reverse micelles. Biochem Eng J 2001;9:67-72.
14. Uehara A, Imai M, Suzuki I. The most favorable condition for lipid hydrolysis by Rhizopus delemar lipase in combination with a sugar-ester and alcohol W/O microemulsion system. Coll Surf A: Physicochem Eng Asp 2008;324:79-85.
15. Dizge N, Aydiner C, Imer DY, Bayramoglu M, Tanriseven A, Keskinler B. Biodiesel production from sunflower, soybean, and waste cooking oils by transesterification using lipase immobilized onto a novel microporous polymer. Biores Tech 2009;100:1983-1991.
16. Sabbani S, Hedenstrom E, Nordin O. The enantioselectivity of Candida rugosa lipase is influenced by the particle size of the immobilising support material Accurel. J Mol Cat B: Enzymatic 2006;42:1-9.
17. Zhou G, Chen Y, Yang S. Comparative studies on catalytic properties of immobilized Candida rugosa lipase in ordered mesoporous rod-like silica and vesicle-like silica. Micropor Mesopor Mat 2009;119:223-229.
18. Naya M, Imai M. Regulation of the hydrolysis reactivity of immobilized Candida rugosa lipase with the aid of a hydrophobic porous carrier. Asia-Pacific J Chem Eng 2012; 7, Issue S1:S157-S165.
19. Lowry RR, Tinsley IJ. Rapid colorimetric determination of free fatty acids. J Am Oil Chemists' Soc 1976;53:470-472.
20. Kyte J and Doolittle R. A simple method for displaying the hydropathic character of a protein. J Mol Biol 1982;157:105- 132.

CHAPTER 17

EVEALUATION OF INORGANIC MATRIXES AS SUPPORTS FOR IMMOBILIZATION OF MICROBIAL LIPASE

H.F. Castro, M.L.C.P. Silva, and G.L.J.P Silva

Department of Chemical Engineering, School of Chemical Engineering of Lorena,

ABSTRACT

Candida rugosa **was immobilized by physical adsorption on several inorganic supports using hexane as coupling medium. The enzymatic activities of the different derivatives were determined by both hydrolysis of olive oil and esterification of n-butanol with butyric acid. The results were compared to previous data obtained by using a controlled porous silica matrix. The goal was to contribute in searching inexpensive supports for optimum lipase performance. All supports examined exhibited good properties for binding the enzyme lipase. Zirconium phosphate was the best support, giving the highest percentage of protein fixation (86%) and the highest retention of lipase activity after immobilization (34%). The operational stability performance for niobium oxide derivative was improved by previously activated the support with silane and glutaraldehyde. Thermal stabilities were also examined by thermal gravimetric analysis (TG).**

INTRODUCTION

Lipases (EC 3.11.3) are versatile catalysts with potential applications in a number of industrial processes. Important uses include their addition to detergents, the manufacture of food ingredients, pitch control in the pulp industry, and biocatalyst of stereoselective transformations (de Castro and Anderson, 1995; Ghandi, 1997; Jarger and Reetz, 1998).

In view of the current high costs of lipases, the possibility of regenerating and reusing the enzyme would be an attractive feature of biocatalysis. Immobilization may protect enzyme to some extent from solvent denaturation. It helps in maintaining homogeneity of enzymes in the reaction media since it avoids aggregation of enzyme particles. In addition to the ease of handling, immobilized enzymes are well suited for the use in continuous packed-bed reactors (de Castro and Anderson, 1995; Balcão *et al.*, 1996).

Support materials with high mechanical strength are desirable particularly in stirred systems. The presence of solvents may necessitate support with high chemical resistance. Various microorganisms grow readily on many substrates catalyzed by lipases which call for microbially resistant supports. The supports should allow the effective utilization of

Citation: H.F. Castro, M.L.C.P. Silva, and G.L.J.P Silva, Evaluation of Inorganic Matrixes as Supports for Immobilization of Microbial Lipase, http://dx.doi.org/10.1590/S0104-66322000000400048

the enzyme by having the enzyme molecule accessible to the substrates. Since the loss of activity is a certainty over time, the support material should be easy to be regenerated with active enzymes (Yahya *et al.*, 1998).

Lipase has been immobilized by several methods, namely adsorption, crosslinking, adsorption followed by crosslinking, covalent attachment and physical entrapment (Balcão *et al.*, 1996). However, the insoluble carriers used are almost always polymeric resins, natural polymeric derivatives, organogels and fibers with limited capacity to reuse the immobilization matrix and therefore creating disposal problems of organic materials. Inorganic carriers are more expensive than their organic commercial counterparts but have the advantage of being reusable which in some circumstances can decrease the immobilization support costs (Fonseca *et al.*, 1993).

Among inorganic supports, silica based carriers and its derivatives with well defined pore size are popular matrixes for immobilization of several enzymes (Trevisan and Mei, 1992; Ison *et al.*, 1994; Zanin *et al.*, 1996; Baron *et al.*, 1996). This support has also been used successfully for the immobilization of microbial lipase (Soares *et al.*, 1999), however its high cost (US$ 67/gram, Sigma) may impose some economical limitations if scale-up is envisaged.

The search for an inexpensive support has motivated our group to undertake this work dealing with the selection of potential inorganic matrices which can be recognized by lipases, at molecular level, as solid surfaces. We investigated the feasibility of using metal phosphates and oxides (zirconium, thorium, tungsten, and niobium) which have important ion exchange properties and showed to be effective as an adsorbent for cations from dilute aqueous or non-aqueous solutions and color removal from industrial wastes (Clearfield, 1982; Clearfield and Thakur, 1980; Thakur and Clearfield, 1981; Watanabe *et al.*, 1974). In this preliminary study, amorphous zirconium phosphate and niobium oxide (amorphous and crystalline) synthesized according to the methodology previously established in our Institution (Medeiros *et al.*, 1996; Serafim *et al.*, 1996) were evaluated as supports for immobilizing *Candida rugosa* lipase.

MATERIALS AND METHODS

Materials

Commercial *Candida rugosa* lipase (Type VII) and bovine serum albumin (BSA) were purchased from Sigma Chemical Co. (St Louis, MO, United States). The lipase was a crude preparation with a nominal specific activity of 835 units. mg^{-1} solid and 10% protein based on the Lowry method protein assay (Lowry *et al.*, 1954). Low acidity olive oil (Carbonel) was purchased at a local market. Substrates for esterification reactions were dehydrated, with 0.32 cm molecular sieves (aluminum sodium silicate, type 13 X-BHD Chemicals, Toronto, Canada), previously activated in an oven at 350°C for 6 hours. Solvents were standard laboratory grade. Alcohol, organic acid and other reagents were purchased either from Aldrich Chemical Co. (Milwaukee, WI, USA) or Sigma Chemical Co. (St Louis, MO, USA).

Supports

Crystalline hydrous niobium oxide (CNO), amorphous hydrous niobium oxide (ANO) and amorphous zirconium phosphate (ZrP) were prepared following methodology previously established (Medeiros *et al.*, 1996, Serafim *et al.*, 1996). Details of the characterization procedures for the textural and ion exchange properties of the solids have been published elsewhere (Medeiros *et al.*, 1996, Serafim *et al.*, 1996).

Crystalline hydrous niobium oxide was obtained by precipitation from the corresponding niobium metallic in an acidic mixture solution (HF/HNO_3) followed by addition of ammonium hydroxide. Amorphous hydrous niobium oxide was prepared by alkaline fusion with K_2CO_3 at 1000°C for 6 hours, followed by addition of hot water and nitric acid 1.0 M. Amorphous zirconium phosphate (ZrP) was obtained by precipitation from the corresponding $ZrOCl_2.H_2O$ in an H_3PO_4 aqueous solution followed by addition of ammonium hydroxide. All supports were dried at 100°C and the main particle size was between 75 and 180 mm. The surface areas, S_{BET} in $m^2.g^{-1}$, determined by nitrogen adsorption from BET method were 18, 110 and 234 $m^2.g^{-1}$ for CNO, ANO and ZrP, respectively.

Immobilization Procedure

Lipase was immobilized by physical adsorption on the selected supports according to the procedure previously described for Celite (de Castro *et al.*, 1999). Alternatively, lipase was also covalently bound on crystalline hydrous niobium oxide (CNO) previously treated with g-aminopropyltriethoxysilane (g-APTS), followed by reaction of the pretreated beads with glutaraldehyde solution, according to the procedure described for controlled pore silica (Soares *et al.*, 1999). For each gram of dry support, 0.25 grams of lipase was used. The enzyme was dissolved in 10 ml of 100 mM of sodium phosphate buffer (pH 7.0) and mixed with the support under low stirring during 2 hours at room temperature. After this, 10 ml of hexane was added to the mixture enzyme-support and coupling took place overnight at 4°C. The derivative was filtered (Whatman filter paper 41) and thoroughly rinsed with hexane. Analyses of the hydrolytic activities carried out on initial and spent lipase solutions and immobilized preparations were also used to determine the coupling yield (h%) according to the following expression:

$$\eta\% = \frac{\text{Overall activity of the immobilized enzyme}}{\text{Overall activity of the initial enzyme solution}} \times 100 \qquad (1)$$

Protein Assay

Protein was determined according to Lowry *et al.*(1954) using bovine serum albumin (BSA) as a standard. The amount of bound protein was determined indirectly from the difference between the amount of protein introduced into the coupling reaction mixture and the amount of protein in the filtrate and in the washing solutions.

Moisture Content

The moisture content of the immobilized preparations was determined by Karl Fisher titration (Mettler DL18).

Hydrolysis Assay

Hydrolytic activities of free and immobilized lipase were assayed by the olive oil emulsion method according to the modification proposed by Soares *et al.* (1999). The substrate was prepared by mixing 50 ml of olive oil with 50 ml of gum arabic solution (7% w/v). The reaction mixture containing 5 ml of the emulsion, 2 ml of 100 mM sodium phosphate buffer (pH 7.00) and either free (1 ml, 5 $mg.ml^{-1}$) or immobilized (250 mg) lipase was incubated for 10 min at 37°C. The reaction was stopped by the addition of 10 ml of acetone-ethanol solution (1:1). The liberated fatty acids were titrated with 25 mM potassium hydroxide solution in the presence of phenolphthalein as an indicator. One unit (U) of enzyme activity was defined as the amount of enzyme which liberated 1 mmol of free fatty acid per min under the assay conditions.

Esterification Assay

Reaction systems consisted of heptane (20 ml), n-butanol (250 mM), butyric acid (300 mM) and immobilized lipase (1.0 gram, dry weight). The mixture was incubated at 37°C for 24 hours with continuous shaking at 150 rpm. The product formed was determined by gas chromatography using a 6 ft 5% DEGS on Chromosorb WHP, 80/10 mesh column (Hewlett Packard, Palo Alto, CA, USA) and hexanol as internal standard. Activity was expressed as mmol of butyl butyrate formed per minute per gram of dry support.

Operational Stability

The operational stability of the immobilized enzyme was assayed by using 1.0 gram of immobilized lipase in successive batches performed under the same conditions as described for esterification activity assay. Twenty-four hours after starting each batch, the immobilized lipase was removed from the reaction medium and rinsed with hexane, in order to extract any substrate or product eventually retained in the matrix. One hour later (length of time required for the solvent to evaporate) the immobilized derivative was introduced in a fresh medium.

Thermal Analysis

Thermal gravimetric analysis (TG-DTG) was performed in a Shimadzu thermogravimetric instrument, TGA-50 model. Samples weighting 10 mg were examined at heating rates of 10°C $.min^{-1}$ in a dry nitrogen flux from 30 to 600°C.

RESULTS AND DISCUSSION

Immobilization of Lipase onto Inorganic Materials by Adsorption

Three materials (see material and methods for abbreviations) were tested as support matrixes for the immobilization of *Candida rugosa* lipase by adsorption. In this method, the enzyme was bound to the support without activation using hexane as coupling medium. The results obtained are shown in Table 1.

The amount of protein adsorbed on each support was in the range of 16 to 21 mg/ gram support (dry weight) which gave 64 to 86% protein recovered. After filtration, all derivatives had similar moisture content (25%). The best result with respect to both protein (86%) and catalytic activity (34%) recovered after immobilization was obtained with zirconium phosphate (ZrP). Further information on the catalytic properties of the derivatives was obtained by performing esterification assays (last row in Table 1). These results indicated that ZrP had the most favourable configuration for immobilization of *Candida rugosa* lipase leading to preparations with high activities to perform both hydrolysis and esterification reactions.

Table 1: Adsorption recovery, moisture content, coupling yield and catalytic activities of candida rugose lipase immobilized onto different inorganic supports

Support	Adsorbed protein (mg/g)	Bound Protein (%)	Moisture content (%)	Coupling Yield (%)	Hydrolytic Activity ($\mu mol.mg^{-1}.min^{-1}$)	Synthetic Activity ($\mu mol .g^{-1}.min^{-1}$)
CNO	17.4	69.5	25.9	12.6	26.5	350
ANO	16.0	64.1	23.9	27.3	50.5	210
ZrP	21.6	86.4	24.6	34.2	31.0	389

Lipase loading: 244 $units.g^{-1}$ support
Protein loading: 25 mg. g^{-1} support
Esterification activities are expressed as µmol butyl butyrate per minute per gram dry support by following the rate of product formation from the reaction mixture containing butanol and butyric acid.

Recycle Potential of Adsorbed Lipase Preparations

The operational stability of the adsorbed lipase derivatives was determined during successive batch esterification reactions at 37°C (Table 2). Thus after the first assay (about 24 hours), each derivative was recovered, washed with hexane and reused with fresh substrate mixture at the same conditions as the first experiment. After an initial loss of activity which is probably due to inactivation lost or loss of adsorbed or only weakly stabilized lipase, the ZrP lipase derivative gave stable performance activity over 5 reaction cycles (half life of 136 hours). However, both niobium oxide (crystalline and amorphous) supports showed poor operational stability resulting in high activity lost (over 75%) after five recycles.

Table 2: Repeated use of immobilized lipase derivatives in batch reaction to form butyl butyrate

Support material	Esterification activity (µmol.$g^{-1}$$min^{-1}$)					half life (hours)
	Cycle 1	Cycle 2	Cycle 3	Cycle 4	Cycle 5	
CNO	354	258	190	102	50	43
ANO	157	114	80	51	42	39
ZrP	388	258	260	258	208	136

Physical adsorption method is relatively easy and simple for immobilization and leads to minimal structural change of the enzyme during the immobilization process. However, bonding forcing such as hydrogen bonds, Van der Walls forces, and/or hydrophobic interactions are not strong and cause desorption of enzymes from the support during repeated reactions. To overcome this, it is recommended to use a porous support material so that suitable amount of enzyme can be spread on a surface area without conformational changes. Therefore, the increased stability of immobilized lipase on ZrP might result from the improved enzyme retention in this support which has the highest surface area (234 m^2.g^{-1}). From these results, zirconium phosphate may be selected as the most favorable support matrix for immobilizing *Candida rugosa* lipase.

Immobilization by Covalent Binding

The direct coupling of lipase on niobium oxide crystalline (CNO) structure as first proposed gave the lowest operational stability (activity dropped 85% after five recycles). To overcome this limitation another experiment was performed in which lipase was immobilized according to the procedure described for controlled porous silica (Soares *et al.*, 1999). In this method the lipase was covalently bound on CNO previously activated with g-aminopropyltriethoxysilane (g-APTS), followed by transformation of the aminoalkyl groups attached to the niobium oxide surfaces into Schiff□s bases by treatment with glutaraldehyde.

Using such a method, both hydrolytic and synthetic activities were enhanced (Table 3) and even more important than that, higher operational stability was obtained (Figure 1). Such results were favorable comparable to those described by Soares *et al.* (1999). This suggested that silica matrix could be also replaced by CNO with great advantage in the process costs.

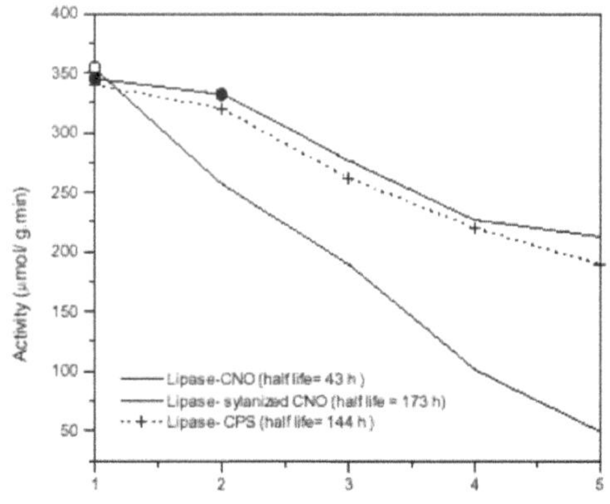

Figure 1: Batch operational stability test of immobilized lipase derivatives. Esterification assay was carried out with substrate containing 250 mM butanol and 300 mM butyric acid in heptane. Initial esterification activities were: 350 µmol.g^{-1}.min^{-1} for the lipase adsorbed onto CNO, 486 µmol.g^{-1}.min^{-1} for the lipase covalent binding onto sylanized CNO and 340 µmol.g^{-1}.min^{-1} for the lipase covalent binding onto sylanized controlled porous silica (CPS).

Table 3: Immobilization yield of lipase binding on inactivated and sylanized CM) compared to the immobilization yield reported for lipase binding on controlled porous silica

Support	Coupling yield (%)	Hydrolytic Activity (µmol.mg^{-1}.min^{-1})	Esterification Activity (µmol .g^{-1}.min^{-1})
CNO	12.6	26.5	350
Activated CNO	13.9	33.9	486
Activated CPS*	18.0	51.9	340

*data from Soares *et al.* (1999).

Thermal Stability

The thermal stability of enzymes is one of the important criteria for long-term and commercial application. The activity of immobilized enzyme is known to be more resistant against heat than that native state. Thermal gravimetric analysis (TG) provides an important tool for thermal stability studies of macromolecules (Airoldi and Oliveira, 1992; Aly, 1996). This technique has enabled us to determine the temperature range at which a heated sample undergoes a major conformational change by means of monitoring the thermal weight loss profile. In the case of free lipase and immobilized lipase derivatives such temperature range can be related to the protein unfolding and thus to the enzyme denaturation. Thermal degradation temperatures were determined for free lipase, all tested supports (crystalline hydrous niobium oxide (CNO); amorphous hydrous niobium oxide (ANO) and amorphous zirconium phosphate (ZrP) and the resulting immobilized lipase derivatives (lipase-CNO, lipase-ANO and lipase-ZrP). Thermal weight loss results are showed in Table 4 and examples of TG curves are given inFigures 2 to 4.

Table 4: Thermogravimetric analysis fur supports, free lipase and immobilized lipase derivatives

Material	Temperature range (°C)	Thermal degradation temperature* (°C)	Weight loss (%)
Support			
CNO	30-400	49	14.0
ANO	30-200 200-400	53	10.5 4.9
ZrP	30-400	111	11.0
Lipase preparation			
Free lipase	30-180	52	4.3
CNO-lipase	30-400	52	15.4
Sylanized CNO-lipase	30-200 200-400	54	13.6 5.0
ANO-lipase	30-400	50	13.0
ZrP-lipase	30-400	65	23.2

* Temperature at which a heated sample undergoes a major weight loss.

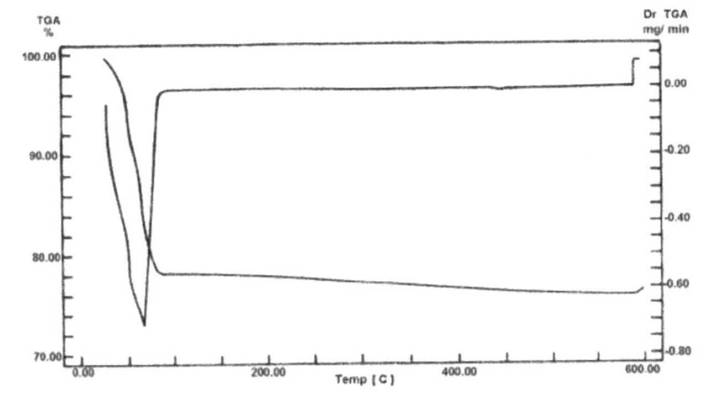

Figure 2: thermal analysis of free lipase

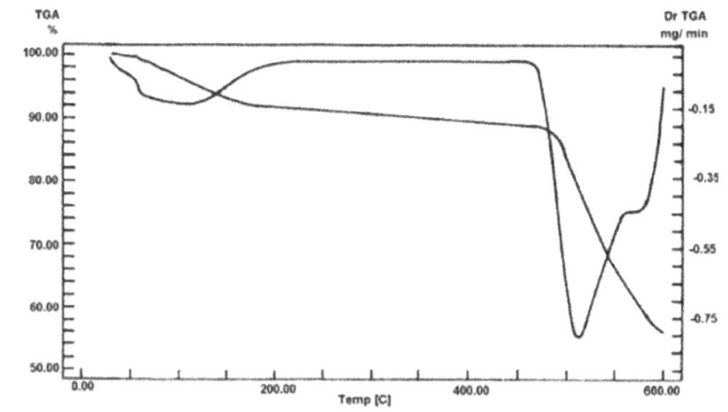

Figure 3: Thermal analysis of zirconium phosphate (ZrP) support

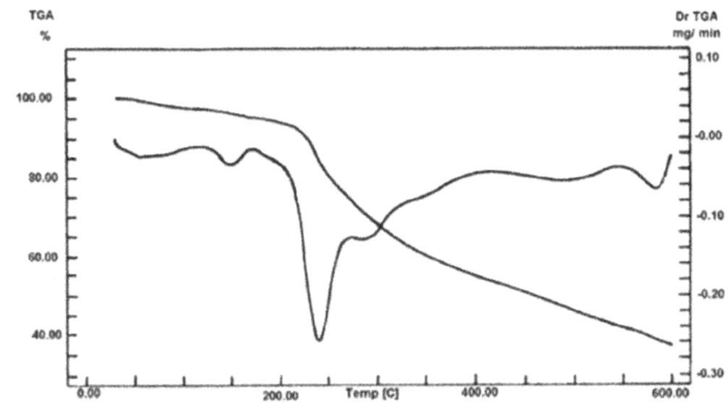

Figure 4: Thermal analysis of lipase immobilized on ZrP

Figure 2 shows the TG curve for free lipase which was characterized by two weight loss peaks. In the first one, at temperature range from 30 to 180°C, characterized by a low weight loss (4.3%) due to the dehydration of the interstitial water containing in the free lipase sample. From 180 to 600°C, continuous weight loss was observed indicating a complete decomposition of the organic structure of lipase.

The amorphous zirconium phosphate TG curve (Figure 3) shows only one weight loss peak (about 11%), at temperature from 30 to 400°C. This was also associated to the dehydration of the interstitial water. Over 500°C, complete condensation of the mono hydrogen phosphate groups occurred as expected for this material under such temperature (Airoldi and Oliveira, 1992). The thermogravimetric curve for ANO support (figure not shown) indicated a large weight lost (10.5%) at temperature lower than 200°C and a small one (4.9%) from 200 to 400°C. The former weight loss was due to the elimination of the free water and the latter to the water that it held in the support pores (Aly, 1996). The TG curve profile for CNO support showed distinct shape, only one step of 14% weight loss at temperature range of 30-400°C was detected also due to the elimination of interstitial water.

The TG curve for the lipase immobilized on ZrP (Figure 4) indicated that maximum rate of weight loss (22.0%) occurred in the range 30 to 100°C, probably due to the elimination of both hydrated water and organic structure of lipase. The thermogravimetric curve for the lipase immobilized on ANO (figure not shown) indicated weight loss of 13% in one step at temperature range of 30 to 400°C. The same behavior was observed for the lipase immobilized on CNO with a weight loss of 15.4% at the same temperature range. The thermogravimetric curve for the lipase immobilized on sylanized CNO showed weight loss in two steps. The first was characterized by a weight loss (13.6%) at 30 to 200°C due to the elimination of hydration water and the second step was characterized by 5% weight loss at 200 to 400°C corresponding to water loss that is held in the pores.

The results indicated that upon immobilization, the profile curves for all lipase derivatives shifted towards higher temperatures suggesting that a strong interaction between enzyme and all tested supports occurred which enhanced the conformation stability of the native form. Further investigation on the thermal stability of such derivatives is under progress.

CONCLUSIONS

Over the last 15 years significant advances have been made in the use of enzymes in non-conventional media. These advances are ushering in a new age in the application of enzymes to organic syntheses. Our efforts in this area have focused on the development of suitable support materials for biotransformations in non-conventional media. To understand the phenomena which lead to enzyme inactivation in these systems, we have turned to various thermodynamic approaches to characterize the supports and understand fundamentals aspects of enzyme stability. In this work we have demonstrated that cheap metal inorganic supports are biocompatible with lipases rendering immobilized derivatives with similar or better characteristics than those obtained with controlled porous silica.

ACKNOWLEDGMENTS

This work was supported by research grants from the Brazilian Research Council (CNPq) and FAPESP.

REFERENCES

1. Airoldi, C. and de Oliveira, S.F. Preparation, Properties and Ionic Exchange Behavior of Titanium (IV) Arsenophsphate, J. Braz. Chem. Soc., 3, No.1/2, 47-41 (1992).
2. Aly, H.M. The Al-13-Phosphatoantimonic Acid Synthesis and Ion Exchange Properties. Solvent Extraction and Ion Exchange, 14, No.5, 947-954 (1996).
3. Balcão, V.M.; Paiva, A.L and Malcata F.X. Bioreactors with Immobilized Lipases: State of the Art. Enzyme Microbial Technology, 18, 392- 416 (1996).

4. Baron, M.; Florencio, J.A.; Zanin, G.M.; Ferreira, A.G.; Ennes, R. and Fontana, J.D. Difructose Anhydride - forming Bacterial Inulinase II and Fructogenic Fungal Inulase I: Free and Immobilized Forms. Applied Biochemistry and Biotechnology, 57/58, 605-615 (1996).

5. Clearfield, A. Inorganic Ion Exchange Materials, Florida, USA, CRC Press, Inc.Boca Raton, 74 (1982).

6. Clearfield, A. and Thakur, D.S. The Acidity of Zirconium Phosphate in Relation to their Activity in the Dehydration of Cyclohexanol. Journal of Catalysis, 65,185-194 (1980).

7. de Castro, H.F. and Anderson, W.A. Fine Chemicals by Biotransformation using Lipases. Química Nova, 18, No. 5, 544-554 (1995).

8. de Castro, H.F.; Soares, C.M.F.; Oliveira, P.C. and Zanin, G.M. Immobilization of Porcine Pancreatic Lipase on Celite for Application in the Synthesis of Butyl Butyrate in Nonaqueous System. Journal of American Oil Chemist□s Society, 76, No.1, 147-152 (1999).

9. Fonseca, L.P.; Cardoso, J.P. and Cabral, J.M.S. Immobilization Studies of an Industrial Penicillin Acylase Preparation on a Silica Carrier. Journal of Chemical Technology and Biotechnology, 58, 27-37 (1993).

10. Gandhi, N.N. Application of Lipase. Journal of American Oil Chemist□s Society, 74, 621-634 (1997).

11. Ison, A.P.; Macrae, A.R. and Smith, C.G. Mass Transfer Effects in Solvent-Free Fat Interesterification Reactions: Influences on Catalyst Design. Biotechnology and Bioengineering, 43, 122-130 (1994).

12. Jarger, K.E. and Reetz, M.T. Microbial Lipases Form Versatile Tolls for Biotechnology. TIBTECH, 16, 396-403 (1998).

13. Lowry, O.H.; Rosebrough, N.J.; Farr, A.L. and Randall, R.J. Protein Measurement with the Folin Phenol Reagent. Journal of Biological Chemistry, 193, 263-275 (1954).

14. Medeiros, F.F.P; Nono, M.C. A; M.L.C.P.; Silva G.L.P. and Serafim, M.J.S. Obtenção e Caracterização de Fosfato de Zircônio Amorfo e Cristalino para uso como Trocador Iônico, Anais do 12° Congresso Brasileiro de Engenharia e Ciência dos Materiais, Águas de Lindóia-S.P, 500-503 (1996).

15. Serafim, M.J.S.; Nono, M.C. A.; Silva M.L.C.P. and Silva, G.L.P. Síntese e Caracterização do Óxido de Nióbio V Hidratado para Aplicação como Trocador Iônico, Anais do 12° Congresso Brasileiro de Engenharia e Ciência dos Materiais, Águas de Lindóia-S.P, 105-108 (1996).

16. Soares, C.M.F.; de Castro, H.F.; Zanin, G.M. and de Moraes, F.F. Characterization and Utilization of *Candida rugosa* Lipase Immobilized on Controlled Pore Silica. Applied Biochemistry and Biotechnology, 77/ 79, 745-756 (1999).

17. Thakur, D.S. and Clearfield, A. Cyclohexanol Dehydration over Zirconium Phosphates of varying Crystallinity, Journal of Catalysis, 69, 230-233 (1981).

18. Watanabe, Y.; Matsumura, Y.; Izumi, Y. and Mizutani, Y. Synthesis of Methyl Isobutyl Ketone from Acetone and Hydrogen Catalyzed by Paladium Supported on Zirconium Phosphate, Bulletin of Chemical Society Japan, 42, No. 2, 2922-2925 (1974).

19. Trevisan, H.C. and Mei, L.H.I. Supports for Immobilization. An. Acad. Bras. Ci., 64, No.2, 111-116 (1992).

20. Yahya, A.R.M.; Anderson, W.A. and Moo-Young, M. Ester Synthesis in Lipase-catalyzed Reactions. Enzyme and Microbial Technology, 23, 438-450 (1998).

21. Zanin, G.M. and de Moraes, F.F. Modeling Cassava Starch Saccharification with Amyloglucosidase. Applied Biochemistry and Biotechnology, 45/46, 627-640 (1996).

CHAPTER 18

GLUCOSE OXIDASE DIRECTLY IMMOBILIZED ONTO HIGHLY POROUS GOLD ELECTRODES FOR SENSING AND FUEL CELL APPLICATIONS

Hendrik du Toit and Mirella Di Lorenzo

University of Bath, Department of Chemical Engineering, Bath, BA2 7AY, UK

ABSTRACT

The successful implementation of redox-enzyme electrodes in biosensors and enzymatic biofuel cells has been the subject of extensive research.

For high sensitivity and high energy-conversion efficiency, the effective electron transfer at the protein-electrode interface has a key role. This is difficult to achieve in the case of glucose oxidase, due to the fact that for this enzyme the redox centre is buried inside the structure, far from any feasible electrode binding sites.

This study reports, a simple and rapid methodology for the direct immobilisation of glucose oxidase into highly porous gold electrodes. When the resulting electrode was tested as glucose sensor, a Michaelis-Menten kinetic trend was observed, with a detection limit of 25 μM. The bioelectrode sensitivity, calculated against the superficial surface area of the bioelectrode, was of 22.7 ± 0.1 μA mM^{-1} cm^{-2}.

This glucose oxidase electrode was also tested as an anode in a glucose/O_2 enzymatic biofuel cell, leading to a peak power density of 6 μW cm^{-2} at a potential of 0.2 V.

INTRODUCTION

The functional immobilisation of redox enzymes, such as laccase and glucose oxidase, onto electrode material surfaces is of keen interest for sensors and biofuel cells development.

Enzymes are the most common bioreceptor molecules used in biosensors due to their extremely high specificity that leads to minimal risk of false positive responses. The implementation of enzymes in biofuel cells allows for the development of membrane-less and compartment-less devices, which not only can be easily miniaturised, but can also be used in situations where it is not feasible to separate the fuel and oxidant [1].

For these applications, the achievement of efficient electron transfer between the enzyme active centre and the electrode is critical. Usually a mediated electron transfer (MET) mechanism is required. This might involve the use of

Citation: Hendrik du Toit, Mirella Di Lorenzo, Glucose Oxidase Directly Immobilized onto Highly Porous Gold Electrodes for Sensing and Fuel Cell applications, Electrochimica Acta, Volume 138, 20 August 2014, Pages 86-92, ISSN 0013-4686, http://dx.doi.org/10.1016/j.electacta.2014.06.074.

small redox active particles and polymers as electron carriers (mediators), such as organic dyes, ferrocene and its derivatives, modified vitamin complexes, and conducting salts [2]. If the mediator is in solution their diffusion to the electrode surface allows for a more rapid electron transfer compared to the direct transfer from the enzyme itself [3]. Alternatively, the mediators can be polymerised directly onto the electrode surface or co-immobilised with the reacting enzymes to further enhance the rate of electron transfer [4] and [5].

However, the use of redox active electron carriers can have several drawbacks, such as short lifespans, poor biocompatibility, risk of leaching away from the electrode surface, potential toxicity. Consequently, the achievement of a direct electron transfer (DET) process is preferred.

Major advancements have come as far back as the 1980s from examining transition metal rich enzymes, such as laccase (LAC), capable of catalysing the reduction of oxygen to water through DET [6]. This is because LAC contains several copper centres that allow the electrons transport through the enzyme redox centre to the electrode surface. Consequently, the achievement of DET with LAC-immobilised electrodes is now well established [7], [8] and [9].

In the case of glucose oxidase (GOx), DET is more difficult to achieve, due to the fact that the GOx redox centre is buried inside the enzyme structure, and is far from any feasible electrode binding sites. To achieve efficient electron transfer, the use of GOx has been often combined with mediator compounds, of which ferrocene is the most common [10]. Most of the GOx immobilisation protocols reported, while effective, are usually very expensive, due to the reagents required. These protocols are often very laborious, involving multi-steps in the immobilization procedure that can be sources of experimental errors [11], [12], [13] and [14]. Moreover the resulting bioelectrodes can be unstable and inefficient with limited opportunity for practical implementations, due to the leaching of the mediator implemented and the dependence of the electron transfer process on the capability of the mediator to be rapidly oxidised and reduced [15].

In recent years, some progress has been reported when using nanostructured electrodes, such as carbon nanotubes (CNTs), as electron acceptors. Due to their size and shape, these electrodes are able to intertwine with the enzyme and come in very close proximity to the enzyme FAD centre. Successful DET of GOx immobilised onto CNTs-based electrodes has been reported [16], [17] and [18]. High surface area microelectrodes are also preferred in the context of an increasing trend towards the miniaturization of bioelectrochemical devices for applications such as implantable healthcare devices. However, the not yet fully addressed potential long-term toxicity of CNTs leads to controversial opinions on the feasibility of using CNTs for *in vivo* or cutaneous applications [19], [20], [21], [22] and [23].

Nano-porous gold (nPG) electrodes (porous gold electrodes with a pore size distribution limited to the nanometre range) are considered a very promising alternative for the development of new generation bioelectrochemical devices with implantable capability. These non-toxic electrodes have remarkable properties, such as high conductivity, large surface area, three-dimensional open porosity, and biocompatibility [24]. In this context, an even more promising material for the production of high sensitivity biosensors is highly porous gold (hPG). While retaining the morphology observed with nPG, hPG electrodes present large micro-pores that are lined with nano-pores themselves [25]. As a consequence, hPG electrodes have a very wide pore size distribution, leading to extremely large surface areas and hence larger current densities in comparison to conventional nPG.

A new and rapid method of producing hPG electrodes by direct electrodeposition of porous gold films onto gold electrodes was recently reported [26]. These electrodes were characterised by a 3D foam-like structure, with a wide pore size distribution (ranging from 10 nm to 30 μm), and a roughness factor (calculated in terms of electrochemically effective surface area) approximately 10^3 times higher than polished gold. The hPG electrodes showed excellent glucose electrooxidation activity with a detection limit as low as 5 μM [26]. However, the high specificity required for some applications, such as implantable biofuel cell devices where the fuel (e.g. glucose) and the oxidant (e.g. oxygen) are fed to the system as a mixture, demands the implementation of enzymatic electrodes [27].

The large surface area of hPG electrodes and their complex morphology make them an ideal support for enzyme immobilisation at high loadings. This hypothesis is encouraged by the successful production of GOx-immobilised nPG electrodes recently reported. In these cases the GOx immobilisation protocols involved the nPG functionalization with thiol-linker molecules or with conductive polymers, such as poly(3,4-ethylenedioxythiophene), to enhance the electron transfer process [24], [28] and [29].

This study reports for the first time, an efficient, simple, cost-effective, and rapid method for the functional immobilisation of GOx onto hPG surfaces. The immobilisation protocol does not require any electrode pre-treatments with linker molecules or polymers and it is simple to reproduce. In particular, GOx is immobilised onto the hPG surface *via* a one-step electrochemical adsorption process in a phosphate buffer with no additional chemicals. The use of the resulting GOx-immobilised hPG electrode is tested for glucose sensing and for energy harvesting in a glucose/O_2 enzymatic biofuel cell.

EXPERIMENTAL

Materials

Glucose Oxidase (GOx) from *Aspergillus niger*, Laccase (LAC) from *Trametes versicolor*, and all other reagents used were of analytical grade and purchased from Sigma-Aldrich. All aqueous solutions were prepared with reverse osmosis purified water. Gold disk electrodes (2 mm diameter), saturated calomel electrodes (SCE) and platinum counter electrodes were purchased from IJCambria Ltd. Gold disk electrodes were polished between uses with a 0.3 micron aluminium oxide polish (Buehler).

All analytical experiments were performed in phosphate buffered saline (PBS). This was prepared with the following constituents: 137 mM NaCl, 2.7 mM KCl, 10 mM Na_2HPO_4, 2 mM KH_2PO_4. The pH of this solution was then adjusted to 7 with the drop wise addition of 1 M solutions of HCl and NaOH.

Deposition of highly Porous Gold (hPG) onto gold electrodes

The hPG was fabricated with a two-step potential process as previously described [26]. Briefly, the gold disk electrodes were immersed in an aqueous electrolyte consisting of 0.1 M $HAuCl_4$ and 1 M NH_4Cl. Gold was then potentiostatically deposited in two steps. Firstly, the system was set at a working potential of -0.7 V (vs. SCE) for 5 s. Afterwards, the potential was stepped down to a value of -4.0 V (vs. SCE) for 15 s.

Immobilisation of GOx onto hPG electrodes

GOx was electrochemically adsorbed onto the prepared hPG disk electrodes by conducting a total of 6 CV scans between 0.42 V and 0.60 V (vs. SCE) at a scan rate of 1 mV s^{-1}, in a PBS solution containing 0.45 mg ml^{-1} GOx (approximately 8 U ml^{-1} as per activity rating of manufacturer).

As a term of comparison of performance, GOx was also immobilised by absorption. In this case, the hPG electrodes were incubated with the GOx solution in PBS (0.45 mg ml^{-1}) for 1 hour at room temperature, without conducting any CV scans. In both cases, the GOx-hPG electrodes were then thoroughly rinsed three times with PBS to remove any weakly bonded enzyme, and stored in PBS at 4 °C until used.

The amount of GOx immobilised onto the hPG electrodes, was estimated by performing a kinetic assay (provided by Megazyme Ltd.) of the enzyme solution before and after the immobilisation procedure and assuming no enzyme losses during the process.

Immobilisation of LAC onto hPG electrodes

The surface of the hPG electrodes was modified with a layer of amino-phenyl groups through a simple two-stage process. Firstly, a layer of nitro-phenyl groups was attached by performing two reductive CV scans at 100 mV s^{-1} from 0.6 V to -0.6 V (vs. SCE), in the presence of an acetone electrolyte containing 2 mM p-nitrophenyldiazonium tetrafluoroborate and 100 mM tetrabutylammonium tetrafluoroborate. In the second stage, the nitro groups were exchanged for amino groups by conducting two reductive CV scans at 50 mV s^{-1} from 0.0 V to -1.4 V (vs. SCE), in an aqueous electrolyte containing 10% v/v EtOH and 0.1 M KCl. Afterwards, the electrodes were transferred to a PBS solution containing 80 U ml^{-1} of LAC. The electrodes were kept in this solution overnight at 4 °C in order to facilitate the covalent linkage between the amine functional groups on the gold, and the carboxylate groups on LAC. Finally, the electrodes were rinsed three times with PBS and stored in PBS at 4 °C until required.

Electrochemical Characterisation

All electrochemical processes were conducted using the Autolab PGSTAT128 N (Metrohm, UK) potentiostat. Cyclic voltammetry (CV) and amperometry tests were performed in a three-electrode electrochemical set-up with a SCE reference electrode and a platinum rod counter electrode. CV scans were performed at a scan rate of 1 mV s^{-1} in the potential range of 0.42 - 0.60 V (vs. SCE) in order to determine the optimum potential for the amperometric analysis (this range incorporates the normal oxidation potential of H_2O_2 on polished gold). CV analyses were repeated over a period of three days to determine the stability of the GOx-hPG electrodes.

The amperometry tests were performed at a fixed potential (0.52 V) during the step-wise increase of glucose concentration. After each concentration increase, the solution was agitated for 20 s and then left for 270 s to allow the amperometric response under convectional mixing to stabilise. The average amperometric response over a period of 10 s was then reported. Three replicates were performed per each experiment. In each case, a glucose concentration range of 5 μM - 50 mM was considered.

Biofuel Cell Setup

The biofuel cell consisted of a glass vial (28 ml in volume) containing a GOx_{abs}-hPG electrode and a LAC-hPG electrode as anode and cathode respectively. The electrode spacing was approximately of 0.5 cm. The system was fed with 12 ml of an aerated PBS solution containing 27.8 mM glucose at pH 7. The potential difference (ΔV) between the two electrodes was measured in open circuit mode until steady state conditions were achieved. Afterwards, the cell potential was measured under a range of different external resistances applied to the system.

RESULTS AND DISCUSSION

GOx-hPG electrodes

The purpose of this study was to investigate the possibility to develop a more rapid and reproducible immobilization protocol of GOx onto hPG electrodes with respect to those so far reported that involve the use of thiol groups or conductive polymers [24] and [28]. This was achieved by running a series of CV scans of a hPG electrode in a GOx buffer solution, with the aim of increasing enzyme loading and affinity through the use of electrostatic forces.

The performance of the resulting bioelectrode was compared with a hPG electrode with GOx immobilised by absorption, as described in Section 2.3. Fig. 1 compares the CV curves produced with the two GOx-hPG electrodes in the presence of glucose. In both cases, no mediators were added to the electrolyte solution. As shown, when the GOx-hPG electrode was prepared electrochemically (GOx_{ads}-hPG electrode) a more prominent response to glucose was observed compared to the electrode produced by absorption (GOx_{abs}-hPG electrode). The CV scan of the GOx_{ads}-hPG electrode shows in particular a shoulder peak between 0.5 V and 0.525 V vs. SCE.

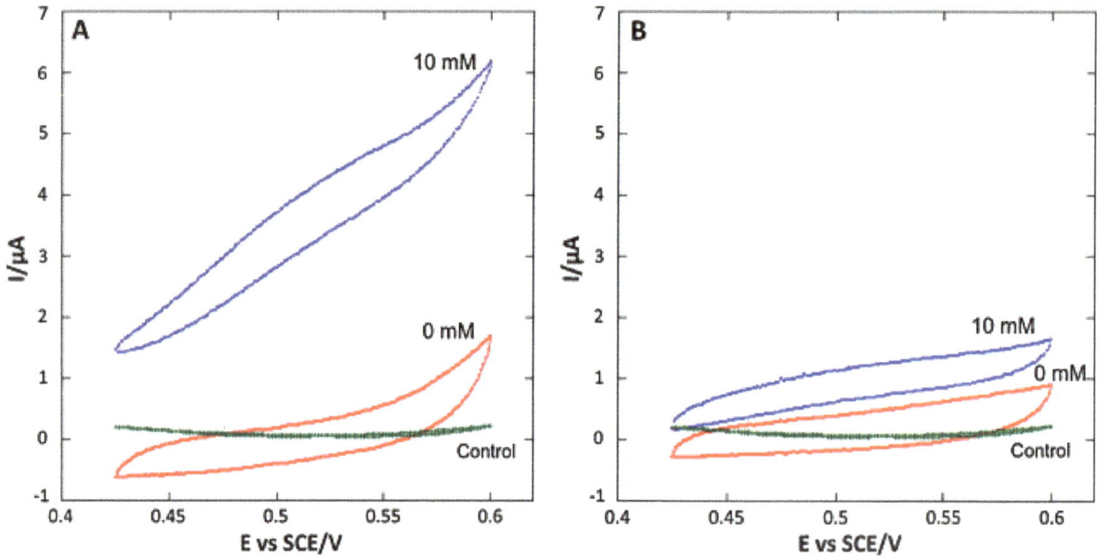

Figure 1: Electrochemical response to glucose in PBS of the hPG electrodes. A: CV scans of the GOx_{ads}-hPG electrode in the absence and presence of 10 mM of glucose. B: CV scans of the GOx_{abs}-hPG electrode in the absence and presence of 10 mM of glucose. The control was performed with the hPG electrode (no GOx) in the presence of 10 mM of glucose. The scan rate for each test was 1 mV s^{-1}.

The difference in the CV curves for the two electrodes suggests that a different type of linkage between the enzyme and hPG electrode surface occurs in the two cases. The immobilisation mechanism achieved during the CV scans in the presence of GOx is not merely *absorption*, but an electrochemically driven physical *adsorption*. Since GOx is anionic at pH 7.0, and since a relatively high positive scan range is used, the CV scans are likely to promote an electrostatic attraction between the gold surface and the free enzyme. In contrast, when the electrodes are simply placed in contact with the enzymatic solution, absorption occurs into the pores of the electrode. Consequently, the resulting electrode is less stable as GOx can easily leach out.

The CV scans might actively draw the enzyme to the surface of the electrode, thus significantly increasing the loading. For the case of electrochemical adsorption in fact an average of 31.7% reduction of activity in solution after the immobilisation process was observed. On the other hand, the amount of enzyme immobilised by absorption was so low that no significant changes were observed with the activity tests in solution prior and after the incubation with the hPG electrode.

The amperometric response of the two GOx-hPG electrodes to glucose (10 mM) was tested and compared at a potential of 0.52 V vs SCE (the potential at which the electrodes showed the highest sensitivity to glucose).

This potential is within the 0.4 - 0.6 V range of values previously reported for the oxidation peak of H_2O_2 on polished gold [11] and [24]. H_2O_2 is formed through the oxidation of glucose by GOx and, since its formation is related to the amount of glucose in solution (see Equation (1)), many glucose sensors rely on the resultant oxidation of H_2O_2 on gold to determine the concentration of glucose in a sample [30].

$$Glucose + O_2 + H_2O \xrightarrow[GOx]{} Gluconic\ \ acid + H_2O_2 \quad (1)$$

The amperometric tests confirmed that the GOx_{ads}-hPG electrode performed better than the GOx_{abs}-hPG electrode. As reported in Table 1, the response of the GOx_{ads}-hPG electrode is in fact approximately four times higher than the GOx_{abs}-hPG electrode, with a sensitivity of 22.7 ± 0.1 μA mM^{-1} cm^{-2}, calculated against the superficial surface area of the bioelectrode (0.03 cm^2). In comparison, the amperometric response of a blank hPG electrode to glucose at 0.52 V is negligible. This is because the current peak related to glucose electrooxidation by the non-enzymatic hPG electrode occurs at a potential of 0.15 V [26].

Table 1: Comparison of the amperometric response at 0.52 V vs SCE of the GOx-hPG electrode to 10 mM of glucose, according to the immobilisation methodology. Standard deviation refers to three replicates.

Electrode	I/μA	Sensitivity/μA mM^{-1} cm^{-2}
hPG	0.07 ± 0.03	
GOx_{abs}-hPG	1.6 ± 0.2	2.42 ± 0.05
GOx_{ads}-hPG	6.10 ± 0.12	22.7 ± 0.1

The use of the GOx_{ads}-hPG electrode as glucose sensor

The amperometric response of the GOx_{ads}-hPG electrode to increasing concentrations of glucose at a constant potential of 0.52 V (vs SCE) was investigated. A typical Michaelis-Menten kinetic trend was observed. This was characterised by a linear increase in the current output up to a concentration of 10 mM as shown in Fig. 2. The detection limit the system was found to be 25 μM, which is lower compared to other GOx electrodes reported previously [28] and [31]. At least three replicates were performed using different electrodes, showing a good reproducibility of the electrode performance with a maximum standard deviation factor of 0.38 (at 5 μM of glucose).

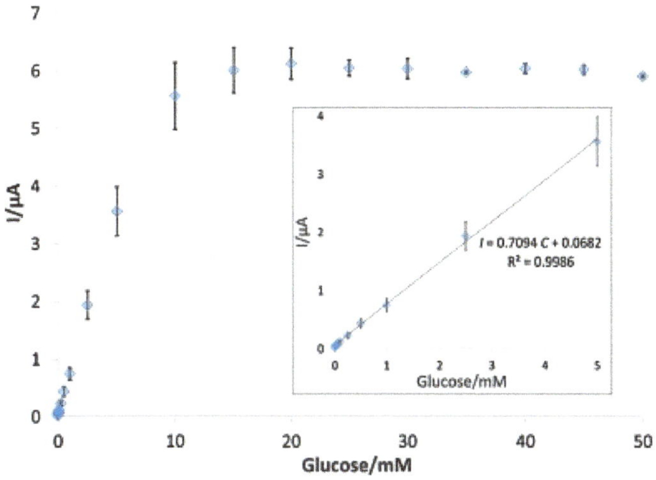

Figure 2: Amperometric tests (at 0.52 V vs SCE) of the GOx_{ads}-hPG electrode in the presence of various glucose concentrations in the electrolyte (range of 5 μM - 50 mM). Error bars refer to three replicates.

The enzymatic affinity towards glucose was estimated in terms of the apparent Michaelis-Menten constant (K_m^{app}), and was calculated by using the electrochemical version of the Lineweaver-Burk equation of enzyme kinetics (Equation (2)) [13].

$$\frac{1}{i} = \frac{1}{i_{max}} + \frac{K_M^{app}}{i_{max}C} \qquad (2)$$

Where i is the steady-state current observed after the addition of glucose; $imax$ is the maximum current under the saturated concentration of glucose; c is the glucose concentration. The reciprocal of the current was plotted versus the reciprocal of the glucose concentration in the range of 0.005- 10 mM. This gave a value of K_m^{app} equal to 6.3 ± 0.7 mM, which was calculated from the slope ($K_m^{app}/imax$) and the intercept ($1/imax$) of the plot. This value is much lower than the one reported for the native GOx from *Aspergillus niger* in solution (27 mM) [32]. This indicates that the immobilised GOx onto hPG electrodes has high enzymatic activity and higher affinity for glucose than the soluble enzyme with a low diffusion barrier. Although this value is slightly higher than the value of 1.5 mM recently reported for a GOx/chitosan/ferrocene carbon nanotubes electrode [13], it is, however, comparable with the value (6.3 mM) reported for a ferrocene-modified multiwalled carbon nanotube [12], and much lower than other reported GOx electrodes [33] and [34].

Table 2 compares the sensitivity to glucose of the GOx_{ads}-hPG electrode with other GOx electrodes created by using much more complex immobilisation procedures [12], [13],[14], [21], [35] and [36]. As reported, the GOx_{ads}-hPG electrode sensitivity is greater than the majority of the other methods reviewed, including some that boast direct electron transfer between the FAD centre of GOx and the electrode surface.

Table 2: Comparison of the sensitivity to glucose of several GOx electrodes. *Standard deviation refers to three replicates. **Methods which claim to achieve direct electron transfer.

Electrode	Sensitivity/μA mM^{-1} cm^{-2}	E vs SCE/V	Reference
GOx_{ads}-hPG *	22.7 ± 0.1	0.52	This Study
Naf/GOx-NPG/GCE	0.697	0.40	[14]
Fc/GOx-MWCNT/Chi-BSA- AuE	7.77 ± 0.08	0.17	[13]
GOx-Fc/MWCNT/Chi-GCE	25	0.31	[12]
Gel/GOx-WMCNT/GCE **	2.47	- 0.48	[35]
AuNPs/GOx/Biofilm-MWCNT/GCE **	16.6	- 0.40	[36]
PDDA/AuNPs-MWCNT/GE **	29.72	- 0.49	[21]

Naf: Nafion; GCE: glassy carbon electrode; Fc: Ferrocene; MWCNT: multi-walled carbon nano-tubes; Chi: Chitosan; BSA: bovine serum albumin; AuE: gold electrode; Gel: gelatine; AuNPs: gold nano-particles; PDDA: poly(diallyldimethylammonium chloride); GE: graphene electrode

Stability

The GOx_{ads}-hPG electrodes were stored in PBS at 4 °C and tested again at regular intervals of time to investigate their stability. In particular, Fig. 3A compares the CV scans obtained immediately after the electrode fabrication, and after 24 and 48 hours, respectively.

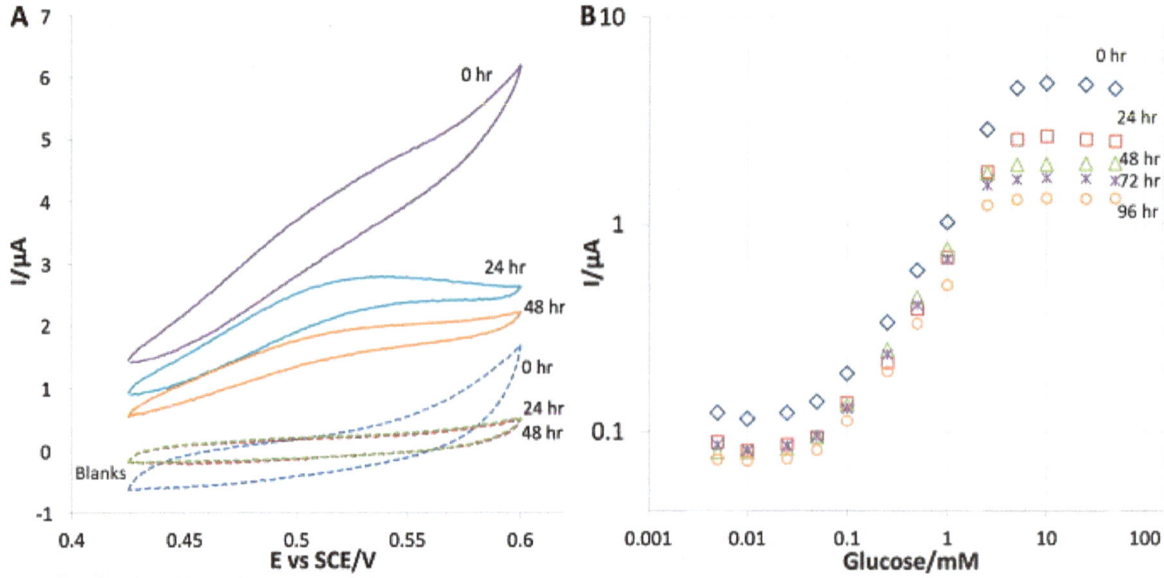

Figure 3: Stability tests. A: CV scans of the GOx_{ads}-hPG electrode after 0, 24, and 48 hours for a glucose concentration of 10 mM. The blanks were performed in the absence of glucose in the electrolyte. B: Amperometric response of the GOx_{ads}-hPG electrode to increasing concentration of glucose over a period of five days of storage.

A reduction in the current signal was observed within the three days, with the largest variation occurring after 24 hours of storage. After the first 24 hours of storage, a 35% reduction in the current peak at 0.52 V was observed. Interestingly, after the first 24 hours of storage, a marked change in the blank (e.g. CV test in the absence of glucose) was also observed, while no further changes occurred in the following days. The reduction in the response observed was lower after 48 hours, with a 25% decrease in the current peak at 0.52 V with respect to data collected at 24 hours.

This suggests that the initial variation in the activity of the bioelectrodes might be caused by a reduction in the hPG sensitivity towards H_2O_2, rather than simply to enzyme deactivation and/or leaching. The reduction in the electrochemical activity of nPG structures with the time has been previously reported and attributed to a coarsening process [37]. This hypothesis was supported by stability tests carried out on the hPG electrode with no enzyme immobilised. When the hPG electrode was tested in PBS after one day, a change in the CV curve was observed, which was minimal after another day of storage.

The stability of the GOx_{ads}-hPG electrodes in terms of amperometric response was also investigated during a period of time of five days. As shown in Fig. 3B, the linearity range and the detection limit did not change. However, the magnitude of the amperometric response diminished by a maximum factor of 3.60 over the 5 days tested. In considering practical applications, the GOx_{ads}-hPG electrode response should be normalised and calibrated prior to use to achieve concurrent data sets. After data normalisation, the variance observed between data sets obtained on different days for concentrations of glucose between 50 μM and 5 mM was minimal.

GOx electrode in enzymatic biofuel cell

The functional immobilisation of LAC was achieved by activating the hPG surface with amino-phenyl groups as previously reported [38]. The steps in the immobilisation protocols are reported in Fig. 4. Firstly, nitrophenyl groups were electrochemically attached to the gold by conducting two reductive CV scans in the presence of p-nitrophenyldiazonium salt. In order to limit the production of polyaniline (by the polymerisation of the phenyl groups through hydrolysis), and to promote the formation of a monolayer, this stage was conducted in an acetone-based electrolyte. The CV scans conducted show two peaks (at 400 mV and -200 mV vs. SCE) on the first scan that greatly decrease on the second scan (Figure S3). These peaks are attributed to the reduction of the diazonium salt on the different planar surfaces of gold and crystalline gold as previously suggested for the case of an acetonitrile electrolyte [39]. Since the phenyl-gold bond formed is not oxidised back in this scan range, the large decrease in the magnitude of the peaks on the second scan suggests that the surface is quickly becoming saturated with the nitro-phenyl groups [8], [39] and [40].

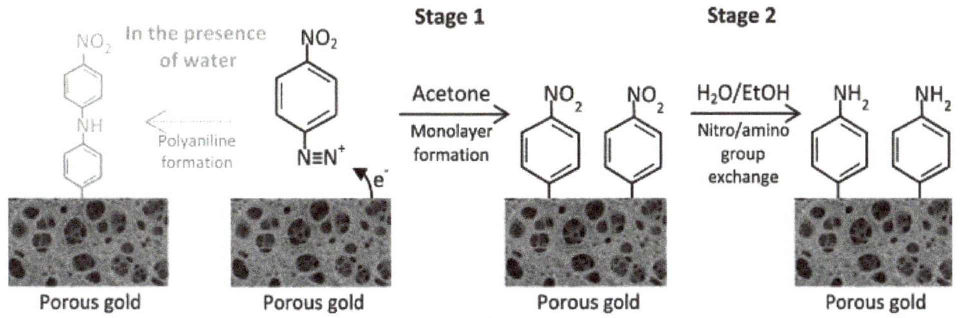

Figure 4: Schematic of gold surface modification prior to LAC immobilisation.

During the second stage, the treated electrodes were immersed in a 10% ethanol and water based electrolyte and two more reductive CV scans were conducted in a much more negative scan range (0.0 V and -1.4 V vs. SCE). This facilitated the exchange of oxygen for protons on the nitro groups resulting in the production of amino and hydroxylamino groups. The voltammogram of the first scan shows a very clear peak at -820 mV which is established as the potential at which nitrophenyl groups are irreversibly reduced to aminophenyl groups [8]. This peak is not evident on the second scan though a small redox peak is observed at -700 mV (Figure S4). This results suggests the presence of hydroxylamine groups amongst the amine groups [41].

The GOx_{ads}-hPG and the LAC-hPG electrodes were subsequently tested as anode and cathode respectively of a glucose/oxygen enzymatic biofuel cell. The electrodes were immerged in an aerated PBS solution containing 27.8 mM of glucose, and the potential difference of the cell was monitored by means of a potentiostat (Fig. 5). An open circuit voltage of 0.58 V was observed, comparable with the value previously reported [27]. Subsequently, the cell was polarised by connecting the electrodes to a range of external resistors. Fig. 5B shows the cell voltage and the power density as a function of the current density. As shown, the peak power density was of 6 μW cm^{-2} at 0.2 V vs SCE. This result, although obtained with a very simple design, is comparable with recently reported miniature enzymatic biofuel cells [27] and [42].

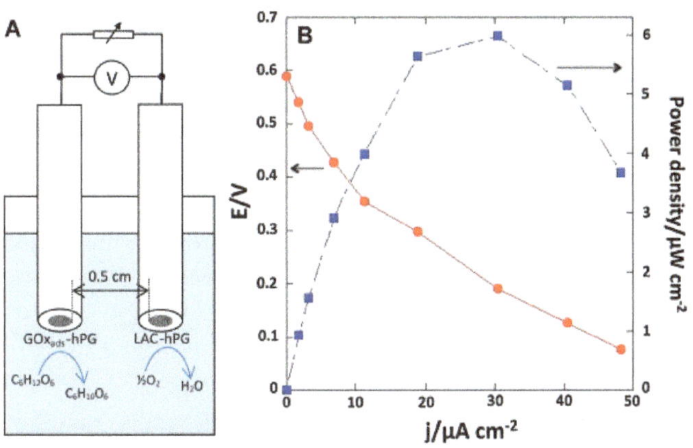

Figure 5: A: Scheme of the enzymatic biofuel cell set up. B: Polarisation and power density curve of the GOx_{ads}-hPG and LAC-hPG biofuel cell. The power and the current density refer to the total superficial surface area of the electrodes used (0.03 cm^2).

CONCLUSIONS

Glucose oxidase was immobilised onto highly porous gold electrodes by electrochemical adsorption, with a rapid one-step protocol that did not involve the use of expensive and/or harsh reagents. The GOx_{ads}-hPG electrode exhibits Michaelis-Menten kinetics with a wide linearity range between 50 μM and 10 mM of glucose, and a detection limit of 25 μM.

The sensitivity, calculated against the total superficial surface area of the bioelectrode was of 22.7 ± 0.1 μA mM^{-1} cm^{-2}. After five days of storage the bioelectrode amperometric response to glucose was 3.6 times lower compared to the initial value. However, the linearity range and the detection limit were maintained.

Preliminary experiments encourage the implementation of the GOx_{ads}-hPG electrode as the anode of an enzymatic biofuel cell. When the electrode was coupled with a LAC-hPG electrode as a cathode, a peak power density of 6 μW cm^{-2} at 0.2 V vs SCE was achieved. Future work will be regarding the implementation of the enzyme hPG electrodes in a micro biofuel cell for healthcare applications.

ACKNOWLEDGMENTS

The authors acknowledge: Prof Frank Marken, University of Bath, for valuable discussions; the Royal Society Research Grant (RG110344) and EPSRC(EP/K031597/1) for funding.

REFERENCES

1. M. Falk, V. Andoralov, M. Silow, M.D. Toscano, S. Shleev, Miniature Biofuel Cell as a Potential Power Source for Glucose-Sensing Contact Lenses, Anal Chem 85 (2013) 6342–6348.
2. A. Chaubey, B.D. Malhotra, Mediated biosensors, Biosens Bioelectron 17 (2002) 441–456.
3. R.A. Bullen, T.C. Arnot, J.B. Lakeman, F.C. Walsh, Biofuel cells and their development, Biosens Bioelectron 21 (2006) 2015–2045.
4. M. Shaolin, X. Huaiguo, Bioelectrochemical characteristics of glucose oxidase immobilized in a polyaniline film, Sensor Actuat B-Chem 31 (1996) 155–160.
5. Y. Degani, A. Heller, Direct electrical communication between chemically modified enzymes and metal electrodes. I. Electron transfer from glucose oxidase to metal electrodes via electron relays, bound covalently to the enzyme, J Phys Chem 91 (1987) 1285–1289.
6. M.J. Moehlenbrock, S.D. Minteer, Extended lifetime biofuel cells, Chem Soc Rev 37 (2008) 1188–1196.
7. C. Vaz-Dominguez, S. Campuzano, O. Rüdiger, M.

Pita, M. Gorbacheva, S. Shleev, V.M. Fernandez, A.L. De Lacey, Laccase electrode for direct electrocatalytic reduction of O2 to H2O with high-operational stability and resistance to chloride inhibition, Biosens Bioelectron 24 (2008) 531–537.

8. M. Pita, C. Gutierrez-Sanchez, D. Olea, M. Velez, C. Garcia-Diego, S. Shleev, V.M. Fernandez, A.L. De Lacey, High Redox Potential Cathode Based on Laccase Covalently Attached to Gold Electrode, J Phys Chem C 115 (2011) 13420–13428.
9. S. Timur, N. Pazarlıolu, R. Pilloton, A. Telefoncu, Thick film sensors based on laccases from different sources immobilized in polyaniline matrix, Sensor Actuat B-Chem 97 (2004) 132–136.
10. J. Wang, Glucose Biosensors: 40 Years of Advances and Challenges, Electroanalysis 13 (2001) 983–988.
11. Y. Xiao, F. Patolsky, E. Katz, J.F. Hainfeld, I. Willner, Plugging into Enzymes: Nanowiring of Redox Enzymes by a Gold Nanoparticle, Science 299 (2003) 1877–1881.
12. J.D. Qiu, W.M. Zhou, J. Guo, R. Wang, R.P. Liang, Amperometric sensor based on ferrocene modified multiwalled carbon nanotube nanocomposites as electron mediator for the determination of glucose, Anal Biochem 385 (2009) 264–269.
13. A. Fatoni, A. Numnuam, P. Kanatharana, W. Limbuta, C. Thammakhet, P. Thavarungkul, A highly stable oxygen-independent glucose biosensor based on a chitosan-albumin cryogel incorporated with carbon nanotubes and ferrocene, Sensor Actuat B-Chem 185 (2013) 725–734.
14. H. Qiu, L. Xue, G. Ji, G. Zhou, X. Huang, Y. Qu, P. Gao, Enzyme-modified nanoporous gold-based electrochemical biosensors, Biosens Bioelectron 24 (2009) 3014–3018.
15. J. Kim, H. Jia, P. Wang, Challenges in biocatalysis for enzyme-based biofuel cells, Biotechnol Adv 24 (2006) 296–308.
16. A. Guiseppi-Elie, C. Lei, R.H. Baughman, Direct electron transfer of glucose oxidase on carbon nanotubes, Nanotechnology 13 (2002) 559.
17. D. Ivnitski, B. Branch, P. Atanassov, C. Apblett, Glucose oxidase anode for biofuel cell based on direct electron transfer, Electrochem Commun 8 (2006) 1204–1210.
18. Y. Liu, M. Wang, F. Zhao, Z. Xu, S. Dong, The direct electron transfer of glucose oxidase and glucose biosensor based on carbon nanotubes/chitosan matrix, Biosens Bioelectron 21 (2005) 984–988.
19. N. German, A. Ramanaviciene, J. Voronovi, A. Ramanavicius, Glucose biosensor based on graphite electrodes modified with glucose oxidase and colloidal gold nanoparticles, Microchim Acta 168 (2010) 221–229.
20. H.-F. Cui, K. Zhang, Y.-F. Zhang, Y.-L. Sun, J. Wang, W.-D. Zhang, J.H.T. Luong, Immobilization of glucose oxidase into a nanoporous TiO2 film layered on metallophthalocyanine modified vertically-aligned carbon nanotubes for efficient direct electron transfer, Biosens Bioelectron 46 (2013) 113–118.
21. Y. Yu, Z. Chen, S. He, B. Zhang, X. Li, M. Yao, Direct electron transfer of glucose oxidase and biosensing for glucose based on PDDA-capped gold nanoparticle modified graphene/multi-walled carbon nanotubes electrode, Biosens Bioelectron 52 (2014) 147–152.
22. A.J. Thorley, T.D. Tetley, New perspectives in nanomedicine, Pharmacol Therapeut 140 (2013) 176–185.
23. M. Bottini, S. Bruckner,K. Nika, N. Bottini, S. Bellucci,A. Magrini,A. Bergamaschi, T. Mustelin, Multi-walled carbon nanotubes induce T lymphocyte apoptosis, Toxicol Lett 160 (2006) 121–126.
24. L.Y. Chen, T. Fujita, M.W. Chen, Biofunctionalized nanoporous gold for electrochemical biosensors, Electrochim Acta 67 (2012) 1–5.
25. S. Cherevko, C.-H. Chung, Direct electrodeposition of nanoporous gold with controlled multimodal pore size distribution, Electrochem Commun 13 (2011) 16–19.
26. H. du Toit, M. Di Lorenzo, Electrodeposited highly porous gold microelectrodes for the direct electrocatalytic oxidation of aqueous glucose, Sensor Actuat BChem 192 (2014) 725–729.
27. M. Falk, V. Andoralov, M. Silow, M.D. Toscano, S. Shleev, Miniature biofuel cell as potential power source for glucose sensing contact lenses, Anal Chem 85 (2013) 6342–6348.
28. X. Xiao, M.e. Wang, H. Li, P. Si, One-step fabrication of bio-functionalised nanoporous gold/poly(3,4-ethylenedioxythiophene) hybrid electrodes for amperometric glucose sensing, Talanta 116 (2013) 1054–1059.
29. M. Hakamada, M. Takahashi, M. Mabuki, Enzyme electrodes stabilized by monolayer-modified nanoporous Au for biofuel cells, Gold Bull 45 (2012) 9–15.
30. F. He, Y. Tang, M. Yu, S. Wang, Y. Li, D. Zhu, Fluorescence-Amplifying Detection of Hydrogen Peroxide with Cationic Conjugated Polymers, and Its Application to Glucose Sensing, Adv Funct Mater 16 (2006) 91–94.
31. G. Wang, N.M. Thai, S.-T. Yau, Preserved enzymatic activity of glucose oxidase immobilized on an unmodified electrode, Electrochem Commun 8 (2006) 987–992.
32. M.J. Rogers, K.G. Brandt, Interaction of d-glucal with Aspergillus niger glucose oxidase, Biochemistry 10 (1971) 4624–4630.
33. J.C. Vidal, E. Garcia, J.R. Castillo, Electropolymerization of pyrrole and immobilization of glucose oxidase in a flow system: influence of the operating conditions on analytical performance, Biosens Bioelectron 13 (1998) 371–382.
34. X. Chen, J.B. Jia, S.J. Dong, Organically modified sol–gel/chitosan composite based glucose biosensor, Electroanalysis 15 (2003) 608–612.
35. A.P. Periasamy, Y.-J. Chang, S.-M. Chen, Amperometric glucose sensor based on glucose oxidase immobilized on gelatin-multiwalled carbon nanotube modi- fied glassy carbon electrode, Bioelectrochemistry 80 (2011) 114–120.
36. H. Zhang, Z. Meng, Q. Wang, J. Zheng, A novel glucose

biosensor based on direct electrochemistry of glucose oxidase incorporated in biomediated gold nanoparticles–carbon nanotubes composite film, Sensor Actuat B-Chem 158 (2011) 23–27.

37. S. Cattarin, D. Kramer, A. Lui, M.M. Musiani, Preparation and Characterization of Gold Nanostructures of Controlled Dimension by Electrochemical Techniques, J Phys Chem C 111 (2007) 12643–12649.

38. P. Olejnik, B. Palys, A. Kowalczyk, A.M. Nowicka, Orientation of Laccase on Charged Surfaces. Mediatorless Oxygen Reduction on Amino- and CarboxylEnded Ethylphenyl Groups, J Phys Chem 116 (2012) 25911–25918.

39. A. Benedetto, M. Balog, P. Viel, F. Le Derf, M. Sallé, S. Palacin, Electro-reduction of diazonium salts on gold: Why do we observe multi-peaks? Electrochim Acta 53 (2008) 7117–7122.

40. P.A. Brooksby, A.J. Downard, Electrochemical and Atomic Force Microscopy Study of Carbon Surface Modification via Diazonium Reduction in Aqueous and Acetonitrile Solutions, Langmuir 20 (2004) 5038–5045.

41. N. Mano, A.Kuhn, Cation induced amplification ofthe electrocatalytic oxidation of NADH by immobilized nitro-fluorenone derivatives, J Electroanal Chem 498 (2001) 58–66.

42. T. Beneyton, I. Putu Mahendra Wijaya, C. Ben Salem, A.D. Griffiths, V. Taly, Membraneless glucose/O2 microfluidic biofuel cells using covalently bound enzymes, Chem Commun 49 (2012) 1094–1096.

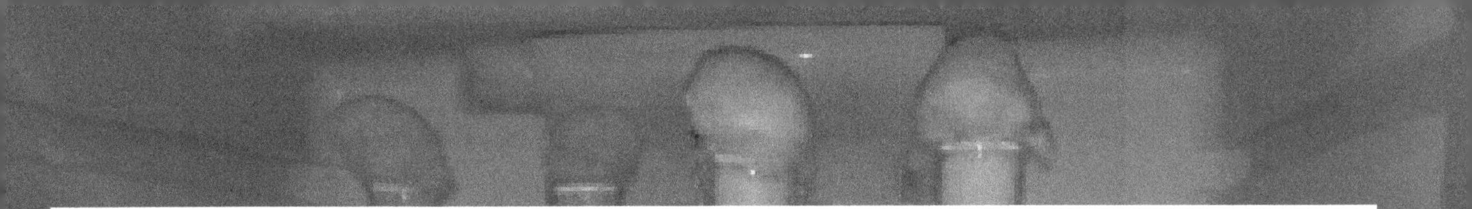

CHAPTER 19

HIGH SENSITIVITY ELECTROCHEMICAL CHOLESTEROL SENSOR UTILIZING A VERTICALLY ALIGNED CARBON NANOTUBE ELECTRODE WITH ELECTROPOLYMERIZED ENZYME IMMOBILIZATION

Anurat Wisitsoraat, Chanpen Karuwan, Krongkamol Wong-ek, Ditsayut Phokharatkul, Pornpimol Sritongkham, and Adisorn Tuantranont

Nanoelectronics and MEMS Laboratory, National Electronics and Computer Technology Center, 112 Thailand Science Park, Pahol Yothin Rd., Klong Luang, Pathumthani 12120, Thailand

ABSTRACT

In this report, a new cholesterol sensor is developed based on a vertically aligned CNT electrode with two-step electrochemical polymerized enzyme immobilization. Vertically aligned CNTs are selectively grown on a 1 mm^2 window of gold coated SiO_2/Si substrate by thermal chemical vapor deposition (CVD) with gravity effect and water-assisted etching. CNTs are then simultaneously functionalized and enzyme immobilized by electrochemical polymerization of polyaniline and cholesterol enzymes. Subsequently, ineffective enzymes are removed and new enzymes are electrochemically recharged. Scanning electron microscopic characterization indicates polymer-enzyme nanoparticle coating on CNT surface. Cyclic voltammogram (CV) measurements in cholesterol solution show the oxidation and reduction peaks centered around 450 and −220 mV, respectively. An approximately linear relationship between the cholesterol concentration and the response current could be observed in the concentration range of 50–300 mg/dl with a sensitivity of approximately 0.22 μA/mg·dl^{-1}, which is considerably higher compared to previously reported CNT bioprobe. In addition, good specificity toward glucose, uric acid acetaminophen and ascorbic acid have been obtained. Moreover, sensors have satisfactory stability, repeatability and life time. Therefore, the electropolymerized CNT bioprobe is promising for cholesterol detection in normal cholesterol concentration in human blood.

INTRODUCTION

Cholesterol is an essential lipid for human body. The desired total plasma cholesterol for an individual is less than 5.2 mM (200 mg/dL) and it poses a potential health threat when the level is greater than 6.2 mM (240 mg/dL). Excessive plasma cholesterol causes poor cardiovascular conditions, such as atherosclerosis and hypertension, which can lead to coronary heart disease, myocardial and cerebral infarction. The total cholesterol in all lipoprotein fractions is normally determined using a colorimetric assay specific for cholesterol [1]. In the process, cholesterol esters are acted upon by cholesterol esterase (ChEs) to release free cholesterol and then cholesterol oxidase (ChOx) catalyzes the reaction to generate H_2O_2, which is reacted to yield quinoneimine dye. The absorbance of the dye is proportional to H_2O_2, hence the cholesterol concentration.

Citation: Anurat Wisitsoraat, Chanpen Karuwan, Krongkamol Wong-ek, Ditsayut Phokharatkul, Pornpimol Sritongkham and Adisorn Tuantranont, High Sensitivity Electrochemical Cholesterol Sensor Utilizing a Vertically Aligned Carbon Nanotube Electrode with Electropolymerized Enzyme Immobilization, doi:10.3390/s91108658

The disadvantages of this technique include instability of color reagent, variable reactivity of ester, corrosive nature of reagents, poor specificity and unreusable reagents.

Electrochemical detection techniques emulate the colorimetric assay by immobilizing ChEs and ChOx onto electrode surfaces to liberate cholesterol and subsequently generate H_2O_2, which is measured amperometrically [2]. Its advantages include rapid analysis, reusability, thermal stability, linearity and good specificity. However, electrochemical biosensors still need further development to improve performances in term of sensitivity, reproducibility and life time. Several approaches including functionalization of the working electrode surface by polymer or nanomaterials have been proposed to enhance the performance of electrochemical biosensors [3,4].

Carbon nanotubes (CNTs) have been shown to promote electron transfer reactions for the redox of important biomolecules [5-8]. In addition, vertical alignment of the CNT nanoeletrodes is preferred over other schemes because the open end of a CNT is expected to show faster electron transfer rate (ETR). This characteristic along with their high surface area and conformal compatibility with biomolecules makes CNTs particularly attractive for high-sensitivity biochemical sensing systems. However, the performance of CNTs based electrochemical biosensor is also critically dependant on effectiveness of enzyme immobilization on CNT structure. Conventional approaches for enzyme immobilization on amperometric bioprobes include direct adsorption [9] and entrapment in polymeric films [10]. Physical adsorption technique suffers from protein desorption due to changes in temperature, pH and ionic strength while the enzyme entrapment in the lattice of a polymer matrix or membrane provides relatively better enzyme retention.

Recently, vertically aligned CNT electrode with enzyme entrapment in polyvinyl alcohol (PVA) by spin coating has been reported [8]. Nevertheless, the disadvantages of the reported technique are low enzyme entrapment efficiency and life-time due to PVA's low physical entrapment capacity, low conductivity and partial water solubility. Electrochemical enzyme entrapment with conducting polymer should offer much higher enzyme retention capacity and better electron transfer to CNTs [12-16]. There have been some reports on biosensors based on CNTs-enzyme-polymer composites prepared by electrochemical polymerization [17-22]. However, there have been few reports that perform enzyme immobilization by electrochemical polymerization on vertically aligned CNTs electrode, which can offer significantly improved sensor's performance and reliability. In this work, cholesterol bioprobe is developed based on vertically aligned CNTs with enzyme immobilization in polyaniline (PANI) using two-step electrochemical process.

EXPERIMENTAL SECTION

The structure for electrode fabrication is shown in Figure 1. First, SiO_2 (400 nm), Cr (50 nm) and Au (500 nm) were successively sputtered on <100> Si substrates. Next, aluminium oxide (10 nm) and stainless steel (SS) catalyst (5 nm) were sequentially sputtered to prepare for CNT synthesis. Titanium dioxide was then sputtered via shadow masking on the gold layer over a defined region, which excludes active sensing area (1 mm^2) and electrical contact region. The aluminium oxide and titanium dioxide layers were deposited by reactive sputtering at a pressure of 3×10^{-3} mbar of 1:5 Ar/O_2 gas mixtures while other metallic layers were deposited in Ar gas at the same pressure.

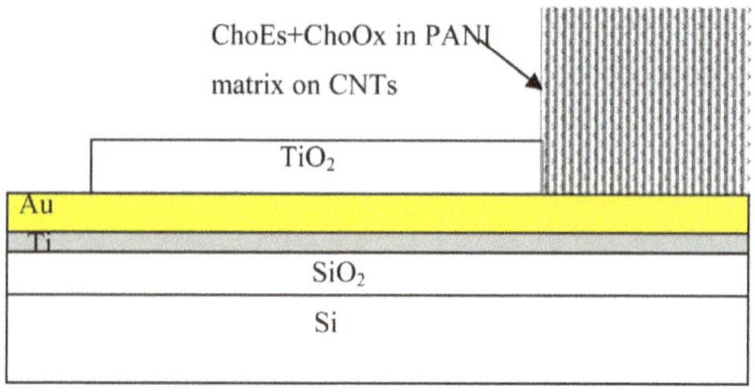

Figure 1: Structure of the CNT based cholesterol bioprobe.

Vertically aligned carbon nanotubes were grown by thermal chemical vapor deposition (CVD) with gravity effect and water-assisted etching. The catalyst layers on substrates were placed upside down along gravitational field on an alumina carrier in a horizontal furnace thermal CVD system. The CNT synthesis was conducted at atmospheric pressure and growth temperature of 700 °C. During CNT growth, acetylene was flown for 1.5 minute and hydrogen to acetylene volume flow ratio was 4.3:1 with hydrogen flow of 1,935 sccm and acetylene flow of 450 sccm. In the course of CNT growth, in-situ-water-assisted etching was employed to remove undesired amorphous carbon formation from random acetylene decomposition. In water etching process, 300 ppm of water vapor was introduced by water bubbling through Ar gas for 3 minutes while acetylene gas was turned off. CNTs growth and water-assisted etching were repeatedly performed for three cycles and the total time was 13.5 minutes. CNTs› functionalization and enzyme immobilization in PANI matrix were then conducted as per the following protocol.

Reagents for biosensing

Analytical-grade bio-reagents used in this study: cholesterol (95%), cholesterol oxidase (ChOx, *Pseudomonas fluorescens* 25 U/mg), cholesterol esterase (ChEs, *Pseudomonas fluorescens* 100 U/mg), horseradish peroxidase (HRP, 5,000 U), potassium ferrocyanide, trehalose, Triton X-100 were purchased from Sigma-Aldrich. Aniline monomer and phosphate buffer saline (PBS) solution was purchased from Fluka (USA).

Surface functionalization and enzyme immobilization

Functionalization of the CNT based electrode was done by electrochemical polymerization on the active zone of the working electrode in a solution of 0.2 M aniline monomer. Prior to aniline polymerization, CNT was refluxed in 1 M nitric acid for 2 hours at room temperature to remove catalyst and clean CNT surface. The aniline solution was prepared by dissolving 4.8 mL of aniline monomer and 1 mL of HCl in 100 mL of PBS solution (pH 7.0) with constant stirring for 10 minutes at room temperature. The final pH of aniline solution is around 2.5. Electrochemical polymerization was performed at ~ 20 °C by a standard three electrode electrochemical system (Metrohm electrochemical workstation with BAS C3 voltammetric cell). The working electrode (WE) was the CNT based electrode while the counter electrode (CE) was a Pt wire and all the potentials were controlled and measured as referred to an Ag|AgCl|KCl (sat.) reference electrode (RE). During polymerization, the CNT electrode was biased at a constant voltage of 0.6 V with respect to RE for 30 minutes. The polyaniline matrix formed on CNTs introduced NH group on CNT surface, which greatly helped immobilizing various biological analytes, particularly proteins [12,13].

Enzyme immobilization was then followed by a two-step electrochemical procedure. The first electrochemical process was conducted in the mixture of 2 mL of cholesterol enzyme and 2 mL of 0.2 M aniline monomer. The pH of the mixture was measured to be about 3. The cholesterol enzyme was prepared by dissolving 1.5 mg ChEs, 2.0 mg ChOx, 1.8 mg HRP, 8.0 mg potassium ferrocyanide (redox mediator) and 6.0 mg trehalose (enzyme stabilizer) in 0.3 mL PBS solution. The PANI-CNT electrode was biased at a constant voltage of 0.6 V with respect to RE for 30 minutes. Due to the acidic condition during aniline polymerization, cholesterol enzymes were inactive or even denatured. Thus, ineffective enzymes were removed by etching in stirred 6 M HCl solution for 2 hours, leaving cavities in and on PANI matrix. The PANI layer was then reduced by biasing at constant voltage of −0.2 V for 20 minutes in PBS solution to remove any residual anions from acidic etching. New enzymes were electrochemically recharged into PANI matrix by biasing the electrode in cholesterol enzyme in PBS solution (pH 6.0) at a constant voltage of 0.6 V with respect to RE for 30 minutes. After immobilization, the sensor chip was stored in a refrigerator for 24 h and thoroughly rinsed in PBS before running each test.

Electrochemical measurements

Electrochemical measurement of cholesterol was performed by the same electrochemical system. Due to cholesterol's water insolubility, the stock cholesterol solution was prepared by dissolving 40 mg cholesterol in 1 mL Triton X-100 and 9 mL 0.1M PBS solution (pH 7.0) and stirring with a magnetic spin bar just before the measurement to obtain a homogeneous solution. Cholesterol solution with different concentration was then prepared for testing by proper dilution of the stock solution. Cholesterol bioprobe was immersed in 0.1 M PBS solution and cyclic voltammograms (CV) were run at several scan rates for different concentrations of cholesterol. The voltage window for CV was from −1.0 to +1.0 V because the redox peaks appeared within this potential window.

RESULTS AND DISCUSSION

The surface morphology of the sensing area of CNT based biochip was examined using scanning electron microscopy (SEM). Scanning electron microscopic images of vertically aligned CNTs before and after polyaniline-enzyme immobilization,is shown in Figures 2(a) and (b), respectively. The cross section seen in the images confirms that CNTs are indeed vertically aligned, with a vertical height of ~50 μm. In addition, it can be seen that vertically aligned CNTs are uniformly coated by polymer-enzyme nanoparticles on CNT surface. Figure 3(a) shows high magnification top view SEM image of CNT electrodes after electrochemical polyaniline-enzyme coating. It can be seen the size of polyaniline-enzyme nanoparticles is in the range between 20 and 50 nm. In addition, transmission electron microscopic (TEM) characterization as shown in Figure 3(b) verifies that CNTs are multiwalls with diameters and numbers of walls in the range of 10–25 nm and 8–15 respectively.

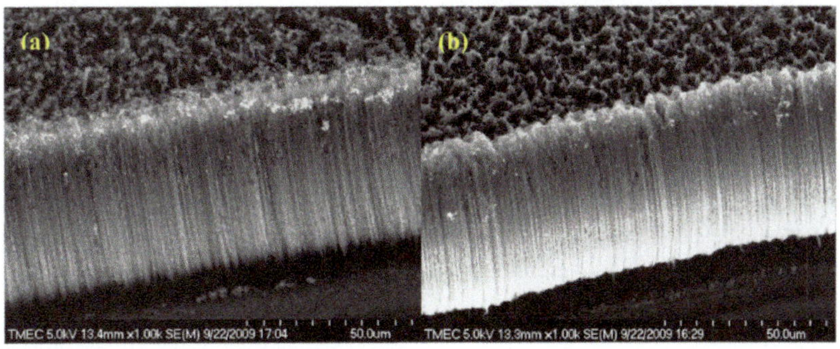

Figure 2: Typical SEM micrograph of CNT electrode (a) before and (b) after electrochemical polyaniline-enzyme coating.

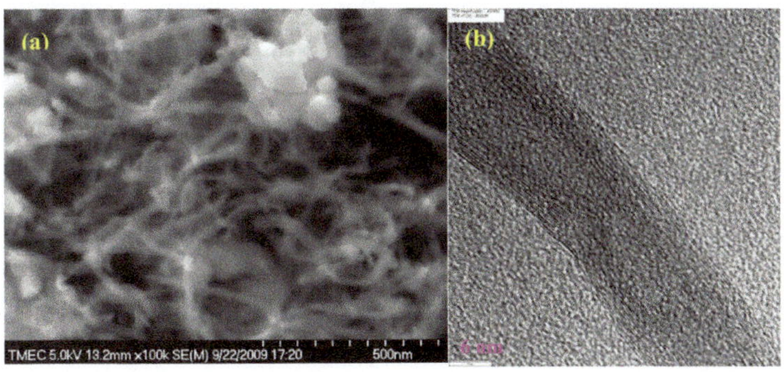

Figure 3: High magnification (a) top view SEM image of CNT electrodes after electrochemical polyaniline-enzyme coating and (b) TEM image of CNT extracted from the synthesized vertically aligned CNTs.

The chemical property of the CNT bioprobe was characterized by Fourier Transform Infrared Spectroscopy (FTIR) using a Perkin Elmer System—2000R FTIR spectrophotometer. The FTIR spectrum of the CNT bioprobe is shown inFigure 4. The observed functional groups confirm successful polyaniline formation and cholesterol enzyme immobilization. The PANI, ChOx, and ChEs exhibit common characteristic bands at around 2,852, 1,565, and 1,300 cm^{-1}, which are the C–H bending vibration out of the plane of the para-disubstituted benzene rings, the C–C stretching band in benzenoid rings, the C–N stretching mode. In addition, PANI have distinctive peaks at 1,143 and 796 cm^{-1}, the vibration band of the anion (HCl–PAni), and N–H vibration band of aniline, respectively. Likewise, ChOx and ChEs have distinct absorption peaks at approximately 1,424, 2,867 and 3,524 cm^{-1}, which correspond to the CH_3, C–CH_3, OH vibration bands.

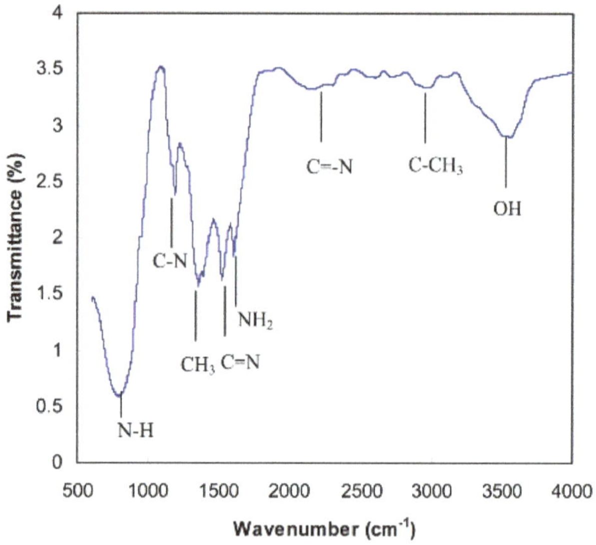

Figure 4: Typical FT-IR spectrum of the CNT electrode after electro-chemical polyaniline-enzyme coating.

CV curves for gold based and CNT based cholesterol bioprobe at a cholesterol concentration of 300 mg·dL^{-1} and 100 mV scan rate are shown in Figure 5. The gold based bioprobe was prepared by the same two-step electrochemical procedure as that was done for CNT based bioprobe. It is evident that CNT based bioprobe show much higher current sensitivity than gold based cholesterol bioprobe. The result confirms the enhancement of electrochemical detection of cholesterol by vertically aligned CNT structures. However, this result is in contrast to Lee's report [22] that Pd sputtered screen-printed carbon electrodes (SPCEs) exhibited higher electrochemical sensitivity than CNT modified SPCEs. The contradiction can be explained from the fact that the substrate surfaces for sputtered thin film coating are significantly different.

In Lee's work, screen-printed carbon electrodes (SPCEs) were used but standard SiO_2/Si substrates were utilized in this work. The surfaces of SPCEs were very rough and pitted while SiO_2/Si substrates were very smooth. The Pd sputtering caused better enhancement than CNTs modification on SPCEs because thin Pd sputtering could directly increase electron transfer rate without losing surface area whereas CNTs modification would reduce electrode surface area as Nafion/CNTs filled into pores of SPCE surface. In our work, vertically aligned CNTs that were directly grown on gold sputtered SiO_2/Si electrode significantly increased specific surface area and electron transfer rate.

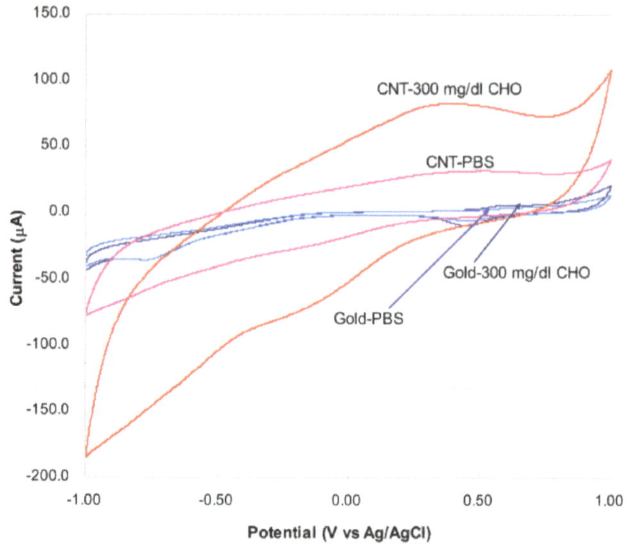

Figure 5: Cyclic voltammograms for gold and CNT based cholesterol bioprobe at 300 mg/dL concentration and 100 mV scan rates coating.

From Figure 5, the anodic and the cathodic peaks at scan rate of 0.1 V/s are centered on 397 and 526 mV, respectively. This could be ascribed to the electrocatalytic reduction of H_2O_2 by horseradish peroxidase following cholesterol release and decomposition by ChEs and ChOx, respectively [11], as illustrated in the following set of reactions:

cholesterol ester → fatty acid + cholesterol (cholesterol esterase) (1)

cholesterol → cholest-4-ene-3 one + H_2O_2 (cholesterol oxidase) (2)

$H_2O_2 + 2[Fe(CN)_6]^{4-} + 2H^+ \longrightarrow 2[Fe(CN)_6]^{3-} + 2H_2O$ (horseradish peroxidase) (3)

CV curves for cholesterol bioprobe at a cholesterol concentration of 300 mg.dl^{-1} with different scan rates are shown in Figure 6. It can be noticed from the given set of CV curves that magnitude of the anodic and cathodic peaks are different for different scan rates. The anodic peaks shift from 490 mV to 540 mV as the scan rate increases from 10 mV/s to 300 mV/s while the cathodic peaks change from −250 mV to −333 mV. Moreover, it can be seen that the redox reaction is not reversible since the peak current ratio between anodic and cathodic (i_{pa}/i_{pc}) is not equal to one for all scan rates. This implies that the electron transfer rate in the chemical reactions, which are coupled to the redox process or to the adsorption of either reactants or products that might occur during the scanning process, is slow relative to the voltage scan rate. Furthermore, the oxidation peak amplitude is found to increase linearly with the square root of scan rate as shown in the inset of Figure 6, indicating that the current is limited by semi-infinite diffusion of cholesterol on the CNTs electrode.

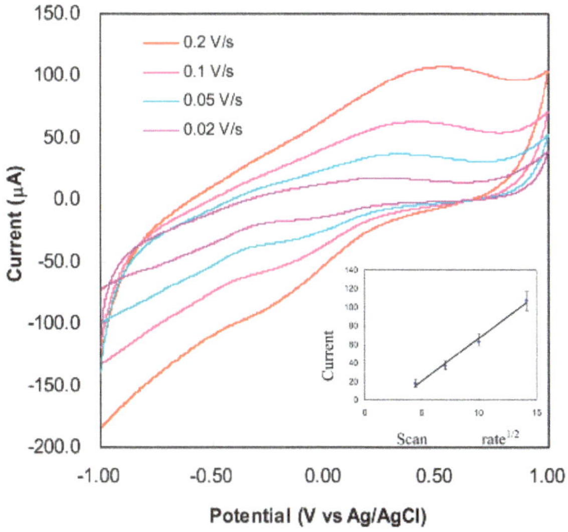

Figure 6: Cyclic voltammograms for cholesterol bioprobe at 300 mg/dL concentration with different scan rates. Inset: Peak current (oxidation peaks) as a function of the square root of scan rate. The error bars indicates the variation from five independent measurements.

Five independent measurements were made to assess repeatability and stability of the sensors. The error bars shown in the inset of Figure 6 demonstrate that the sensor has good repeatability and stability of less than 10% under different scan rates.

Figure 7 represents the set of waveforms obtained for different concentrations of cholesterol at a specific scan rate (100 mV/s). The difference in the peak current would be the detection capacity for the sensor in a cholesterol concentration at this working electrode surface of 1 mm^2 active area. It can be seen that the reduction and oxidation peaks become clearly pronounced as the cholesterol concentration increases from 100 to 300 mg/dL. The oxidation and reduction peaks centered around 450 and −250 mV, respectively. In addition, the peaks are more evident and located at higher voltage as scan rate increases at a fixed concentration.

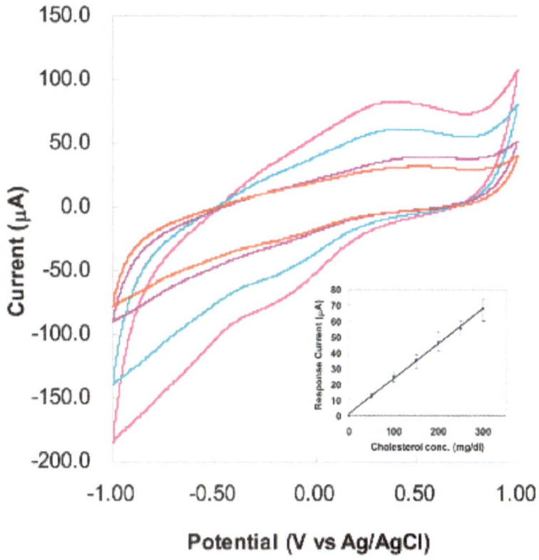

Figure 7: Cyclic voltammograms for cholesterol bioprobe at 100 mV/s scan rate with different cholesterol concentrations. Inset: Amperometric response for bioprobe (oxidation peaks) as a function of cholesterol concentration. The error bars indicates the variation from five independent measurements.

The amperometric response (inset of Figure 7) is the amplitude of oxidation peaks above background from CV curves. An approximately linear relationship between the analyte concentration and the response current of the CNT based electrode can be observed between 50 and 300 mg/dL concentration with a sensitivity of approximately 0.22 $\mu A/mg \cdot dL^{-1}$, which is considerably higher compared to previously reported CNT bioprobe [8]. The electrochemical sensing performance of the electrode can be attributed to the combination of CNTs properties and efficient enzyme immobilization by two step electrochemical process. Since the response current appeared to be in linear relation with the cholesterol concentration, so the enzyme catalytic reaction of cholesterol is the first-order reaction. The error bars shown in the inset of Figure 7 demonstrate that the sensor has good repeatability and stability of less than 15% over the whole concentration range.

Specificity is one of the most critical issues for biosensors to be used in real environment. To evaluate specificity of the sensors, interferences by four common electroactive species including ascorbic acid, glucose, acetaminophen and uric acid were tested at their normal higher concentration level in serum.

Figure 8 demonstrates interference peaks of ascorbic acid (1 mg/dL), glucose (150 mg/dL), acetaminophen and uric acid (7 mg/dL) along with the peak of cholesterol (200 mg/dL). It can be seen that ascorbic acid and glucose give very low interference signals, while acetaminophen and uric acid produce considerable interference. Nevertheless, the interference from these analytes can be minimized by setting the working electrode potential at its peak value (~0.4 V) since the peaks of interfering analytes locate at different potentials except glucose which has similar peak location but its interference is very low. Finally, long-term stability of the sensor was assessed. The sensors were stored dry at 4 °C and tested every day. It was found that the sensitivity was dropped by 15% after one month due to natural enzyme degradation. Thus, the sensors have a satisfactory life time. Therefore, the present sensor is promising for clinical diagnostics after the calibration to be done with real blood samples.

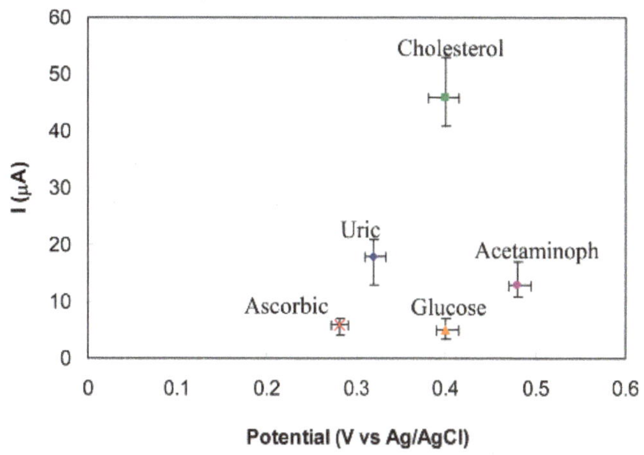

Figure 8: Oxidation peak current vs. corresponding peak potential of cholesterol (200 mg/dL), ascorbic acid (1 mg/dL), glucose (150 mg/dL), acetaminophen (5 mg/dL) and uric acid (7 mg/dL). The error bars indicates the variation for five independent measurements.

CONCLUSIONS

In conclusion, electropolymerized vertically aligned carbon nanotubes have been developed to enhance electrochemical sensing of gold coated Si electrode. The fabricated sensors have been characterized for cholesterol detection in the concentration range between 0 and 300 mg/dL cholesterol concentration by standard cyclic voltammogram measurement. An almost linear relationship between the analyte concentration and the response current of the CNT based electrode could be observed between 100 and 300 mg/dL concentration with a high current sensitivity. In addition, good specificity toward glucose, uric acid acetaminophen and ascorbic acid have been obtained. Moreover, sensors have satisfactory stability, repeatability and life time. Thus, the electropolymerized CNT bioprobe can effectively be used for cholesterol detection in normal range of cholesterol concentration in human blood. In addition, this developed scheme has the potentiality to integrate an array of sensors for lab-on-a-chip systems due to its compatibility with the standard Si micro-fabrication technology.

REFERENCES AND NOTES

1. Allain, C.C.; Poon, L.S.; Chan, C.S. Enzymatic determination of total serum cholesterol. *Clin. Chem.* 1974, *20*, 470–475.
2. McCreery, R.L. *Electroanalytical Chemistry*, 1st ed.; Marcel Dekker: New York, NY, USA, 1991; pp. 221–228.
3. Bescombes, J.L.; Cosnier, S.; Labbe, P. Polyphenol oxidase-catechol an electroenzymatic model system for characterizing the performance of matrices for biosensors. *Talanta* 1996, *43*, 1615–1619.
4. Yeh, J.I.; Lazareck, A.; Kim, J.H.; Xu, J.; Du, S. Peptide nanowires for coordination and signal transduction of peroxidase biosensors to carbon nanotube electrode arrays. *Biosens. Bioelectron.* 2007, *23*, 568–574.
5. Ohba, T.; Kaneko, K. Internal surface area evaluation of carbon nanotube with GCMC simulation-assisted N_2adsorption. *J. Phys. Chem. B* 2002, *106*, 7171–7176.
6. Li, G.; Liao, J.M.; Hu, G.Q.; Ma, N.Z.; Wu, P.J. Study of carbon nanotube modified biosensor for monitoring total cholesterol in blood. *Biosens. Bioelectron.* 2005, *20*, 2140–2144.
7. Guiseppi-Elie, A.; Lei, C.; Baughman, R.H. Direct electron transfer of glucose oxidase on carbon nanotubes. *Nanotechnology* 2002, *13*, 559–564.
8. Roy, S.; Vedala, H.; Choi, W.B. Vertically aligned carbon nanotube probes for monitoring blood cholesterol. *Nanotechnology* 2006, *17*, 14–18.
9. Ferri, T.; Poscia, A.; Santucci, R. Direct electrochemistry of membrane-entrapped horseradish peroxidase. Part I A voltammetric and spectroscopic study. *Bioelectrochem. Bioenergy* 1998, *44*, 177–181.
10. Su, X.; Zong, Y.; Richter, R.; Knoll, W. Enzyme immobilization on poly(ethylene-co-acrylic acid) films studied by quartz crystal microbalance with dissipation monitoring. *J. Colloid Interface Sci.* 2005, *287*, 35–42.
11. Kan, J.; Pan, X.; Chen, C. Polyaniline–uricase biosensor prepared with template process. *Biosensors Bioelectron.*2004, *19*, 1635–1640.
12. Mu, S.; Xue, H. Bioelectrochemical characteristics of glucose oxidase immobilized in a polyaniline film. *Sensors Actuators B* 1996, *3*, 155–160.
13. Trojanowich, M.; Krawczyk, T. Krawczynsky vel Electrochemical biosensors based on enzymes immobilized in electropolymerized films. *Microchim. Acta* 1995, *121*, 167–181.
14. Losada, J.; Garcia Armada, M.P. A glucose amperometric sensor based on covalent immobilization of glucose oxidase in poly-2-aminoaniline film via chloranil on platinized platinum electrode. *Electroanalysis* 1997, *9*, 1416–1421.
15. Garjonyte, R.; Malinauskas, A. Amperometric glucose biosensors based on Prussian Blue—and polyaniline–glucose oxidase modified electrodes. *Biosensors Bioelectron* 2000, *15*, 445–451.
16. Wang, J.; Musameh, M. Carbon-nanotubes doped polypyrrole glucose biosensor. *Analytica Chimica Acta* 2005,*539*, 209–213.
17. Chen, G.; Shaffer, M.; Colbey, D.; Dixon, G.; Zhou, W.; Fray, D.; Windle, A. Carbon nanotube and polypyrrole composites coating and doping. *Adv. Mater.* 2000, *12*, 522–526.
18. Wei, Z.; Wan, M.; Lin, T.; Dai, L. Polyaniline nanotubes doped with sulfonated carbon nanotubes made via a self-assembly process. *Adv. Mater.* 2003, *15*, 136–139.
19. Wang, J.; Musameh, M.; Lin, Y. Solubilization of carbon nanotubes by nafion towards the preparation of amperometric biosensors. *J. Am. Chem. Soc.* 2003, *125*, 2408–2409.
20. Rubianes, M.D.; Rivas, G.A. Carbon nanotubes paste electrode. *Electrochem. Commun.* 2003, *5*, 689–694.
21. Luong, J.; Hrapovic, S.; Wang, D.; Bensebaa, F.; Simard, B. Solubilization of multiwall carbon nanotubes by 3-aminopropyltriethoxysilane towards the fabrication of electrochemical biosensors with promoted electron transfer. *Electroanalysis* 2004, *16*, 132–139.
22. Lee, C.-H.; Wang, S.-C.; Yuan, C.-J.; Wen, M.-F.; Chang, K.-S. Comparison of amperometric biosensors fabricated by palladium sputtering, palladium electrodeposition and Nafion/carbon nanotube casting on screen-printed carbon electrodes. *Biosen. Bioelectron.* 2007, *22*, 877–884.

CHAPTER 20

COMPARATIVE BIOCHEMICAL CHARACTERIZATION OF SOLUBLE AND CHITOSAN IMMOBILIZED β-GALACTOSIDASE FROM KLUYVEROMYCES LACTIS NRRL Y1564

Ariosvana Fernandes Lima[1], Kenia Franco Cavalcante [2], Maria de Fátima Matos de Freitas [2], Tigressa Helena Soares Rodrigues [2], Maria Valderez Ponte Rocha [2], Luciana Rocha Barros Gonçalves [2]

[1]Instituto Federal de Educac¸ão, Ciência e Tecnologia do Ceará, Campus Limoeiro do Norte, Brazil

[2]Departamento de Engenharia Química, Universidade Federal do Ceará, Campus do Pici, Bloco 709, CEP 60455-760, Fortaleza, CE, Brazil

ABSTRACT

An investigation was conducted on the production of β-galactosidase (β-gal) by different strains of *Kluyveromyces*, using lactose as a carbon source. The maximum enzymatic activity of 3.8 ± 0.2 U/mL was achieved by using *Kluyveromyces lactis* strain NRRL Y1564 after 28 h of fermentation at 180 rpm and 30 °C. β-gal was then immobilized onto chitosan and characterized based on its optimal operation pH and temperature, its thermal stability and its kinetic parameters (K_m and V_{max}) using o-nitrophenyl β-d-galactopyranoside as substrate. The optimal pH for soluble β-gal activity was found to be 6.5 while the optimal pH for immobilized β-gal activity was found to be 7.0, while the optimal operating temperatures were 50 °C and 37 °C, respectively. At 50 °C, the immobilized enzyme showed an increased thermal stability, being 8 times more stable than the soluble enzyme. The immobilized enzyme was reused for 10 cycles, showing stability since it retained more than 70% of its initial activity. The immobilized enzyme retained 100% of its initial activity when it was stored at 4 °C and pH 7.0 for 93 days. The soluble β-gal lost 9.4% of its initial activity when it was stored at the same conditions.

INTRODUCTION

The enzymatic hydrolysis of lactose by β-galactosidase (β-gal) plays an important role in the processing of dairy products, such as the production of milk containing low concentrations of lactose, the prevention of crystallization in dairy products, and the use of galactosyltransferase for synthesizing galacto-oligosaccharides [1] and [2].

Kluyveromyces strains have been extensively studied for their production of β-gal, especially *Kluyveromyces lactis* and *Kluyveromyces marxianus*, which possess an ability to assimilate lactose when lactose is used as a carbon source [3]. The yeast *K. lactis* is a source of β-galactosidases (β-d-galactohydrolase, EC 3.2.1.23) [4] and is of great commercial importance, with the complex mixture of enzymes in Lactozym 3000 L HP G being the enzymatic preparation that is most widely used [5].

As it is relatively costly to produce the pure enzyme, an alternative to reduce the production cost would be to use more affordable culture media such as whey, which is the liquid

Citation: Ariosvana Fernandes Lima, Kenia Franco Cavalcante, Maria de Fátima Matos de Freitas, Tigressa Helena Soares Rodrigues, Maria Valderez Ponte Rocha, Luciana Rocha Barros Gonçalves, Comparative biochemical characterization of soluble and chitosan immobilized β-galactosidase from Kluyveromyces lactis NRRL Y1564, Process Biochemistry, Volume 48, Issue 3, March 2013, Pages 443-452, ISSN 1359-5113, http://dx.doi.org/10.1016/j.procbio.2013.02.002.

portion that is formed upon the coagulation of casein in milk when cheese is produced. Milk consists of approximately 85 to 95% whey by volume, and whey contains 55% of the nutrients found in milk [6]. Cheese whey is treated and transformed into various food products. Among those products, whey powder accounts for a large fraction of the whey that is processed, therefore maintaining the quality of fresh whey for a longer period of time, facilitating manipulation and transport [7].

The main disadvantage of using soluble enzymes as industrial catalysts is their poor stability in aqueous solutions against extreme pH, temperature, high ionic strengths, among others [8]. Various approaches have been proposed to solve this problem, including enzyme immobilization, the use of glycols and other stabilizing additives, reticulation with bifunctional agents, the use of enzymes from thermophilic organisms, or a chemical modification of the enzyme [9] and [10]. The immobilization of enzymes is usually advantageous in numerous ways: the biocatalyst can be reused, the reaction volume can be reduced because the immobilized enzyme can be used at higher concentrations, there is an ease of operational control, it is easier to control the process parameters when the reactor is operated continuously, and the product can be more easily separated from the immobilized enzyme because the immobilized enzyme is not soluble in the reaction medium [11]. Pessela et al. [12] studied the immobilization of β-gal from *Escherichia coli* on different supports and observed that the recovery activity was dependent on the immobilization support, which may be associated with the different levels of enzyme distortion caused by immobilization. It has also been reported [13] that the cross-linking of the support with glutaraldehyde may enhance its hydrophobicity, which may affect the enzyme catalytic activity.

This work aims to study how *Kluyveromyces* strains can be used to produce β-gal from an agro-industrial by-product such as whey powder. It also aims to investigate the immobilization of the enzyme onto chitosan and determine its properties such as the optimal operating pH and temperature, the thermal stability of the enzyme, the thermal denaturation constant, the half-life (at 40 °C, 50 °C and 60 °C) and the kinetic parametersK_m and V_{max} of the enzyme β-gal from *K. lactis* strain NRRL Y1564.

Chitosan is the deacetylated form of chitin, the second most abundant polymer in nature after cellulose. It is a low cost, renewable, biodegradable natural product, which has very good biocompatibility, low toxicity, chemically inert and high hydrophilicity [14] and [15]. This biopolymer has a similar molecular structure to cellulose, differing only in its functional groups. Hydroxyl groups (OH) are present on the general structure of these biopolymers, but the main difference between them is the presence of amino groups (NH_2) in the structure of chitosan. Different configurations of chitosan can be obtained in the process of deacetylation of chitin and these configurations may be used in the process of enzyme immobilization. However, hydrogels and membranes are the most used forms due to the obtained physical changes, such as increasing the pore diameter of the support, suitable for the process of immobilization of enzymes. Different protocols may be employed for the immobilization of enzymes in chitosan, such as adsorption, encapsulation and covalent bounding. Among the different methods, the covalent immobilization onto glutaraldehyde activated chitosan is the method of choice due to the presence of amino groups in the structure of chitosan that react with this agent in mild conditions close to neutrality [14].

MATERIALS AND METHODS

Microorganisms

Three *K. marxianus* strains (LAMI CE 025; CCA 510; ATCC 36907 – CCT André Tosello) and two *K. lactis* strains (NRRL Y1564 and NRRL 4087) were tested. Stock cultures were maintained at 4 °C in YPD (yeast extract, peptone and dextrose) agar plates containing 10 g/L of yeast extract, 20 g/L of peptone and 20 g/L of dextrose.

Materials

Powdered chitosan was purchased from Polymar Ind. Ltda (Ceará, Brazil). Bovine serum albumin (BSA), glutaraldehyde solution Grade II 25% and the synthetic substrates *o*-nitrofenil-β-d-galactopyranoside (99%, ONPG) and o-nitrophenol (99%, ONP) were purchased from Sigma–Aldrich (St. Louis, USA). Dinitrosalicylic acid (DNS) was obtained from Vetec (São Paulo, Brazil). Glicose PAP Liquiform, the enzymatic kit for glucose determination, was obtained from Labtest Diagnóstica (Minas Gerais, Brazil). Whey powder was kindly provided by Alibra Ingredientes Ltda (Campinas, Brazil). All other analytical grade reagents were purchased from Synth (São Paulo, Brazil) and Vetec (São Paulo, Brazil).

Pre-treatment of the whey

The whey powder was diluted in distilled water to obtain a lactose concentration of 50 g/L and heated to 90 °C. Commercial lactic acid was then added to the whey solution under mild agitation to reach pH 4.8. Agitation was then interrupted and the solution was left to sit for several minutes to settle its contents. Whey was then filtered using qualitative filter paper [16].

Culture medium preparation

A synthetic medium was prepared according to a previous study [17], comprising 50 g/L of lactose, 5 g/L of KH_2PO_4, 1.2 g/L of $(NH_4)_2SO_4$, 0.4 g/L of $MgSO_4 \cdot 7H_2O$ and 1 g/L of yeast extract in 1 L of potassium phosphate buffer (0.2 M) at pH 5.5. A complex medium, described as "deproteinized whey" in this study, was formulated by adding 1 g/L of yeast extract to the pre-treated whey, where the pH was adjusted to 6.0 using a 4 mol/L solution of KOH. Both types of culture medium were placed in 250 mL Erlenmeyer flasks and sterilized at 115 °C for 15 min.

Inoculum preparation

The *Kluyveromyces* strains were initially grown in YPD agar plates (containing 10 g/L yeast extract, 20 g/L of peptone, 20 g/L of dextrose, 20 g/L of agar) and incubated at 30 °C for 24 h. Three colonies were then transferred into separate 250-mL Erlenmeyer flasks containing 50 mL of synthetic medium each [17]. The flasks were incubated at 30 °C and shaken at 180 rpm for 24 h. The optical density (OD) was then determined at a wavelength of 620 nm using a spectrophotometer and OD was adjusted to 1.0.

Batch cultivation conditions

All assays were conducted in 250-mL Erlenmeyer flasks containing 50 mL of culture medium that were shaken on a TE240 rotary shaker (Tecnal, São Paulo, Brazil) at 180 rpm and 30 °C [17]. Experiments were initiated by inoculating the culture medium with inocula to a concentration of 10% (v/v) and then maintaining the culture at 30 °C with shaking at 180 rpm. Batch cultivations lasted 24 h, when to the selection of lactose-consuming strains, and 30 h, when studying the effect of incubation time on the batch production of β-gal by the *K. lactis* NRRL Y1564 strain cultivated in deproteinized whey. Approximately 1 mL of sample was collected from each flask to determine the cell growth, the pH and the lactose concentration. The assays were performed in duplicate for each different microorganism.

The volumetric enzyme activity in terms of U/mL is defined as the enzyme activity per unit volume of culture medium, and it was used to investigate the production kinetics of β-gal[18]. The specific enzyme activity in terms of U/g was calculated by dividing the enzymatic activity by the mass of cells harvested prior to the extraction of the enzyme [18] and [19].

Selection of β-gal-producing yeasts

A screening of lactose-utilizing among six non-pathogenic *Kluyveromyces* strains was conducted in a synthetic culture medium. The criteria used for screening and microorganism selection were cell growth and lactose uptake. The strains that demonstrated both growth and lactose uptake were then cultivated in the complex medium to select yeasts that could produce β-gal using the lactose present in whey.

Extraction of the enzyme β-galactosidase from the culture medium

β-Galactosidase is an intracellular enzyme and has to be released from the cells; hence, cell autolysis was performed according to a previous literature report [20] with some modifications. Approximately 20 mL of the sample was collected and centrifuged at 10,000 × *g* for 15 min at 4 °C. The cell pellets obtained from centrifugation were washed once with 20 mL of cold distilled water and centrifuged again under the same conditions. After washing, the cell pellets were re-suspended in a potassium phosphate buffer at pH 6.6 containing 0.1 mM of $MnCl_2 \cdot 4H_2O$ to obtain a cell density of 2.9 mg/mL. Ethanol and toluene were added to the buffer at concentrations of 10% (v/v) and 2% (v/v) respectively to permeabilize the cells in 125 mL Erlenmeyer flasks, which were shaken at 150 rpm for 20 min at 37 °C. The suspension of the permeabilized cells was then centrifuged at 10,000 × *g*, for 15 min at 4 °C. The cell-free supernatant containing the crude enzyme extract was then collected and refrigerated.

Preparation and modification of supports used in enzyme immobilization

In this work, chitosan was used as a support for enzyme immobilization [21]. Powdered chitosan was solubilized in 2% (v/v) acetic acid to a concentration of 2.0 wt.% and homogenized for 30 min before 20 mL of this solution was placed in a 100 mL-glass beaker at 50 °C and left to stabilize for several minutes under mechanical agitation. After the stabilization of the chitosan solution, 30 mL of a KOH 0.5 M solution was added to the reactor, and the reactor was then further agitated for 30 min. Support activation was conducted by adding glutaraldehyde to the solution of chitosan coagulated with KOH to a concentration of 0.8% (v/v). After 30 min of activation at 50 °C, the particles were immediately washed with distilled water at room temperature to remove excess glutaraldehyde and then dried under a vacuum.

The immobilization of β-gal onto glutaraldehyde-activated chitosan

The immobilization of β-gal that was produced by *K. lactis* NRRL Y1564 was carried out in a batch reactor under mild agitation at 25 °C by mixing the enzyme in a 50 mM potassium phosphate buffer at pH 7.0, containing 0.1 mM manganese chloride and the previously activated support, for different contact times between the enzyme and the support (5, 10 and 24 h). In all experiments, the initial quantity of enzyme added to the buffer was 10 mg of protein/g support. To determine the immobilization yield, samples of supernatant were withdrawn and their hydrolytic activities were measured. A blank assay was also conducted to evaluate the possibility of enzyme deactivation in the buffer while it was undergoing immobilization. Therefore, a solution of β-gal in phosphate buffer with 0.1 mM of added manganese chloride was prepared in a separate reactor as a control experiment, but there were no chitosan supports placed in that reactor.

Immobilization yield

After measuring the initial (*Ati*) and final (*Atf*) enzyme activities in the supernatant of the immobilization suspension and in the reference suspension, the remaining enzyme activity in the supernatant (*Ati* − *Atf*) and the immobilization yield (IY) could be calculated according to Eq. (1)[22]. The reference suspension is a control assay that was performed in the absence of support to discard any possibility of loss of enzyme activity during immobilization.

$$IY_{imob}(\%) = \frac{At_i - At_f}{At_i} \times 100 \qquad (1)$$

Recovery activity (Atr)

The theoretical activity (*Att*) of immobilized β-gal onto the chitosan support could be calculated (Eq. (2)) as a product of the enzymatic activity offered per unit mass of support (At_{off}) and the immobilization yield, where the immobilization yield can be obtained from Eq. (1). After measuring the activity of the immobilized enzyme per unit mass of support (*Atd*), the recovery activity (*Atr*) could be calculated according to Eq. (3).

$$At_t = IY \times At_{off} \qquad (2)$$

$$At_r = \frac{At_d}{At_t} \times 100 \qquad (3)$$

The optimal operating pH and temperature of β-gal produced by the *K. lactis*NRRL Y1564 strain

The effects of pH and temperature on the activities of soluble and immobilized β-gal produced by the *K. lactis* NRRL Y1564 strain were evaluated. Samples were incubated at a pH ranging between 5.5 and 8.0 and a constant temperature of 37 °C to evaluate the effect of pH on enzymatic activity. The effect of temperature on enzymatic activity was determined at a constant pH of 6.6 [23] in a temperature range of 25 and 60 °C. In this work, the optimal operating pH and temperature refers to the maximum enzymatic activity.

An evaluation of the thermal stabilities of soluble and immobilized β-gal produced by the *K. lactis NRRL Y1564 strain*

Soluble and immobilized β-gal was placed in separate 50 mM sodium phosphate buffer solutions at pH 8.6 and kept at 3 different temperatures of 40, 50 and 60 °C for more than 4 h. Samples were periodically withdrawn and their residual activities were evaluated by analyzing their ability to hydrolyze ONPG. A first order deactivation model [24] was fitted to the experimental data using Origin 8.1 software (OriginLab Corporation, Northampton, USA). The knowledge of the exponential decay model and its parameters allows for the determination of the deactivation constants (*Kd*) and the half-lives ($t_{1/2}$) of the biocatalysts. Stabilization factors (SF) were obtained as the ratio between the half-lives of the immobilized β-gal and the soluble β-gal, see Eq. (4). The deactivation energy was estimated based on the Arrhenius equation [25].

$$SF = \frac{t_{1/2\,\text{immobilized}}}{t_{1/2\,\text{soluble}}} \tag{4}$$

Calculation of cultivation parameters

The specific cell growth rates (μX), lactose uptake rate (μS), and the specific product formation rates (μP) were estimated according to the methodology described by Schmidell et al. [26] according to Eqs. (5) and (6). The relative magnitudes of these parameters provide an important classification of the fermentation process, where product formation may be associated, partly linked, or not linked at all to cell growth.

$$\mu_X = \frac{1}{x}\frac{dX}{dt} \tag{5}$$

$$\mu_P = \frac{1}{x}\frac{dP}{dt} \tag{6}$$

The volumetric productivity was calculated as the ratio of the component concentration to the fermentation time (t), see Eqs. (7) and (8):

$$P_X = \frac{X}{t} \tag{7}$$

$$P_1 = \frac{C_1}{t} \tag{8}$$

where, PX is the volumetric productivity of biomass, P_1 is the volumetric productivity of β-gal, X is biomass concentration and C_1 is volumetric enzyme activity.

The substrate conversion, which in the case of this study was lactose, was calculated by Fogler [27] according to Eq. (9):

$$\Delta S(\%) = \frac{S_0 - S}{S_0} \times 100 \tag{9}$$

where S_0 refers to the initial substrate concentration and S refers to the substrate concentration in the samples taken for analysis at each time interval.

Determination of the kinetic parameters in the hydrolysis of *o*-nitrofenyl-β-d-galactopyranoside

The kinetic parameters of ONPG hydrolysis were determined in the range of 0.1–7.5 mM in 50 mM potassium phosphate buffer containing 0.1 mM $MnCl_2 \cdot 4H_2O$ at 37 °C. The Michaelis–Menten constant (K_m) and the maximum reaction velocity (V_{max}) were calculated by fitting a non-linear Michaelis–Menten model to the experimental data using Origin 8.1 software. The catalytic efficiency of the immobilized enzyme was calculated based on a previous report [28], but with slight modifications:

$$\text{catalytic efficiency}(\%) = \frac{(V_{max}/K_m)_{\text{Immob}}}{(V_{max}/K_m)_{\text{soluble}}} \times \frac{[E_0]_{\text{Immob}}}{[E_0]_{\text{soluble}}} \times 100 \tag{10}$$

where E_0 refers to the initial enzyme activity.

The estimation of thermodynamic parameters

To determine the thermodynamic parameters of the immobilized enzyme, the enthalpy of inactivation (ΔH^*), the Gibbs free energy of inactivation and the entropy of inactivation (ΔS^*) were calculated according to Ustok et al. [25].

Reuse of the immobilized β-gal

The reuse of the immobilized enzyme was assayed by using 1 g of the biocatalyst in successive batches of ONPG hydrolysis. The assay conditions were the same as those described in Section 2.19.2. At the end of each batch, the support was removed from the reaction medium, washed with phosphate buffer to remove any remaining substrate or product, dried under a vacuum and subjected to a fresh batch of ONPG for hydrolysis. The residual activity was calculated as the activity of the immobilized enzyme measured after each cycle relative to the activity of the immobilized enzyme during the first cycle.

Storage stability of the immobilized β-gal

Biocatalyst samples were also stored at 4 °C in a 50 mM potassium phosphate buffer at pH 7.0, and their residual hydrolytic activities were determined after 93 days of storage to evaluate the storage stability of the soluble and immobilized enzyme.

Lactose hydrolysis by the soluble and immobilized β-gal

The hydrolysis of lactose by the soluble and immobilized β-gal was performed based on previous reports [12] and [28], but with slight modifications. Lactose hydrolysis was performed using pure lactose at 2% in 50 mM potassium phosphate buffer solution, pH 7.0, containing 0.1 mM $MnCl_2 \cdot 4H_2O$. The reaction was conducted in a 100 mL-glass beaker at 40 °C for 360 min under orbital agitation (150 rpm). Experiments were initiated by contacting 50 mL of lactose solution with 1 g or 1.3 mL of immobilized (3.8 U/g) or soluble enzyme (2.9 U/mL), respectively. Samples of supernatant were withdrawn and glucose release was measured using an enzymatic kit (Glicose PAP Liquiform) for glucose determination. The assays were performed in triplicate for each different biocatalyst.

Analytical methods

Biomass concentration

Biomass concentration was determined by measuring the optical density of the samples (O.D.) at 620 nm using a Spectronic spectrophotometer (Genesys, Series 20, USA) and converted into concentration (g/L) according to the standard curve that was developed for the specific microorganism.

Enzyme activity

The hydrolytic activity of soluble and immobilized β-gal was determined using *o*-nitrophenyl-β-d-galactopyranoside as a substrate at pH 6.6 and 37 °C, according to the methodology described in literature previous report [20] but with some modifications. A 50 µL enzyme sample was added to a test tube containing a 2 mL solution of 1.25 mM of ONPG, mixed into a 50 mM potassium phosphate buffer solution containing 0.1 mM $MnCl_2 \cdot 4H_2O$, and maintained at 37 °C for 5 min. The immobilized enzyme activity was measured by adding 0.05 g of immobilized enzyme to the test tube. The reaction was stopped and 0.5 mL of 1 M sodium carbonate was added. The *o*-nitrophenol product concentration was measured at a UV wavelength of 420 nm. One unit (U) of β-gal is defined as the quantity of enzyme that liberates 1 µmol of *o*-nitrophenol/min under the test conditions.

Lactose and ethanol concentrations

Lactose and ethanol concentrations were measured using high performance liquid chromatography (Waters, Milford, MA, USA), with a Waters refractive index detector (Model 2414) and an Aminex HPX-87H column at 65 °C. The eluent was 5 mM H_2SO_4, and it was pumped through the column at a flow rate of 0.5 mL/min. The injection volume of the samples was 20 µL and the analysis time for each sample was 30 min.

Reducing sugars concentration

The concentration of the total reducing sugars was determined colorimetrically by the dinitrosalicylic acid (DNS) method [29]. An appropriate calibration curve was used to convert absorbance to concentration.

Protein concentration

The protein concentration was determined according to the Bradford assay [30] at a wavelength of 595 nm using bovine serum albumin (BSA) as the standard.

RESULTS AND DISCUSSION

Selection of lactose-consuming strains

Literature reports [17] and [31] have shown that yeasts such as *K. marxianus* and *K. lactis*produce β-gal when lactose is used as the carbon source for cultivation. Therefore,*Kluyveromyces* strains were screened for their ability to uptake lactose, where five strains were grown in a synthetic medium containing 10 g/L of lactose as shown in Table 1.

Table 1: The cell growth and the lactose uptake of the different strains of *Kluyveromyces* cultivated in the synthetic medium containing 10 g/L of lactose at 30 °C and an agitation of 180 rpm after 24 h.

Microorganism	Cell growth	Lactose consump-tion
K. marxianus CE 025	+	–
K. marxianus CCA 510	+	–
K. marxianus ATCC 36907	+	+
K. lactis NRRL Y1564	+	+
K. lactis NRRL Y4087	+	+

Response: (+) positive (cell growth or lactose uptake); (–), negative (no measurable cell growth or lactose uptake).

All strains were able to grow in the synthetic medium as shown in Table 1. Nevertheless, only three of the five strains evaluated exhibited lactose uptake, namely the *K. marxianus*ATCC 36907 strain, the *K. lactis* NRRL Y1564 strain, and the *K. lactis* NRRL Y4087 strain. These three strains were selected to assess their β-gal production capabilities when grown on whey. The *K. marxianus* CE 025 and CCA 510 strains showed no lactose uptake and were therefore excluded from this study.

The selection of β-gal-producing yeasts from deproteinized whey

Table 2 shows the cell growth, the lactose uptake and the enzyme activity of β-gal after 24 h of cultivating the *K. lactis* NRRL Y1564 strain, the *K. lactis* NRRL Y4087 strain, and the *K. marxianus* ATCC 36907 strain in a deproteinized whey medium. The *K. lactis*NRRL Y1564 strain showed the highest cell growth, consuming approximately 2.2 times more lactose than the *K. lactis* NRRL Y4087 strain. However, the *K. marxianus* ATCC 36907 strain showed practically no lactose uptake when cultivated in the whey medium.

Table 2: The cell growth, the lactose uptake and the production of β-gal by *K. lactis* NRRL Y1564, *K. lactis* NRRL Y4087 and *K. marxianus* ATCC 36907 strains cultivated in deproteinized whey at 30 °C and an agitation of 180 rpm for 24 h. Volumetric enzyme activity was defined as the hydrolytic β-gal activity/mL of culture medium while specific enzyme activity was calculated by dividing the hydrolytic activity by the amount of cell harvested before enzyme extraction. The initial lactose concentration (S_0) was 50 g/L.

Microorganism	Initial biomass (g/L)	Final biomass (g/L)	Final lactose concentration (g/L)	Volumetric enzyme activity[a](U/mL)	Specific enzyme activity[a](U/g)
K. lactis NRRL Y1564	0.27	2.97	11.50	0.75	251.72
K. lactis NRRL Y4087	0.27	1.35	20.05	0.09	68.96
K. marxianus ATCC 36907	0.16	1.97	50.00	0.33	168.96

[a]Enzyme activity was determined at room temperature (27 °C).

Of the yeasts used in this study, the *K. lactis* NRRL Y1564 strain was selected because it had a volumetric enzyme activity that was 72.6% greater than the enzyme produced by the *K. lactis* NRRL Y4087 strain and 67.1% more activity than the enzyme produced by the *K. marxianus* ATCC 36907 strain.

Other authors [20] cultivated various strains of *K. lactis* and *K. marxianus* in a medium containing 10 g/L lactose as a carbon source. The values that they obtained for enzyme activity ranged between 0.7 and 1.3 U/mL for *K. lactis* and ranged between 0.7 and 1.0 U/mL for *K. marxianus*. Braga et al. [31] studied seven strains of *Kluyveromyces* and found that the *K. marxianus* ATCC 16045 strain was the best producer of β-gal in culture medium containing agro-industrial waste materials such as whey and water from parboiled rice (left over). The highest enzymatic activity achieved was 10.4 U/mL when whey was used as a culture medium, with concentrations of 120 g/L lactose, 5 g/L yeast extract, 15 g/L peptone, 15 g/L $(NH_4)_2SO_4$ and 30 g/L water that was left over from parboiled rice at pH 4.0.

The effect of incubation time on the batch production of β-gal by the *K. lactis* NRRL Y1564 strain cultivated in deproteinized whey

Fig. 1 shows the cell growth profile, the lactose uptake, the volumetric enzyme activity, and the ethanol concentration when the *K. lactis* NRRL Y1564 strain was cultivated in deproteinized whey that was supplemented with yeast extract. A maximum enzyme activity of 3.8 ± 0.2 U/mL and a maximum volumetric productivity of 0.14 U/mL/h was obtained after 28 h of cultivation. Other authors [18] also observed a peak in the enzyme activity and productivity when β-gal was produced from the *K. marxianus* CCT 7082 strain. The highest concentration of biomass was 4.3 g/L, which was achieved after 30 h of cultivation when the lactose was completely consumed. On the other hand, the maximum biomass productivity of approximately 0.035 g/L/h was achieved after 12–14 h of cultivation. During the same period, approximately 100% of the substrate was consumed, indicating that the final lactose concentration in the medium was less than 1 g/L. Ethanol was also produced as a result of cell cultivation, reaching a concentration of 16.9 ± 3.5 g/L after 25 h of cultivation. Other authors [32] have studied the production of β-gal by *K. lactis* and observed that the volumetric enzymatic activity increased during the exponential growth phase. They also observed the production of ethanol and mentioned that it is always a fermentation product when lactose is metabolized by *K. lactis*.

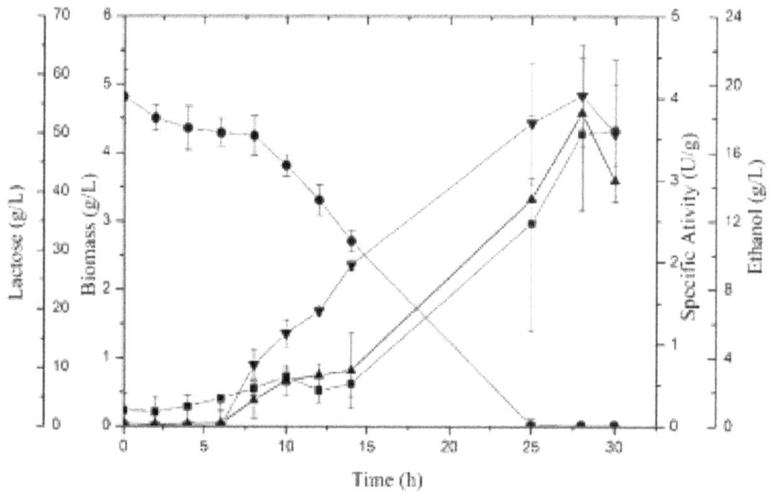

Figure 1: Cultivation of a *K. lactis* NRRL Y1564 batch culture in deproteinized whey at 30 °C and 180 rpm as a function of time: (■) biomass (g/L); (●) lactose (g/L); (▲) enzyme activity (U/mL) and (▼) ethanol concentration (g/L).

It can be observed through a comparison of cell growth and enzyme production as shown in Fig. 1 that β-gal production started early in the exponential phase and ran parallel to cell growth kinetics during that phase for up to 28 h, indicating that it was associated with cell growth. Furthermore, the similarity between the μX and the μP curves as shown in Fig. 2suggests that ethanol is a primary metabolite produced by this strain when lactose is fermented [33]. According to Gaden [34], processes that show only one maximum for the specific rate curves, where the curves are virtually coincident, are usually known as growth-associated processes. It is therefore possible to say that the synthesis of β-gal by the *K. lactis* NRRL Y1564 strain in this experiment is associated with populational growth when cultivated in a medium containing lactose as the sole carbon source. Various studies [20], [35] and [36] performed on enzyme production have made a common observation that the highest enzyme activity per unit of cell concentration is achieved at the beginning of the stationary phase and declines thereafter, even though lactose is still present in the medium.

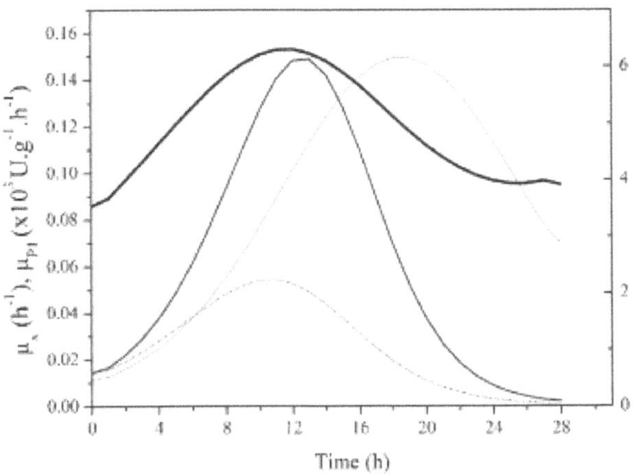

Figure 2: Specific growth (μX), lactose uptake rate (μS), β-gal production (μP_1) and ethanol production (μP_2) rates by a *K. lactis* NRRL Y1564 strain cultivated in deproteinized whey at 30 °C and 180 rpm: (—) μX (h^{-1}); (▬▬) μS (h^{-1}); (- · · - · -) μP_1 (U/g h) and (- - - -) μP_2 (h^{-1}).

Other authors [17] and [20] working with strains of *K. lactis* (NRRL Y 1564 and NRRL Y 8279) grown on a synthetic medium containing lactose, observed that the peaks in enzymatic activity were obtained after 24 h of fermentation with values ranging between 0.5 and 1.5 U/mL, which are lower than the 3.5 U/mL that has been obtained in this present work after 28 h of fermentation. Matheus and Rivas [37] optimized the β-gal production by *K. lactis* using deproteinized whey as a fermentation medium, which contained 4.85% lactose supplemented with 0.15% yeast extract and 0.1% $(NH_4)_2SO_4$. The enzyme activity varied between 1.29 and 7.43 U/mL, depending on the operational conditions.

Immobilization of β-gal onto chitosan activated with glutaraldehyde

The immobilization of β-gal onto chitosan activated with glutaraldehyde was investigated in this work, with the aim of obtaining a biocatalyst possessing operational and thermal stabilities. Table 3 shows the results of the effect of contact time between the enzyme and the support on the derivative activity (*Atd*), the recovery activity (*Atr*), and the immobilization yield (IY). No expressive loss of enzyme activity was observed in the reference suspension during immobilization.

Table 3: Immobilization parameters of β-gal on chitosan activated with glutaraldehyde: derivative (*Atd*) and recovery activity (*Atr*), immobilization yield (IY), enzymatic activities of the soluble enzyme in the supernatant before (At_0) and after (*Atf*) immobilization. Immobilization conditions: 9 mL of enzyme solution mixed with 50 mM potassium phosphate buffer containing 0.1 mM manganese chloride at pH 7.0 and mixed with 1 g of support at 25 °C. Enzyme load: 10 mg protein/g support.

Contact time (h)	At_0 (U/g)	*Atf* (U/g)	*Atd* (U/g)	*Atr* (%)	IY (%)
5	6.40	2.26	0.32	7.7	64.7
10	6.15	1.13	0.62	12.4	81.6
24	5.39	0.03	0.53	9.9	99.5

Table 3 shows that higher immobilization yields were achieved with an increased contact time. These results indicate that the first bonds between enzyme molecules and activated support are built up quickly. Nevertheless, the next bonds take more time to be formed since the protein itself becomes a barrier to enzyme molecules diffusion [38]. Thus, longer contact times are needed to overcome the intra-particle diffusion effects, caused by the presence of the enzyme inside the pores, and promote more enzyme molecules to link to the support.

The *Atd* and the *Atr* were observed to increase for 10 h (see Table 3). After 10 h, there was no further increase in enzyme and recovery activity despite the increased enzyme load on the support. Therefore, a contact time of 10 h was selected as the optimal contact time for further immobilization assays.

Afterwards, soluble and immobilized β-gal samples were stored in potassium phosphate buffer at a pH 7.0 and a temperature 4 °C for more than 3 months. The immobilized enzyme retained 100% of its initial activity after being stored for 93 days. In contrast, the soluble β-gal experienced a 9.4% reduction in its activity when stored under the same conditions for the same period of time. This is in accordance with the results that have been published in the literature [8], [15], [39] and [40]. Dwevedi and Kayastha [15] immobilized β-gal obtained from peas onto chitosan and observed no activity loss of the immobilized enzyme for up to 40 days. Ansari and Husain [39] immobilized β-gal obtained from *Aspergillus oryzae* onto a support constructed out of concanavalin A-Celite 545 and observed that the enzyme retained 78% of its initial activity after 2 months of storage while its soluble enzyme counterpart retained only 40% of its initial activity. Bayramoglu et al.[40] reported that β-gal immobilized on magnetic poly-(glycidyl methacrylate ethylene dimethyl methacrylate) beads retained 59% activity after 2 months of storage, while the free enzyme experienced a complete loss of activity after 5 weeks of storage. Other authors [8] stored the soluble and the immobilized β-gal from *K. lactis* at 5 °C in a Tris–HCl buffer. They observed that the soluble enzyme retained 46% of its activity after seven days, while the immobilized enzyme retained 60% after the same period of storage time. After 14 days, soluble enzyme lost more than 80% of its initial activity, while the immobilized enzyme only lost 59% of its initial activity.

The effects of temperature and pH on the activity of soluble and immobilized β-gal

The hydrolysis of ONPG was used as a model reaction to study the effects of temperature and pH on the properties of soluble and immobilized β-gal. The effect of temperature was investigated at temperatures ranging from 25 to 60 °C as shown in Fig. 3, at a constant pH of 6.6. The immobilized enzyme possessed a higher activity at a temperature range of 30–37 °C in comparison with the soluble enzyme. However, the immobilized enzyme exhibited a lower activity at temperatures above 45 °C. Furthermore, while the optimal operating temperature for the soluble enzyme was 50 °C, the optimal operating temperature for the immobilized enzyme was only 37 °C. This result is comparable to what has been published in a different study [8], where the immobilized enzyme is better at maintaining its activity at lower temperatures but loses it faster when subjected to temperatures above 37 °C in comparison with the soluble enzyme. The change in the optimal temperature toward a lower value is unexpected, but this change has also been observed by other authors [8], [41] and [42]. One possible explanation is that the impurities contained in the crude extract (soluble enzyme) may stabilize the enzyme molecule, protecting it against thermal inactivation. Those impurities may have been removed from the nearby of the enzyme after immobilization, probably reducing enzyme stability.

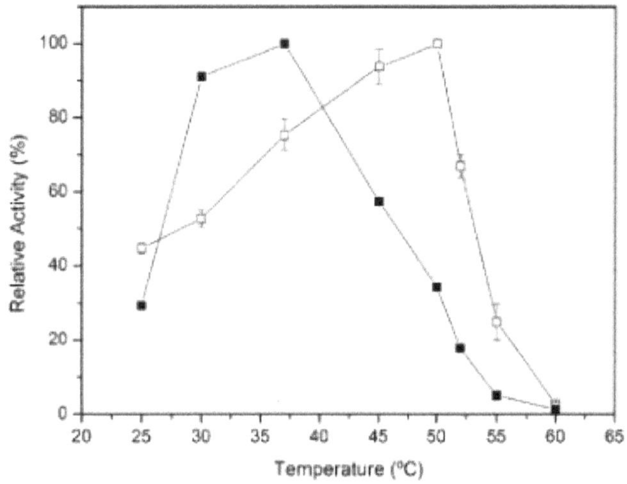

Figure 3: Temperature effects on the hydrolytic activity of soluble (□) and immobilized (■) β-gal when hydrolyzing a 1.25 mM ONPG solution in 50 mM of potassium phosphate buffer supplemented with 0.1 mM $MnCl_2$ at pH 6.6.

Fig. 4 shows the pH dependence of the immobilized enzyme activity at a constant temperature of 37 °C in comparison with the activity of the soluble enzyme. The soluble enzyme exhibited a low activity at the pH of approximately 5.5. The immobilized enzyme, on the other hand, retained 60% of its activity at the same pH, indicating that the immobilization helped to increase the stability of the enzyme. The higher stability of the immobilized β-gal may be explained by the new microenvironment provided by the chitosan matrix. At high H^+ concentrations, the amino groups of chitosan are protonated, attracting OH^- ions, resulting in a higher local pH in comparison with the bulk solution [43]. Furthermore, immobilized enzyme has a broader optimum pH range when compared to the soluble enzyme. In the case of the soluble enzyme, the maximum activity was achieved at pH 6.5, while the immobilized enzyme exhibited a plateau of maximum activity from pH 6.5 to pH 7.0. The results from the temperature and pH changes on the enzymatic activity, suggests that the immobilized enzyme possesses a maximum catalytic activity at a lower temperature in comparison with the soluble enzyme, and it is believed that this change in the optimum temperature was neither promoted by interactions between the enzyme and the support nor any conformational changes of the enzyme molecules. However, it is highly possible that the micro-environment in the vicinity of the support may be very different from the bulk solution environment, which may explain the pH stabilization observed for the immobilized enzyme. Immobilization may promote the stability of enzymes against extreme reaction conditions that are deleterious to soluble enzymes [44].

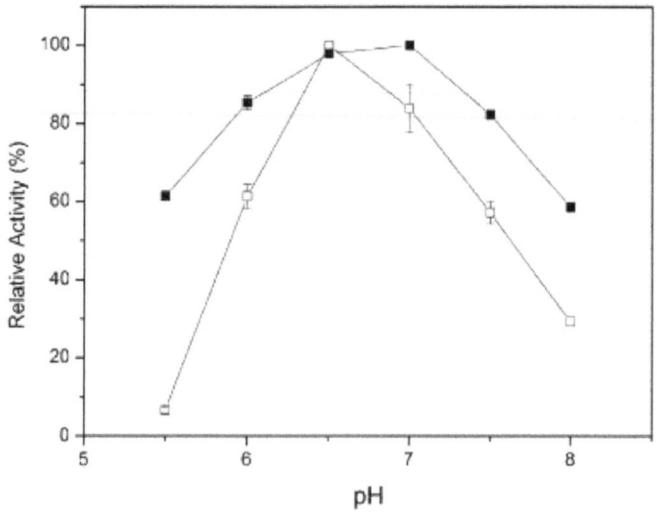

Figure 4: pH effects on the on hydrolytic activity of soluble (□) and immobilized (■) β-gal at 37 °C when hydrolyzing a 1.25 mM ONPG solution in 50 mM of potassium phosphate buffer supplemented with 0.1 mM $MnCl_2$ at pH 6.6.

The influence of immobilization on ONPG hydrolysis kinetics

K_m and V_{max} were estimated by using a non-linear regression of the Michaelis–Menten equation to fit the experimental data obtained for the initial rates of ONPG hydrolysis. The results obtained for soluble and immobilized β-gal are shown in Fig. 5. For the soluble enzyme, K_m was approximately 3.34 ± 0.3 mM and V_{max} was approximately 1.78 ± 0.07 mM/min with R^2 value of 0.997. For the immobilized enzyme, K_m was approximately 3.68 ± 1.84 mM and V_{max} was approximately 3.38 ± 0.77 mM/min with anR^2 value of 0.95. It can be observed that the K_m values for the immobilized enzyme and the soluble enzyme are similar when taking into account experimental errors, indicating that immobilization did not reduce any substrate access to the enzyme and that there are no apparent diffusion limitations. The maximum velocity of the immobilized β-gal was double that of the soluble enzyme, although enzyme loads in the assays were similar, where there was 0.04 U of soluble enzyme and 0.03 U of immobilized enzyme. The catalytic efficiency of the immobilized enzyme in this work was approximately 130% that of the soluble enzyme. Some authors [8], [45] and [46] have discussed the effect of immobilization on enzymes and have concluded that the impact is not always negative. For instance, Goto et al. [46] observed that the immobilization of β-glucosidase caused a decrease in the K_m value, but the V_{max} value was increased.

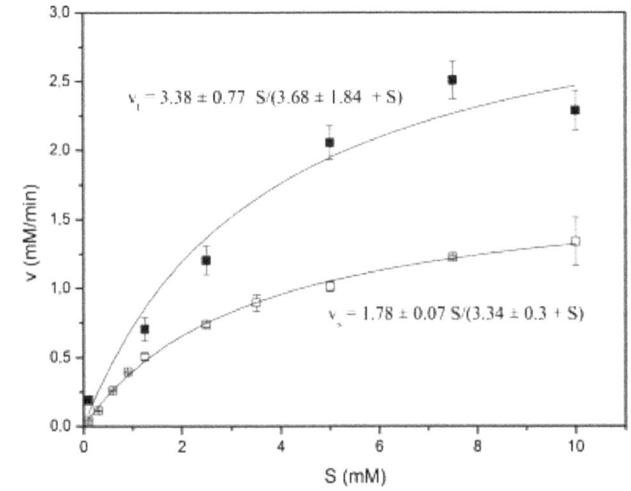

Figure 5: Influence of substrate concentration on the initial rates of ONPG hydrolysis by (□) soluble and (■) immobilized β-gal from *K. lactis* at 37 °C. The lines represent the fitting of the non-linear Michaelis–Menten model to the experimental data. v_s is the rate of ONPG hydrolysis catalyzed by soluble enzyme and v_I is the rate of ONPG hydrolysis catalyzed by the immobilized enzyme.

The kinetic parameters estimated in this work are similar to those that have been reported by other researchers. Zhou and Chen [23] determined the K_m and V_{max} for β-gal from the *K. lactis* Maxilact LX 5000 strain in soluble form and when it was immobilized on graphite. For the soluble β-gal, K_m was 1.74 mM and V_{max} was 77.45 mmol ONP/min mg, while they were K_m was 9.34 mM and V_{max} was 8.75×10^{-3} mmol ONP/min mg for the immobilized enzyme. The authors noted that an increase in K_m is related to the difficulty experienced by the substrate when coming into contact with the enzyme, which was not observed in this work.

Thermal stability and estimation of inactivation energy

The thermal stability of β-gal was studied at temperatures of 40, 50 and 60 °C, and the results can be observed in Fig. 6. For each thermal stability assay, the thermal deactivation constant, the half-life ($t_{1/2}$) and the stabilization factor were determined as shown in Table 4.

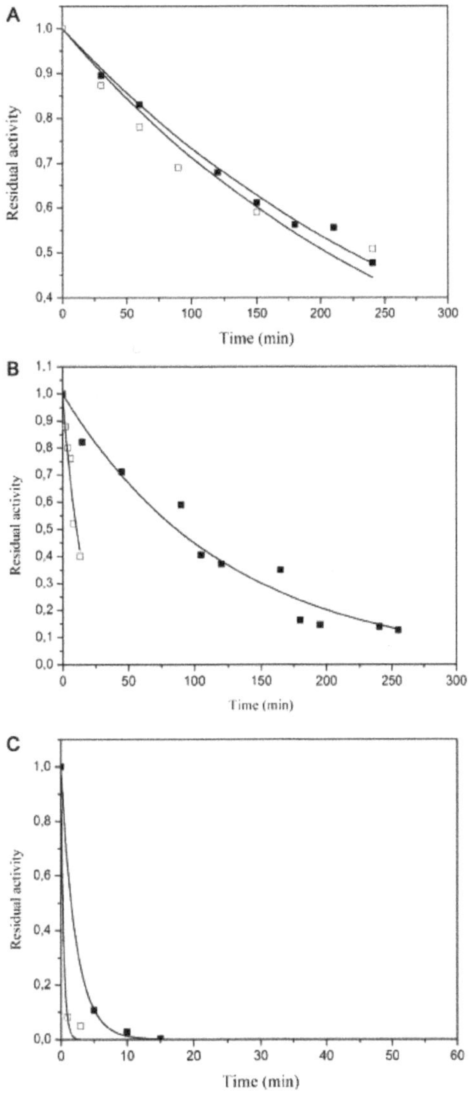

Figure 6: Thermal deactivation profiles for soluble (□) and immobilized (■) β-gal at pH 6.6 at temperatures of (a) 40 °C, (b) 50 °C and (c) 60 °C. The lines represent the trend of the first-order deactivation model.

Table 4: Experimental values for the thermal deactivation constant (K_d), half-life ($t_{1/2}$) and the Stabilization Factor (SF) of soluble and immobilized β-gal.

Temperature	β-Galactosidase	1st-order model		$t_{1/2}$ (min)	(SF)
		K_d	R^2		
40 °C	Soluble	0.0034 ± 0.000	0.9474	203.9	1.1
	Immobilized	0.0031 ± 0.000	0.9921	223.6	
50 °C	Soluble	0.0661 ± 0.006	0.9447	10.5	8.3
	Immobilized	0.0080 ± 0.001	0.9650	86.6	
60 °C	Soluble	2.4906 ± 0.412	0.9961	0.3	5.6
	Immobilized	0.4414 ± 0.015	0.9997	1.6	

In Fig. 6(A), it can be observed that the soluble and immobilized enzyme showed similar deactivation profiles at 40 °C. Both biocatalysts showed the same stability, retaining approximately 50% of their initial activities after 2 h of exposition. However, the immobilized enzyme showed an increased stability at 50 °C (Fig. 6(B)). The soluble enzyme lost all its activity even before it had been incubated for 50 min, while the immobilized β-gal still retained more than 70% of its initial activity. The soluble and immobilized enzymes showed a similar deactivation profile at 60 °C, where both types of enzymes lost all their activities before 20 min of incubation, as shown in Fig. 6(C).

A low stabilization factor (SF) was achieved at 40 °C as shown in Table 4, since half-lives of soluble and immobilized enzyme were very close, 203.87 and 223.60 min, respectively. Nevertheless, a higher SF was observed when the incubation temperature was increased. The immobilized enzyme showed the highest stabilization factor at 50 °C, being approximately 8.3 times more stable than the soluble enzyme. These results reinforce the idea that the immobilized β-gal may be stabilized better in the new microenvironment provided by the chitosan matrix, as has been discussed previously. Although immobilized enzyme was approximately 5.6 times more stable than the soluble enzyme at 60 °C, half-lives were reduced when the enzyme was exposed to drastic temperatures. For the immobilized enzyme, for instance, it decreased from 223.6 to 1.6 min, which means it is suffers a rapid deactivation at 60 °C.

According to the literature [25], the estimation of the deactivation thermodynamic parameters provides important information on the probable mechanism of enzyme denaturation. In this work, the deactivation energy calculated by the Arrhenius equation between the temperatures of 40–60 °C for soluble β-gal and immobilized β-gal were 285.7 kJ/mol and 213.6 kJ/mol, respectively. These values are in the range that has been estimated for many microbial enzymes (167–293 kJ/mol) [24].

Table 5 presents the changes in enthalpy (ΔH^*) and entropy (ΔS^*) calculated using the transition state theory for the thermal deactivation of the soluble β-gal and the immobilized β-gal at a temperature range of 40–60 °C. The thermodynamic parameter values estimated in this work are of the same order of magnitude as the values that have been previously reported [25]. It can be observed that ΔH^* slightly decreases with an increase in the temperature while ΔS^* remains constant. Furthermore, the ΔS^* values are close to zero, suggesting that thermal deactivation did not cause significant changes to the tertiary structure of the enzyme [25].

Table 5: Thermodynamic parameter values of thermal inactivation of soluble and immobilized β-gal at 40–60 °C.

T (°C)	ΔH (kJ/mol)		ΔG (kJ/mol)		ΔS (kJ/mol/K)	
	Soluble	Immobilized	Soluble	Immobilized	Soluble	Immobilized
40	283.09	211.00	91.62	91.86	0.61	0.38
50	283.01	210.92	86.66	92.33	0.61	0.37
60	282.92	210.84	79.37	84.16	0.61	0.38

According to another study [47], a decrease in the ΔS^* and the ΔH^* values indicates the thermostabilization of the enzyme. By comparing the ΔS^* and the ΔH^* values of the soluble and immobilized enzymes as presented in Table 5, it can be observed that immobilization was able to promote the thermal stability of β-gal because the ΔS^* and the ΔH^* values are shown to decrease upon immobilization. The authors explain that the decrease in ΔS^* may be promoted by decreasing the flexibility of an external loop, thereby stabilizing the modified enzyme [47].

Reuse of the immobilized -gal

Fig. 7 shows a reuse study of the immobilized β-gal, using the subsequent hydrolysis of ONPG as a model reaction. In this work, the immobilized enzyme was reused for 10 cycles with a good stability as more than 70% of the initial activity was retained after the tenth cycle. In this work, one cycle is defined as a batch of ONPG hydrolysis at 37 °C and pH 7.3, lasting 5 min. The result achieved may be considered satisfactory in comparison with the results obtained by other researchers [48], [49] and [50]. Verma et al. [48]immobilized β-gal from *K. lactis* on functionalized silicon dioxide nanoparticles and used the same model reaction to evaluate the immobilized enzyme at 40 °C over 11 cycles, where the biocatalyst retained approximately half of its initial activity at the end of the last cycle. Tardioli et al. [49] observed that the immobilized enzyme had a higher stability than the soluble enzyme. After four cycles of lactose hydrolysis lasting 20 min each at 40 °C and pH 7, the immobilized enzyme retained 83% of its initial activity. Other authors [50]used composite microspheres of tamarind gum and chitosan as a support but only retained 53% of its initial activity after 9 cycles of ONPG hydrolysis at 37 °C and pH 7.3.

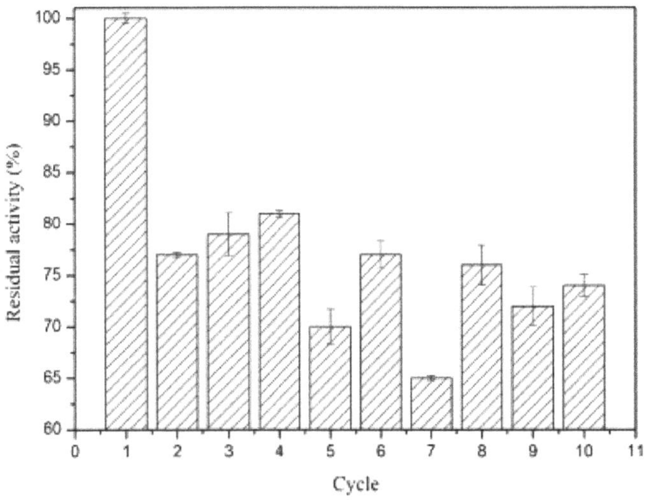

Figure 7: Reuse and stability of immobilized β-gal when hydrolyzing ONPG at 37 °C.

Lactose hydrolysis by the soluble and immobilized β-gal

Fig. 8 shows the hydrolysis of lactose catalyzed by the soluble and immobilized β-gal performed in a batch reactor and at 40 °C. Soluble enzyme hydrolyzed lactose linearly until it reaches a value of 71% at 210 min. After that, the reaction slowed down until almost termination, reaching a conversion higher than 80% after 330 min. This result shows the capability of β-gal from *K. lactis* NRRL Y1564 to catalyze the lactose hydrolysis into galactose and glucose. When immobilized enzyme was used, a higher lactose conversion (79%) was achieved at the beginning of the assay (up to 210 min). Afterwards, the hydrolysis of lactose catalyzed by the immobilized β-gal also showed a slow down and lactose conversion reached a value over 80%, which was similar to the value achieved when using the soluble enzyme. The improvement of the reaction course when lactose hydrolysis was catalyzed by the immobilized β-gal was also observed by other authors [12]. They noticed that the immobilization of a thermophilic β-galactosidase on Sepabeads supports decreased product inhibition. They attributed the improvement of the reaction courses to higher inhibition constants of the immobilized enzyme.

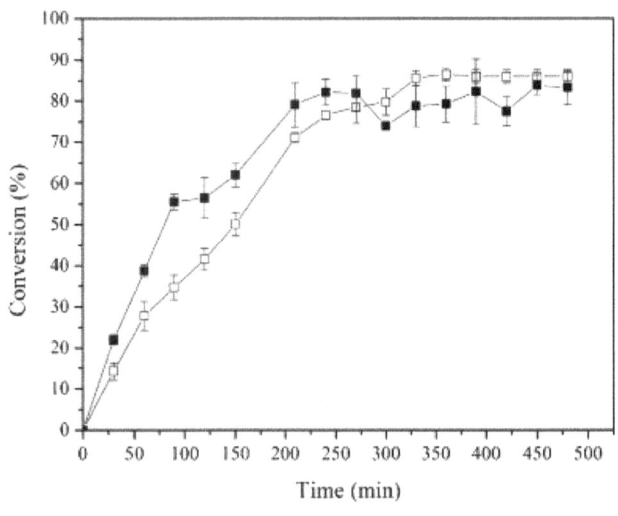

Figure 8: Time course of lactose hydrolysis catalyzed by soluble (□) and immobilized (■) β-gal.

CONCLUSIONS

In this work, a culture medium prepared by using deproteinized whey supplemented with yeast extract was efficiently used for the production of β-gal through the cultivation of*Kluyveromyces* strains. Among the strains evaluated, the *K. lactis* NRRL Y1564 strain was selected because it showed the highest volumetric enzyme activity. Furthermore, the kinetic results suggest that the synthesis of β-gal by *K. lactis* NRRL Y1564 is associated with cell growth. Chitosan activated with glutaraldehyde is a suitable alternative low cost support for β-gal immobilization, providing the immobilized enzyme with higher thermal, operational and storage stabilities in comparison with the soluble enzyme. After immobilization, the optimal temperature was lowered and a broader optimal pH operating range was observed in comparison with the soluble enzyme. The kinetics of ONPG hydrolysis catalyzed by soluble and immobilized β-gal suggests that there is no conformational change experienced by the enzyme during immobilization. The higher stability of immobilized β-gal was most likely a result of the new microenvironment provided by the chitosan matrix. Last but not least, the immobilization of β-gal not only to improved enzyme stability, but also improved the performance of the biocatalyst in the hydrolysis of lactose, probably by altering the kinetic parameters of the enzyme (inhibition constants).

ACKNOWLEDGEMENTS

The authors kindly acknowledge the financial support given by Conselho Nacional de desenvolvimento Científico e Tecnológico (CNPq). The authors also thank Dr. Vania M.M. Melo (UFC) and Dr. Sandra Regina Ceccato-Antonini (UFSCar) for the strains of *K. marxianus* LAMI CE 025 and *K. marxianus* CCA 510, respectively.

REFERENCES

1. Jurado E, Camacho F, Luzón G, Vicaria JM. Kinetic models of activity for - galactosidases: influence of pH, ionic concentration and temperature. Enzyme Microb Technol 2004;34:33–40.
2. Hsu CA, Yu RC, Lee SL, Chou CC. Cultural condition affecting the growth and production of -galactosidase by Bifidobacterium longum CCRC 15708 in a jar fermenter. Int J Food Microbiol 2007;116:186–9.
3. Lane MM, Morrissey JP. Kluyveromyces marxianus: a yeast emerging from its sister's shadow. Fungal Biol Rev 2010;24:17–26.
4. Tello-Solis SR, Jimenez-Guzman J, Sarabia-Leos C, Gomez-Ruiz L, CruzGuerrero AE, Rodriguez-Serrano G, et al. Determination of the secondary structure of Kluyveromyces lactis -galactosidase by circular dichroism and its structure–activity relationship as a function of the pH. J Agric Food Chem 2005;53:10200–4.
5. Martínez-VillaluegaC, Cobas-CardelleA, CorzoN, Olano A. Optimization of conditions for galactooligosaccharide synthesis during lactose hydrolysis by -galactosidase from Kluyveromyces lactis (Lactozym 3000 L HP G). Food Chem 2008;107:258–64.
6. Guimarães PMR, Teixeira JA, Domingues L.

Fermentation of lactose to bioethanol by yeasts as part of integrated solutions for the valorisation of cheese whey. Biotechnol Adv 2010;28:375–84.

7. Siso MIG. The biotechnological utilization of cheese whey: a review. Bioresour Technol 1996;57:1–11.
8. Jochems P, Satyawali Y, Van Roy S, Doyen W, Diels L, Dejonghe W. Characterization and optimization of -galactosidase immobilization process on a mixed-matrix membrane. Enzyme Microb Technol 2011;49: 580–8.
9. Ladero M, Santos A, Garcia LJ, Carrascosa VA, Pessela CCB, Garcia-Ochoa F. Studies on the activity and the stability of -galactosidases from Thermus sp. Strain T2 and from Kluyveromyces fragilis. Enzyme Microb Technol 2002;30:392–405.
10. Giacomini C, Irazoqui G, Batista-Viera F, Brena MB. Influence of the immobilization chemistry on the properties of immobilized -galactosidases. J Mol Catal B: Enzym 2001;11:597–606.
11. Brahim S, Narinesingh D, Guiseppi-Elie A. Kinetics of glucose oxidase immobilized in p(HEMA)-hydrogel microspheres in a packed-bed bioreactor. J Mol Catal B: Enzym 2002;18:69–80.
12. Pessela BCC, Dellamora-Ortiz G, Betancor L, Fuentes M, Guisan JM, Fernandez-Lafuente R. Modulation of the catalytic properties of multimeric -galactosidase from E. coli by using different immobilization protocols. Enzyme Microb Technol 2007;40:310–5.
13. Gupta KC, Jabrail FH. Effects of degree of deacetylation and cross-linking on physical characteristics, swelling and release behavior of chitosan microspheres. Carbohydr Polym 2006;66:43–54.
14. Mendes AA, Oliveira PC, Castro HF, Giordano RLC. Application of chitosan as support for immobilization of enzymes of industrial interest. Quim Nova 2011;34:831–40.
15. Dwevedi A, Kayastha AM. Optimal immobilization of -galactosidase from Pea (PsBGAL) onto Sephadex and chitosan beads using response surface methodology and its applications. Bioresour Technol 2009;100: 2667–75.
16. Santiago PA, Marquez LDS, Cardoso VL, Ribeiro EJ. Estudo da produc¸ ão de -galactosidase por fermentac¸ ão de soro de queijo com Kluyveromyces marxianus. Cienc Tecnol Aliment 2004;24: 567–72.
17. Manera AP, Ores JC, Ribeiro VA, Burket CAV, Kalil SJ. Optimization of the culture medium for the production of -galactosidase from Kluyveromyces marxianus CCT 7082. Food Technol Biotechnol 2008;46:66–72.
18. Alves FG, Maugeri F, Burkert JFM, Kalil SJ. Maximization of -galactosidase production: a simultaneous investigation of agitation and aeration effects. Appl Biochem Biotechnol 2010;160:1528–39.
19. Cortes G, Trujillo-Roldán MA, Ramırez OT, Galindo E. Production of - galactosidase by Kluyveromyces marxianus under oscillating dissolved oxygen tension. Process Biochem 2005;40:773–8.
20. Inchaurrondo VA, Yantorno OM, Voget CE. Yeast growth and -galactosidase production during aerobic bath cultures in lactose-limited synthetic medium. Process Biochem 1994;29:47–54.
21. Budriene S, Gorochovceva N, Romaskevic T, Yugova VL, Miezeliene A, Dienys G, et al. -galactosidase from Penicillium canescens. Properties and immobilization. Cent Eur J Chem 2005;3:95–105.
22. Silva JA, Macedo GP, Rodrigues DS, Giordano RLC, Gonc¸ alves LRB. Immobilization of Candida antarctica lipase B by covalent attachment on chitosan-based hydrogels using different support activation strategies. Biochem Eng J 2012;60:16–24.
23. Zhou QZK, Chen XD. Effects of temperature and pH on the catalytic activity of the immobilized -galactosidase from Kluyveromyces lactis. Biochem Eng J 2001;9:33–40.
24. Shuler ML, Kargi F. Bioprocess engineering: basic concepts. 2nd ed. Upper Saddle River, NJ: Prentice Hall; 2002.
25. Ustok FI, Tari C, Harsa S. Biochemical and thermal properties of - galactosidase enzymes produced by artesanal yoghurt cultures. Food Chem 2010;119:1114–20.
26. Schmidell W, Lima UA, Aquarone E, Borzani W. Biotecnologia industrial, vol. 2. São Paulo: Edgard Blucher; 2001.
27. Fogler HS. Elements of chemical reaction engineering. 4th ed. Upper Saddle River, NJ: Prentice Hall; 2005.
28. Neri DFM, Balcão VM, Carneiro-da-Cunha MG, Carvalho Jr LB, Teixeira JA. Immobilization of -galactosidase from Kluyveromyces lactis onto a polysiloxane–polyvinyl alcohol magnetic (mPOS–PVA) composite for lactose hydrolysis. Catal Commun 2008;9:2334–9.
29. Miller GL. Use of dinitrosalicylic acid reagent for determination of reducing sugar. Anal Chem 1959;31:426–8.
30. Bradford MM. A rapid and sensitive method for the quantitation of microgram quantities of proteins utilizing the principle of protein-dye binding. Anal Chem 1976;72:248–54.
31. Braga ARC, Gomes PA, Kalil SJ. Formulation of culture medium with agro-industrial waste for -galactosidase from Kluyveromyces marxianus ATCC 16045. Food Bioprocess Technol 2011:1–11, http://dx.doi.org/10.1007/s11947-011-0511-0.
32. OrnelasAP, SilveiraWB, Sampaio FC, Passos FML. The activity of -galactosidase and lactose metabolism in Kluyveromyces lactis cultured in cheese whey as a function of growth rate. J Appl Microbiol 2008;104:1008–13.
33. Neves LCM, Oliveira KS, Kobayashi MJ, Penna TCV, Converti A. Biosurfactant production by cultivation of Bacillus atropheus ATCC 9372 in semidefined glucose/casein-based medium. Appl Microbiol Biotechnol 2007;136–140:539–54.
34. Gaden ELJ. Fermentation process kinetics. J Biochem Microbiol Technol Eng 1959;1:413–9.
35. Mahoney RR, Nickerson TA, Whitaker JR. Selection

of strains, growth conditions and extraction procedures for optimal production of lactase from Kluyveromyces lactis. J Dairy Sci 1974;58:1620–9.

36. Dickson RC, Markin JS. Physiological studies of -galactosidase induction in Kluyveromyces lactis. J Bacteriol 1980;142:777–85.

37. Matheus AOR, Rivas N. Production and partial characterization of betagalactosidase from Kluyveromyces lactis grown in deproteinized whey. Arch Latinoam Nutr 2003;53:194–201.

38. Gonc¸ alves LRB, Ferreira ALO, Fernandez-Lafuente R, Guisan JM, Giordano RC, Giordano RLC. Influence of mass transfer limitations on the enzymatic synthesis of -lactam antibiotics catalyzed by penicillin G acylase immobilized on glioxilagarose. Bioprocess Biosyst Eng 2008;31:411–8.

39. Ansari SA, Husain Q. Lactose hydrolysis from milk/whey in batch and continuous processes by concanavalin A-Celite 545 immobilized Aspergillus oryzae -galactosidase. Food Bioprod Process 2011, http://dx.doi.org/10.1016/j.fbp.2011.07.003.

40. Bayramoglu G, Tunali Y, Arica MY. Immobilization of beta-galactosidase onto magnetic poly (GMA-MMA) beads for hydrolysis of lactose in bed reactor. Catal Commun 2007;8:1094–101.

41. Delcheva G, Dobrev G, Pishtiyski I. Performance of Aspergillusniger B03 - xylosidase immobilized on polyamide membrane. J Mol Catal B: Enzym 2008;54:109–15.

42. Adikane HV, Thakar DM. Studies of penicillin g acylase immobilization using highly porous cellulose-based polymeric membrane. Appl Biochem Biotechnol 2010;160:1130–45.

43. Wentworth DS, Skonberg D, Donahue DW, Ghanem A. Application of chitosanentrapped -galactosidase in a packed-bed reactor system. J Appl Polym Sci 2003;91:1294–9.

44. Tischer W, Wedekind F. Immobilized enzymes: methods and applications. Top Curr Chem 1999;200:96–126.

45. Mazzei R, Giorno L, Piacentini E, Mazzuca S, Drioli E. Kinetic study of a biocatalytic membrane reactor containing immobilized -glucosidase for the hydrolysis of oleuropein. J Membr Sci 2009;339:215–23.

46. Goto M, Kawakita H, Uezu K, Tsuneda S, Saito K, Goto M, et al. Esterification of lauric acid using lipase immobilized in the micropores of a hollow-fiber membrane. J Am Oil Chem Soc 2006;83:209–13.

47. Rashid MH, Siddiqui KS. Thermodynamic and kinetic study of stability of the native and chemically modified -glucosidase from Aspergillus niger. Process Biochem 1998;33:109–15.

48. Verma ML, Barrow CJ, Kennedy JF, Puri M. Immobilization of -d-galactosidase from Kluyveromyces lactis on functionalized silicon dioxide nanoparticles: characterization and lactose hydrolysis. Int J Biol Macromol 2012;50:432–7.

49. Tardioli PW, Vieira DC, Adriano WS, Mendes AA, Giordano RC, Giordano RLC. Immobilization of -galactosidase from Kluyveromyces fragilis. New Biotechnol 2009;25:S164.

50. Zhang ZJ, Zeng J, Zhang S, Wang Y, Wang Y. Immobilization of -galactosidase on tamarind gum and chitosan composite microspheres. J Bioact Compat Polym 2006;21:415–32.

CHAPTER 21

HIGH STABILITY OF IMMOBILIZED β-D-GALACTOSIDASE FOR LACTOSE HYDROLYSIS AND GALACTOOLIGOSACCHARIDES SYNTHESIS

Manuela P. Klein[1,2], Lucas P. Fallavena [2], Jéssie da N. Schöffer [2], Marco A.Z. Ayub, Rafael C. Rodrigues [2], Jorge L. Ninow[1], and Plinho F. Hertz [2]

[1]Chemical and Food Engineering Department, Federal University of Santa Catarina, ZC 88040-900 Florianópolis, SC, Brazil

[2] Enzymology Laboratory, Food Science and Technology Institute, Federal University of Rio Grande do Sul, Av. Bento Gonc¸ alves 9500, P.O. Box 15090, ZC 91501-970 Porto Alegre, RS, Brazil

ABSTRACT

***β*-d-Galactosidase from** Kluyveromyces lactis **was immobilized on glutaraldehyde-activated chitosan and used in a packed-bed reactor for the continuous hydrolysis of lactose and the synthesis of galactooligosaccharides (GOS). The biocatalyst was tested for its optima pH and temperature, thermal stability in the presence of substrate and products, and operational stability. Immobilization increased the range of operational pH and temperature, and the enzyme thermal stability was sharply increased in the presence of lactose. Almost complete lactose hydrolysis was achieved for both milk whey and lactose solution at 37 °C at flow rates up to 2.6 mL min^{-1}. Maximal GOS concentration of 26 g L^{-1} was obtained at a flow rate of 3.1 mL min^{-1}, with a productivity of 186 g L^{-1} h^{-1}. Steady-state operation for 15 days showed the reactor stability concerning lactose hydrolysis.**

INTRODUCTION

Lactose (β-d-galactopyranosyl-(1→4)-d-glucopyranose), the main sugar of milk, can be hydrolyzed by β–d-galactosidase liberating d-glucose and d-galactose, making possible the consumption of milk and other dairy products by lactose-intolerant people (Haider & Husain, 2009). In the presence of highly concentrated lactose, this enzyme can also produce galactooligosaccharides (GOS), by transferring galactosyl residues to lactose molecules. The GOS produced (for example, Gal (β 1→4) Gal (β 1→4) Glc), usually has the structure Galn–Glc, where n indicates the degree of polymerization, which is typically 1–5 (Gosling et al., 2010 and Mussatto and Mancilha, 2007). GOS are non-digestible oligosaccharides, which are used as prebiotics food ingredients. The regular consumption of GOS can promote the growth and metabolism of intestinal bifidobacteria, microorganisms that are associated with positive health effects (as the reduction of the level of cholesterol, anticarcinogenic properties, and production of vitamins), when applied in human and other animals diets (Grosova, Rosenberg, & Rebros, 2008). Therefore, the enzyme β-d-galactosidase, in its free or immobilized forms, has an important application in the food industry for lactose hydrolysis and for GOS synthesis.

Citation: Manuela P. Klein, Lucas P. Fallavena, Jéssie da N. Schöffer, Marco A.Z. Ayub, Rafael C. Rodrigues, Jorge L. Ninow, Plinho F. Hertz, High stability of immobilized β-d-galactosidase for lactose hydrolysis and galactooligosaccharides synthesis, Carbohydrate Polymers, Volume 95, Issue 1, 5 June 2013, Pages 465-470, ISSN 0144-8617, http://dx.doi.org/10.1016/j.carbpol.2013.02.044.

For the industrial-scale applications of immobilized enzymes, where large amounts of biocatalyst are required, the immobilization protocol must be simple and preferably make use of inexpensive materials as supports (Garcia-Galan, Berenguer-Murcia, Fernandez-Lafuente, & Rodrigues, 2011). Moreover, concerning food applications, nontoxicity and biocompatibility are also required. In this context, chitosan [(1→4)-2-amino-2-deoxy-β-d-glucan], which is a natural polyaminosaccharide derived from chitin [(1→4)-2-acetamido-2-deoxy-β-d-glucan], a by-product of the seafood processing industry, is both safe and cheap (Krajewska, 2004 and Muzzarelli, 1980). Another unique property of chitosan is its low propensity for microbial contamination, a problem associated with the immobilization of *β*-d-galactosidase for lactose hydrolysis (Benhabiles et al., 2012). Chitosan is soluble in acid solutions, and can be precipitated in alkaline pH, thus chitosan particles can be readily obtained dripping an acid chitosan solution into an alkaline coagulation solution (Krajewska, 2004). Prior to enzyme immobilization, chitosan particles can also be easily activated using various agents such as glutaraldehyde (Muzzarelli, 1980), genipin (Chiou et al., 2007 and Muzzarelli, 2009), glycidol or epichlorohydrin (Rodrigues, Mendes, Adriano, Gonçalves, & Giordano, 2008) in order to generate specific groups for enzyme attachment.

Reactors configuration for continuous operations is another important aspect in designing industrial enzymatic processes. Packed-bed reactors (PBRs) are widely and successfully used in many industrial processes. Some of the advantages of using this type of reactor are the reuse of the enzyme without the need of a prior separation; the continuous production of the desired product; the easiness of handling substrates of low solubility by the use of large volumes containing low concentrations of substrate; and the long-term runs and industrial-scale operations. Therefore, PBRs are more cost effective than batch operations (Chang et al., 2007 and Halim et al., 2009).

The main objective of this research was to set up a packed-bed reactor filled with chitosan-immobilized β-d-galactosidase for the continuous hydrolysis of lactose and the synthesis of galactooligosaccharides. The immobilized enzyme was also characterized for its operational optima pH and temperature, and biocatalyst thermal stability, assayed in the presence of substrates and products. Process performance was evaluated in terms of lactose hydrolysis and GOS synthesis as a function of flow rate of substrate feeding. Finally, continuous lactose hydrolysis was carried out in order to evaluate the operational stability of the PBR under steady-state operation.

MATERIALS AND METHODS

Materials

Whey powder was obtained from a local supplier (Elegê Laticínios S.A., Teutônia, Brazil),Kluyveromyces lactis β-d-galactosidase was the liquid formulation of Maxilact LX 5000, with a declared activity of ≥5000 NLU/g (DSM Food Specialties, Heerlen, Netherlands). Chitosan (from shrimp shells, ≥75% deacetylated), o-nitrophenyl-β-d-galactopyranoside (ONPG), d-glucose, d-galactose, lactose, and raffinose (β-d-fructofuranosyl α-d-galactopyranosyl-(1→6)-α-d-glucopyranoside) were obtained from Sigma–Aldrich (St. Louis, USA). A d-glucose determination kit was purchased from Labtest Diagnóstica SA (São Paulo, Brazil). All solvents and other chemicals were of analytical grade.

Methods

Preparation of β-d-galactosidase immobilized on chitosan macroparticles

β-d-galactosidase was covalently immobilized on glutaraldehyde-activated chitosan macroparticles as described in a previous work (Klein et al., 2012). The amount of protein attached on chitosan macroparticles was 46.2 mg g^{-1} of dry support, presenting an activity of 230 U g^{-1} of dry support. The chitosan-β-d-galactosidase derivative was stored at 7 °C in 0.1 M of phosphate–potassium buffer (pH 7.0) containing $MgCl_2$ 1.5 mM (activity buffer).

Activity of free and immobilized β-d-galactosidase

Free β-d-galactosidase activity was assayed by incubating the diluted enzyme (50 µL) with 0.5 mL of activity buffer containing ONPG (10 mM) at 37 °C during 2 min. For immobilized β-d-galactosidase, 1.5 mg (dry support) was incubated with 1 mL of activity buffer containing ONPG (10 mM) at 37 °C during 2 min. The reactions were stopped with the addition of borate buffer pH 9.8. Released o-nitrophenol (ONP) was spectrophotometrically determined at 415 nm. One unit of β-d-galactosidase activity (U) was defined as the amount of enzyme that hydrolyze 1 µmol of ONPG to o-nitrophenol per minute under the conditions previously stated. Protein content of the enzyme solutions was determined by the Lowry assay.

Optima pH and temperature for free and immobilized β-d-galactosidase

The optimum operational pH of β-d-galactosidase activity was studied by monitoring enzyme activity of both free and immobilized preparations in 0.1 M of phosphate–potassium buffer (pH 5.7–8.0) containing $MgCl_2$ 1.5 mM, at 37 °C, while the optimum temperature was determined by measuring the activity between 10 °C and 70 °C at pH 7.

Thermal stability of the immobilized β-d-galactosidase

Thermal stability of the immobilized enzyme was evaluated incubating the biocatalyst in sealed tubes in a thermostatically controlled water bath at 55 °C. Thermal stability was performed under the following conditions: activity buffer, different concentrations of lactose (50 g L^{-1} and 400 g L^{-1}), d-glucose (200 g L^{-1}), d-galactose (200 g L^{-1}), or d-glucose plus d-galactose (200 g L^{-1} of each sugars). Samples were withdrawn at different time intervals and placed in ice bath to stop the thermal inactivation instantaneously. The remaining enzyme activity was determined as described above.

Packed-bed reactor setup

The column type packed-bed reactor (12 cm × 3 cm) was composed of a water-jacketed glass column, flow-rate controller, and water bath. The reactor was packed with 4 g (dry weight) of chitosan-immobilized β-d-galactosidase 2 mm diameter macroparticles, corresponding to approximately 920 U of β-d-galactosidase total activity. The substrate solution was fed through the bottom of the column using a peristaltic pump. The topside and the bottom of the column were fitted with a sintered glass disc (4 mm thick). The total volume of the packed-bed reactor was 29 mL.

Lactose hydrolysis

The hydrolysis of lactose was performed by flowing through the reactor either buffered lactose solution or whey, both containing 50 g L^{-1} of sugar. The lactose solution was prepared in activity buffer, while the whey solution was prepared by suspension of whey powder in distilled water and adjusting the pH to 7. Previously to starting the reaction, the packed-bed reactor was washed with activity buffer. Flow rates were tested from 0.26 mL min^{-1} to 3.4 mL min^{-1}. Samples were collected after the steady state was reached and analyzed for d-glucose formation. Previously to the d-glucose quantification, samples of hydrolyzed whey were diluted and filtered through a 0.22 μm acetate cellulose membrane in order to remove proteins that may cause interference in the analysis. The hydrolysis was carried out at two different temperatures, 37 and 7 °C, in order to simulate real industrial conditions and to avoid possible microbial contaminations.

Continuous synthesis of galactooligosaccharides (GOS)

GOS synthesis was performed by flowing through the reactor buffered lactose solution at high concentration (400 g L^{-1}). Flow rates were tested from 1 to 15 mL min^{-1} for GOS synthesis at 37 °C. Samples were collected after steady state was reached and the reaction product was analyzed by HPLC.

Analytical procedures

Lactose and products from the transgalactosylation reaction (GOS, d-galactose and d-glucose) were analyzed by HPLC (Shimadzu, Tokyo, Japan) equipped with refractor index and Aminex HPX-87C (300 mm × 7.8 mm) column. Ultra-pure water was used as eluting solvent at a flow rate of 0.6 mL min^{-1}, at 85 °C. The concentration of saccharides was calculated by interpolation from external standards. Standards were used for lactose, d-glucose, and d-galactose. GOS concentration was calculated as raffinose equivalents from an external raffinose standard, as described by Gosling, Stevens, Barber, Kentish, and Gras (2011). The commercial product Vivinal GOS® (Friesland Foods, Netherlands) was used to compare the retention time of the obtained GOS, and its yield (%) was defined as the percentage of GOS produced compared with the weight of initial lactose in the reaction medium.

Operational stability of the packed-bed reactor during lactose hydrolysis

The operational stability of the PBR under steady-state regime was evaluated during 15 continuous days at 37 °C using a flow rate of 2.6 mL min^{-1} containing 50 g L^{-1} of lactose. Aliquots were periodically collected and analyzed for d-glucose formation.

RESULTS AND DISCUSSION

Properties of the immobilized β-d-galactosidase

Optima pH and temperature

The effect of pH on the relative activity of the free and immobilized β-d-galactosidase was evaluated in the range of 5.7–8.0 (Fig. 1A). After immobilization on chitosan macroparticles, the optimum pH of β-d-galactosidase was enhanced from 6.5 to a broader range between 6.5 and 8.0. Even at acidic pH, the enzyme activity was remarkably enhanced, with more than 70% remaining at pH 5.5, compared to less than 25% for the free form, suggesting the strong stabilization effect on enzyme molecules, provided by the covalent binding to the support. Bayramoglu, Tunali, and Arica (2007), in their studies with Escherichia coli β-d-galactosidase immobilized onto magnetic poly(GMA–MMA) beads, found a shift of optimum pH of the enzyme from 7.5 to 7 after immobilization. The authors reported that this effect was probably due to the basic nature of the amino functionalized surface of the magnetic beads. Similarly, β-d-galactosidase from K. lactis immobilized by adsorption on a mixed-matrix membrane containing zirconium dioxide presented a shift in the optimum pH from 6.5 to 7 (Jochems et al., 2011). The authors suggested that this change was due to the microenvironment in the vicinity of the membrane. The support, indeed, can change the pH value around the enzyme catalytic site by changing the concentration of the charged species (e.g. hydrogen and hydroxyl ions) in respect to the bulk solution (De Maio et al., 2003).

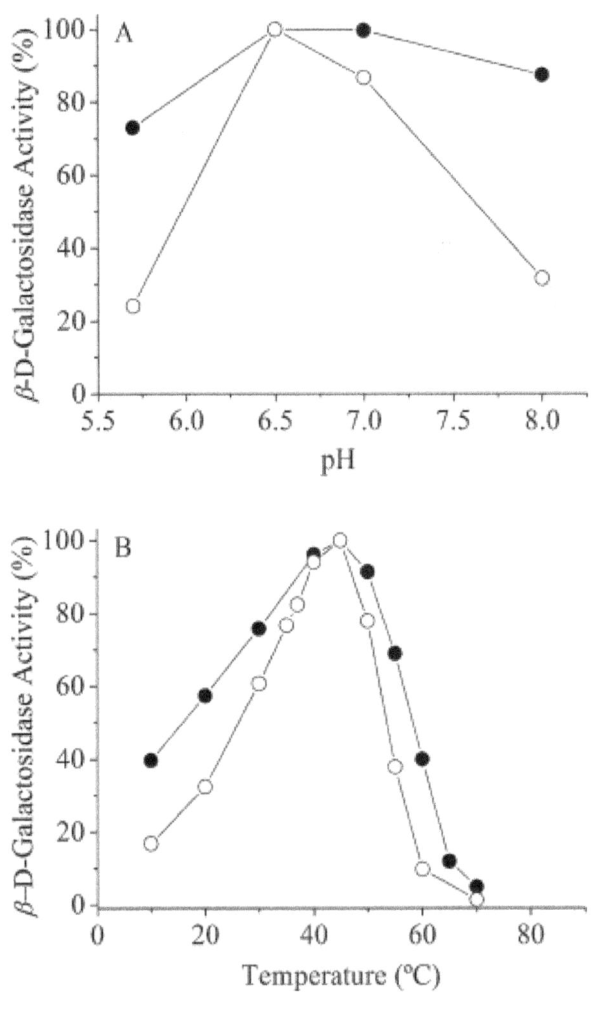

Figure 1: Effect of pH (A) and temperature (B) on the activity of free β-d-galactosidase (○) and chitosan-immobilized β-d-galactosidase (●).

Changes in activity related to temperature of immobilized β-d-galactosidase were investigated in the range of 10–70 °C (Fig. 1B). The free and immobilized forms presented maximum activities at 45 °C, while the immobilized enzyme was more active in a wider range of temperatures. Similarly to the changes in pH, the wider range of temperatures with higher activities can be attributed to the effect of immobilization, the protection offered by the chitosan macroparticles to the enzyme. Song, Lee, Kang, and Kim (2010) reported similar results and a broader range of activity was observed for pH and temperature after immobilization of K. lactis β-d-galactosidase pretreated with lactose on the surface of silica gel using glutaraldehyde as linker. According to Pereira-Rodríguez et al. (2012), the tetramer of β-d-galactosidase from K. lactis is an assembly of dimers, with higher dissociation energy for the dimers than for its tetrameric structure, causing an equilibrium between the dimeric and tetrameric forms of the enzyme when in solution. The binding of the enzyme to the support can prevent the dissociation of subunits of either forms (Fernandez-Lafuente, 2009), consequently causing it structure to be stabilized. Although immobilized enzymes are heterogeneous catalysts, the widened range for pH and temperature can configure another advantage of using immobilized enzymes over the use of their free forms by allowing their applications under different conditions, improving industrial uses.

Thermal stability

Thermal inactivation is an important, limiting factor for prolonged use of enzymes in industrial processes. Inactivation in the presence of substrate and products (reactive conditions) should be evaluated, since this reflects the reaction in an enzyme reactor. Fig. 2 depicts the thermal stability of chitosan-immobilized β-d-galactosidase under different reactive conditions (substrate type and concentration) at 55 °C.

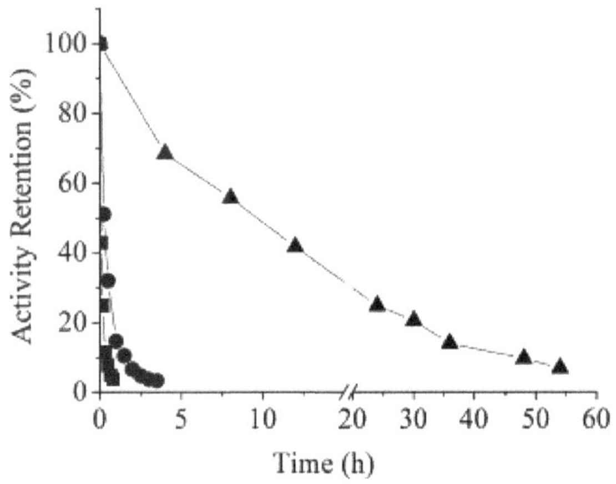

Figure 2: Thermal stability at 55 °C of chitosan-immobilized β-d-galactosidase in the presence of 400 g L^{-1}lactose solutions (▲), 50 g L^{-1} lactose solutions (●), and activity buffer (■).

Activity of the immobilized enzyme under nonreactive conditions (in activity buffer) was approximately 43% after 5 min of incubation at 55 °C, which is consistent with the fact thatK. lactis β-d-galactosidase is relatively thermolabile. According to Jurado, Camacho, Luzon, and Vicaria (2004), experiments conducted at 30 °C and 35 °C, it was found that this β-d-galactosidase preserved practically all its activity after 3 h of reaction, while at higher temperatures it was virtually deactivated.

In order to assay operational conditions, it was evaluated the thermal stability of the immobilized enzyme in the presence of lactose 50 g L^{-1}, which is the average concentration of lactose present in milk and whey. In the same way, thermal stability test was carried out in presence of whey and lactose 400 g L^{-1} (the concentration used in the GOS synthesis). The results showed that in 50 g L^{-1} of lactose the enzyme retained around 51% of its activity after 15 min at 55 °C, while in 400 g L^{-1} the activity was considerably improved (100% after 15 min and approximately 55.8% after 8 h of incubation) at the same temperature, suggesting that lactose stabilizes the enzyme. According to Jurado et al. (2004), Jurado, Camacho, Luzon, and Vicaria (2005) andLadero, Santos, and Garcia-Ochoa (2006), lactose (and possibly d-galactose) can stabilize the overall structure of the enzyme by the coupling of the substrate into the active site.

The products generated during the reaction can also play some important role in the enzyme stability. Binding reaction products and other ligands to a protein is a simple way to stabilize its conformation, which is a widely used strategy for stabilizing proteins and enzymes during various operations including enzyme purification and enzyme immobilization (Illanes, Altamirano, & Zuniga, 1996).

Then, in an effort to accurately investigate the effect of substrate and products in the enzyme stability, the thermal stability in the presence of products (d-glucose and d-galactose) was analyzed. As can be seen in Fig. 3, when only d-glucose (200 g L^{-1}) was present, the chitosan-immobilized β-d-galactosidase presented around 50% of its activity after 11 min of incubation; the stability was not significantly increased compared to the inactivation under nonreactive conditions (4.8 min). For K. lactis β-d-galactosidase, d-glucose is known to be a non-competitive inhibitor ($K_i^{gli} = 794$ mM) (Cavaille & Combes, 1995), which means that lactose and d-glucose will independently bind at different sites of the enzyme.

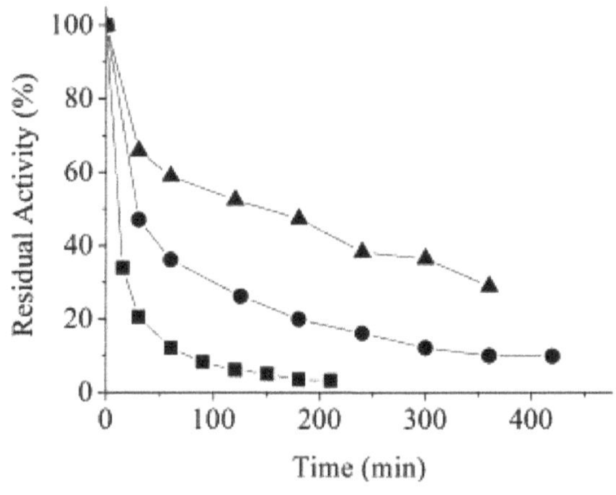

Figure 3: Thermal stability at 55 °C of chitosan-immobilized β-d-galactosidase in the presence of mixture of d-glucose and d-galactose (each 200 g L^{-1}) (▲), 200 g L^{-1}d-galactose solution (●), and 200 g L^{-1}d-glucose solution (■).

When inactivation was performed in presence of d-galactose (200 g L^{-1}), the β-d-galactosidase residual activity was about 47% after 30 min of incubation, which was higher than that obtained when d-glucose was used in the same concentration. d-Galactose is a competitive inhibitor for this enzyme ($K_i^{gli} = 42\,mM$) (Cavaille & Combes, 1995) and has a much higher inhibitory power than d-glucose. Since the only difference between d-glucose and d-galactose lies on the binding site to the protein, it can be suggested that the protector effect promoted by d-galactose comes from the binding of this sugar to the active site of the enzyme. It was recently reported the three-dimensional structure of K. lactis β-galactosidase and the complex structure of the molecule when d-galactose is at the active site, showing that a tryptophan residue, responsible for the binding of d-glucose in the active site of β-galactosidase from E. coli, is missing from the active site of K. lactis β-galactosidase (Pereira-Rodríguez et al., 2012).

When inactivation was carried out in the presence of d-glucose and d-galactose, both at 200 g L^{-1}, the enzyme stability was improved with 47% of activity still remaining after 180 min of incubation at 55 °C, suggesting the stabilizing effect of osmolytes such as sugars, amino acids, and trehalose (Sampedro, Cortes, Munoz-Clares, Fernandez, & Uribe, 2001). Stabilization would be achieved by an increase in the surface tension of the solution, causing the exclusion of the saccharides from the protein domain and reducing backbone movements away from the fully folded protein form (Bromberg et al., 2008,Butler and Falke, 1996, Lin and Timasheff, 1996 and Sampedro et al., 2001). The stabilizing effect also depends on the type of carbohydrate used (Sola-Penna & Meyer-Fernandes, 1998).

The analyze of the sugars during inactivation in the presence of lactose 400 g L^{-1} showed that lactose was present during all experiment, with the enzyme probably constantly hydrolyzing lactose up to the moment that it was inhibited by the high concentration of produced d-galactose. After 54 h of inactivation at 55 °C, 63 g L^{-1} of lactose still remained.

Effect of flow rate on lactose hydrolysis in a packed-bed reactor

Since lactose proved to have the major protecting effect on this β-d-galactosidase, operations under conditions in which the enzyme is constantly catalyzing lactose hydrolysis and avoiding product inhibition could be interesting to test. The use of PBRs for lactose hydrolysis and GOS synthesis could be advantageous over batch operations. Substrates (lactose solution or whey) were continuously pumped at different flow rates (0.26–3.4 mL min^{-1}) through the PBR. Fig. 4 shows the lactose conversion at 37 °C and 7 °C for whey and lactose solution at various flow rates.

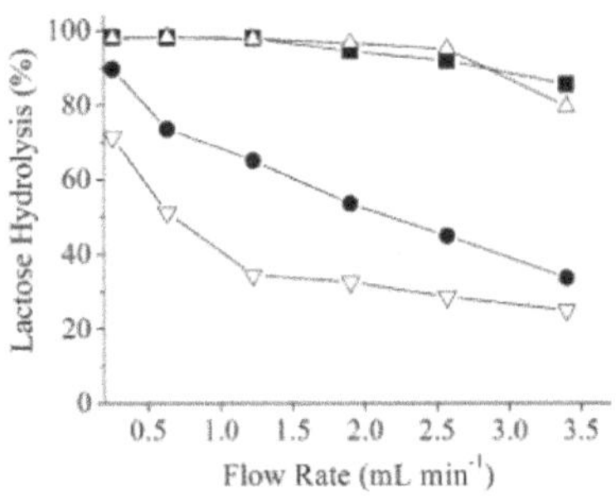

Figure 4: Effect of the flow rate on the lactose hydrolysis by chitosan-immobilized β-d-galactosidase under different conditions: whey at 37 °C (▵), 50 g L^{-1} buffered lactose solution at 37 °C (■), whey at 7 °C (▿), and 50 g L^{-1} buffered lactose solution at 7 °C (●).

At 37 °C, for both lactose solution (50 g L^{-1}) and whey, more than 90% of lactose hydrolysis was reached at a flow rate of 2.6 mL min^{-1} (residence time, 11.3 min). Ansari and Husain (2010), reported 95% of lactose hydrolysis during a 10-days steady-state operation of a PBR filled with β-d-galactosidase adsorbed on concanavalin A-cellulose. However, their operation was conducted at slow flow rate of 0.166 mL min^{-1}. A pilot scale module (108 cm × 14 cm), packed with K. lactis β-d-galactosidase immobilized on cotton fabric, resulted in 30.23% hydrolysis of lactose from milk, at 37 °C and residence time of 11.8 min (Li, Zhou, & Chen, 2007).

Increasing the substrate flow rate to 3.4 mL min^{-1} caused the lactose conversion to drop to 86 and 80% for lactose solutions (50 g L^{-1}) and whey, respectively. This result can be explained by the short residence time of the substrate inside the reactor.

In order to simulate industrial conditions and reduce the possibility of microbial contamination, lactose hydrolysis was also carried out at 7 °C. As expect, the degree of lactose hydrolysis was reduced for all flow rates when compared with the reaction at 37 °C. For lactose solution (50 g L^{-1}), approximately 90% of lactose was hydrolyzed at a flow rate of 0.26 mL min^{-1}, while for whey 72% of hydrolysis was achieved under same conditions. It is possible that, at 7 °C, the difficulty of mass transfer caused by whey proteins is more pronounced than at 37 °C, which explain the lower lactose conversion observed.

Continuous synthesis of galactooligosaccharides

Fig. 5 shows the changes in the concentration of d-glucose, d-galactose, lactose, and GOS as a function of the substrate flow rate, under steady state operation. The maximum GOS concentration of 26 g L^{-1} was achieved using a substrate flow rate of 3.1 mL min^{-1}(residence time of 9.4 min), corresponding to a yield of 6.5% in GOS and 58% of lactose conversion. At lower flow rates (1–2.6 mL min^{-1}), lower concentrations of GOS were obtained, probably due to the hydrolysis of the synthesized GOS, because the longer residence times (11–29 min) allows for its subsequent hydrolysis by the enzyme. Higher than 3.1 mL min^{-1} flow rates (5–15 mL min^{-1}) causes the residence time to be insufficient for best synthesis, reducing GOS concentration.

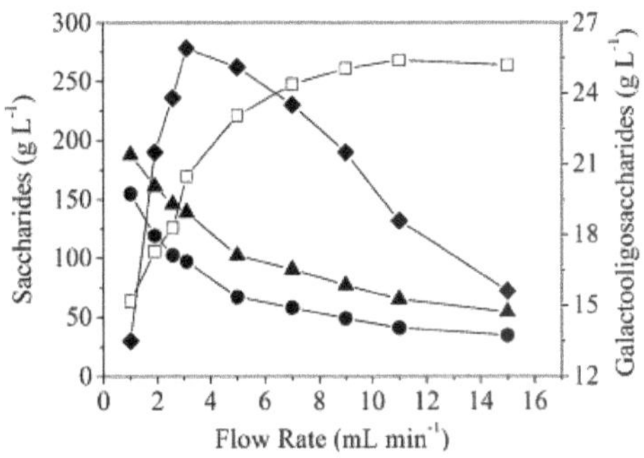

Figure 5: Effect of flow rate on GOS synthesis by chitosan-immobilized β-d-galactosidase, using 400 g L^{-1}buffered lactose solution, pH 7 at 37 °C. d-Glucose (▲), d-galactose (●), lactose (□), and galactooligosaccharides (♦).

Comparatively, Nakkharat and Haltrich (2007) reported 16 g L^{-1} of GOS in a packed-bed reactor containing β-d-galactosidase from Talaromyces thermophilus immobilized on Eupergit C, with 50% lactose conversion and 200 g L^{-1} of initial lactose concentration.Neri et al. (2009), worked with β-d-galactosidase from Aspergillus oryzae immobilized on magnetic polysiloxane-polyvinyl alcohol, obtaining 103.4 g L^{-1} of GOS, with lactose conversion of 47% and 500 g L^{-1} of lactose, in a batch reaction. Although K. lactis β-d-galactosidase is known to show low transgalactosylation activity and thermal stability during the production of galactooligosaccharides compared to A. oryzae enzyme (Park & Oh, 2009), when applied in a PBR, the immobilized enzyme was able to work continuously, with high productivities of galactooligosaccharides.

The GOS productivity in the PBR related to the operational flow rate is shown in Fig. 6. It increased to a maximum of 484.5 g L^{-1} h^{-1} at 15 mL min^{-1}. Comparatively, Shin, Park, and Yang (1998), reported oligosaccharide productivity of 4.4 g L^{-1} h^{-1} with lactose solutions of 100 g L^{-1} in PBR with chitosan-immobilized Bullera singularis β-d-galactosidase (970 U g^{-1} resin). The higher GOS productivity so far reported, 6000 g L^{-1} h^{-1}, was obtained by Albayrak and Yang (2002), using A. oryzae β-d-galactosidase immobilized on cotton cloth by polyethyleneimine and applied in a PBR fed with lactose (400 g L^{-1}) at 2.5 mL min^{-1}.

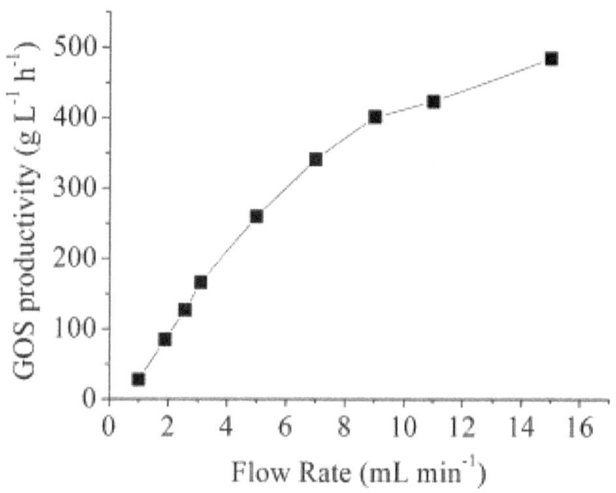

Figure 6: Effect of flow rate on GOS productivity. Experiments were performed using 400 g L^{-1} buffered lactose solution, pH 7 at 37 °C.

Stability of immobilized β-d-galactosidase in the packed-bed reactor

The operational stability of a system is an important parameter in an industrial process, since it directly affects the costs (Nie, Xie, Wang, & Tan, 2006). Fig. 7 shows the operational stability of the immobilized enzyme in the PBR. The reactor was operated continuously at 37 °C at a flow rate of 2.6 mL min^{-1}, mantaining 90% of lactose hydrolysis for longer than 15 days.

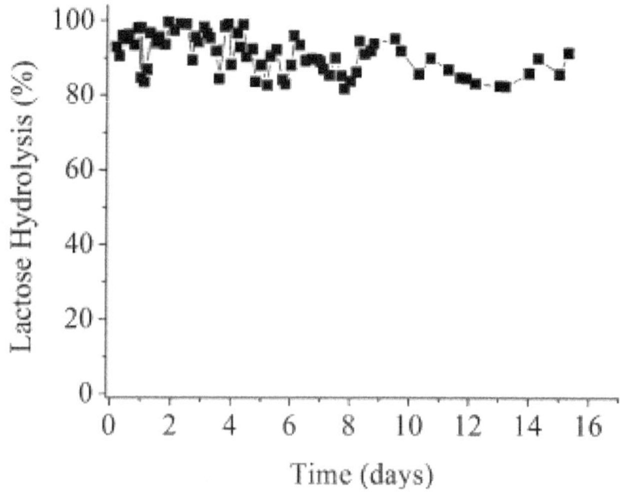

Figure 7: Operational stability of the PBR filled with chitosan-immobilized β-d-galactosidase, operated continuously using 50 g L^{-1} buffered lactose solution, pH 7 at 37 °C and flow rate of 2.6 mL min^{-1}.

CONCLUSIONS

One of the main objectives in immobilized enzyme technology is to increase the enzyme stability, thus allowing the obtained derivative to be repeatedly used, inclusive at different process conditions, such as in continuous reactors. The use of PBRs for lactose hydrolysis and GOS synthesis may replace batch reactors, with several cost reductions and operation advantages, including reduced reaction inhibition by substrate and products. In the present study, using chitosan macroparticles, a relatively low cost and easily accessible support, improvements in the operational range of pH and temperature of the enzyme were observed as a consequence of the immobilization process. Furthermore, for the first time it was clearly shown that, the combination of continuous flow with a high content of lactose can sharply increase the stability of K. lactis β-d-galactosidase. Thus, Maxilact LX 5000, which is generally used for lactose hydrolysis as a free enzyme, could be advantageously employed in its covalent immobilized form to the hydrolysis of lactose and production of GOS in a continuous PBR.

ACKNOWLEDGEMENTS

The authors wish to thank Coordenação de Aperfeiçoamento de Pessoal de Nível Superior (CAPES), for the scholarships of the first and third authors. This work was funded by the Conselho Nacional de Desenvolvimento Científico e Tecnológico (CNPq)and Fundação de Amparo a Pesquisa do Rio Grande do Sul (FAPERGS).

REFERENCES

1. Albayrak, N., & Yang, S. T. (2002). Immobilization ofbeta-galactosidase on fibrous matrix by polyethyleneimine for production of galacto-oligosaccharides from lactose. Biotechnology Progress, 18(2), 240–251.
2. Ansari, S. A., & Husain, Q. (2010). Lactose hydrolysis by galactosidase immobilized on concanavalin A-cellulose in batch and continuous mode. Journal of Molecular Catalysis B: Enzymatic, 63(1–2), 68–74.
3. Bayramoglu, G., Tunali,Y., & Arica,M.Y.(2007). Immobilization of beta-galactosidase onto magnetic

poly(GMA–MMA) beads for hydrolysis of lactose in bed reactor. Catalysis Communications, 8(7), 1094–1101.

4. Benhabiles, S., Salah, R., Lounici, H., Drouiche, N., Goosen, M. F. A., & Mameri, N. (2012). Antibacterial activity of chitin, chitosan and its oligomers prepared from shrimp shell waste. Food Hydrocolloids, 29(1), 48–56.

5. Bromberg, A., Marx, S., & Frishman, G. (2008). Kinetic study of the thermal inactivation of cholinesterase enzymes immobilized in solid matrices. Biochimica Et Biophysica Acta: Proteins and Proteomics, 1784(6), 961–966.

6. Butler, S. L., & Falke, J. J. (1996). Effects of protein stabilizing agents on thermal backbone motions: A disulfide trapping study. Biochemistry, 35(33), 10595–10600.

7. Cavaille, D., & Combes, D. (1995). Characterization of beta-galactosidase from Kluyveromyces lactis. Biotechnology and Applied Biochemistry, 22, 55–64.

8. Chang, S.-W, Shaw, J.-F., Yang, C.-K., & Shieh, C.-J. (2007). Optimal continuous biosynthesis of hexyl laurate by a packed bed bioreactor. Process Biochemistry, 42(9), 1362–1366.

9. Chiou, S.-H., Hung, T.-C., Giridhar, R., & Wu, W.-T. (2007). Immobilization of lipase to chitosan beads using a natural cross-linker. Preparative Biochemistry & Biotechnology, 37(3), 265–275

10. De Maio, A., El-Masry, M. M., Portaccio, M., Diano, N., Di Martino, S., Mattei, A., et al. (2003). Influence ofthe spacer length on the activity of enzymes immobilised on nylon/polyGMA membranes. Part 1. Isothermal conditions. Journal of Molecular Catalysis B: Enzymatic, 21(4–6), 239–252.

11. Fernandez-Lafuente, R. (2009). Stabilization of multimeric enzymes: Strategies to prevent subunit dissociation. Enzyme and Microbial Technology, 45(6–7), 405–418.

12. Garcia-Galan, C., Berenguer-Murcia, A., Fernandez-Lafuente, R., & Rodrigues, R. C. (2011). Potential of different enzyme immobilization strategies to improve enzyme performance. Advanced Synthesis & Catalysis, 353(16), 2885–2904.

13. Gosling, A., Stevens, G. W., Barber, A. R., Kentish, S. E., & Gras, S. L. (2010). Recent advances refining galactooligosaccharide production from lactose. Food Chemistry, 121(2), 307–318.

14. Gosling, A., Stevens, G. W., Barber, A. R., Kentish, S. E., & Gras, S. L.(2011). Effect ofthe substrate concentration and water activity on the yield and rate of the transfer reaction of beta-galactosidase from Bacillus circulans. Journal of Agricultural and Food Chemistry, 59(7), 3366–3372.

15. Grosova, Z., Rosenberg, M., & Rebros, M. (2008). Perspectives and applications of immobilised beta-galactosidase in food industry—Areview. Czech Journal of Food Sciences, 26(1), 1–14.

16. Haider, T., & Husain, Q. (2009). Immobilization of beta galactosidase from Aspergillus oryzae via immunoaffinity support. Biochemical Engineering Journal, 43(3), 307–314.

17. Halim, S. F. A., Kamaruddin, A. H., & Fernando, W. J. N. (2009). Continuous biosynthesis of biodiesel from waste cooking palm oil in a packed bed reactor: Optimization using response surface methodology (RSM) and mass transfer studies. Bioresource Technology, 100(2), 710–716.

18. Illanes, A., Altamirano, C., & Zuniga, M. E. (1996). Thermal inactivation of immobilized penicillin acylase in the presence of substrate and products. Biotechnology and Bioengineering, 50(6), 609–616.

19. Jochems, P., Satyawali, Y., Van Roy, S., Doyen, W., Diels, L., & Dejonghe, W. (2011). Characterization and optimization of beta-galactosidase immobilization process on a mixed-matrix membrane. Enzyme and Microbial Technology, 49(6–7), 580–588.

20. Jurado, E., Camacho, F., Luzon, G., & Vicaria, J. M. (2004). Kinetic models of activity for beta-galactosidases: Influence of pH, ionic concentration and temperature. Enzyme and Microbial Technology, 34(1), 33–40.

21. Jurado, E., Camacho, F., Luzon, G., & Vicaria, J. M. (2005). Kinetic and enzymatic adsorption model in a recirculation hollow-fibre bioreactor. Bioprocess and Biosystems Engineering, 28(1), 27–36.

22. Klein, M. P., Nunes, M. R., Rodrigues, R. C., Benvenutti, E. V., Costa, T. M. H., Hertz, P. F., et al. (2012). Effect of the support size on the properties of -galactosidase immobilized on chitosan: Advantages and disadvantages of macro and nanoparticles. Biomacromolecules, 13(8), 2456–2464.

23. Krajewska, B.(2004).Application of chitin- and chitosan-based materials for enzyme immobilizations: A review. Enzyme and Microbial Technology, 35(2–3), 126–139.

24. Ladero, M., Santos, A., & Garcia-Ochoa, F. (2006). Kinetic modelling of the thermal inactivation of an industrial beta-galactosidase from Kluyveromyces fragilis. Enzyme and Microbial Technology, 38(1–2), 1–9.

25. Li, X., Zhou, Q. Z. K., & Chen, X. D. (2007). Pilot-scale lactose hydrolysis using - galactosidase immobilized on cotton fabric. Chemical Engineering and Processing: Process Intensification, 46(5), 497–500.

26. Lin, T. Y., & Timasheff, S. N. (1996). On the role of surface tension in the stabilization of globular proteins. Protein Science, 5(2), 372–381.

27. Mussatto, S. I., & Mancilha, I. M. (2007). Non-digestible oligosaccharides: A review. Carbohydrate Polymers, 68(3), 587–597.

28. Muzzarelli, R. A. A. (1980). Immobilization of enzymes on chitin and chitosan. Enzyme and Microbial Technology, 2(3), 177–184.

29. Muzzarelli, R. A. A. (2009). Genipin-crosslinked chitosan hydrogels as biomedical and pharmaceutical aids. Carbohydrate Polymers, 77(1), 1–9.

30. Nakkharat, P., & Haltrich, D. (2007). beta-Galactosidase from Talaromyces thermophilus immobilized on to Eupergit C for production of galactooligosaccharides

during lactose hydrolysis in batch and packed-bed reactor. World Journal of Microbiology & Biotechnology, 23(6), 759–764.

31. Neri, D. F. M., Balcão, V. M., Costa, R. S., Rocha, I. C. A. P., Ferreira, E. M. F. C., Torres, D. P. M., et al. (2009). Galacto-oligosaccharides production during lactose hydrolysis by free Aspergillus oryzae -galactosidase and immobilized on magnetic polysiloxane-polyvinyl alcohol. Food Chemistry, 115(1), 92–99.

32. Nie, K., Xie, F., Wang, F., & Tan, T. (2006). Lipase catalyzed methanolysis to produce biodiesel: Optimizationofthe biodiesel production. Journal of Molecular Catalysis B: Enzymatic, 43(1–4), 142–147.

33. Park, A.-R., & Oh, D.-K. (2009). Galacto-oligosaccharide production using microbial -galactosidase: Current state and perspectives. Applied Microbiology and Biotechnology, 85(5), 1279–1286.

34. Pereira-Rodríguez, Á., Fernández-Leiro, R., González-Siso, M. I., Cerdán, M. E., Becerra, M., & Sanz-Aparicio, J. (2012). Structural basis of specificity in tetrameric Kluyveromyces lactis -galactosidase. Journal of Structural Biology, 177(2), 392–401.

35. Rodrigues, D. S., Mendes, A. A., Adriano, W. S., Gonc¸alves, L. R. B., & Giordano, R. L. C. (2008). Multipoint covalent immobilization of microbial lipase on chitosan and agarose activated by different methods. Journal of Molecular Catalysis B: Enzymatic, 51(3–4), 100–109.

36. Sampedro, J. G., Cortes, P., Munoz-Clares, R. A., Fernandez, A., & Uribe, S. (2001). Thermal inactivation of the plasma membrane H+-ATPase from Kluyveromyces lactis. Protection by trehalose. Biochimica Et Biophysica Acta: Protein Structure and Molecular Enzymology, 1544(1–2), 64–73.

37. Shin, H. J., Park, J. M., & Yang, J. W. (1998). Continuous production of galacto-oligosaccharides from lactose by Bullera singularis beta-galactosidase immobilized in chitosan beads. Process Biochemistry, 33(8), 787–792.

38. Sola-Penna, M., & Meyer-Fernandes, J. R.(1998). Stabilization againstthermal inactivation promoted by sugars on enzyme structure and function: Why is trehalose more effective than other sugars? Archives of Biochemistry and Biophysics, 360(1), 10–14.

39. Song, Y. S., Lee, J. H., Kang, S. W., & Kim, S. W. (2010). Performance of -galactosidase pretreated with lactose to prevent activity loss during the enzyme immobilisation process. Food Chemistry, 123(1), 1–5.

CHAPTER 22 NOVEL HIGH-PERFORMANCE METAGENOME β-GALACTOSIDASES FOR LACTOSE HYDROLYSIS IN THE DAIRY INDUSTRY

Sarah Erich[1], Beatrice Kuschel[1], Thilo Schwarz [1], Jacob Ewert [1], Nico Böhmer [1], Frank Niehaus [2], Jürgen Eck [2], Sabine Lutz-Wahl [1], Timo Stressler [1], and Lutz Fischer[1]

[1]University of Hohenheim, Institute of Food Science and Biotechnology, Department of Biotechnology and Enzyme Science, Garbenstr. 25, 70599 Stuttgart, Germany

[2]B.R.A.I.N. Biotechnology Research and Information Network AG, Darmstaedter Str. 34-36, 64673, Zwingenberg, Germany

ABSTRACT

The industrially utilised β-galactosidases from *Kluyveromyces* spp. and *Aspergillus* spp. feature undesirable kinetic properties in praxis, such as an unsatisfactory lactose affinity (K_M) and product inhibition (K_I) by galactose. In this study, a metagenome library of about 1.3 million clones was investigated with a three-step activity-based screening strategy in order to find new β-galactosidases with more favourable kinetic properties. Six novel metagenome β-galactosidases (M1–M6) were found with an improved lactose hydrolysis performance in original milk when directly compared to the commercial β-galactosidase from *Kluyveromyces lactis* (GODO-YNL2). The best metagenome candidate, called "M1", was recombinantly produced in *Escherichia coli* BL21(DE3) in a bioreactor (volume 35 L), resulting in a total β-galactosidase M1 activity of about 1100 μkat$o_{NPGal,37\ °C}$ L^{-1}. Since milk is a sensitive and complex medium, it has to be processed at 5–10 °C in the dairy industry. Therefore, the β-galactosidase M1 was tested at 8 °C in milk and possessed a good stability ($t_{1/2}$ = 21.8 d), a desirably low apparent $K_{M,lactose,8\ °C}$ value of 3.8 ± 0.7 mM and a high apparent $K_{I,galactose,8\ °C}$ value of 196.6 ± 55.5 mM. A lactose hydrolysis process (milk, 40 nkat$_{lactose}$ mL$_{milk,8\ °C}^{-1}$) was conducted at a scale of 0.5 L to compare the performance of M1 with the commercial β-galactosidase from *K. lactis*(GODO-YNL2). Lactose was completely (>99.99%) hydrolysed by M1 and to 99.6% (w/v) by *K. lactis* β-galactosidase after 25 h process time. Thus, M1 was able to achieve the limit of <100 mg lactose per litre milk, which is recommended for dairy products labelled as "lactose-free".

Citation: Sarah Erich, Beatrice Kuschel, Thilo Schwarz, Jacob Ewert, Nico Böhmer, Frank Niehaus, Jürgen Eck, Sabine Lutz-Wahl, Timo Stressler, Lutz Fischer, Novel high-performance metagenome β-galactosidases for lactose hydrolysis in the dairy industry, Journal of Biotechnology, Volume 210, 20 September 2015, Pages 27-37, ISSN 0168-1656, http://dx.doi.org/10.1016/j.jbiotec.2015.06.411.

INTRODUCTION

β-Galactosidases (E.C. 3.2.1.23) catalyse the hydrolysis of terminal non-reducing β-d-galactose residues in β-d-galactosides. One of the main applications is lactose hydrolysis in milk and milk products, where the β-galactosidase cleaves lactose into d-glucose and d-galactose (Pereira-Rodríguez et al., 2012). The enzymatic hydrolysis of lactose is an important biotechnological process in the food industry. The technological and sensorial characteristics of food can be improved by hydrolysing lactose. The hydrolysis of lactose in ice cream, for instance, improves the creaminess significantly. Moreover, the monosaccharides formed by lactose hydrolysis increase the sweetness of the products (Mlichová and Rosenberg, 2006). The enzyme is used in the dairy industry to produce lactose-free milk products, which is a suitable approach to overcome the lactose intolerance problem affecting humans worldwide (Harju et al., 2012). The authorities in Scandinavian countries set the limit of residual lactose content for products labelled as "lactose-free" to below 10 mg lactose per 100 g of product (<0.01% (w/w)). The GDCh (Gesellschaft Deutscher Chemiker e.V.) in Germany also recommends similar criteria for this type of lactose-free declaration (GDCh, 2005). According to these criteria, the lactose content of products with this label must be lower than 100 $mg_{lactose}$ kg^{-1} or L^{-1} of the dairy product (<0.01%). Thus, β-galactosidases should be at hand which possess industrially suitable properties at reasonable costs.

A considerable number of studies have been published over the past few decades investigating β-galactosidases from different sources for lactose hydrolysis (Panesar et al., 2006). Generally, the β-galactosidases utilised in the dairy industry were obtained from *Kluyveromyces* spp. or *Aspergillus* spp. and applied for industrial applications mainly as soluble, or rarely as immobilised, enzyme preparations (Husain, 2010, Novalin et al., 2005 and Oliveira et al., 2011).

The kinetic properties of β-galactosidases utilised in the food industry still do not fit the processing conditions required optimally (Ansari and Satar, 2012 and Harju et al., 2012). They are all, for example, significantly inhibited by the hydrolysis-generated product d-galactose (K_I value is too low), whereas the substrate affinity to lactose is not beneficial (K_M value is not low enough). Therefore, it is of great economic interest to discover β-galactosidases with improved processing characteristics for their application in milk systems.

Only 0.1–1% of the bacterial species in nature are calculated to be cultivable using conventional methods (Handelsman, 2005). Thus, an auspicious source for finding novel β-galactosidases with industrially desirable properties is the metagenome. Metagenomics has been in the spotlight since the 1990s (Handelsman et al., 1998). Briefly, only the DNA is isolated from environmental samples (e.g. soil) instead of living microorganisms. It is cut into shape and cloned into well-known vector/host systems (e.g.*Escherichia coli*) allowing the expression of enzymes from the non-cultivable organisms. This so-called metagenomic approach has proven to be an efficient tool for the discovery of novel biocatalysts (Gans et al., 2005 and Lorenz and Eck, 2004). Metagenomic libraries have been screened for a wide range of enzymes, such as lipases, proteases, dehydrogenases, nitrile hydratases and oxidoreductases (Nacke et al., 2011 and Steele et al., 2008). A novel metagenome-derived β-galactosidase was found by Wang et al. (2010), who screened a metagenomic library and found a new β-galactosidase which they characterised and tested for lactose hydrolysis in milk at temperatures between 4 and 25 °C over a time period of only 1 h. Thus, no evidence could be presented that this metagenome enzyme was really able to achieve a lactose hydrolysis down to 100 $mg_{lactose}$ L^{-1} milk in a reasonable time.

In literature, the best lactose conversion yields by using free or immobilized enzymes were in the range of 70–99% (Cieśliński et al., 2005, Horner et al., 2011, Nakagawa et al., 2007 and Pan et al., 2010). However, the lactose content in milk of less than 100 $mg_{lactose}L^{-1}$ could not be achieved with these hydrolysis yields, since milk contains about 45–50 $g_{lactose}$ L^{-1} and a 99% hydrolysis yield would result in 450–495 $mg_{lactose}$ L^{-1}.

In our study, a metagenomic library from German soil samples was screened for β-galactosidase activity. Novel β-galactosidases were found using a three-step activity-based screening strategy and were expressed in *E. coli*. The best metagenome β-galactosidases were tested for lactose hydrolysis in milk at 8 °C and compared to a commercial reference β-galactosidase from *K. lactis* called GODO-YNL2 (Shusei Co, Ltd., Tokyo, Japan), since this enzyme is often utilised in the dairy industry today. The most promising metagenome β-galactosidase was then characterized concerning process-relevant and kinetic parameters. The enzyme activities and the kinetic parameters were determined with the natural substrate lactose at 8 °C, because it is known that the synthetic substrate *o*-nitrophenyl-β-d-galactopyranoside (*o*NPGal) leads to dissimilar enzyme activities and kinetic parameters (Hildebrandt et al., 2009).

MATERIALS AND METHODS

Chemicals

All chemicals used were of analytical reagent grade and purchased from Merck (Darmstadt, Germany), Sigma–Aldrich (Steinheim, Germany), Gerbu (Heidelberg Wieblingen, Germany) or Carl-Roth (Karlsruhe, Germany).

Water was purified by reverse osmosis and passed through a Millipore Mili-Q unit. The commercial reference GODO-YNL2 (Shusei Co, Ltd., Tokyo, Japan) containing the β-galactosidase from *Kluyveromyces lactis* was used for comparison studies.

T4 DNA-ligase and hexokinase/glucose-6-phosphate-dehydrogenase were purchased from Roche Diagnostics (Mannheim, Germany). All restriction enzymes were acquired from New England Biolabs (Frankfurt, Germany). HotStar HiFidelity Polymerase was purchased from Qiagen (Hilden, Germany).

The chromatography resin BioFox 40 IDA_{low} and the Bioline HR glass column (300 × 10 mm) were kind gifts from KNAUER Wissenschaftliche Geräte GmbH (Berlin, Germany).

Isolation of metagenomic DNA

Genomic DNA from different sampling sites near Zwingenberg (Germany) was prepared, as described in Gabor et al. (2012).

Metagenome library construction

High-copy plasmid-based metagenome libraries were constructed from these habitat samples. Therefore, DNA fragments of 2–10 kb were obtained by partially digesting metagenomic DNA with *Alu*I. After sizing by agarose gel electrophoresis and electroelution, fragments were ligated into a *Sma*I-digested dephosphorylated pBCS-PvegII vector. This vector (Supplemental S1) is a derivative of the commonly used pBC16 plasmid (Bernhard et al., 1978). Modifications of this vector include the replacement of the tetracycline by a chloramphenicol resistance marker, the addition of a pUC ori and the introduction of the strong constitutive vegII promoter of *Bacillus subtilis* up-stream of the*Nde*I cloning site to allow the initiation of transcription in both *B. subtilis* and *E. coli*. Ligation mixtures were used to transform *E. coli* DH5α (Fisher Scientific) by electroporation. Altogether, 5340 Mb DNA was isolated from several soil samples. The average insert size was 4 kb, which results in a total theoretical number of about 1335,000 clones for the screening.

Screening for β–galactosidase activity

A three-step activity-based screening strategy was applied to the metagenome clones (Fig. 1).

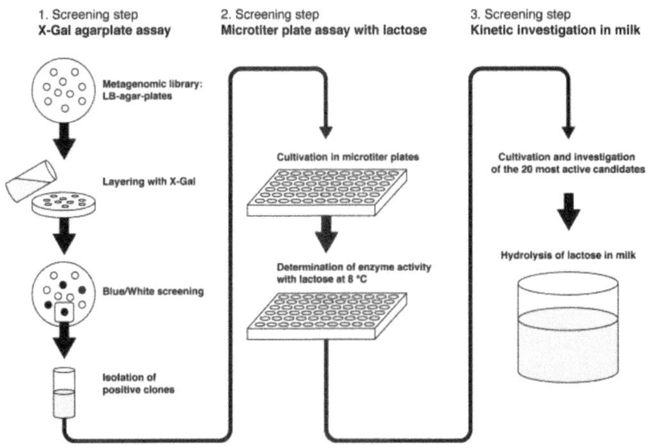

Figure 1: Scheme of the three-step screening strategy for novel metagenome β-galactosidases.

First screening step using X-Gal agar plates

As described in patent EP2530148 (A1) (Niehaus and Eck, 2012), the metagenome clones were cultivated on LB agar plates containing 12.5 μg mL^{-1} chloramphenicol and 80 μg mL^{-1} 5-bromo-4-chloro-3-indoxyl-β-d-galactopyranoside (X-Gal) for the first screening step. Positive clones with presumed β-galactosidase activity showed a blue colour on the agar plates, which resulted from the hydrolysis of X-Gal.

Second screening step with lactose as a substrate

In a second screening step, the positive clones from the first screening step were selected, cultivated and disrupted, as described in patent EP2530148 (A1) (Niehaus and Eck, 2012). The β-galactosidase activity was measured with lactose as a substrate at 8 °C. Therefore, a lactose solution (160 μL; 0.2 M) in potassium phosphate buffer (0.1 M, pH 6.75) was combined with 40 μL cell-free extract. The reaction was stopped by transferring 95 μL reaction mixture to 190 μL perchloric acid (1 M) after 1 h. The sample was centrifuged (740 × *g*) at 4 °C for 10 min after neutralisation with KOH (2 M). The d-glucose released was determined with the hexokinase/glucose-6-phosphate-dehydrogenase assay (HK assay) by photometric measurement (340 nm) in microtiter plates, following the manufacturer's protocol for the d-glucose/d-fructose test kit (R-Biopharm AG, Darmstadt, Germany; product code 10 139 106 035).

Third screening step: Enzymatic lactose hydrolysis in milk

The lactose hydrolysis in milk ("Schwarzwälder H-Weidemilch, 3.8%" obtained from Schwarzwaldmilch GmbH, Germany) was performed with an operating volume of 30 mL at 8 °C in 100 mL flasks. Equal amounts of β-galactosidase activity (0.83 $nkat_{lactose,8\ °C}mL^{-1}$) from the twenty most active metagenome enzymes from the second screening step and GODO-YNL2 (reference) were added to milk which contained 46 ± 0.5 g L^{-1}lactose and 0.1% (w/v) sodium azide (for the prevention of microbial growth). Samples (0.5 mL) were taken every 24 h. The enzymatic reaction in the samples was stopped with 1 mL of perchloric acid (1 M). After centrifugation of the samples at 7000 × *g* and 4 °C for 10 min (Eppendorf centrifuge 5417R), the samples were neutralised with potassium hydroxide (2 M) and cooled on ice for 10 min. Finally, the samples were filtered through a 0.45 μm PTFE filter and analysed by HPLC (high performance liquid chromatography; see Section 2.12).

Analysis of β-*galactosidase* genes

The genes of the six best metagenomes of β-galactosidase (M1 to M6) were sequenced. Database homology research was performed using the "Basic Local Alignment Search Tool" (BLAST, http://blast.ncbi.nlm.nih.gov) for analysis. Furthermore, the pairwise alignment of the metagenome was performed by using the ClustalW2 programme (http://www.genome.jp/tools/clustalw/). The nucleotide sequences of M1 to M6 are available in the GenBank database (M1: KM651981, M2: KM651982, M3: KM651983, M4: KM651984, M5: KJ792469; M6: KJ792470). In addition, the sequences for M1–M4 (M1: 3J_33, M2: 3J_37, M3: 3J_11, M4: 186B1/330C6) have already been published in patent EP2530148 (A1) (Niehaus and Eck, 2012).

Subcloning of the best metagenome -galactosidase M1

In order to obtain a higher amount of enzyme, the best metagenome β-galactosidase M1 was subcloned into the expression vector pET20b. The construction of the expression vector was performed using *E. coli* XL1-blue and standard molecular biology techniques (Sambrook and Russell, 2001). The M1 β-galactosidase gene was amplified by polymerase chain reaction (PCR) using the template pBCS-3J-33(M1) with the primers M1_for_*Nde*I (5′-agagtctgcatatgcgacaaaagcttgttta-3′) and M1_rev_*Xho*I (5′-actaatctcgagaatggtgcgaaacgtaaag-3′). The PCR product was digested with *Nde*I and*Xho*I and ligated to a *Nde*I/*Xho*I-treated pET20b vector with a six-histidine tag using T4-Ligase to generate pET20b-M1.

Production of M1 in a stirred tank reactor

E. coli BL21(DE3) was transformed with pET20b-M1 prior to the recombinant expression. The cultivation was realized using 2YT medium (16 g peptone, 5 g NaCl, and 10 g yeast extract per L) with d-glucose (2% w/v) and ampicillin (100 μg mL^{-1}) in a techfors bioreactor (Infors AG, Bottmingen, Switzerland; operating volume 35 L) at 500 rpm, air gassing *vvm* = 2 (pO_2 > 30%) and pH 7.0, controlled with 12.5% (v/v) NH_4OH and 0.66 M H_3PO_4. The cells were cultivated at 37 °C up to an OD_{600} of 10. Then, the culture broth was cooled to 30 °C and isopropyl-β-d-1-thiogalactopyranoside (IPTG; 0.4 mM) was added to induce the expression of the recombinant protein. Cells were harvested by centrifugation (8000 × *g*, 4 °C, 10 min) after 15.5 h. These cells were used for cell-disruption and purification, as described below.

During cultivation, the bio dry mass was determined gravimetrically. Therefore, samples were centrifuged (10,000 × *g*, 4 °C, 5 min), washed twice with saline and dried at 40 °C and 10 mbar in a RVC 2-33 IR vacuum centrifuge from Christ (Osterode, Germany). The residual d-glucose concentration of the medium was determined with the hexokinase/glucose-6-phosphate-dehydrogenase assay (see Section 2.4.2). The enzyme activity was determined with *o*NPGal as a substrate (see Section 2.9.1).

The expression and purification (see Section 2.8) of M1 was analysed by SDS-PAGE using 8% and 12% polyacrylamide gels, according to the method of Laemmli (1970). The molecular weight marker Protein Ladder (10–250 kDa, New England Biolabs Inc., Ipswich, MA, USA) was used as a reference. Gels were stained with Coomassie Brilliant Blue. The protein concentrations were determined according to the method of Bradford (1976) using bovine serum albumin as the standard (quadruplicate measurement).

Purification of the metagenome β-galactosidase M1

After cell-disruption, the β-galactosidase M1 was purified by fractionated ammonium sulphate precipitation. Additionally, M1 was further purified using a Ni^{2+}-affinity chromatography step for the determination of the kinetic constants and the influence of d-glucose and d-galactose on the enzyme activity.

Cell-disruption

A cell suspension (30% w/v) was made in potassium phosphate buffer (100 mM, pH 6.75) with $MgCl_2$ (5 mM). This suspension was conducted through a bead mill (Dyno®-mill, containing 0.1–0.2 mm glass beads). Afterwards, the suspension was centrifuged (13,000 × *g*, 4 °C, 20 min). The cell-free extract (supernatant) obtained was used for purification of the β-galactosidase M1.

Ammonium sulphate precipitation

In the first purification step, ammonium sulphate was added to the cell-free extract by stirring at 4 °C to obtain a salt saturation of 30% (w/v). After centrifugation (13,000 × *g*, 4 °C, 20 min), the pellet was discarded and more ammonium sulphate was added to the supernatant to obtain a final salt saturation of 45% (w/v). After further centrifugation (13,000 × *g*, 4 °C, 20 min), the supernatant was discarded and the pellet was dissolved in potassium phosphate buffer (100 mM, pH 6.75) with $MgCl_2$ (5 mM) and dialysed (dialysis tubing cellulose membrane, Sigma Aldrich, Steinheim, Germany) against this buffer at 8 °C for 24 h. This enzyme preparation was used for lactose hydrolysis in milk at 8 °C and for the biochemical characterisation of the M1 (optimal pH, optimal temperature).

Ni^{2+}-affinity chromatography

The ammonium sulphate pellet precipitated was dissolved in potassium phosphate buffer (100 mM, pH 6.75) with NaCl (150 mM) and dialysed against this buffer, as described above. The dialysed solution was used for further purification using an Äkta FPLC system (Äkta FPLC, GE Healthcare, Germany). The flow rate was 1 mL min^{-1}. Eluted protein was detected at 280 nm.

A volume of 20 mL was injected via a Superloop™ (GE Healthcare, Germany) onto a Ni^{2+}-charged BioFox 40 IDA_{low} (KNAUER, Berlin) filled column (Bioline HR glass column (300 × 10 mm), KNAUER, Berlin; 1 column volume (CV) = 4.7 mL) for each chromatography run (*n* = 4). The chromatography resin had been equilibrated previously with buffer A (potassium phosphate buffer (100 mM, pH 6.75) with NaCl (150 mM)). Unbound protein was washed out with buffer A over eight CV. Bound protein was eluted by a linear gradient from 0 to 50% with buffer B (buffer A plus imidazole (500 mM)) over three CV and was held for a further two CV. Finally, the column was cleaned by 100% buffer B over two CV. Fractions (1 mL) were collected during the chromatography process and tested for β-galactosidase activity (*o*NPGal-assay; see Section 2.9.1). Active fractions were pooled and desalted to potassium phosphate buffer (100 mM, pH 6.75) with $MgCl_2$ (5 mM) using PD-10 columns (GE Healthcare, Germany). This enzyme preparation was used for the determination of the kinetic constants and the influence of d-glucose and d-galactose on the enzyme activity.

Assays for β-*galactosidase* activity determination

The β-galactosidase activity was determined by different methods, depending on the needs of the particular experiment. One katal was defined as the amount of enzyme that catalyses the release of 1 mol *o*-nitrophenol from *o*NPGal (*o*-nitrophenyl-β-d-galactopyranoside) per s or the release of 1 mol d-glucose from lactose per s. All assays were performed at least in triplicate.

β–Galactosidase activity using oNPGal as a substrate

During the bioreactor cultivation, the β-galactosidase activity of the cell-free extract was measured at 405 nm with the substrate *o*NPGal at 37 °C. The reaction was performed in the kinetic mode of a spectral photometer containing a temperature-controlled cuvette (Ultrospec 3000, Amersham Bioscience, Freiburg, Germany). An amount of 0.6 mL*o*NPGal solution (50 mM) in potassium phosphate buffer (100 mM, pH 6.75) with $MgCl_2$(5 mM) was mixed with 0.5 mL of the same buffer and preincubated at 37 °C for 5 min. The reaction was started with 0.1 mL enzyme solution.

The *o*NPGal assay was modified and performed on a smaller scale prior to the kinetic measurements. A volume of 100 μL *o*NPGal solution (50 mM) in potassium phosphate buffer (100 mM, pH 6.75) with $MgCl_2$ (5 mM) was mixed with 80 μL of the same buffer and preincubated at 37 °C for 5 min. The reaction was started with 20 μL enzyme solution and incubated at 37 °C for 120 s. The reaction was terminated by adding 200 μL precooled Na_2CO_3-solution (1 M). After centrifugation (13,000 × *g*, 4 °C, 5 min), the absorption was measured at 405 nm.

The effect of d-glucose and d-galactose on the enzyme activity of purified M1 and GODO-YNL2 was determined with a modified *o*NPGal assay. Synthetic milk ultra filtrate (SMUF), according to Pessela et al. (2003), was used to simulate the salt composition of the milk. A volume of 100 μL *o*NPGal solution (50 mM) in SMUF was mixed with 80 μL SMUF containing 25–500 mM of d-glucose, d-galactose or a mixture of both sugars (equimolar: e.g. MIX-50 containing 50 mM d-glucose and 50 mM d-galactose) and preincubated at 37 °C for 5 min. The assay was performed as described above.

β-*Galactosidase activity using lactose as a substrate*

The lactose assays were performed at different temperatures, depending on the particular experiment. A lactose stock solution (400 mM in $H_2O_{bidest.}$) was used. This solution (600 μL) was mixed with 500 μL potassium phosphate buffer (100 mM, pH 6.75; standard buffer) and $MgCl_2$ (5 mM), and preincubated at the respective temperature for 5 min. The reaction was started with 100 μL enzyme solution. After 90 s, a 100 μL sample of the reaction mixture was taken and transferred into 150 μL of perchloric acid (1 M) in order to stop the enzymatic reaction. After centrifugation (13,000 × *g*, 4 °C, 10 min), 160 μL of the supernatant was added to 55 μL potassium hydroxide (2 M). After a second centrifugation step (13,000 × *g*, 4 °C, 10 min), the d-glucose of the clear supernatant released was quantified enzymatically (HK assay, see Section 2.4.2). This enzyme assay was used for the determination of the kinetic parameters. It is the recommended official method of the "Lebensmittel- und Futtergesetzbuch" (LFGB) in Germany

Investigations of pH, temperature properties and kinetic constants

The pH and temperature profiles, temperature stabilities and kinetic constants of the β-galactosidases were investigated with lactose as a substrate.

Optimal pH, optimal temperature and temperature stability

The β-galactosidase activity was measured in a pH range between 5 and 9 and in a temperature range between 8 and 60 °C. All buffers for the pH profile had a concentration of 100 mM. The following buffers were used: citrate buffer (pH 5–6), potassium phosphate buffer (pH 6–8) and Tris–HCl buffer (pH 8–9). The temperature profile was determined with potassium phosphate buffer (100 mM, pH 6.75) containing $MgCl_2$(5 mM). The assay temperature ranged from 8 to 60 °C. In addition, the enzyme solutions were incubated at two different temperatures (8 °C and optimal temperature) for 48 h to determine the temperature stability. The residual enzyme activities of the samples were measured at 37 °C. The data were fitted to first-order plots and analysed, with the first-order rate constants (*k*) determined by linear regression of ln residual activity versus the incubation time. The activity half-life ($t_{1/2}$) was calculated using the following equation:$t_{1/2} = \ln 2/k$.

Determination of kinetic parameters

The Michaelis–Menten kinetic parameters (K_M and K_I) of the metagenome β-galactosidases M1 (purified) and GODO-YNL2 were determined at 8 °C. The kinetic parameters were calculated by non-linear regression fitting with the Enzyme Kinetics Module of SigmaPlot 12.5 (Systat Software, Inc,. San Jose, CA).

For determining the apparent K_M value in buffer, the lactose concentrations in the reaction mixture ranged from 5 to 200 mM. For determining the apparent K_I value, the galactose concentrations ranged from 100 to 400 mM. Each galactose concentration was tested at five different lactose concentrations (5–200 mM).

The apparent kinetic parameters K_M and K_I were also investigated in diluted milk. A dilution was needed to obtain various appropriate lactose concentrations in the range of 5–62.5 mM in the final reaction mixture. The milk was diluted with H_2O_{bidest}. The d-galactose concentration for the determination of the K_I value ranged between 100 and 400 mM. For this, d-galactose was added to the milk. All kinetic experiments were carried out with β-galactosidase activity ranging between 10 and 20 $nkat_{lactose,8\ °C}\ mL_{total\ sample}^{-1}$.

Comparison of the lactose hydrolysis in milk of M1 and GODO-YNL2 on a larger scale

Large-scale lactose hydrolyses were performed in a table bioreactor (Infors AG, Bottmingen, Switzerland) using the metagenome β-galactosidase M1 and GODO-YNL2. The bioreactor was coupled to a cooling circulator (Julabo F12, JULABO Labortechnik GmbH, Seelbach, Germany) and contained an operating volume of 500 mL. The lactose hydrolysis in milk was investigated at 8 °C over 24 h with an enzyme activity equal to 40 $nkat_{lactose,8\ °C}\ mL^{-1}$. After 24 h, the temperature was raised to 40 °C for one hour to increase the activity of the β-galactosidases. When this temperature step was included, a residue lactose concentration below 0.1 g L^{-1} was achieved. Samples (0.5 mL) were taken after 0, 1, 4, 10, 24, 24.5 and 25 h and prepared as described above (see Section2.4.3).

Quantification of lactose via HPLC

HPLC analysis was carried out on a Spectra System (Thermo Fisher Scientific, Dreieich, Germany) coupled with an evaporative light scattering detector (ELSD, SEDEX 75 SEDERE, France). Three REZEX Ca^{2+} columns (300 × 7.8 mm, 5 μm; Phenomenex Aschaffenburg, Germany) were used in sequence for the resolution of the sugars. Water was used as an isocratic mobile phase. The flow rate was set to 0.5 mL min^{-1}, the injection volume was 20 μL and the total runtime was 55 min.

The HPLC method of Erich et al. (2012) was used for the quantification of lactose at contents below 100 mg L^{-1}. The limit of detection for lactose with this method (UV detection) was estimated at 2 mg L^{-1} and the limit of quantification was 6 mg L^{-1}.

RESULTS

Screening of the metagenomic library for β-galactosidase activity

The screening started with more than one million metagenome clones (theoretical estimation), which were cultivated on LB agar plates containing X-Gal. A blue colour around a bacterial colony indicated potential β-galactosidase activity. This amounted to 354 single colonies. After the second screening step, using lactose as a substrate, the amount of β-galactosidase activity-positive clones was reduced to 97. The third screening step was intended to investigate the performance of the novel metagenome β-galactosidases for their lactose hydrolysis in milk at 8 °C. As described above, equal amounts of β-galactosidase activity (0.83 $nkat_{lactose,8\ °C}$ mL^{-1}) from the twenty most active metagenome enzymes from the second screening step and GODO-YNL2 (reference) were added to milk (30 mL). Six metagenome candidates (M1–M6) showed a better lactose hydrolysis performance in milk than the β-galactosidase reference GODO-YNL2. They achieved lower levels of remaining lactose after a conversion time of 250 h (Fig. 2). The other fourteen candidates (M7–M20; not illustrated in Fig. 2) demonstrated a poorer performance than GODO-YNL2. The host strain *E. coli* DH5α of the metagenome enzymes was cultivated under the same conditions and no background β-galactosidase activity was detected with lactose as a substrate at 8 °C.

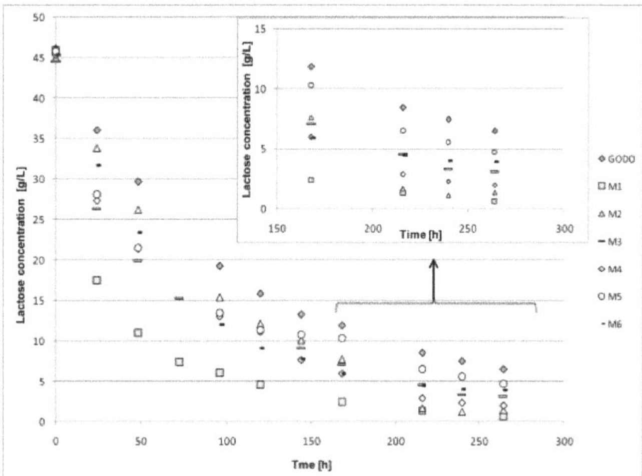

Figure 2: Hydrolysis of lactose in milk (30 mL scale) with six metagenome β-galactosidases (M1–M6) and the commercial enzyme preparation GODO-YNL2 at 8 °C.

Analysis of the β-galactosidase genes of the best candidates

A BLASTx search in the databases of NCBI (http://blast.ncbi.nlm.nih.gov) exposed that M1–M4 are members of the family 2 glycosyl hydrolases and M5 and M6 are members of the family 1 glycosyl hydrolases. The most similarity to M1 is shown by the β-galactosidase sequence from *Paenibacillus wynnii* (strain JDR-2) with an identity of 77% (M1). M2 and M3 show the highest identity of 97% to the next neighbours, which are the β-galactosidase from *Paenibacillus* sp. FSL H7-0357 (M2) and from *Paenibacillus* sp. FSL R5-192 (M3). The next neighbour to M4 is the β-galactosidase from *Paenibacillus vortex* (74% identity) and to M5, the β-glucosidase from *Paenibacillus stellifer* (73% identity). The sequence of M6 shows an identity of 63% to the next neighbour,*Acaryochloris* sp. CCMEE5410 β-galactosidase. The pairwise alignment of the amino acid sequences of M1–M4 depicts identities between 42.9% (M2/M3) and 66.3% (M1/M4), whereas M5 and M6 depict an identity of 44.3%. However, the novel M1–M6 sequences show quite low identity to other metagenome β-galactosidases, such as ZD410 (highest identity of 17.2% with M6; Wang et al., 2010) and Gal308 (highest identity of 18.6% with M2; Zhang et al., 2013), respectively.

Production of the metagenome β-galactosidase M1

Since M1 showed the most favourable properties in the comparative milk hydrolysis, only this candidate was investigated further. To characterise and confirm the suitability of M1 for lactose hydrolysis in milk, a larger amount of the enzyme was produced using the vector/host system pET20b-M1/*E. coli* BL21(DE3). The cultivation was carried out in 2YT medium with 2% (w/v) glucose in a bioreactor with 35 L working volume. The cultivation is shown in Fig. 3. The recombinant *E. coli* pET20b-M1 was initially grown at 37 °C for 2.5 h. Induction was initiated with IPTG at an OD_{600} of 10 and the temperature was lowered to 30 °C in order to avoid inclusion body formation. The highest β-galactosidase activities were measured in the stationary phase (range: 963.0 ± 1.9 – 1,089 ± 3 $\mu kato_{NPGal,37\ °C}L_{culture}^{-1}$). The cells were harvested after a cultivation time of 15.5 h.

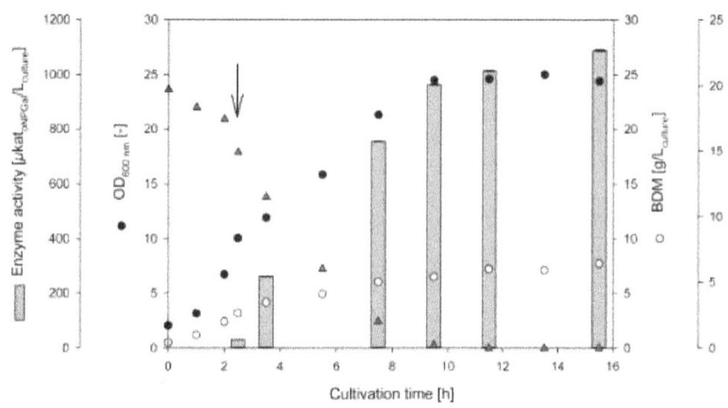

Figure 3: Cultivation of *E. coli* BL21(DE3) pET20b-M1 for metagenome-β-galactosidase M1 production. Bioreactor at 35 L scale, arrow indicates induction with IPTG and temperature shift; enzymatic activity was measured with *o*NPGal at 37 °C, BDM = Bio Dry Mass.

The expression of M1 was investigated by gel electrophoresis experiments (SDS-PAGE). The PAGE gel with samples of the cell-free extracts of *E. coli* BL21(DE3) wild-type and *E. coli* BL21(DE3) pET20b-M1 at various cultivation times is shown in Fig. 4. The overexpressed M1 appeared as a clearly visible protein band with an apparent mass of about 120 kDa (see Fig. 4). This result is in agreement with the in silico mass predicted of 119.77 kDa (including His_6-tag). No protein band of that molecular mass was seen on the SDS-PAGE performed with the protein pellet after cell disruption (data not shown). Thus, no formation of inclusion bodies by the β-galactosidase M1 was observed.

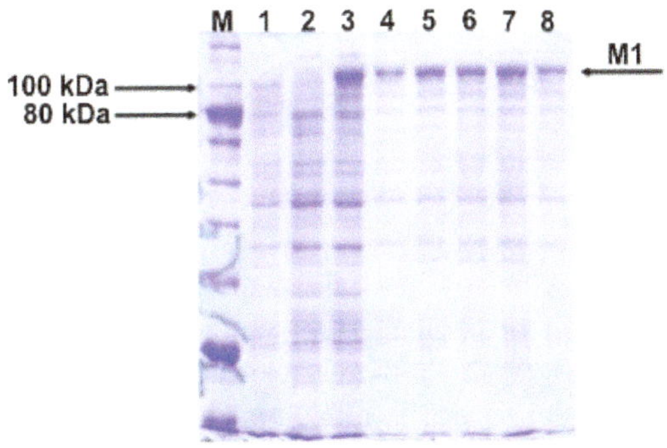

Figure 4: SDS-PAGE analysis of the expression of the metagenome-β-galactosidase M1 in *E. coli* BL21(DE3). 12% gel; M: Protein Ladder 10–250 kDa; 1: *E. coli* BL21(DE3) wild-type (reference); 2: *E. coli*BL21(DE3) pET20b-M1 before induction; 3–8: *E. coli* BL21(DE3) pET20b-M1 1, 5, 7, 9, 11 and 13 h after induction; Coomassie stained.

Purification of M1 and comparison with GODO-YNL2

Partial purification by ammonium sulphate precipitation

An enzyme purification using column chromatography is not common for applications in the food industry due to the high costs of this process step (Stressler et al., 2015). Therefore, the β-galactosidase M1 was partially purified only for the later experiments concerning food process-relevant characteristics, such as optimal pH and temperature, and temperature stability. This partially purified enzyme preparation was consistently used for the lactose hydrolysis in milk on a larger scale. The volumetric activity of M1 determined was 4700 ± 100 $nkat_{lactose,30\ °C}$ mL^{-1} (705 ± 14 $nkat_{lactose,8\ °C}$ mL^{-1}) after a fractionated ammonium sulphate precipitation of the cell-free extract. The total protein content was 42 ± 1.7 mg mL^{-1}. By comparison, the activity of the commercial β-galactosidase GODO-YNL2 was 65,600 ± 980 $nkat_{lactose,30\ °C}$ mL^{-1}(11,808 ± 12 $nkat_{lactose,8\ °C}$ mL^{-1}) and its protein content was 48 ± 1.7 mg mL^{-1}. Thus, GODO-YNL2 is a highly concentrated β-galactosidase preparation (see Fig. 5). It is about 14-fold higher in volumetric activity than the β-galactosidase M1 preparation produced after salt precipitation.

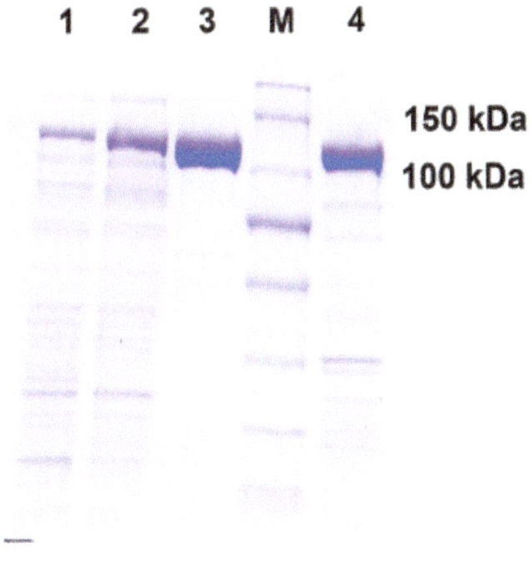

Figure 5: SDS-PAGE analysis of the purification of M1 and comparison to GODO-YNL2. 8% gel; M: Protein Ladder 10–250 kDa; 1: M1 after cell disruption (cell-free extract); 2: M1 after ammonium sulphate precipitation; 3: M1 after purification with Ni^{2+}-affinity chromatography; 4: GODO-YNL2 (commercial enzyme preparation); Coomassie stained.

Due to the fact that the host strain used, *E. coli* BL21(DE3), is not β-galactosidase negative, its background activity was determined. Therefore, *E. coli* BL21(DE3) wild type was cultivated and processed as described for the M1 production. No background β-galactosidase activity was measured with lactose as a substrate at 8 °C. A negligible background β-galactosidase activity of 100 ± 9 $nkat_{lactose}$ mL^{-1} was detectable at 30 °C (2% of the M1 activity at this temperature).

Further purification of M1 by Ni^{2+}-affinity chromatography

In contrast to an application in the food industry, a protein should be purified before the functional properties (Camper and Viola, 2009), such as enzyme kinetics (Stressler et al., 2014), can be investigated. Here, M1 was further purified by Ni^{2+}-affinity chromatography, as seen in Fig. 5. This figure shows that the purified enzyme preparation of M1 had a higher purity than industrially applied GODO-YNL2. The specific enzyme activity of M1 after purification increased about sixfold up to 99.12 $nkat_{lactose,8\ °C}$ $mg_{protein}^{-1}$. However, the specific activity of GODO-YNL2 (246 $nkat_{lactose,8\ °C}$ $mg_{protein}^{-1}$) was still about 2.5 times higher than the specific activity of the purified metagenome β-galactosidase M1.

Determination of process-relevant parameters of M1 and GODO-YNL2

The pH and temperature profiles of enzymes determine their application area in the food industry. Thus, these properties were ascertained for the metagenome β-galactosidase M1 and were compared with the commercial enzyme preparation GODO-YNL2. All data were determined with lactose as the substrate. The results are shown in Table 1 . Both enzymes tested possessed an optimal pH between 6.5 and 7.0, which enables them for an application in milk or sweet whey (both: pH 6.7). Regarding the optimal temperature, M1 was most active at 37 °C. GODO-YNL2 showed a slightly higher optimal temperature (40 °C). Since industrial lactose hydrolysis in milk is carried out at low temperatures (below 10 °C) and the β-galactosidases must be thermally inactivated afterwards at pasteurisation temperatures (above 65 °C), the optimal temperature of M1 was appropriate for food applications.

Table 1: Optimal pH, optimal temperature and temperature stability (half-life $t_{1/2}$) of the novel metagenome β-galactosidase M1 and the commercial β-galactosidase preparation GODO-YNL2. (Determination in buffer with the substrate lactose, experiments were done in triplicate, standard deviation of the enzyme activities determined <5%.

		Temperature properties		
			Half-life $t_{1/2}$ [d]	
	Optimal pH [–]	Optimal temp. [°C]	At optimal temp.	At 8 °C
GODO-YNL2	6.5	40	5.0	30.3
M1[a]	7	37	3.0	21.8

[a] Partially purified (ammonium sulphate precipitation).

The stability of the -galactosidases was measured at 8 °C and at their particular optimal temperature (see Table 1), respectively. M1 showed a sufficient temperature stability ($t_{1/2}$ 21.8 days at 8 °C and $t_{1/2}$ 3 days at 37 °C). However, the commercial enzyme GODO-YNL2 was more stable ($t_{1/2}$ 30.3 days at 8 °C and $t_{1/2}$ 5 days at 40 °C). The residual activities of M1 and GODO-YNL2 were also examined at 8 °C after 48 h. Here, M1 and GODO-YNL2 showed a residual enzyme activity of 85% and 90%, respectively.

In summary, all results suggest that both enzyme preparations are suitable for lactose hydrolysis in milk.

Lactose hydrolysis in milk on a larger scale

Both β-galactosidase preparations M1 and GODO-YNL2 were compared under industrial-like conditions in a scale of 0.5 L milk concerning their ability to hydrolyse lactose (see Fig. 6). M1 was used as a partially purified enzyme preparation due to the higher resemblance to commercial preparations and, therefore, increased relevance for industrial applicability. An equal amount of β-galactosidase activity was used (40 $nkat_{lactose,8\ °C}\ mL_{milk}^{-1}$). The initial lactose concentration of the milk was 46 ± 0.5 g L^{-1}. The β-galactosidase M1 exhibited a better performance in lactose hydrolysis from the beginning (see Fig. 6). After 60 min, 75% of the milk lactose had already been hydrolysed with M1 compared to 67% with GODO-YNL2. However, the total conversion of lactose after 25 h was most important (see small illustration in Fig. 6). In comparison to M1 leaving lactose amounts well below 50 mg L^{-1}, GODO-YNL2 still left about 180 mg lactose L^{-1}. Thus, the recommended level for "lactose-free" labelling (below 100 mg L^{-1}) was reached with the novel β-galactosidase M1 (below the limit of quantification, see Erich et al., 2012). A higher enzyme amount than 40 $nkat_{lactose,8\ °C}\ mL_{milk}^{-1}$ would be needed to reach this desired result with the commercial reference GODO-YNL2.

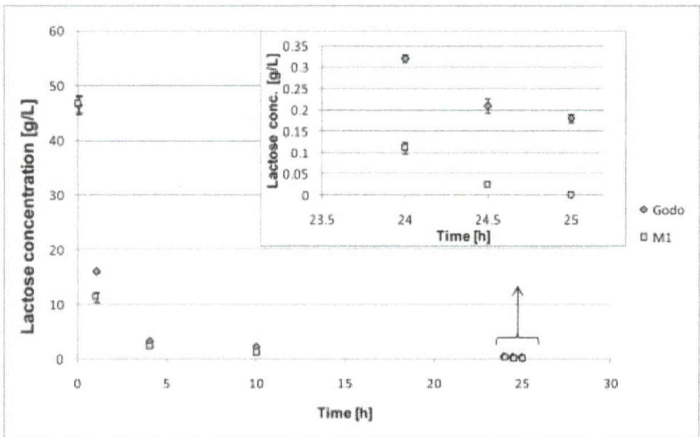

Figure 6: Hydrolysis of lactose in milk (500 mL scale) with the metagenome β-galactosidase M1 and the commercial β-galactosidase GODO-YNL2 at 8 °C for 25 h.

Comparison of M1 and GODO-YNL2 concerning kinetic parameters and the influence of the lactose hydrolysis products on the enzyme activity

Although M1 and GODO-YNL2 showed nearly the same pH, temperature and stability characteristics, the performance of M1 during the lactose hydrolysis was better. A possible reason could have been the particular kinetic properties of the enzymes.

Kinetics

The apparent kinetic parameters K_M and K_I (product inhibition by d-galactose) for both β-galactosidases were determined with lactose in buffer and in diluted milk at 8 °C. The diluted milk was chosen as an additional test matrix to simulate the behaviour of the enzymes when applied in the food industry. The results are presented in Table 2. Interestingly, the apparent kinetic parameters for M1 and GODO-YNL2 depended on the medium applied (diluted milk or buffer system) and displayed an opposing trend for each enzyme. The apparent K_M value decreased in the case of purified M1 from 14.3 ± 1.6 mM in buffered medium to 3.8 ± 0.7 mM in diluted milk. Contrarily, in the case of GODO-YNL2, an increase of the apparent K_M value was seen from 20.4 ± 1.9 mM in buffered medium to 30.9 ± 3.7 mM in diluted milk. Looking at the K_I values in buffer and diluted milk, M1 again performed more favourably than GODO-YNL2. In the case of GODO-YNL2, with an unfavourable low apparent K_I of 32.4 ± 3.9 mM in diluted milk, the product inhibition by d-galactose is already noticeable after about 25% lactose hydrolysis (35 mM d-galactose). Although the apparent K_I value for M1 decreased from 364.4 ± 49.6 mM in buffer to 196.6 ± 55.5 mM in diluted milk, the apparent K_I values for M1 were still higher than the theoretically possible d-galactose concentration in milk after total lactose hydrolysis (approx. 140 mM = 48 g L^{-1}). Thus, the product inhibition of M1 in milk restricts the lactose hydrolysis only marginally and only at a high lactose degree of hydrolysis towards the end of the process. However, it should be mentioned that these results explain the shift of the kinetic parameters in a milk-like environment. Further studies, using a more appropriate dilution medium than water (e.g. SMUF-buffer), should be carried out in the future.

Table 2: Apparent kinetic parameters K_M and K_I of the novel metagenome β-galactosidase (M1) and GODO-YNL2 (commercial β-galactosidase). (Experiments were carried out in triplicate, and data were calculated by non-linear regression fitting with the Enzyme Kinetics Module of SigmaPlot 12.5, Systat Software, Inc,. San Jose, CA).

		Buffer[a] [8 °C]	Diluted milk[b] [8 °C]
$K_{M,\ lactose}$ [mM]	M1[c]	14.3 ± 1.6	3.8 ± 0.7
	GODO-YNL2	20.4 ± 1.9	30.9 ± 3.7
$K_{I,\ galactose}$ [mM]	M1[c]	364.4 ± 49.6	196.6 ± 55.5
	GODO-YNL2	34.3 ± 2.9	32.4 ± 3.9

[a] Potassium phosphate buffer (100 mM, pH 6.75) with $MgCl_2$ (5 mM).
[b] H-Schwarzwaldmilch (3.8% fat) diluted with H_2O.
[c] Purified (ammonium sulphate precipitation and Ni^{2+}-affinity chromatography).

Investigation of product inhibition by d-glucose and d-galactose

In addition to d-galactose, the effect of released d-glucose on the enzyme activity of M1 and GODO-YNL2 during hydrolysis was also of interest. Due to the experimental setup (analysis), it was not possible to measure the influence of d-glucose with the natural substrate lactose. Therefore, *o*NPGal was used for the determination of β-galactosidase activity in the presence of d-glucose. In addition, the influence of d-galactose and the influence of an equimolar concentration of both d-glucose and d-galactose was tested. The influence of M1 and GODO-YNL2 on the enzyme activity is shown in Fig. 7. The enzyme activity of GODO-YNL2 decreased with increasing d-galactose concentration (150 mM: 81.1 ± 0.3% activity) as seen in Fig. 7A. This is in accordance with the results displayed previously. By the addition of d-glucose, a minor activating effect of the enzyme activity was observed. However, this was only seen at high d-glucose concentrations. The equimolar mixture of both sugars showed that the activating effect of d-glucose resulted in a lesser reduction of the enzyme activity compared to the reduction resulting from the exclusive influence of d-galactose. However, as expected, the inhibiting effect of d-galactose was dominant. In the case of M1, the results were quite different. No inhibition was visible up to a concentration of 200 mM d-galactose. Furthermore, the presence of d-glucose showed a much stronger activating effect on the enzyme activity of M1 (at 150 mM d-glucose: 118.6 ± 2.1% activity) compared to GODO-YNL2. No further effect on the enzyme activity (at MIX-150: 116.1 ± 0.3% activity) was observed for the equimolar mixture of d-glucose and d-galactose, compared to the exclusive addition of d-glucose.

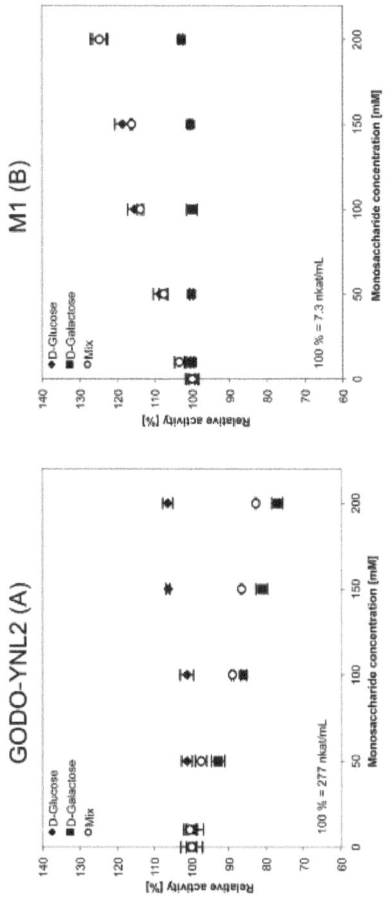

Fig. 7:Effect of added d-glucose, d-galactose or equimolar mixture of both sugars on the activity of GODO-YNL2 (A) and purified M1 (B). Enzymatic activity was measured with *o*NPGal at 37 °C; experiments were carried out in triplicate.

In conclusion, the kinetic data presented promoted the finding that M1 performed better in milk lactose hydrolysis than GODO-YNL2.

DISCUSSION

Metagenome screening

The screening strategy is the most important factor when looking for novel enzymes with desirable properties. The general saying "you get what you screen for" (Schmidt-Dannert and Arnold, 1999) means, in the case of novel β-galactosidases for the food/dairy industry, that one should, firstly, utilise lactose as a screening substrate and, secondly, verify the novel enzymes under process conditions (medium, pH, temperature). The first screening step was carried out with an artificial substrate (X-Gal) for β-galactosidases, since the high number of metagenome clones (more than one million) needed to be managed. This principle was also successfully proven in other metagenome β-galactosidase screenings (Wang et al., 2010Zhang et al., 2013). Nevertheless, it is likely that an unknown number of metagenome β-galactosidases could not be discovered because X-Gal did not fit into their active site, as the indoxyl residue is quite different in shape and functionality compared to natural hexose units. On the contrary, the fact that X-Gal cleaving metagenome enzymes do not all act on lactose was measured in our second screening step. In fact, the percentage of true β-galactosidases hydrolysing lactose was only 27.4% (97 from 354). The third screening step challenged the positive candidates under industrial process conditions (milk, 8 °C), and six candidates (about 6% of the true β-galactosidases) featured promising properties when compared to the commercial reference GODO-YNL2 (see Fig. 2). Thus, the three-step screening strategy proposed accomplished its objective.

Since screening step two was carried out at a pH of 6.75, similar to milk, the resulting metagenome enzyme M1 showed an optimal pH of 7 (see Table 1). The optimal neutral pH and the optimal temperature- of the β-galactosidase (37 °C) are quite common for mesophilic β-galactosidases reported in the literature (Ansari and Satar, 2012 and Mahoney, 2002). The temperature stabilities of β-galactosidases are often investigated for only a few hours in the literature (Hoyoux et al., 2001, Hu et al., 2007 and Wang et al., 2010). Thus, a comparison with literature data does not seem appropriate. Nevertheless, the M1 β-galactosidase was compared with GODO-YNL2 in our study at a process-relevant temperature of 8 °C over 48 h (Table 1). The extrapolated half-lives for both β-galactosidases (M1 21.8 days, GODO-YNL2 30.3 days) were more than sufficient for an industrial process, since the industrial lactose hydrolysis in milk operates no longer than 48 h.

Hydrolysis of lactose in milk

The milk lactose hydrolysis performance of M1 in milk was directly compared to GODO-YNL2 on a scale of 0.5 L. The hydrolysis was carried out at 8 °C for 24 h. Subsequently, the milk was heated for 1 h to 40 °C, similar to what would have been done in the dairy industry for pasteurisation. Under these conditions, the M1 β-galactosidase hydrolysed >99.9% of the lactose in milk after 25 h, whereas GODO-YNL2 (industrially utilised reference) reached 99.6% (w/v) after the same time. The metagenome β-galactosidase ZD410, which was investigated by Wang et al. (2010), could hydrolyse 4.2% of the lactose in milk at 4 °C after 60 min of incubation. Unfortunately, they did not investigate it for a longer period of time. In the study by Hoyoux et al. (2001), lactose hydrolysis was also performed for short time intervals. It was reported that a β-galactosidase from*Pseudoalteromonas haloplanktis* hydrolysed 33% of the lactose in milk at 4 °C after 50 min. Longer time intervals were investigated by Pan et al. (2010). In this study, the hydrolysis conversion of lactose in milk was 73% with a β-galactosidase from*Lactobacillus acidophilus* at 10 °C after 30 h. Thus, this hydrolysed milk sample still had a residual lactose content of 13.5 g L^{-1}. In other studies, cold-active β-galactosidases were also investigated for lactose hydrolysis in milk (Cieśliński et al., 2005 and Nakagawa et al., 2007). With the recombinant cold-adapted β-galactosidase from Antarctic bacterium*Pseudoalteromonas* sp. 22b, immobilised on chitosan beads, Cieśliński et al. (2005) achieved a lactose hydrolysis conversion of 90.5% at 15 °C after 28 h. The β-galactosidase activities of four commercial preparations were investigated by Horner et al. (2011) at 2 °C. Here, they obtained a maximum lactose hydrolysis conversion with GODO-YNL2 of 99.08% after 24 h.

Kinetic parameters and effects of d-glucose and d-galactose on the enzyme activity

Although the metagenome β-galactosidase M1 and GODO-YNL2 showed similar biochemical characteristics, the performance during the lactose hydrolysis in milk was better for M1. To find an explanation for this, the apparent kinetic parameters K_M and K_I of M1 and GODO-YNL2 were investigated in buffer and in milk diluted with water at 8 °C. Both M1 and GODO-YNL2 exhibited product inhibition by d-galactose in buffer and diluted milk. The apparent K_I of M1 obtained was higher than the maximum theoretical hydrolysis yield of d-galactose in milk (approx. 140 mM = 48 g L^{-1}). Wang et al. (2010)discovered the metagenome β-galactosidase ZD 410. However, the kinetics (K_M values) were only measured with various synthetic nitrophenyl-monosaccharides (chromogenic substrates) and, thus, were not comparable with our results using lactose. Zhang et al. (2013) reported a novel metagenome β-galactosidase Gal308 (optimal temperature with lactose 78 °C, pH 6.8), which possessed product inhibition by d-galactose (K_I = 238 mM) in buffer when *o*-nitrophenyl-β-d-galactopyranoside was used as the substrate. Zhang et al. (2013) also tested Gal308 for milk lactose hydrolysis at 65 °C over 60 min. At the end, a 93.6% hydrolysis rate was observed. A comparison with M1 is not possible from their data due to the different conditions and a shorter time-dependent course of observation. Comprehensive kinetic data from β-galactosidases of culturable strains, such as *K. lactis,Aspergillus niger* and others, were published by Mahoney (1997). The *K. lactis* -galactosidase, which should be similar to our reference GODO-YNL2, was reported to possess a $K_{M,lactose}$ of 12–17 mM and a $K_{I,}$d$_{-galactose}$ of 42 mM. This is in accordance with our data for GODO-YNL2 (see Table 2). In our work, the apparent K_M/K_I ratio for GODO-YNL2 is 0.6 (in buffer, 8 °C) and 1.0 (milk, 8 °C). It was calculated as Mahoney (1997) did for the evaluation of the hydrolysis potential for *K. lactis* (0.38 in buffer, 35 °C). Again, the data of Mahoney (1997) and ours in buffer correspond to each other, but the different temperatures should be kept in mind. The metagenome β-galactosidase M1 had favourably low apparent K_M/K_I ratios of 0.04 (in buffer, 8 °C) and 0.02 (milk, 8 °C). The rather large difference between M1 and GODO-YNL2, with M1 having apparent K_M/$K_I$ratios up to 50 times lower than GODO-YNL2, highlights the favourable properties of M1, especially for usage in milk. In summary, the apparent kinetic constants $K_{M,lactose}$ and $K_{I,}$d$_{-galactose}$ of both β-galactosidases were highly dependent on the medium and, with the aspect of lactose hydrolysis in milk, they have to be determined in milk-like systems in order to allow a more realistic estimation for an industrial utilisation. No inhibitory effect was determined for M1 and GODO-YNL2 in the presence of d-glucose up to a concentration of 200 mM (Fig. 7; substrate: *o*NPGal). Even a slight activating effect was determined in the case of GODO-YNL2 and a more significant activating effect for M1. Therefore, the d-glucose released during the lactose hydrolysis should not negatively influence the performance of the enzymes. Thus, the d-glucose released during lactose hydrolysis will most probably not have a noticeable negative effect on M1 during lactose hydrolysis, especially in milk.

CONCLUSIONS

In conclusion, the theoretical total number of about 1.3 million metagenome clones were screened for novel β-galactosidases with potential applications in the dairy industry. The third screening step, since it allowed quantitative evaluation of the performance of the positive candidates in the target reaction environment, is especially highly recommended. The metagenome bore six novel β-galactosidases with previously unknown gene sequences which performed very promisingly in milk lactose hydrolysis when compared to an industrially utilised reference enzyme. The apparent kinetic constants of the most promising metagenome candidate M1 (K_M, K_I) were dependent on the environment and were of empiric character. M1 showed the most favourable kinetic properties in diluted milk. This novel metagenome β-galactosidase M1 was able to completely hydrolyse milk lactose in 25 h (>99.9% conversion).

ACKNOWLEDGMENTS

The authors would like to thank Wolfgang Claaßen (University of Hohenheim, Germany) for his support during the cultivation and cell-disruption experiments. We also thank Karin Maierhofer (University of Hohenheim) for her highly appreciated support in the metagenome screening experiments. We express our thanks to KNAUER Wissenschaftliche Geräte GmbH (Berlin, Germany) for providing the chromatography resin and glass column. This research project was supported by the Federal Ministry for Economic Affairs and Energy (via AiF) and the FEI (Forschungskreis für Ernährungsindustrie e.V., Bonn): Project AiF 15,801N.

REFERENCES

1. Ansari, S.A., Satar, R., 2012. Recombinant -galactosidases—past, present and future: a mini review. J. Mol. Catal. B Enzym. 81, 1–6.
2. Bernhard, K., Schrempf, H., Goebel, W., 1978. Bacteriocin and antibiotic resistance plasmids in Bacillus cereus and Bacillus subtilis. J. Bacteriol. 133, 897–903.
3. Bradford, M.M., 1976. A rapid and sensitive method for the quantitation of microgram quantities of protein utilizing the principle of protein dye binding. Anal.

Biochem. 72, 248–254

4. Camper, D.V., Viola, R.E., 2009. Fully automated protein purification. Anal. Biochem. 393, 176–181.
5. Ciesli ′ nski, H., Kur, J., Białkowska, A., Baran, I., Makowski, K., Turkiewicz, M., 2005. ′ Cloning, expression, and purification of a recombinant cold-adapted -galactosidase from antarctic bacterium Pseudoalteromonas sp. 22b. Protein Expr. Purif. 39, 27–34.
6. Erich, S., Anzmann, T., Fischer, L., 2012. Quantification of lactose using ion-pair RP-HPLC during enzymatic lactose hydrolysis of skim milk. Food Chem. 135, 2393–2396.
7. Gabor, E., Niehaus, F., Aehle, W., Eck, J., 2012. Zooming in on metagenomics: molecular microdiversity of subtilisin carlsberg in soil. J. Mol. Biol. 418, 16–20.
8. Gans, J., Wolinsky, M., Dunbar, J., 2005. Computational improvements reveal great bacterial diversity and high metal toxicity in soil. Science 309, 1387–1390.
9. GDCh (Gesellschaft Deutscher Chemiker e.V.), 2005. Positionspapier der Lebensmittelchemischen Gesellschaft zu den Angaben laktosefrei und laktosearm. Lebensmittelchemie, 59.
10. Handelsman, J., 2005. Sorting out metagenomes. Nat. Biotechnol. 23, 38–39.
11. Handelsman, J., Rondon, M.R., Brady, S.F., Clardy, J., Goodman, R.M., 1998. Molecular biological access to the chemistry of unknown soil microbes: a new frontier for natural products. Chem. Biol. 5, 245–249.
12. Harju, M., Kallioinen, H., Tossavainen, O., 2012. Lactose hydrolysis and other conversions in dairy products: technological aspects. Int. Dairy J. 22, 104–109.
13. Hildebrandt, P., Wanarska, M., Kur, J., 2009. A new cold-adapted beta-d-galactosidase from the Antarctic Arthrobacter sp. 32c—gene cloning, overexpression, purification and properties. BMC Microbiol. 9, 151.
14. Horner, T.W., Dunn, M.L., Eggett, D.L., Ogden, L.V., 2011. -Galactosidase activity of commercial lactase samples in raw and pasteurized milk at refrigerated temperatures. J. Dairy Sci. 94, 3242–3249.
15. Hoyoux, A., Jennes, I., Dubois, P., Genicot, S., Dubail, F., Franc¸ ois, J.M., Baise, E., Feller, G., Gerday, C., 2001. Cold-adapted -galactosidase from the Antarctic psychrophile Pseudoalteromonas haloplanktis. Appl. Environ. Microbiol 67, 1529–1535.
16. Hu, J.M., Li, H., Cao, L.X., Wu, P.C., Zhang, C.T., Sang, S.L., Zhang, X.Y., Chen, M.J., Lu, J.Q., Liu, Y.H., 2007. Molecular cloning and characterization of the gene encoding cold-active -galactosidase from a psychrotrophic and halotolerant Planococcus sp. L4. J. Agric. Food Chem. 55, 2217–2224.
17. Husain, Q., 2010. -Galactosidases and their potential applications: a review. Crit. Rev. Biotechnol. 30, 41–62.
18. Laemmli, U.K., 1970. Cleavage of structural proteins during the assembly of the head of bacteriophage T4. Nature 227, 680–685
19. Lorenz, P., Eck, J., 2004. Screening for novel industrial biocatalysts. Eng. Life Sci. 4, 501–504.
20. Mahoney, R.R., 1997. Lactose: enzymatic modification. In: Fox, P.F. (Ed.), Advanced Dairy Chemistry. Chapman & Hall, London, pp. 77–125.
21. Mahoney, R.R., 2002. -Galactosidase. In: Whitaker, J.R., Voragen, A.G.J., Wong, D.W.S. (Eds.), Handbook of Food Enzymology. CRC Press, Boca Raton, FL, pp. 823–828.
22. Mlichová, Z., Rosenberg, M., 2006. Current trends of -galactosidase application in food technology. J. Food Nutr. Res. 45, 47–54.
23. Nacke, H., Will, C., Herzog, S., Nowka, B., Engelhaupt, M., Daniel, R., 2011. Identification of novel lipolytic genes and gene families by screening of metagenomic libraries derived from soil samples of the German biodiversity exploratories. FEMS Microbiol. Ecol. 78, 188–201.
24. Nakagawa, T., Ikehata, R., Myoda, T., Miyaji, T., Tomizuka, N., 2007. Overexpression and functional analysis of cold-active -galactosidase from Arthrobacter psychrolactophilus strain F2. Protein Expr. Purif. 54, 295–299.
25. Niehaus, F., Eck, J., 2012. Novel beta-galactosidases useful for the production of lactose depleted milk products. Eur. Pat. Appl., EP2530148.
26. Novalin, S., Neuhaus, W., Kulbe, K.D., 2005. A new innovative process to produce lactose-reduced skim milk. J. Biotechnol. 119, 212–218.
27. Oliveira, C., Guimarães, P.M.R., Domingues, L., 2011. Recombinant microbial systems for improved -galactosidase production and biotechnological applications. Biotechnol. Adv. 29, 600–609, http://dx.doi.org/10.1016/j. biotechadv.2011.03.008
28. Pan, Q., Zhu, J., Liu, L., Cong, Y., Hu, F., Li, J., Yu, X., 2010. Functional identification of a putative beta-galactosidase gene in the special lac gene cluster of Lactobacillus acidophilus. Curr. Microbiol. 60, 172–178.
29. Panesar, P.S., Panesar, R., Singh, R.S., Kennedy, J.F., Kumar, H., 2006. Microbial production, immobilization and applications of -d-galactosidase. J. Chem. Technol. Biotechnol. 81, 530–543.
30. Pereira-Rodríguez, Á., Fernández-Leiro, R., González-Siso, M.I., Cerdán, M.E., Becerra, M., Sanz-Aparicio, J., 2012. Structural basis of specificity in tetrameric Kluyveromyces lactis -galactosidase. J. Struct. Biol. 177, 392–401.
31. Pessela, B.C.C., Mateo, C., Fuentes, M., Vian, A., Garcıa, J.L., Carrascosa, A., V, Guisán, ′ J.M., Fernández-Lafuente, R., 2003. The immobilization of a thermophilic -galactosidase on Sepabeads supports decreases product inhibition. Enzyme Microb. Technol. 33, 199–205, http://dx.doi.org/10.1016/S0141-0229(03)120-0
32. Sambrook, J., Russell, D.W., 2001. Molecular Cloning: A Laboratory Manual. Cold Spring Harbor Laboratory Press, Cold Spring Harbor, N.Y.
33. Schmidt-Dannert, C., Arnold, F.H., 1999. Directed evolution of industrial enzymes. Trends Biotechnol. 17, 135–136.
34. Steele, H.L., Jaeger, K.E., Daniel, R., Streit, W.R.,

2008. Advances in recovery of novel biocatalysts from metagenomes. J. Mol. Microbiol. Biotechnol. 16, 25–37.

35. Stressler, T., Eisele, T., Baur, C., Wangler, J., Kuhn, A., Fischer, L., 2015. Extraular peptidases from insect- and compost-associated microorganisms: screening and usage for wheat gluten hydrolysis. Eur. Food. Res. Technol., http://dx.doi. org/10.1007/s00217-015-2452-8 (in-press).

36. Stressler, T., Eisele, T., Kranz, B., Fischer, L., 2014. PepX from Lactobacillus helveticus: automated multi-step purification and determination of kinetic parameters with original tripeptide substrates. J. Mol. Catal. B Enzyme 108, 103–110.

37. Wang, K., Li, G., Yu, S.Q., Zhang, C.T., Liu, Y.H., 2010. A novel metagenome-derived beta-galactosidase: gene cloning, overexpression, purification and characterization. Appl. Microbiol. Biotechnol. 88, 155–165.

38. Zhang, X., Li, H., Li, C.-J., Ma, T., Li, G., Liu, Y.-H., 2013. Metagenomic approach for the isolation of a thermostable -galactosidase with high tolerance of galactose and glucose from soil samples of Turpan Basin. BMC Microbiol. 13, 237.

CHAPTER 23

STUDY OF IMMOBILIZED CANDIDA RUGOSA LIPASE FOR BIODIESEL FUEL PRODUCTION FROM PALM OIL BY FLOW MICROCALORIMETRY

J.C. Moreno-Piraja`n [1] and Liliana Giraldo [2]

[1] Facultad de Ciencias, Departamento de Quı´mica, Grupo de Investigacio´n en So´lidos Porosos y Calorimetrı´a, Universidad de los andes, Carrera 1 No 18 A 10, Edifcio Q Oficina 837, Bogota´, Colombia

[2] Facultad de Ciencias, Departamento de Quı´mica, Universidad Nacional de Colombia, Universidad de los andes, Carrera 30 No 45-00, Ciudad Universitaria, Bogota´, Colombia

ABSTRACT

Enzymatic transesterification of palm oil with methanol and ethanol was studied. Of the four lipases that were tested in the initial screening, lipase Candida Rugosa (CR) resulted in the highest yield of mono alkyl esters. Lipase CR was further investigated in immobilized form within an activated carbon as support. The activated carbon was prepared by activation physical. Using the immobilized lipase CR, the effects of water and alcohol concentration, enzyme loading and enzyme thermal stability in the transesterification reaction were investigated. The optimal conditions for processing 50 g of palm oil were: 37 °C, 1:14.5 oil/methanol molar ratio, 1.0 g water and 500 mg lipase for the reactions with methanol, 35 °C, 1:15.0 oil/ethanol molar ratio, 1.0 g water, 500 mg lipase for the reactions with ethanol, and 35 °C, 1:10.0 oil/*n*-butanol molar ratio, 1.0 g water, 500 mg lipase for the reactions with ethanol. Subject to the optimal conditions, methyl and ethyl esters formation of 70 and 85 mol% in 1 h of reaction were obtained for the immobilized enzyme reactions. The flow microcalorimetry is an important and novel techniques is used in evaluation of biodiesel production.

INTRODUCTION

Vegetable oils have attracted attention as a potential renewable source for the production of an alternative source of fuel for petroleum based diesel due to continuous huge world-wide consumption of petroleum reserves and environmental consequences of exhaust gases from diesel engines. Methyl and ethyl esters of fatty acids, also known as biodiesel, are considered nontoxic, biodegradable, and an excellent replacement for petroleum diesel. Biodiesel's cetane number, energy content, viscosity and other diesel characteristics are also similar to those of petroleum-based diesel fuel (Mittelback and Tritthart, 1988). Moreover, biodiesel is essentially sulfur free and is considered ultra-sulfur fuel quality and the diesel engines fueled by biodiesel emit significantly fewer particulates, hydrocarbons, and carbon monoxide, carbon dioxide than those operating on conventional fossil fuel diesel. Emissions of NOx, however, are slightly higher than those of diesel engines operating on conventional diesel fuels (Schumacher et al., 1996 and Ali et al., 1995). Transesterification of vegetable oils and animal fats with significant free FFA (free

Citation: J.C. Moreno-Pirajàn, Liliana Giraldo, Study of immobilized candida rugosa lipase for biodiesel fuel production from palm oil by flow microcalorimetry, Arabian Journal of Chemistry, Volume 4, Issue 1, January 2011, Pages 55-62, ISSN 1878-5352, http://dx.doi.org/10.1016/j.arabjc.2010.06.019.

fatty acids) contents are regarded a well established industrial process for the production of fatty acid alkyl esters also known as biodiesel. Biodiesel is well miscible with diesel and reduces significantly air pollution emission including sulfur and currently produces commercially in significant amounts in Europe, USA and some Asian countries all over the world using renewable oil feedstock sources. For the product and process development of biodiesel, enzymatic transesterification has been suggested to produce a high purity product with an economic, environment friendly process at mild reaction conditions. The enzyme cost being the main hurdle can be overcome by immobilization. Immobilized enzyme, which has been successfully used in various fields over the soluble counterpart, could be employed in biodiesel production with the aim of reducing the production cost by reusing the enzyme. The conventional biodiesel technology involves the use of an inorganic base or acid catalyst at or near the boiling temperatures of the triglyceride/alcohol mixture. The removal of catalyst is through neutralization and eventual separation of salt from the product esters, which is difficult to achieve. Moreover, the physiochemical synthesis schemes often result in poor reaction selectivity and may lead to undesirable side reactions. Enzymatic conversion of triglycerides has been suggested as a realistic alternative to the conventional physiochemical methods (give the relative reference). Enzymatic transesterification of triglycerides offers an environmentally more attractive option to the conventional process. However, the high cost of the enzymes often makes the enzymatic processes economically unattractive. The key step in enzymatic process lies in successful immobilization of the enzyme, which will allow for its recovery and reuse (Balcao et al., 1996). Immobilization is the most widely used method for achieving stability in lipases and to make them more attractive for industrial use (Cowan, 1996, Clark, 1994 and Kawakami and Yoshida, 1996). Common immobilization techniques include physical adsorption onto a solid support (e.g. Bosley and Pielow, 1997), covalent bonding to a solid support (e.g. Walt and Agayn, 1994) and physical entrapment within a polymer matrix support (e.g. Pizarro et al.,1997). Entrapment of lipase entails capture of the lipase within a matrix of a polymer (Hartmeier, 1985). The immobilized lipase by entrapment is much more stable than physically adsorbed lipase and unlike the covalent bonding method, this method uses a relatively simple procedure and at the same time the immobilized lipase maintains its activity and stability (Kennedy and Melo, 1990). A variety of methods have been used for trapping lipases in a polymer matrix (Bickerstaff, 1997). Entrapment of enzymes in an inorganic polymer matrix is one method that has received a considerable attention in recent years. This method which was pioneered by Avnir et al. (1994) is based on sol–gel process. The application of the sol–gel material in the immobilization of lipases is well documented (e.g.Reetz, 1997). A substantial collection of research on the enzymatic transesterification of triglycerides has focused on free enzyme reactions with and without organic solvents.Nelson et al. (1996) studied the lipase-catalyzed transesterification of triglycerides in hexane, using different lipases and a variety of triglycerides and alcohols. The lipase from Mucor miehi (Lipozyme IM-20) was found to be the most effective in converting tallow oil to their respective mono alkyl esters with primary alcohols, whereas the lipase from Candida antarctica was found to be most suitable for reacting with secondary alcohols, giving branched alkyl esters. Wu et al. (1999) employed response surface methodology (RSM) to optimize reaction parameters such as temperature, time, level of lipase, and molar ratio of reactants in the Pseudomonas cepacia lipase and Candida antarctica lipase catalyzed transesterification reaction of restaurant grease to mono alkyl ethylesters using 95% of ethanol. Results showed a synergistic effect when the two lipases were used in sequence which surpassed the RSM predictions. Abigor et al. (2000) studied the lipase-catalyzed transesterification of two Nigerian lauric oils, palm kernel oil and coconut oil, by transesterification of oils with different alcohols using PS-30 lipase as catalyst. In the conversion of palm kernel oil to mono alkyl esters, ethanol resulted in the highest conversion of 72%. Other alcohols tested and their corresponding transesterification conversions included: *t*-butanol 62%, 1-butanol 42%, *n*-propanol 42%, and iso-propanol 24%, while only 15% methylesters was observed using methanol.Kaieda et al. (2001) studied the effect of methanol and water concentrations on the methanolysis of soybean oil using different lipases in a solvent free system. Lipase from Pseudomonas cepacia showed the greatest methanol resistance among the tested lipases. This lipase also exhibited the highest catalytic activity toward the transesterification reaction, even in the absence of water, methylester content reached 32% after 50 h of reaction. Cao et al. (1992) reported the esterification and transesterification reactions catalyzed by Porcine pancreatic lipase adsorbed on glass, acetone precipitated on porous glass, kieselghur, aluminum oxide and agar beads in organic solvents. Results showed the lipase adsorbed on kieselguhr and agar beads with the highest activity. Shimada et al. (1999) and Watanabe et al. (2000) used immobilized Candida antarctica (Novozyme 435) for the conversion of vegetable oil to biodiesel. Results showed incomplete methanolysis of vegetable oil which was attributed to the inactivation of the enzyme. Stepwise addition of methanol prevented this inactivation and conversions in excess of 98% were obtained. Samukawa et al. (2000) investigated the effect of the preincubation of immobilized Candida antarctica lipase (Novozyme 435) in methyloleate and soybean oil prior to the biodiesel production. Results indicated a much faster rate of methanolysis for the preincubated lipase. Hsu et al. (2001) developed a novel procedure for the immobilization of lipase from Pseudomonas cepacia within a phyllosilicate sol–gel matrix. In this process phyllosilicate clay saturated with sodium ions was suspended in water and then exchanged with alkylammonium ions by the addition of cetyltrimethyl ammonium chloride. This mixture was then used in the entrapment of Pseudomonas cepacia with tetramethoxysilane (TMOS) as the polymerization precursor. The immobilized enzyme so prepared was then used in the transesterification of tallow and grease where conversions in excess of 95% were reached. In the present study, lipase PS from candida rugosa was entrapped within an activated carbon matrix, prepared by lignocellulosic waste. The immobilized enzyme (Candida Rugosa) was used in the transesterification of palm oil with methanol and ethanol. In addition to this, we proposed a study of microcalorimetric as a technique alternative for analysis of process of transesterification.

METHODS

Materials

All lipases were purchased by Amano Enzyme (Nagoya, Japan), and Meito Sangyo (Nagoya, Japan). Palm oil was purchased to Federación Nacional de Cultivadores de Aceite de Palma (Colombia). Methanol, ethanol and *n*-butanol were purchased from Fisher Scientific (Pittsburgh, PA). Bis-(trimethylsilyl) trifluroacetamide (BSTFA, derivative grade), 1,2,3-tridecanoylglycerol (tricaprine, 95%) and pyridine. Hexane (GC grade) was purchased from EM Science (Gibbstown, NJ). Tetramethoxysilane (TMOS, 95%), iso-butyltrimethoxysilane (iso-BTMS, 97%), and sodium fluoride (NaF) were purchased from the Aldrich Chemical Company (Milwaukee, WI).

Lipase immobilization

A specified amount of lipase Candida Rugosa (optimally 1 g) was measured into a flask and 10 ml of water was added. The mixture was stirred at 200 rpm using a magnetic stirrer for about 15 min. To this mixture, 0.5 ml of a 0.5 M NaF solution and the activated carbon was added. The flask was removed from the stirrer and left sealed at room temperature for 24 h. The flask was then uncapped and was incubated in a water bath at 37 °C for about 48 h. The powder was washed with 100 ml of distilled water in a 250 ml flask for 1 h at a mixing speed of 500 rpm. The mixture was then filtered. The activated carbon with the lipases was dried again at 37 °C for 48 h. The granular carbon was stored at 4 °C until use. Based on the degree of immobilization tests, about 95% of the enzyme was immobilized in this procedure. The actual enzyme loading was determined at 500 mg of lipase Candida Rugosa 10 g of activated carbon.

Reaction setup and optimization

The reactions were carried out in a constant temperature water bath, under which a Barnstead Thermolyne Haake DC10-W46/B was placed to agitate the reaction mixture. Water was circulated into the bath/circulator which, via an external probe, was able to control the temperature of the bath to within ±0.01 °C. A standard set of conditions was used as the baseline in the optimization studies. These conditions were recommended byNoureddini et al. (2002) in the literature. The initial conditions were 10 g palm oil, 1 g methanol (methanol to oil molar ratio was 5.0), 1.0 g water, 5 g immobilized lipase Candida Rugosa, 37 °C, 800 rpm, and 1 h reaction time. In the reactions with ethanol an*n*-butanol, too, 1.0 g of water and 1 g of ethanol and/or *n*-butanol (ethanol and *n*-butanol to oil molar ratio too 5.0) were used under otherwise identical conditions. In the optimization studies, also the reaction parameters were maintained at the same time. For example, when the effect of molar ratio was investigated, the rest of the reaction conditions were maintained at: 10 g palm oil, 1.0 g methanol, 1.0 g water, 5 g immobilized lipase Candida Rugosa I (CR), 800 rpm, and 1 h reaction time. Lipase Candida Rugosa was used in all of the optimization studies.

Degree of immobilization tests

The immobilized enzyme was washed with water and after filtration; about 100 ± 5 ml of supernatant was collected. This supernatant may potentially contain free enzyme, partially hydrolyzed precursors, alcohols, and soluble oligomers. To quantify the amount of enzyme in the supernatant, a calibration curve relating the formation of free fatty acids as a function of free enzyme loading was constructed, which was based on the hydrolysis of soybean oil. Details about this work may be found elsewhere (Noureddini et al., 2002 and Noureddini et al., 2005). By comparing the supernatant from the immobilization wash procedure with this calibration, the amount of enzyme in the supernatant and the degree of immobilization was determined. In order to model the exact reaction conditions, in the calibration studies, a blank gel with no enzyme was prepared. The supernatant, which was collected from washing this gel, was used as the reaction medium in the calibration experiments. Otherwise, the reaction procedure for the calibration experiments was identical to the free enzyme hydrolysis reactions. The enzyme loading was varied in the range of 0.3–30 mg lipase per 1 g of palm oil for this calibration. The calibration curve showed a monotonically increasing activity as the enzyme loading was increased. The activity of the enzyme leveled off beyond the upper limit of 40 mg of enzyme per 1 g of palm oil.

Sampling and analysis

Samples were 0.8–2.0 ml in volume and were collected in 5 ml sample vials. Samples from the reactions were initially heated to ensure enzyme denaturation, freeze dried to remove excess water and alcohol, and finally derivatized with BSTFA. The silylating agent reacts with the carboxyl groups of fatty acids and results in trimethylsilylated fatty acids which are readily separated and quantified. The derivatized samples were then analyzed by a gas chromatograph (GC) to determine the concentration of fatty acid esters, free fatty acids, mono, di, and triglycerides. A Shimadzu (United States) 2010 Series GC system was used. This equipment was equipped with a column suitable that was designed specially for this work. Sample volumes were 1.5 μL, the carrier gas was helium, and the GC was operated in constant flow mode with a flow rate of 10.0 ml/min. A split injector was used with a split ratio of 10:1 and a temperature of 350 °C. The FID detector was operated at 370 °C and used a helium makeup flow to maintain a constant detector flow of 30.0 ml/min. The oven was initially held at 90.0 °C and was then elevate to 200 °C at 20.0 °C/min, to 300 °C at 2.0 °C/min, and finally to 350 °C at 10.0 °C/min. The oven was held at this temperature for 32.65 min before returning to 90.0 °C. Total run time for this method was 64.0 min. Calibration of the GC method was carried out by analyzing standard solutions of mixed glycerol, fatty acid esters, free fatty acids, mono-, di-, and triglycerides. The standards were derivatized in the same fashion as the reaction samples. More details about the sample preparations and analysis procedures are explained elsewhere (Wagner, 1999).

Data analysis

Experimental results are presented in Figure 1, Figure 2, Figure 3, Figure 4, Figure 5 and Figure 6. In these figures, the activity of the lipases is presented by the formation of esters and other reaction products in terms of molar percent of the components. The data presented in Figure 1, Figure 2, Figure 3, Figure 4, Figure 5 and Figure 6 were replicated at least five times. The mean values for the replicated data not are presented in the graphs. In Figure 1, Figure 2, Figure 3, Figure 4, Figure 5 and Figure 6, the standard deviations for fatty acids were approximately ±2% (this not show in the graphics) of the mean values and are not shown in the figures. The data was analyzed by Statistica Software®.

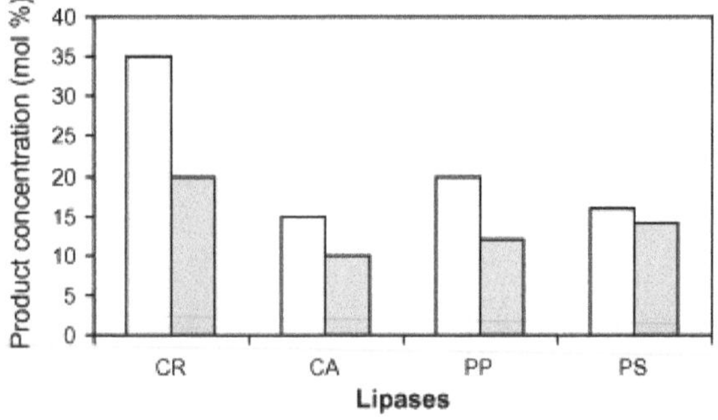

Figure 1: Lipase screening on transesterification of palm oil, a loading 300 mg free lipase, 1.0 g methanol, 1.0 g water, 10 g palm oil, a stirring rate of 800 rpm and a 1-h reaction at 37 °C. (□) Methylesters and (▪) free fatty acids.

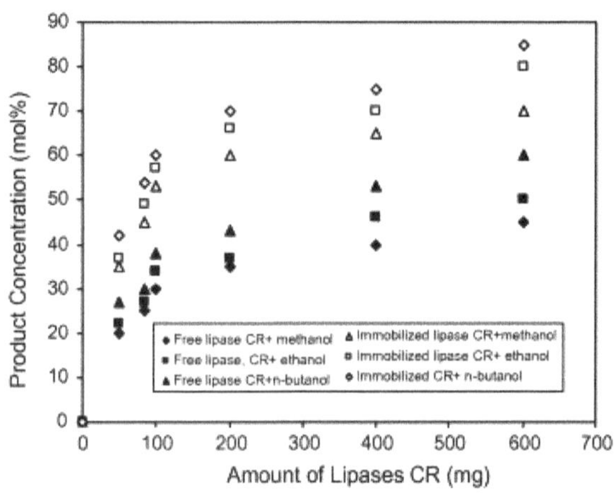

Figure 2: Effect of enzyme loading on transesterification of palm oil 1.0 g water, 10 g palm oil, a stirring rate of 800 rpm and a 1 h reaction at 37 °C. (♦) Free lipase CR, 1 g of methanol. (Δ) Immobilized lipase CR, 1.0 g of methanol. (▪) Free lipase CR, 1 g of ethanol (□) Immobilized lipase CR, 1 g of ethanol. (▴) Free lipase CR, 1 g of *n*-butanol (◊) Immobilized lipase CR, 1 g of *n*-butanol.

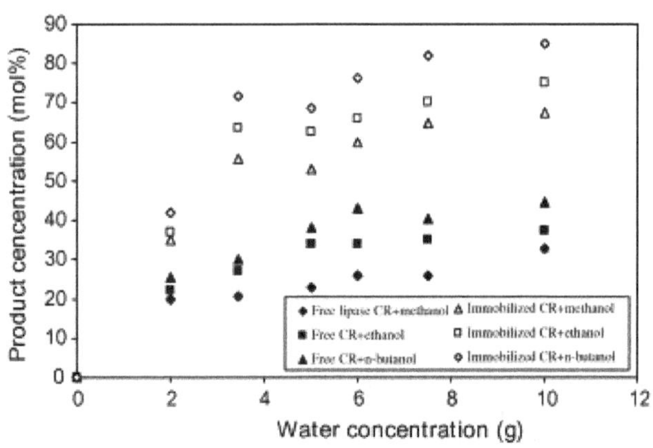

Figure 3: Effect of water concentration on immobilized lipase catalyzed transesterification of palm oil, subject to a loading of 10 of activated carbon and palm oil, a stirring rate of 800 rpm and a 1 h reaction at 37 °C. (♦) Free lipase CR, 1 g of methanol. (Λ) Immobilized lipase CR, 1.0 g of methanol. (▪) Free lipase CR, 1 g of ethanol (□) Immobilized lipase CR, 1 g of ethanol. (▴) Free lipase CR, 1 g of *n*-butanol (◊) Immobilized lipase CR, 1 g of *n*-butanol.

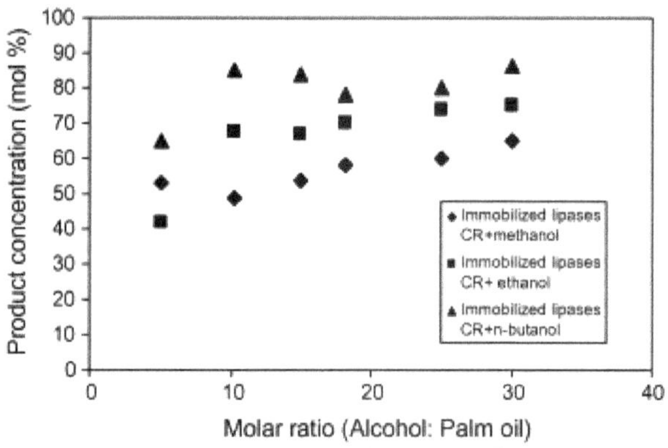

Figure 4: Effect of alcohol concentration on immobilized lipase catalyzed transesterification of palm oil, subject to a loading of 10 g activated of palm oil, a stirring rate of 800 rpm and a 1 h reaction at 37 °C. (♦) Immobilized lipase CR, 1.0 g of methanol. (▪) Immobilized lipase CR, 1 g of ethanol. (▴) Immobilized lipase CR, 1 g of *n*-butanol.

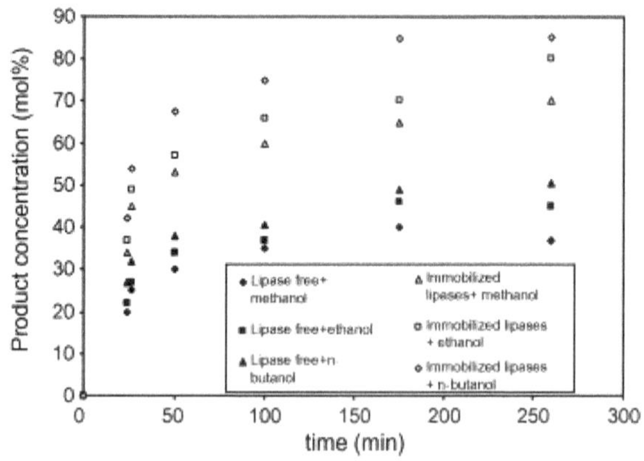

Figure 5: Time course of the transesterification of palm oil, subject to a loading of 10 g of activated carbon, a stirring rate of 800 rpm for at 37 °C. (♦) Free lipase CR, 1 g of methanol. (Δ) Immobilized lipase CR, 1.0 g of methanol. (▪) Free lipase CR, 1 g of ethanol (□) Immobilized lipase CR, 1 g of ethanol. (▴) Free lipase CR, 1 g of *n*-butanol (◊) Immobilized lipase CR, 1 g of *n*-butanol.

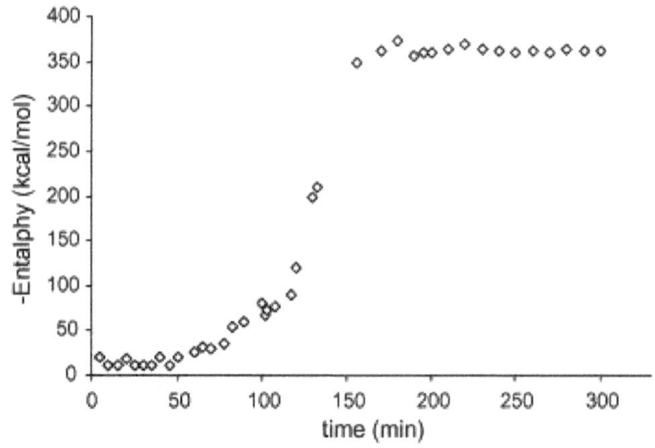

Figure 6: Enthalpy development in the production of biodiesel.

Determination of production biodiesel with flow microcalorimetry

For these measurement a micromacalorimeter was employed which was constructed in our Laboratory (Giraldo et al., 1998). Fig. 7 shows a general scheme of the flow microcalorimeter that has been built. The apparatus consists of:

- An outer heat sink (A) which is an aluminum cylinder (150 mm diameter by 220 mm long) with an inner hole (75 mm diameter) to accommodate the calorimetric units. The cylinder has been cut horizontally so that one half serves as lid, as shown in the figure.
- Two inner heat sinks (B), each of which consists of two aluminum cylinders (75 mm diameter by 25 mm long) that contain two thermopiles and the reaction vessel. The whole set is held in place by three stainless steel screws.
- The thermopiles (Tellurex Corporation, 42 mm by 42 mm by 3.8 mm) (C) work at a maximum intensity of 6 A and a maximum potential of 15.5 V.
- The reference and reaction calorimetric cells (D) are accommodated in aluminum containers that allow a good thermal contact with the thermopiles. Two different reaction vessels were constructed; one is a gold spiral tube (0.2 mm inner diameter) placed inside aluminum parallel plates; a very good thermal contact is assured by filling the empty space with aluminum powder. The other is a silver plate (43 mm by 43 mm by 1.8 mm) with an engraved channel that accommodates a silver tube (0.2 mm diameter). The reference cell differs with the reaction cell in that the former has only one tube for the entrance of liquids. The silver reference cell is identical to the reaction cell. The calibration resistance (99.9 Ω and 0.125 W) resides within the body of each reaction cell. The inner heat sink, the thermopiles, and the flow reaction cell constitute a calorimetric unit. In order to eliminate external heat effects, a similar calorimetric unit (with the reference cell) is connected opposite to the previously described. This twin arrangement is frequently used in microcalorimetry. The heat exchanger (E) is a tin cylinder (75 mm diameter by 60 mm long) that contains two concentric stainless steel spirals through which the calorimetric liquids are driven. The function of the heat exchanger is to minimize the temperature gradients between the solutions, as well as to equilibrate their temperature with that of the reaction cell. The equipment is placed in a chrome steel box hermetically sealed immersed in a water thermostat (75 L capacity) maintained at 25.000 ± 0.005 °C. The whole arrangement is located in a room kept at 25.0 ± 0.2 °C by means of an air thermostat. The potential signal generated by the thermopiles is read by a digital precision multimeter (Hewlett–Packard 3478A) with sensitivity of 0.1 μV that is connected to a personal computer by a GPIB NI-488.2 interface (National Instrument). Data storage and processing are performed by the computer program CETAC developed in this laboratory. The calorimetric liquids are conducted to the reaction cells by means of a peristaltic pump (Master Flex 7520-25) with variable flow rate between 0.2 mL and 42 mL min^{-1}. The system they place in the cells and there is fixed a constant flow of the palm oil and is begun to register the respective signals. Then, is evaluated enthalpic value in function of time.

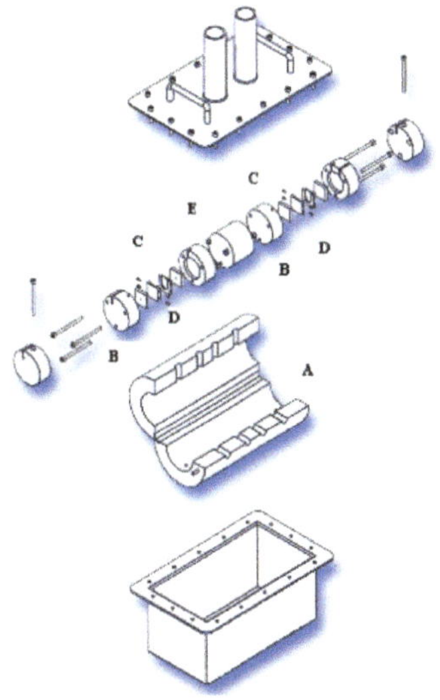

Figure 7: Flow microcalorimeter used for the transesterification of biodiesel study.

RESULTS AND DISCUSSION

Lipase screening

Although lipases generally catalyze the hydrolysis of carboxylic esters, they bring about a range of bioconversion reactions such as esterification, transesterification, acidolysis, and aminolysis. Lipase screening was performed to find the lipase with the best catalytic activity in the transesterification of soybean oil. The most active lipase was then used in the imobilization studies. Four lipases, as listed in Table 1, were screened for their transesterification activity. As the screening experiments were intended for an initial evaluation of the activity of the lipases, they were conducted under a preliminary set of reaction conditions which may not have been the optimum set for all the lipases. In a typical reaction, 300 mg of dry enzyme was added to the mixture of 1 g methanol (5.0 molar ratio of methanol to soybean oil), 1.0 g of water and 10 g of palm oil. The reactions were carried out at 37 °C and according to the reaction setup which was described earlier. The screening results for the tested lipase are presented in Fig. 1. Reaction products are presented as mol% of methyl esters of fatty acids in the reaction mixture.

Table 1: Lipases tested in the transesterification screening

Lipase	Source organism
CR	Candida rugosa
PP	Porcine pancreatic
PS	Pseudomonas cepacia
SP435	Candida antarctica

The formation of free fatty acids is also included in this figure since the presence of water in the reaction medium naturally promotes the competing hydrolysis reaction. As this figure shows, among the tested lipases, lipase Candida Rugosa showed the highest activity toward the transesterification of palm oil with alcohols, specially with the *n*-butanol. Other lipases showed very little or no activity toward the transesterification reaction. After 1 h of reaction with lipase Candida Rugosa, the product contained 37 mol% methyl esters, 5 mol% of fatty acids, 4 mol% of monoglycerides, 52 mol% of diglyceride and 2 mol% of triglycerides.

Enzyme loading

Experiments were performed to determine the effect of enzyme loading on the extent of the transesterification reaction. Enzyme loading in the range of 0–800 mg of free enzyme and 0–5.0 g of the immobilized enzyme were examined in the transesterification of palm oil with methanol, ethanol and *n*-butanol. One gram of immobilized enzyme corresponds to 170 mg free enzyme in these reactions. Based on the concentration of palm oil, the alcohol concentration was at 5.0 molar equivalents for methanol, ethanol and *n*-butanol (1.0 g methanol, 1.0 ethanol and 1.0 g *n*-butanol). Other reaction parameters were as was stated earlier in the reaction setup and optimization section. Reaction results for the formation of methyl and ethyl esters of fatty acids are presented in Fig. 2. For all cases studied, as the enzyme loading was increased there was a sudden surge in the formation of alkyl esters, followed by a slower rate at higher enzyme loadings. This surge was steeper and the formation of alkyl esters was significantly higher for the immobilized enzymes compared with the free enzymes. For the reactions with methanol, the concentration of methylesters reached 45 and 70 mol% for the free and the immobilized lipase, respectively. At the end of the reaction period, the concentration of triglyceride in both system reached negligible levels, while, the formation of fatty acid, mono- and di-glyceride were 5, 7, and 48 mol% for the free enzyme reaction and 5, 20, and 10 mol% for the immobilized enzyme. Similarly, for the reaction with ethanol, the level of ethylesters reached 50 and 80 mol% for the free and the immobilized lipase. At the end of the reaction period, the concentration of triglyceride in both system reached low levels, while, the formation of alkyl-esters were 6, 3, 49 mol% for the free enzyme reaction and 6, 19, and 12 mol% for the immobilized. Also, for the reaction with *n*-butanol, the level of alkyl-esters reached 60 and 85 mol% for the free and the immobilized lipase. At the end of the reaction period, the concentration of triglyceride in both system reached negligible levels, while, the formation of alkyl-esters were 2, 2, 54 mol% for the free enzyme reaction and 4, 16, and 17 mol% for the immobilized This behavior of the immobilized lipase was not consistent with those of other researchers (e.g. Reetz, 1997) and has attributed not only to lipophilic nature of the alkyl group but to activated carbon that for high superficial area allows that a number great of lipases settle and the process is more efficiently. Free alkyl groups in the support-lipase create a lipophilic microenvironment that subsequently interacts with the lipase, triggering a phenomenon similar to a classical interfacial interaction. However, unlike the interfacial activation, which is an interactive process, the alkyl effect is believed to be due to a more favorable lipase conformation caused during the carbon activated-lipse-oil interaction. The lipophilic environments are also believed to facilitate the transport of the organic substrate to the biocatalyst sites in the outer surface of the support and possibly in and out of the matrix.

Effect of water concentration

Lipases possess the unique feature of acting at the interface between an aqueous and an organic phase. Lipase interfacial action is due to the fact that their catalytic activity generally depends on the aggregation of the substrates. Activation of the enzyme involves unmasking and restructuring of the active site through conformational changes of the lipase molecule, which requires the presence of oil–water interface. Lipase activity generally depends on the available interfacial area. With the increased addition of water, the amount of water available for oil to form oil–water droplets increases, thereby, increasing the available interfacial area. However, since lipases usually catalyze hydrolysis in aqueous media, excess water may also stimulate the competing hydrolysis reaction. The optimum water content is a compromise between minimizing hydrolysis and maximizing enzyme activity for the transesterification reaction. The effect of water concentration in the range of 0.02–5.0 g and at constant alcohol concentrations of 5.0, 15.2 and 17.5 molar ratios of methanol, ethanol and *n*-butanol with respect to palm oil were examined (3.9 g methanol, 8.0 g ethanol and 10.0 *n*-butanol). The reactions were carried out according to the reaction setup and optimization section described earlier. Results presented in Fig. 3 indicate very little enzyme activity at low water concentrations which supports the fact that a minimum amount of water is required to activate the enzyme. With the increased addition of water there was a considerable increase in the ester production showing the enhancement in the activity of the enzyme. In the case with methanol, the concentration of methylesters reached 32.7 and 62.4 mol% for the free and the immobilized lipase, respectively. The production maximum is reached with the *n*-butanol 45 and 85 mol% for the free and the immobilized lipase, respectively of about 65 mol% at 0.3 g water.

Effect of alcohol concentration

In the immobilized enzyme transesterification reaction, the reactants initially form a three-phase system (triglyceride/alcohol/support). The reaction is diffusion controlled and poor diffusion between the phases exists. As alkyl esters are formed, they act as a mutual solvent for the reactants and a two phase liquid/solid system results (Noureddini and Zhu, 1997 and Noureddini et al., 2005). However, as the reaction progresses toward completion and the glycerol concentration is increased, a mainly alcohol and glycerol phase separates from the rich alkyl ester phase and a three-phase system forms again. This is more likely at lower initial alcohol concentrations and higher extent of reaction and in the range of alcohol concentration that was investigated this separation did not occur. Alcohol in excess of the stoichiometric molar ratio of 3:1, was used to: (1) ensure higher reaction rates as the transesterification of triglycerides with alcohols consist of three stepwise and reversible reactions and (2) minimize the diffusion limitations specially in this case by the use of activated carbon as support. However, excess alcohol levels may inhibit the enzyme activity and thereby decrease its catalytic activity toward the transesterification reaction. Experiments were performed to optimize the amount of ester production by varying the alcohol concentration. Optimum alcohol requirements were determined for both methanol and ethanol. The amount of alcohol added was varied from 2.7 to 13.7 molar equivalents for methanol and from 5.7 to 26.7 molar equivalents for ethanol, based on the moles of triglycerides. Water concentration was kept constant at the optimum level of 0.5 g for the methanol reactions and at 0.3 g for the ethanol reactions. Results are summarized in Fig. 4. As was expected, an increase in the number of moles of alcohols with respect to the triglycerides resulted in an increase in the production of esters. Ultimately, the formation of esters reached a maximum level and further increases in the alcohol concentrations resulted in a decrease in the formation of esters. The optimum alcohol concentration was determined at 10.4 molar ratio for alls alcohols to palm oil where about 85 mol% of methyl esters was formed with *n*-butanol.

Determination of biodiesel with flow micro-calorimetry

This one is a very versatile technology that in our laboratory we use for several decades to characterize solid porous as the coal activated with excellent results. In this research it has been implemented to evaluate the production of biodiesel under the experimental conditions of this work. For it before I realize a study of production of the biodiesel with regard to the time as one shows in the Fig. 5. These results show an ideal production to 150 min for each of the studied systems.

In these conditions we take to study the respective results we represent time in function enthalpy. The results show themselves in the Fig. 6. This show appears clearly that there is a change abrupt in the value of enthalpy about 150 min.

This shows that the microcalorimetry of flow is an effective and best tool to do the follow-up of the reaction of transesterificación of the oil of palm in the production of the biodiesel. This type of study has not been reported by the literature.

CONCLUSIONS

Enzymatic transesterification of triglycerides offers an environmentally more attractive option to the conventional physiochemical process. The key step in enzymatic processes lies in successful immobilization of the enzyme which will allow for its recovery and reuse. In this study the immobilized enzyme transesterification of soybean oil with methanol and ethanol was investigated. Lipase Candida Rugosa was immobilized by entrapment within a activated carbon which was prepared by activation physical. The immobilized lipase so prepared was consistently more active than the free lipase toward the transesterification of palm oil. The immobilized lipase also proved to be stable and lost little activity when was subjected to repeated uses. The flow microcalorimetry is a technical novel for followed of transesterification of oil palm process.

ACKNOWLEDGEMENTS

The authors express their gratitude to the Departments of Chemistry of Universidad Nacional de Colombia, Universidad de Los Andes (Colombia) and the Master Agreement established between these institutions. Special gratitude is expressed to the Fondo Especial de Investigaciones de la Facultad de Ciencias de la Universidad de Los Andes (Colombia) for its partial financing.

REFERENCES

1. Abigor, R.D., Uadia, P.O., Foglia, T.A., Haas, M.J., Jones, K.C., Okpefa, E., Obibuzor, J.U., Bafor, M.E., 2000. Lipase-catalysed production of biodiesel fuel from some Nigerian lauric oils. Biochemical Society Transactions 28 (6), 979–981.
2. Ali, Y., Hanna, M.A., Leviticus, L.I., 1995. Emissions and power characteristics of diesel engines on methyl soyate and diesel fuel blends. Bioresource Technology 52 (2), 185–195.
3. Avnir, D., Braun, S., Lev, O., Ottolenghi, M., 1994. Enzymes and other proteins entrapped in sol–gel materials. Chemistry of Materials 6 (10), 1605–1614.
4. Balcao, V.M., Paiva, A.L., Malcata, F.X., 1996. Bioreactors with immobilized lipases: state of the art. Enzyme and Microbial Technology 18 (5), 392–416.
5. Bickerstaff, G.F., 1997. Immobilization of Enzymes and Cells. Humana Press, Totowa, NJ.
6. Bosley, A., Pielow, A., 1997. Immobilization of lipases on porous polypropylene: reduction in esterification efficiency at low loading. Journal of American Oil Chemists Society 74 (1), 107–111.
7. Cao, S.G., Liu, Z.B., Feng, Y., Ma, L., Ding, Z.T., Cheng, Y.H., 1992. Esterification and transesterification with immobilized lipase in organic solvent. Applied Biochemistry and Biotechnology 32 (1), 1– 6.
8. Clark, D.S., 1994. Can immobilization be exploited to modify enzyme activity? TIBTECH 12 (11), 439–443.
9. Cowan, D., 1996. Industrial enzyme technology. TIBTECH 14 (6), 177–178.
10. Hartmeier, W., 1985. Immobilized biocatalysts – from simple to complex systems. Trends in Biotechnology 3 (6), 149–153.
11. Hsu, A.F., Jones, K., Marmer, W.N., Foglia, T.A., 2001. Production of alkyl esters from tallow and grease using lipase immobilized in a phyllosilicate sol–gel. Journal of American Oil Chemists Society 78 (6), 585–588.
12. Giraldo, L., Moreno, J.C., Go´mez, A., 1998. A heat conduction flowmicrocalorimeter for solute enthaphy transfer determinations. Design and Calibration. Instrumentation Science and Technology 26 (5), 51–531.
13. Kaieda, M., Samukawa, T., Kondo, A., Fukuda, H., 2001. Effect of methanol and water contents on production of biodiesel fuel from plant oil catalyzed by various lipases in a solvent-free system. Journal of Bioscience and Bioengineering 91 (1), 12–15.
14. Kawakami, K., Yoshida, S., 1996. Thermal stabilization of lipase by sol– gel entrapment in organically modified silicates formed on kieselguhr. Journal of Fermentation and Bioengineering 82 (3), 239–245.
15. Kennedy, J.F., Melo, E.H.M., 1990. Immobilized

enzymes and cells. Chemical Engineering Progress 86 (7), 81–89.

16. Mittelback, M., Tritthart, P., 1988. Diesel fuel derived from vegetable oils. III. Emission tests using methyl esters of used frying oil. Journal of American Oil Chemists Society 65 (7), 1185–1187.
17. Nelson, L.A., Foglia, T.A., Marmer, W.N., 1996. Lipase catalyzed production of biodiesel. Journal of American Oil Chemists Society 73 (8), 1191–1195.
18. Noureddini, H., Zhu, D., 1997. Kinetics of transesterification of soybean oil. Journal of American Oil Chemists Society 74 (11), 1457–1463.
19. Noureddini, H., Gao, X., Joshi, S., Wagner, P.R., 2002. Immobilization of Pseudomonas cepacia lipase by sol–gel entrapment and its application in the hydrolysis of soybean oil. Journal of American Oil Chemists Society 79 (1), 33–39.
20. Noureddini, H., Gao, R.S., Philcana, R.S., 2005. Immobilized of Pseudomonas cepacia lipase for biodiesel fuel production from soybean oil. Bioresource Technology 96 (1), 769–777.
21. Pizarro, C., Fernandez-Torroba, M., Benito, C., Gonzalez-Saiz, J., 1997. Optimization by experimental design of polyacrylamide gel composition as support for enzyme immobilization by entrapment. Biotechnology and Bioengineering 53 (2), 497–506.
22. Reetz, M.T., 1997. Entrapment of biocatalysts in hydrophobic sol–gel material for use in organic chemistry. Advanced Materials 9 (12), 943–954.
23. Samukawa, T., Kaieda, M., Matsumoto, T., Ban, K., Konda, A., Shimada, Y., Noda, H., Fududa, H., 2000. Pretreatment of immobilized Candida antarctica lipase for biodiesel fuel production from plant oil. Journal of Bioscience and Bioengineering 90 (2), 180–183.
24. Schumacher, L.G., Borgelt, S.C., Fosseen, D., Goetz, W., Hires, W.G., 1996. Heavy-duty engine exhaust emission tests using methyl ester soybean oil/diesel fuel blends. Bioresource Technology 57 (1), 31– 36.
25. Shimada, Y., Watanabe, Y., Samukawa, T., Sugihara, A., Noda, H., Fukuda, H., Tominaga, Y., 1999. Conversion of vegetable oil to biodiesel using immobilized Candida antarctica lipase. Journal of American Oil Chemists Society 76 (7), 789–793.
26. Watanabe, Y., Shimada, Y., Sugihara, A., Noda, H., Fukuda, H., Tominaga, Y., 2000. Continuous production of biodiesel fuel from vegetable oil using immobilized Candida antarctica lipase. Journal of American Oil Chemists Society 77 (4), 355–360.
27. Wagner, P., 1999. Sol–gel immobilization of lipase for the hydrolysis of soybean oil. MS Thesis, Faculty of Chemical Engineering, University of Nebraska-Lincoln.
28. Wu, W.H., Foglia, T.A., Marmer, W.N., Phillips, J.G., 1999. Optimizing production of ethyl esters of grease using 95% ethanol by response surface methodology. Journal of American Oil Chemists Society 76 (4), 517–521.
29. Walt, D.R., Agayn, V., 1994. The chemistry of enzyme and protein immobilization with glutaraldehyde. Trends in Analytical Chemistry 13 (10), 425–430.

INDEX

A

Alpha-amylase 99, 100
Amounts of biocatalyst 190
Assimilate lactose 175
Atomic absorption spectroscopy (AAS) 91

B

Bio-derived catalysts 1
Biological analyse 114
Biotechnological industries 1
Bovine serum albumin (BSA) 101
Butyric acid 153, 155, 156

C

Candida rugosa (CR) 209
Carbon electrode (GC) 75
Carbon nanotubes (CNTs) 50, 76, 160, 170
Cassava root 99, 100
Catalytic efficiency 49, 50, 54, 55, 56
Cell free filtrate (CFF) 106
Chemical co-precipitation method 59, 60, 65
Chemical structure 44, 45, 47
Chemical treatments 43
Chemical vapor deposition (CVD) 169, 170
Cholesterol sensor 169
Cyclic voltammetry (CV) 76
Cyclic voltammogram (CV) 169

D

Direct electron transfer (DET) 160

E

Economic interest 198
Energy dispersive X-ray analysis (EDX 91
Energy dispersive X-Ray (EDX) 99, 100
Enzymatic conversion 210
Enzymatic hydrolysis 175
Enzymatic process 43, 47
Enzymatic processes 68
Enzyme cost 210
Enzyme hydrolysis 60
Enzyme immobilization 11, 26, 28
Enzyme molecule 106, 108
Enzyme specificity 1
Ethyl esters of fatty acids 209, 214

F

Flow-rate controller 190
Fluorinated ethylene propylene (FEP) 116
Formaldehyde dehydrogenase (FDH) 33, 34
Fourier transform infrared spectroscopy (FTIR) 91, 99

G

Galactooligosaccharides (GOS) 189, 191
Glucose Oxidase (GOx) 160

H

High investment 1
High level 145, 150
High performance liquid chromatography (HPLC) 85
High value 83
Hydrolytic process 105
Hydrophobic supports 124, 125, 127, 129

I

Immobilization yield (IY) 126, 177, 181
Immobilized enzyme microreactors (IEMR) 113
Industrial processes 50
International unit (IU) 125
Interparticle mesoporous structure (IPMPS) 34
Ion affinity chromatography (IMAC) 114
Ionotropic gelation 67, 68

K

Kinetic constant value 59

L

Laccase (Lac) 89
Light scattering analysis (LSA) 67, 68

M

Magnetic iron oxide 59, 60
Maltose hydrolysis 105
Mediated electron transfer (MET) 159
Membrane systems 12
Mesoporous silica nanoparticles (MSNs) 2
Mesoporous zirconia material (MPZ) 33
Microemulsion-based organogels (MBGs) 135
Milk products 197, 198, 207
Molar ratio 146
Molecular structure 146
Multi-walled carbon nanotubes (MWCNTs) 75, 76

N

Nature catalyze 123
Nicotinamide adenine dinucleotide (NAD) 35

O

Operational performance 135
Optical density (OD) 176

P

Packed-bed reactors (PBRs) 190
Partial purified enzymatic extract (PPE) 50, 53
Pharmaceutical industry 146
physical adsorption 153, 154
Polymerase chain reaction (PCR) 199
Polyvinyl alcohol (PVA) 170
Poor stability 176
Potential applications 153
Potentially carcinogenic 49
Production rate 146

R

Reverse osmosis (RO) 115

S

Scanning electron microscope (SEM) 91, 107, 110
Skin disorder 84
Solvent experiment 136
Square-wave voltammetry (SWV) 76
Ssynthetic medium 176, 179, 181, 186
Substrate reaction time 105, 108
Sustainable source of fuel 83

T

Thermal gravimetric analysis (TG) 153
Thiamine pyrophosphate (TPP) 116
Transmission electron microscopy (TEM) 35, 67, 68, 69
Turnover frequency 12

V

Vegetable oil 209
Vertical bioreactor 43, 44, 47
Volatile organic compounds (VOCs) 33

W

Water-jacketed glass 190
Water pollution 43
Wide range 49, 54
Wide variety 89